Multidisciplinary Applications of Marine Resources

Mohd Rafatullah • Masoom Raza Siddiqui •
Moonis Ali Khan • Riti Thapar Kapoor
Editors

Multidisciplinary Applications of Marine Resources

A Step towards Green and Sustainable Future

Springer

Editors
Mohd Rafatullah
Division of Environmental Technology
Universiti Sains Malaysia
Penang, Malaysia

Masoom Raza Siddiqui
Chemistry Department
King Saud University
Riyadh, Saudi Arabia

Moonis Ali Khan
Chemistry Department
King Saud University
Riyadh, Saudi Arabia

Riti Thapar Kapoor
Centre for Plant and Environmental
Biotechnology
Amity University
Noida, Uttar Pradesh, India

ISBN 978-981-97-5056-6 ISBN 978-981-97-5057-3 (eBook)
https://doi.org/10.1007/978-981-97-5057-3

This Springer imprint is published by the registered company Springer Nature Singapore Pte Ltd.
The registered company address is: 152 Beach Road, #21-01/04 Gateway East, Singapore 189721, Singapore

Foreword

Ocean plays a significant role in energy production, human health and economic prosperity. Blue economy has been widely advocated to safeguard oceans and water resources through sustainable application of marine organisms/biomass for societal benefit all over the globe. Sea or oceans represent safe sources of food, energy and minerals, and they also provide the transport of more than 90% of traded goods worldwide. Seas cover more than 70% of the Earth's surface and contain more than 97% of the Earth's water as well as minerals. The marine environment is a source of a plethora of beneficial organisms which are rich in functional/bioactive compounds and exhibit their remarkable potential for medical, industrial and biotechnological applications. Marine organisms such as micro- and macroalgae, sponges, crustaceans, microbes and fishes are valuable source of functional ingredients such as polysaccharides, vitamins, minerals, pigments, enzymes, proteins and peptides, lipids, polyunsaturated fatty acids, phenolics and other secondary metabolites. Due to the presence of bioactive compounds, marine biomass can be incorporated in the food industry as colorants, natural preservatives, stabilizers, gelling agents and also in nutraceutical and pharmaceutical industries due to their anti-inflammatory, anti-hypertensive, antioxidant, anti-coagulant, anti-diabetic and cardiovascular protective activities.

However, current biorefinery technologies mainly rely on edible crops and freshwater resources. With an ever-growing population, there are concerns for excessive exploitation of the world's limited freshwater and food crop resources for non-nutritional activities. In this context, marine-based biorefinery represents as a promising approach, where seawater, seaweed, kelp, microalgae, yeast and bacteria act as renewable and alternative feedstock due to their tremendous potential for the production of biofuel and high-value products. Marine biomass can be used in bioconversion/fermentation processes for the production of food, feed, biofuel, biogas and high-value chemicals. Due to the fast growth rate of marine biomass, ease of availability and harvesting, low production cost make them suitable candidate for trapping of the solar energy, utilize CO_2 from atmosphere and can be used in marine biorefinery. As compared to terrestrial biomass, seaweed grows much faster up to 50 times and highly efficient in CO_2 fixation. Seaweed can form sea forest and can be used in the production of food, cosmetics, fertilizers, extraction of industrial gums and other chemicals, etc. Biofuels and bio-based products are among the best alternatives to fossil-based fuels and synthetic chemicals due to their capacity for net-negative carbon emissions, which is a vital contribution to the global ambition of a net-zero economy. Marine biorefinery may enhance the overall energy output and economic feasibility as compared to conventional fossil fuels. Therefore, the development of integrated marine biorefinery system with marine resources can be a promising approach in mitigation of climate change as well as addressing the shortage of water, food and energy at global level. This book provides wide information to readers on state-of-the-art of application of marine biomass, diatom and their integrative applications for the development of marine biorefinery, wastewater/industrial effluent treatment, exploitation of marine organisms/biomass in food, pharmaceutical industries and other value-added products.

Bioenergy and Environmental Remediation Laboratory (BERL), Department of Earth Resources and Environmental Engineering
Hanyang University,
Seoul, Republic of Korea

Ramesh Kumar

Preface

Oceans cover 70% of the planet's surface and besides being an immense reservoir of biological life they serve as an important source for transport, commerce, and human sustenance. However, 80% of the oceans remain unexplored. Energy is an essential infrastructure for economic growth and social advancement due to the rapid expansion of global population and industrialization. Using fossil fuels as a primary energy source has resulted in a slew of negative consequences, including climate change, adverse impact on environment and human health. Therefore, in order to address the energy crisis, it is crucial to rely on a sustainable and promising energy source alternative. There is a high demand for alternative fuels because of rising environmental issues and depletion of fossil fuels. Renewable energy and sustainable technologies offer a promising avenue for mitigating environmental impacts while reducing dependence on fossil fuels. Amid the global energy crisis and environmental concerns, marine biomass in its diverse forms has emerged as a promising, green, and cost-effective alternative which has potential for conversion into sustainable energy in alignment with the United Nations sustainable development goals. Marine biomass is a globally abundant renewable resource that has gained significant attention for its potential for the development of wide portfolio of bioproducts such as biofuel, bioethanol, value-added products, and chemicals that are traditionally derived from non-renewable resources. Marine biomass has excellent biological activity, unique chemical texture, and good biocompatibility that can be used for various purposes. Bioactive compounds derived from marine biomass are considered as vital sources of functional ingredients with favorable functions for human health and disease risk reduction. Hence, there has been a growing interest in researching and developing marine-derived functional food ingredients, nutraceuticals, pharmaceuticals, cosmetics, and dietary supplements. The utilization of renewable marine resources to produce carbon materials, such as biofertilizers, biochar, and low-emission fuels, plays an important role in gradually replacing conventional fossil fuel. Application of marine biomass offers ecological viability, low-cost production, and green credentials. After careful evaluation of every aspect of marine resources, it represents a cornerstone of our future global sustainability.

Sankha Chakrabortty and their research team members from multiple institutions such as Kalinga Institute of Industrial Technology, Bhubaneswar and Bankura Christian College, Bankura, West Bengal, India, present a comprehensive review on the production of biofertilizer from marine biomass. Vivekananda Saha and Goutam

Biswas from Panchanan Barma University, Cooch Behar, India highlight the formulation of marine-derived natural cosmetics. John Tsado Mathew et al., an interdisciplinary group of authors from multiple institutions of Nigeria, focus on applications of diatoms in the treatment of industrial effluent which acts as a sustainable tool for food production. Krishna Neeti et al. from the National Institute of Technology Patna, Bihar, India highlight the utilization of diatoms in the development of value-added products. A study by a research group of W.V.Celcia Gnana Rathinam et al. from Annamalai University, Tamil Nadu, India highlights the application of marine resources as a promising source for cosmetic industries. Immanuel Suresh. J and Divyeswari S from the American College, Madurai, Tamil Nadu, India describe the application of seaweeds in the production of bioethanol. Deepshikha Datta and their research group from Brainware University, Kolkata, West Bengal and the National Institute of Technology, Durgapur, India describe valorization of marine waste towards the production of value-added industrial products like bioplastics. They also highlight the application of marine biomass, regulations, and policy framework for carbon neutrality, sustainable environment, and bioeconomy. Anwesha Patel and their research team members from Dayananda Sagar University, Bengaluru, India highlight the potential of marine microalgae in food and feed industry. Asmita Saha et al. from Dayananda Sagar University, Bengaluru focus on seaweed-derived bioactive compounds for pharmaceutical applications. Rekha Rani et al., a group of authors from multiple institutions such as the College of Dairy and Food Technology, Agriculture University, Jodhpur and Dayalbagh Educational Institute, Agra, India and Food Science and Nutrition Department, University of Minnesota, Minnesota, USA, highlight the potential of seaweeds as antioxidants and marine-based food as a drug to fight against infectious diseases. Mardiana Idayu Ahmad and their research team from the School of Industrial Technology, Universiti Sains Malaysia, Penang, Malaysia and Hail University, Saudi Arabia describe extraction and development procedure of biodiesel from marine macroalgae-seaweed as a renewable feedstock. Mohammad Aliff Shakir et al. from the School of Industrial Technology, Universiti Sains Malaysia, Penang, Malaysia focus on the development of bioplastics from marine green algae. Hamza Mohamed Flafel et al., a group of authors from multiple institutions such as the School of Industrial Technology, Universiti Sains Malaysia, Penang, Malaysia, Wadi Al-Shatti University, Brack Al-Shatti, Libya, Amity University Uttar Pradesh, India and King Saud University, Riyadh, Saudi Arabia, highlight the production of biochar from marine algae and its use in the treatment of industrial effluent. C. Madhuri et al. from Yogi Vemana University, Kadapa, Andhra Pradesh, India explain the production of biofuel from macroalgae/seaweed. Indahayu Ab. Rahman et al. from the School of Industrial Technology and Centre for Marine and Coastal Studies, Universiti Sains Malaysia, Penang, Malaysia focus on biomonitoring of mangroves which acts as an indicator of sustainable coastal environment. Harinisri Ram K and Thamarai Selvi B from Sri Ramakrishna College of Arts and Science for Women, Coimbatore, India explore inventive applications of marine resources in the production of cosmetics.

This book presents a snapshot of research covering transformation of marine biomass into valuable products. The intended readers of this book include environmentalists, engineers, scientists, industrial personnel, and policy makers who wish to specialize in utilization of marine biomass-based renewable resources for the production of biofuel and in the development of value-added products such as biofertilizers, biochar, cosmetics, pharmaceuticals, and nutraceuticals. We are very thankful to the authors for contributing chapters and to the numerous learned reviewers who carefully read and gave valuable suggestions on the chapters of this book to improve the quality. We would also like to gratefully acknowledge Springer Editorial Office, especially Dr. Sanchi Bhimrajka and Dr. Naren Aggarwal for bringing us this opportunity and guiding us throughout the publication process. Our sincere thanks go to Ms Muthuneela Muthukumar for her wonderful help and continuous support.

Penang, Malaysia — Mohd Rafatullah
Riyadh, Saudi Arabia — Masoom Raza Siddiqui
Riyadh, Saudi Arabia — Moonis Ali Khan
Noida, Uttar Pradesh, India — Riti Thapar Kapoor

Contents

Editors and Contributors

About the Editors

Mohd Rafatullah I am presently working as an Associate Professor of Environmental Technology in the School of Industrial Technology, Universiti Sains Malaysia (USM), Malaysia. I joined this School in the year 2008 as a Post-Doctoral Fellow. I have completed my education; Ph.D. in Environmental Chemistry, Master of Science in Analytical Chemistry, and Bachelor of Science in Chemistry from Aligarh Muslim University (AMU), India. My research interest is in the areas of environmental water pollutants and their safe removal; preparation of various nanomaterials to protect the environment; water and wastewater treatment; adsorption and ion exchange; microbial fuel cells; advance oxidation process; activated carbons and their electrochemical properties. My contribution was recognized by Guest Editors and Member of Editorial Board of various scientific journals. I am listed in World's Top 2% Scientist by Stanford University and also listed in Top 1% peer reviewer, in Chemistry, Environmental Science, and cross-field on Publons global reviewer, Web of Science. I am life-time Fellow Member of the International Society of Sustainable Developments and member of various professional international societies. I have published several review articles and regular research papers in the journals of international repute and presented my research work in various national and international conferences. I have also attended many workshops and seminars of environmental chemistry. Based on my performance and contribution in research, total citations: >11500 and h index: 47 reported in Scopus database.

Masoom Raza Siddiqui is currently working as professor of Chemistry at King Saud University, Riyadh, Saudi Arabia. Prof. Siddiqui received his bachelor's, master's, and Ph.D. degrees from Aligarh Muslim University, Aligarh, Uttar Pradesh, India. Prof. Siddiqui has 2.5 years of experience in the pharmaceutical R&D in addition to 12 years of academic experience. His research experience includes analytical method development of pharmaceutical, environmental, and food additive and contaminants, bioremediation, and water treatment. Prof. Siddiqui has more than 120 papers to his credit. He also contributed to book chapters and patents. Prof. Siddiqui also served as guest editor in two special issues of sustainability journal, published by MDPI. He also successfully guided master's and Ph.D. students.

Moonis Ali Khan is a full professor working in Chemistry Department, College of Science, King Saud University, Riyadh, Saudi Arabia. He is an interfacial chemist with research focused on the development of advanced materials for energy, environmental, and biomedical applications. His research is also focused on waste conversion to valuables for both energy and environmental applications. Dr. Khan had guided three students for their respective Ph.D. research work. Till date, he has authored more than 195 research/review articles and have three US patents to his name. He had edited one and authored chapters in two books. His current H-index is 47 (on Google Scholar). He is also serving as an editor, guest editor, and editorial board member for many journals. For the last three consecutive years (2021, 2022, and 2023), he has been listed among the World's Top 2% Scientists, a list released by Stanford University.

Riti Thapar Kapoor is Associate Professor in Amity Institute of Biotechnology, Amity University Uttar Pradesh, Noida, India. Dr. Kapoor received her Ph.D. from the University of Allahabad and worked as post-doctoral fellow at Banaras Hindu University, Varanasi, India. Dr. Kapoor has 15 years of teaching and research experience and her area of specialization is environmental biotechnology, bioremediation, wastewater treatment, and abiotic stresses. Dr. Kapoor has published 8 books and over 105 research papers in various journals of national and international repute. Dr. Kapoor has visited 8 countries for participation in different academic programs. Dr. Kapoor has received prestigious travel award from Bill & Melinda Gates Research Foundation under CGIAR project for participation in international training program held at the International Rice Research Institute (IRRI), Manila, Philippines in 2010. She is also recipient of DST travel grant for participation in International Conference held at Sri Lanka in 2013. Dr. Kapoor has been awarded with Teacher's Research Fellowship from the Indian Academy of Sciences, Bengaluru in 2019. She has supervised and mentored a number of research projects sanctioned by different government funding agencies such as DAE, DST, and UPCST. She has successfully supervised three research students for Ph.D. degree besides several (more than 70) master's and graduate students for their dissertation thesis.

Contributors

Indahayu Ab. Rahman Environmental Technology Section, School of Industrial Technology, Universiti Sains Malaysia, Penang, Malaysia

H. P. S. Abdul Khalil Renewable Biomass Transformation Cluster, School of Industrial Technology, Universiti Sains Malaysia, Penang, Malaysia

Bioresource Technology Division, School of Industrial Technology, Penang, Malaysia

Charles Oluwaseun Adetunji Applied Microbiology, Biotechnology and Nanotechnology Laboratory, Department of Microbiology, Edo University Uzairue, Auchi, Edo State, Nigeria

Department of Chemistry, Federal University of Lafia, Lafia, Nasarawa State, Nigeria

Harlina Ahmad Environmental Technology Section, School of Industrial Technology, Universiti Sains Malaysia, Penang, Malaysia

Renewable Biomass Transformation Cluster, School of Industrial Technology, Universiti Sains Malaysia, Penang, Malaysia

Mardiana Idayu Ahmad Environmental Technology Division, School of Industrial Technology, Universiti Sains Malaysia, Penang, Malaysia

Renewable Biomass Transformation Cluster, School of Industrial Technology, Universiti Sains Malaysia, Penang, Malaysia

Shaz Ahmad Government Engineering College, Kaimur, Bhabua, Bihar, India

Sami Alsaadi Environmental Technology Division, School of Industrial Technology, Universiti Sains Malaysia, Penang, Malaysia

Faculty of Science, Biology Department, Hail University, Hail, Saudi Arabia

V. Anu Prasanna Department of Animal Science, Yogi Vemana University, Kadapa, Andhra Pradesh, India

Yakubu Azeh Department of Chemistry, Ibrahim Badamasi Babangida University, Lapai, Niger State, Nigeria

Piyansi Bhangar Department of Biotechnology, Brainware University, Kolkata, West Bengal, India

Fozy Binhweel Environmental Technology Division, School of Industrial Technology, Universiti Sains Malaysia, Penang, Malaysia

Goutam Biswas Department of Chemistry, Cooch Behar Panchanan Barma University, Cooch Behar, India

Soheli Biswas Department of Biotechnology, Brainware University, Kolkata, West Bengal, India

S. Bragadeeswaran Faculty of Marine Sciences, CAS in Marine Biology, Annamalai University, Parangipettai, Tamil Nadu, India

W. V. Celcia Gnana Rathinam Faculty of Marine Sciences, CAS in Marine Biology, Annamalai University, Parangipettai, Tamil Nadu, India

Sankha Chakrabortty School of Chemical Technology, Kalinga Institute of Industrial Technology, Bhubaneswar, India

T. Chandrasekhar Department of Environmental Science, Yogi Vemana University, Kadapa, Andhra Pradesh, India

Bimal Das Department of Chemical Engineering, National Institute of Technology, Durgapur, West Bengal, India

Sudipta Kumar Das Department of Biotechnology, Brainware University, Kolkata, West Bengal, India

Deepshikha Datta Department of Chemistry, Brainware University, Kolkata, West Bengal, India

A. Divyashree Department of Microbiology, School of Basic and Applied Sciences, Dayananda Sagar University, Bengaluru, India

S. Divyeswari Department of Microbiology, The American College, Madurai, Tamil Nadu, India

Widad Fadhullah Environmental Technology Section, School of Industrial Technology, Universiti Sains Malaysia, Penang, Malaysia

Renewable Biomass Transformation Cluster, School of Industrial Technology, Universiti Sains Malaysia, Penang, Malaysia

Hamza Mohamed Flafel Environmental Technology Division, School of Industrial Technology, Universiti Sains Malaysia, Penang, Malaysia

Libyan Center for Studies and Research in Environmental Science and Technology, Brack Al-Shatti, Libya

Faculty of Environment and Natural Resources, Department of Environmental Science, Wadi Al-Shatti University, Brack Al-Shatti, Libya

K. Gunamathy Faculty of Marine Sciences, CAS in Marine Biology, Annamalai University, Parangipettai, Tamil Nadu, India

K. Harinisri Ram Department of Microbiology, Sri Ramakrishna College of Arts and Science for Women, Coimbatore, India

J. Immanuel Suresh Department of Microbiology, The American College, Madurai, Tamil Nadu, India

Abel Inobeme Department of Chemistry, Edo University Uzairue, Auchi, Edo State, Nigeria

Riti Thapar Kapoor Centre for Plant and Environmental Biotechnology, Amity University, Noida, Uttar Pradesh, India

Prithviraj Karak Department of Physiology, Bankura Christian College, Bankura, West Bengal, India

Payal Karmakar Food Science and Nutrition Department, University of Minnesota, St. Paul, MN, USA

Moonis Ali Khan Chemistry Department, King Saud University, Riyadh, Saudi Arabia

Abhishek Kumar Department of Civil Engineering, National Institute of Technology Patna, Patna, India

S. Kumerasan Faculty of Marine Sciences, CAS in Marine Biology, Annamalai University, Parangipettai, Tamil Nadu, India

Deep Laha Department of Chemical Engineering, National Institute of Technology, Durgapur, West Bengal, India

Japareng Lalung Environmental Technology Division, School of Industrial Technology, Universiti Sains Malaysia, Penang, Malaysia

C. Madhuri Department of Biotechnology and Bioinformatics, Yogi Vemana University, Kadapa, Andhra Pradesh, India

Amos Mamman Department of Chemistry, Ibrahim Badamasi Babangida University, Lapai, Niger State, Nigeria

Esha Mandal Department of Biotechnology, National Institute of Technology, Durgapur, West Bengal, India

John Tsado Mathew Department of Chemistry, Ibrahim Badamasi Babangida University, Lapai, Niger State, Nigeria

J. Mohamed Asarudeen Faculty of Marine Sciences, CAS in Marine Biology, Annamalai University, Parangipettai, Tamil Nadu, India

Jyotisikha Mohapatra School of Biotechnology, Kalinga Institute of Industrial Technology, Bhubaneswar, India

Musah Monday Department of Chemistry, Ibrahim Badamasi Babangida University, Lapai, Niger State, Nigeria

Etsuyankpa Muhammad Bini Department of Chemistry, Federal University of Lafia, Lafia, Nasarawa State, Nigeria

Krishna Neeti Department of Civil Engineering, National Institute of Technology Patna, Patna, India

Anwesha Patel Department of Microbiology, School of Basic and Applied Sciences, Dayananda Sagar University, Bengaluru, India

Mohammad Qutob Environmental Technology Division, School of Industrial Technology, Universiti Sains Malaysia, Penang, Malaysia

Mohd Rafatullah Environmental Technology Division, School of Industrial Technology, Universiti Sains Malaysia, Penang, Malaysia

N. Rajesh Department of Biotechnology and Bioinformatics, Yogi Vemana University, Kadapa, Andhra Pradesh, India

Shristi Ram Department of Microbiology, School of Basic and Applied Sciences, Dayananda Sagar University, Bengaluru, India

Rekha Rani Department of Dairy Chemistry, College of Dairy and Food Technology, Jodhpur, Rajasthan, India

K. Riazunnisa Department of Biotechnology and Bioinformatics, Yogi Vemana University, Kadapa, Andhra Pradesh, India

Asmita Saha Department of Microbiology, School of Basic and Applied Sciences, Dayananda Sagar University, Bengaluru, India

Vivekananda Saha Department of Chemistry, Cooch Behar Panchanan Barma University, Cooch Behar, India

Sakshi Government Engineering College, Kaimur, Bhabua, Bihar, India

Sazlina Salleh Centre for Policy Research, Universiti Sains Malaysia, Penang, Malaysia

Centre for Marine and Coastal Studies, Universiti Sains Malaysia, Penang, Malaysia

Sayantan Sarkar Department of Chemical Engineering, National Institute of Technology, Durgapur, West Bengal, India

T. Sasidharan Faculty of Marine Sciences, CAS in Marine Biology, Annamalai University, Parangipettai, Tamil Nadu, India

Wardah Senusi Environmental Technology Division, School of Industrial Technology, Universiti Sains Malaysia, Penang, Malaysia

Elijah Yanda Shaba Department of Chemistry, Federal University of Technology Minna, Minna, Nigeria

Mohammad Aliff Shakir Environmental Technology Division, School of Industrial Technology, Universiti Sains Malaysia, Penang, Malaysia

Ehsan Shalfoh Environmental Technology Division, School of Industrial Technology, Universiti Sains Malaysia, Penang, Malaysia

Masoom Raza Siddiqui Chemistry Department, King Saud University, Riyadh, Saudi Arabia

Bhopal Singh Department of Dairy Technology, Faculty of Sciences, Dayalbagh Educational Institute, Agra, Uttar Pradesh, India

Reena Singh Department of Civil Engineering, National Institute of Technology Patna, Patna, India

Satyabrata Si School of Chemical Technology, Kalinga Institute of Industrial Technology, Bhubaneswar, India

Bhoomika Sridhar Department of Microbiology, School of Basic and Applied Sciences, Dayananda Sagar University, Bengaluru, India

S. Srikavibarathi Faculty of Marine Sciences, CAS in Marine Biology, Annamalai University, Parangipettai, Tamil Nadu, India

B. Thamarai Selvi Department of Microbiology, Sri Ramakrishna College of Arts and Science for Women, Coimbatore, India

B. Visnu Faculty of Marine Sciences, CAS in Marine Biology, Annamalai University, Parangipettai, Tamil Nadu, India

Bio-Fertilizer Synthesis from Marine Biomass: An Eco-Friendly Approach to Sustainable Agriculture

1

Sankha Chakrabortty, Satyabrata Si, Jyotisikha Mohapatra, and Prithviraj Karak

Abstract

Meeting the rising demand for food while reducing the negative effects on the environment caused by traditional farming methods is a major challenge for the agricultural sector worldwide. In addition to contributing to soil and water pollution, the widespread use of synthetic chemical fertilizers in agriculture also significantly depletes nonrenewable resources. Bio-fertilizers are a promising new option in the search for environmentally friendly substitutes. Marine biomass is an alternative and largely untapped source for bio-fertilizer production that has enormous potential for eco-friendly, nutrient-rich formulations. This abstract explores the process, benefits, challenges, and potential future applications of bio-fertilizer synthesis from marine biomass. Macroalgae (seaweeds) and microalgae are two types of marine biomass that together make up a wide range of species with a high nutrient density. These marine organisms are a rich source of nutrients because they absorb them from their environment. Many urgent agricultural and environmental problems could be alleviated if these nutrients could be extracted and processed into bio-fertilizers. Harvesting, pretreatment, extraction, and formulation are just some of the steps involved in making bio-fertilizers from marine biomass. Ecological damage can be avoided during the harvesting phase by collecting macroalgae and microalgae in a sustainable manner. The next step is pretreatment, which involves getting rid of things like salts and

S. Chakrabortty (✉) · S. Si
School of Chemical Technology, Kalinga Institute of Industrial Technology, Bhubaneswar, India
e-mail: sankha.chakrabortty@kiitbiotech.ac.in

J. Mohapatra
School of Biotechnology, Kalinga Institute of Industrial Technology, Bhubaneswar, India

P. Karak
Department of Physiology, Bankura Christian College, Bankura, West Bengal, India

M. Rafatullah et al. (eds.), *Multidisciplinary Applications of Marine Resources*,
https://doi.org/10.1007/978-981-97-5057-3_1

organic matter that are not wanted. Macro and micronutrients like nitrogen, phosphorus, potassium, and trace elements are extracted using methods like enzymatic hydrolysis and bio-refinery processes. Careful formulation of these extracted nutrients into bio-fertilizers makes them compatible with widespread agricultural practices. Bio-fertilizers made from marine biomass have numerous advantages. First, they are a greener option than chemical fertilizers and help cut down on pollution and climate change byproducts. These bio-fertilizers improve crop yields and quality by boosting soil structure, nutrient retention, and microbial activity. Additionally, a blue bio-economy can flourish, providing advantages to coastal communities and encouraging ocean conservation, thanks to the sustainable harvesting of marine biomass. However, there are still barriers to the widespread use of bio-fertilizers made from marine biomass. Extraction and formulation methods must be fine tuned through research and development to achieve maximum efficiency and scalability. Standardizing production processes and guaranteeing the safety and quality of bio-fertilizers call for the establishment of regulatory frameworks. Successful adoption also requires greater publicity and understanding among farmers and other relevant parties. Bio-fertilizers made from marine biomass have bright future prospects. The use of marine biomass for bio-fertilizer production is in line with international targets for environmentally responsible and economically viable food production, as sustainability becomes imperative in agriculture. More study and development in this area could have far-reaching effects on marine ecosystem preservation and the agricultural industry. The synthesis of bio-fertilizer from marine biomass represents a novel and environmentally sound strategy for meeting the challenges of contemporary agriculture. This area has a lot of potentials to improve farming in the future because of its environmental benefits, the chance to increase crop yields, and the backing of a growing blue bio-economy. To overcome obstacles and reach its full potential, however, researchers, policymakers, and stakeholders must work together.

Keywords

Agriculture · Biofertilizer · Environment · Marine biomass · Sustainable

1.1 Introduction

The specter of a burgeoning global population casts a long shadow over the agricultural sector, posing an unprecedented challenge: how to nourish billions while safeguarding the very environment that sustains us. Conventional farming practices, while undeniably productive, have left deep scars on the Earth. The widespread reliance on synthetic fertilizers, for instance, has become a double-edged sword, boosting yields but exacting a heavy toll on soil and water quality, jeopardizing long-term sustainability [1]. In the face of this mounting pressure, the agricultural community urgently seeks solutions that can reconcile productivity with environmental responsibility. Bio-fertilizers, formulated from organic matter and teeming with beneficial microbes, offer a beacon of hope in this quest for sustainable agriculture [2]. Among the diverse candidates for bio-fertilizer production, marine biomass shines brightly as a largely untapped treasure trove. Encompassing a breathtaking array of macroalgae (seaweeds) and microalgae, this marine bounty boasts remarkable nutrient density. These organisms, through their inherent ability to absorb nutrients from their environment, serve as natural storehouses of vital elements like nitrogen, phosphorus, and potassium [3]. Harnessing this inherent richness to create bio-fertilizers presents a potent opportunity to address crucial agricultural and environmental concerns.

Transforming this marine bounty into potent bio-fertilizers involves a meticulous dance of science and sustainability. The journey begins with sustainable harvesting practices, ensuring minimal ecological impact during the collection of macroalgae and microalgae [4]. This is followed by a meticulous pretreatment phase, aimed at removing unwanted elements such as salts and organic matter that could hinder the extraction process. The heart of the process lies in the extraction stage, where various techniques, including enzymatic hydrolysis and bio-refinery processes, are employed to liberate macro and micronutrients. Finally, careful formulation ensures the extracted nutrients are transformed into bio-fertilizers compatible with existing agricultural practices, maximizing their accessibility and impact. The advantages of bio-fertilizers derived from marine biomass extend far beyond their impressive nutrient content. They offer a greener alternative to conventional fertilizers, reducing pollution and mitigating the harmful byproducts associated with climate change. By improving soil structure, nutrient retention, and microbial activity, these bio-fertilizers contribute to enhanced crop yields and quality, creating a virtuous cycle of abundance and environmental well-being. Furthermore, the sustainable harvesting of marine biomass can stimulate the development of a thriving blue bio-economy, generating economic benefits for coastal communities while incentivizing ocean conservation efforts [5]. Imagine vibrant coastal communities bustling with activity, fueled by the responsible harvesting and processing of marine biomass, while pristine ocean ecosystems flourish alongside thriving agricultural landscapes—a powerful testament to the harmonious relationship between humanity and nature. However, the path to widespread adoption of marine-based bio-fertilizers is not without its challenges. Research and development efforts are crucial to optimize extraction and formulation methods, ensuring maximum efficiency and scalability. Robust regulatory frameworks are needed to standardize production processes and guarantee the safety and quality of these bio-fertilizers, fostering trust and confidence among farmers and policymakers [4]. Moreover, increasing awareness and education among farmers and stakeholders is vital for successful adoption. Overcoming these hurdles requires a collaborative effort, bringing together researchers, policymakers, farmers, and consumers in a shared commitment to sustainable agriculture.

Despite these challenges, the future of bio-fertilizers derived from marine biomass is brimming with promise. This approach aligns seamlessly with international aspirations for environmentally responsible and economically viable food production. Continued research and development hold immense potential to transform the agricultural landscape, safeguarding marine ecosystems while bolstering agricultural productivity. This novel and environmentally sound strategy represents a significant step toward overcoming the challenges of contemporary agriculture, paving the way for a more sustainable and prosperous future. As we embark on this exciting journey, let us remember that the vast potential of marine biomass lies not just in its ability to nourish our crops, but also in its power to forge a new paradigm of agriculture—one that is in harmony with nature, ensures food security for all, and paves the way for a thriving blue future.

1.2 Marine Biomass as a Source for Bio-fertilizers

1.2.1 Natural Marine Biomass Used as a Source for Bio-fertilizers

The complex and widely distributed group of photosynthetic organisms known as seaweed is essential to aquatic environments [6]. The aquatic plant kingdom Thallophyta includes seaweeds, commonly referred to as macroalgae, which are regarded as a vital component of marine ecology and lives in coastal waters [7]. There are thought to be 9000 species of macroalgae. Based on the existence of photosynthetic pigment, storage capacity, and other factors, macroalgae can be divided into three primary groups based on components of food products' cell walls and include Rhodophyta (red), Phaeophyta (brown), and Chlorophyta (green) [8]. Seaweeds are typically found clinging to the bottom of rather shallow coastal waters. A new generation of natural organic fertilizers called seaweed extract contains an effective nutrient source that encourages faster seed germination, increased crop yield, and increased crop resistance [9]. Seaweed extracts enhance the biological characteristics of the soil and raise output under biotic and abiotic stress. Brown macroalgae, such as *Ascophyllum nodosum* (L.), are most frequently utilized in agriculture. In addition to *Ascophyllum nodosum*, other brown algae such as *Ecklonia maxima* (Osbeck) Papenfuss, Fucus spp., Laminaria spp., Sargassum spp., and Turbinaria spp. serve as biofertilizers. Mediterranean red algae like Corralina, Green algae, *Pterocladia pinnata* (Hudson) Papenfuss, and *Jania rubens* (L.) J.V. such as *Enteromorpha intestinalis* (L.) Nees, *Cladophora dalmatica* Kützing, and *Ulva lactuca* L. are also employed as a biostimulant for plant growth [5]. Algal extracts and inorganic fertilizers can be combined, which might promote sustainable agricultural productivity [10]. They can be used in various ways, including foliar sprays, soil additives, and on seeds. Demonstrating a broad range of applications, many beneficial effects are found, such as improved germination, radicular system development, better fruit quality, increased leaf area and chlorophyll content, improved crop output, vitality, strong resilience to biotic and abiotic stress, and extended postharvest shelf-life products. Macroalgae have the ability to increase crop yield, improve seed germination, improve soil characteristics, increase growth and quality of crops, and improve abiotic stress tolerance.

Some marine macroalgae are used as biofertilizers by mixing their pulverized form with the soil. They contain unexplored reservoirs of naturally occurring physiologically active chemicals [7]. They serve as abundant sustainable botanical bioresources. Due to their high quantities of organic matter, which enriches soil with nutrients, marine macroalgae are effective biofertilizers. Additionally, when applied in sufficient amounts, they were superior and a more acceptable alternative to chemical and mineral fertilizers [11]. Marine macroalgae have numerous uses as eco-friendly fertilizers in contemporary agriculture and horticulture crop enhancement (in finely powdered form) [12]. The application of seaweed as a soil amendment was effective in enhancing plant growth. Many parts of the world employ seaweed manure to improve agricultural soil [13].

While other marine macroalgae are used as liquid biofertilizers and are sprayed over the soil and plants. Due to their high levels of organic matter, micro and macro elements, vitamins, and other nutrients, liquid fertilizers made from natural sources, such as seaweed, are effective fertilizer alternatives for agricultural crops. Fatty acids and growth regulators are abundant. In agriculture and horticulture, bioactive compounds derived from marine algae are used, and numerous positive results in terms of improvement in productivity and quality have been observed. Seaweed extracts have the potential to replace chemical fertilizers with environmentally friendly liquid biofertilizers, which is essential for organic agricultural techniques leading to sustainable agriculture. A variety of liquid fertilizers from seaweed are applied as a foliar spray.

1.2.2 Waste Marine Biomass, a Source for Bio-fertilizers

Marine trash is typically utilized as a fertilizer in sea areas due to the fact that it includes a significant number of nutrients (nitrogen and phosphorus). Biotransformation is the process by which organic materials can be transformed into complex materials by the activity of bacteria or fungi [14]. Composting is classified as a biotransformation operation. There are four stages that are involved in the process of composting. These stages are as follows: (1) the decomposition phase, (2) the thermophilic phase, (3) the second thermophilic phase, and (4) the maturation phase. The final product that is utilized for the purpose of enhancing the fertility and texture of soil results in an increase in the ability to store carbon, a reduction in the utilization of synthetic fertilizers, and ultimately a reduction in the emission of greenhouse gases. On one side, the chemical in question has a significant potential for use in agricultural settings, and on the other hand, it is recommended that it be stabilized in order to prevent the problems that are associated with the appearance of phytotoxic components [15].

1.2.3 Nutrient Content in Marine Biomass

Marine biomass, a vast and diverse collection of living organisms inhabiting the world's oceans, is an essential component of Earth's biosphere. From microscopic phytoplankton to majestic whales, marine biomass encompasses a wide range of species that play crucial roles in marine ecosystems and global nutrient cycles. Understanding the nutrient content within marine biomass is fundamental to comprehending its ecological significance, its potential for human utilization, and its role in sustaining life on our planet. Research into the nutrient content of marine biomass has revealed the abundance and diversity of essential elements within these organisms. Nitrogen, phosphorus, potassium, calcium, magnesium, and various trace minerals are among the key nutrients found in marine biomass [16, 17]. These nutrients are essential for the growth, development, and reproduction of marine organisms, as well as for the functioning of marine ecosystems as a whole (Fig. 1.1). One of the most abundant and widely studied components of marine biomass is seaweed, also known as marine algae. Seaweeds are renowned for their high nutrient content, particularly in nitrogen and phosphorus. Studies have shown that the nitrogen content of seaweeds can range from 1% to 4% of dry weight, making them a significant source of this essential nutrient [19]. Additionally, seaweeds are rich in other nutrients such as potassium, calcium, and magnesium, which are vital for plant and animal nutrition. Fish and shellfish are another important component of marine biomass, providing valuable nutrients for human consumption. These organisms are rich sources of protein, omega-3 fatty acids, vitamins, and minerals, including calcium and phosphorus. Consuming fish and shellfish as part of a balanced diet

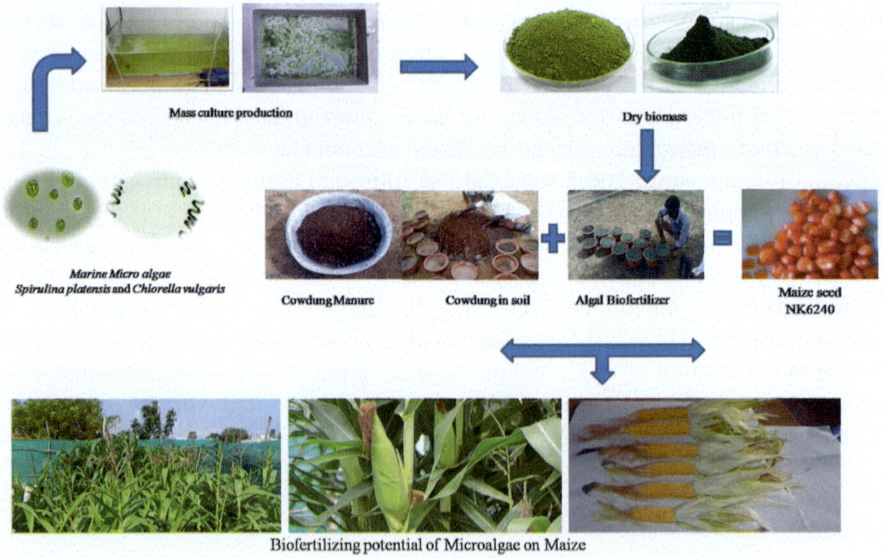

Fig. 1.1 Effect of marine microalgal fertilizer on the growth and yield of maize [18]

can provide numerous health benefits, including cardiovascular health and cognitive function [20]. In addition to macroscopic organisms like seaweeds, marine microorganisms also play critical roles in nutrient cycling and ecosystem functioning. Phytoplankton, for example, are microscopic algae that form the base of the marine food web. These tiny organisms are prolific producers of organic matter and are responsible for a significant portion of the Earth's oxygen production through photosynthesis. Phytoplankton also play a crucial role in the global nitrogen cycle by fixing atmospheric nitrogen into organic compounds, making this essential nutrient available to other marine organisms [21]. Understanding the nutrient content of marine biomass has important implications for various human activities, including fisheries, aquaculture, and marine conservation efforts. Sustainable management of marine resources requires knowledge of nutrient availability and distribution within marine ecosystems. By understanding which nutrients are present in marine biomass and how they are utilized by different organisms, scientists and resource managers can develop strategies to promote ecosystem health and resilience. Furthermore, the nutrient content of marine biomass has implications for human nutrition and food security. Incorporating marine-derived nutrients into the human diet can provide essential vitamins, minerals, and omega-3 fatty acids that are beneficial for health. Additionally, exploring alternative sources of marine biomass for food and feed production, such as seaweeds and marine microorganisms, has the potential to alleviate pressure on traditional fisheries and contribute to sustainable food systems. Marine biomass is a rich and diverse source of essential nutrients that play critical roles in marine ecosystems and global nutrient cycles. Understanding the nutrient content of marine biomass is essential for comprehending its ecological significance, its potential for human utilization, and its role in sustaining life on our planet. Further research into marine biomass nutrition holds promise for addressing challenges related to food security, human health, and environmental sustainability in the future.

1.2.4 Sustainability Considerations in Marine Biomass Harvesting

Sustainability considerations in marine biomass harvesting are paramount due to the delicate balance of marine ecosystems and the potential impact of human activities. Harvesting marine biomass, whether for food, pharmaceuticals, or other purposes, must be conducted in a manner that ensures the long-term health and productivity of oceanic ecosystems. Factors such as overfishing, habitat destruction, bycatch, and pollution can all pose significant threats to the sustainability of marine biomass harvesting practices. Overfishing, in particular, can deplete fish stocks to unsustainable levels, disrupting marine food webs and endangering species. Bycatch, the unintentional capture of nontarget species, can also have detrimental effects on marine biodiversity and ecosystem health. To address these sustainability challenges, various management measures and regulations have been implemented at local, regional, and international levels. These may include fishing quotas, gear restrictions, marine protected areas, and certification schemes such as the Marine Stewardship Council (MSC) for sustainable seafood. Furthermore, advancements in technology and innovation are enabling more sustainable practices in marine biomass harvesting. This includes the development of selective fishing gear to reduce bycatch, the implementation of ecosystem-based management approaches, and the promotion of sustainable aquaculture practices as alternatives to wild harvesting. However, achieving true sustainability in marine biomass harvesting requires a holistic approach that considers ecological, social, and economic factors [22]. Collaboration among governments, industry stakeholders, scientists, and local communities is essential to develop and implement effective management strategies that balance conservation objectives with socio-economic needs. Overall, sustainability considerations in marine biomass harvesting are critical for safeguarding the health and resilience of marine ecosystems, preserving biodiversity, and ensuring the long-term viability of marine resources for future generations.

1.3 Benefits of Bio-fertilizers from Marine Biomass

Since synthetic fertilizers originate from chemicals, their over-application causes eutrophication, which is harmful to both soil and plants [23]. These fertilizers seep into adjacent streams and rivers, contaminating water and harming fish and other aquatic animals. Conversely, plants gradually and safely deliver nutrients to the soil. Many plants used for soil fertilization are initially planted as cover crops to mitigate soil erosion from wind and rain [24, 25]. The enhancement of soil fertility necessitates the maintenance of a healthy topsoil layer. Hence, scientists are currently exploring new technologies to produce eco-friendly fertilizers from living organisms, particularly microbes, which can biodegrade without leaving any residues and are known as biofertilizers [26]. Thus, biofertilizers are natural or organic substances containing living microorganisms, primarily bacteria, fungi, or algae, which enhance soil fertility and promote plant growth by fixing atmospheric nitrogen, solubilizing phosphorus, or producing plant growth-promoting substances. Figure 1.2 shows the pictorial representation of the benefits of bio-fertilizers for soil fertility and crop productivity. Unlike chemical fertilizers, which provide nutrients directly to plants, biofertilizers work symbiotically with plants and soil microorganisms to improve soil health and nutrient availability. There are several types of biofertilizers, each with specific modes of action and benefits. Bio-fertilizers derived from marine biomass offer multiple benefits for sustainable agriculture, including nutrient enrichment, soil health improvement, environmental conservation, and cost effectiveness [27–29]. Their widespread adoption has the potential to contribute to more resilient and environmentally friendly agricultural practices while enhancing food security and livelihoods worldwide [30].

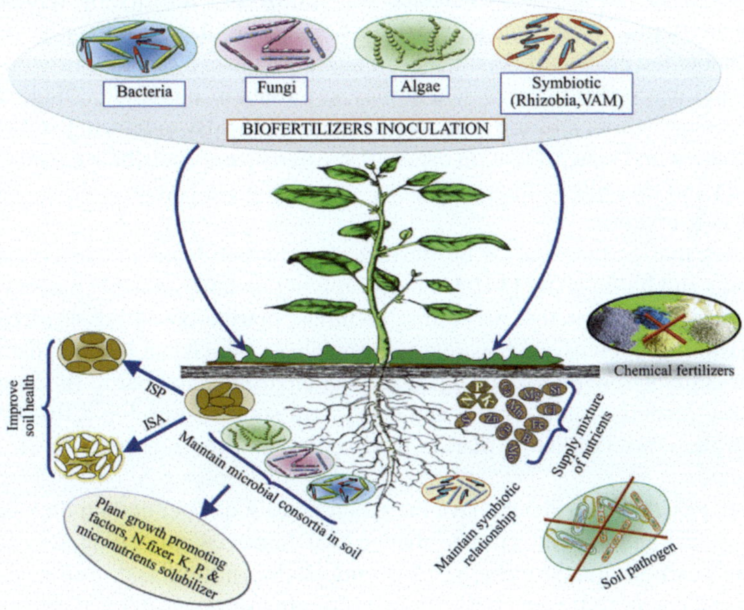

Fig. 1.2 The benefit of biofertilizers toward soil fertility and crop productivity [27]

1.4 Environmental Benefits

Marine-based bio-fertilizers contribute to environmental sustainability and resilience in the face of climate change [23]. Bio-fertilizers derived from marine biomass offer significant environmental benefits, including reduced dependence on chemical fertilizers, decreased soil and water pollution, improved soil health, conservation of marine resources, and carbon sequestration, which are described below.

Marine-based bio-fertilizers provide a natural and sustainable alternative to synthetic chemical fertilizers [31]. By reducing the dependence on chemical inputs, particularly N- and P-based fertilizers derived from fossil fuels, the environmental burden associated with their production, transportation, and application is mitigated [32]. This helps to minimize the release of greenhouse gases, such as CO_2 and N_2O, during fertilizer manufacturing and application, thus contributing to climate change mitigation efforts.

Unlike chemical fertilizers, which can leach into groundwater and surface water bodies, marine-based bio-fertilizers have minimal environmental impact and do not contribute to soil and water pollution [33, 34]. They are biodegradable and non-toxic, posing no risk of contaminating water sources or harming aquatic ecosystems. By promoting nutrient cycling and reducing nutrient runoff, bio-fertilizers help to maintain water quality and protect biodiversity in freshwater and marine environments.

Marine-based bio-fertilizers enhance soil health and fertility by replenishing essential nutrients, improving soil structure, and stimulating microbial activity. They promote the growth of beneficial soil microorganisms, such as nitrogen-fixing bacteria and mycorrhizal fungi, which play key roles in nutrient cycling, organic matter decomposition, and soil structure formation [35]. By enhancing soil organic carbon content and water retention capacity, bio-fertilizers contribute to soil erosion prevention and drought resilience, thereby supporting ecosystem stability and biodiversity conservation.

Utilizing marine biomass for bio-fertilizer production helps to conserve marine ecosystems and biodiversity [34]. Instead of relying on unsustainable practices such as overfishing or destructive harvesting methods, marine-based bio-fertilizers utilize renewable resources, such as seaweed and algae, which can be sustainably harvested or cultivated without causing significant environmental harm. By promoting responsible resource management and ecosystem stewardship, bio-fertilizers contribute to the conservation of marine biodiversity and the preservation of coastal ecosystems.

Marine-based bio-fertilizers have the potential to sequester carbon dioxide from the atmosphere and mitigate climate change [36]. Seaweed cultivation, for example, can act as a carbon sink, absorbing CO_2 during photosynthesis and storing carbon in biomass and ocean sediments [37]. By incorporating seaweed-derived organic matter into agricultural soils, bio-fertilizers enhance soil carbon sequestration and contribute to climate change mitigation efforts. Additionally, the cultivation of marine biomass for bio-fertilizer production can help to reduce greenhouse gas emissions associated with conventional agricultural practices, such as fertilizer production and land-use change.

1.5 Agricultural Benefits

Bio-fertilizers derived from marine biomass offer significant agricultural benefits, including nutrient enrichment, soil conditioning, sustainable nutrient management, improved crop quality, reduced environmental footprint, adaptability to organic farming practices, and cost-effectiveness [37]. By harnessing the natural potential of marine resources, bio-fertilizers contribute to sustainable agriculture, food security, and environmental conservation.

Marine-based bio-fertilizers are rich sources of essential nutrients, including nitrogen, phosphorus, potassium, calcium, magnesium, and micronutrients [34, 37]. These nutrients are readily available in forms easily assimilated by plants, promoting healthy growth, flowering and fruiting. By replenishing soil fertility and correcting nutrient deficiencies, bio-fertilizers support optimal crop development and yield potential [38].

Marine-based bio-fertilizers improve soil structure, texture, and fertility, improving soil tilth and fostering aeration, water retention and root penetration [39, 40]. They contribute organic matter to the soil, increasing microbial activity and nutrient cycling, and reducing soil compaction and erosion. This leads to healthier, more resilient soils that are better able to support vigorous plant growth and withstand environmental stresses.

Utilizing marine-based bio-fertilizers enables farmers to adopt sustainable nutrient management practices, reducing reliance on synthetic chemical fertilizers and minimizing the risk of nutrient leaching, runoff, and pollution [41, 42]. By promoting nutrient cycling and soil organic matter accumulation, bio-fertilizers support long-term soil health and productivity, contributing to sustainable agricultural systems and ecosystem resilience.

The application of marine-based bio-fertilizers can lead to improvements in crop quality, including increased nutrient content, enhanced flavor, color, and aroma and extended shelf life [43, 44]. By providing balanced nutrition and promoting plant vigor and stress resistance, bio-fertilizers help to optimize crop yield and quality, ensuring high marketability and consumer satisfaction.

Marine-based bio-fertilizers have lower environmental impacts compared to synthetic chemical fertilizers [23, 26]. They are biodegradable, nontoxic, and environmentally friendly, posing minimal risks to soil, water, and air quality. By reducing greenhouse gas emissions, nutrient runoff, and chemical residues in the environment, bio-fertilizers contribute to environmental conservation and mitigate climate change.

Marine-based bio-fertilizers are compatible with organic and sustainable farming practices, supporting certification requirements and principles of ecological stewardship [45]. They can be used in organic farming systems as alternatives to synthetic inputs, helping farmers meet organic certification standards and fulfill consumer demand for sustainably produced food.

While initial investment and production costs may vary, marine-based bio-fertilizers offer long-term cost savings for farmers by reducing the need for expensive synthetic fertilizers and agrochemical inputs. By improving soil fertility and crop productivity, bio-fertilizers contribute to sustainable productivity gains and economic viability in agriculture, enhancing farm profitability and livelihoods.

1.6 Challenges in Adoption

The rising adoption of biofertilizers represents a recent effort to enhance crop yields and soil fertility by harnessing their ability to fix atmospheric nitrogen, solubilize insoluble phosphate in the soil biologically, and produce growth hormones and vitamins [28]. Additionally, biofertilizers aid in improving soil structure, texture, and water-holding capacity within the agricultural sector [41]. A key and beneficial aspect of biofertilizers is their significant role in reducing environmental pollution and promoting agroecological sustainability [28]. Biofertilizers are economically advantageous for farmers due to their affordability and their critical role in supplying essential nutrients like nitrogen and phosphorus to crop plants. But there still remain many challenges that need to be addressed for their ease of adoption to the mass community.

1.7 Research and Development Challenges

Addressing the research and development challenges is essential for advancing the adoption of biofertilizers and realizing their potential to contribute to sustainable agriculture, food security, and environmental conservation. Some of the key challenges under this category are as follows:

There is a need to explore and understand the microbial diversity present in different soil types and environments to identify novel strains with desirable traits [46]. Moreover, ensuring the stability and viability of microbial inoculants during production, storage, and application is crucial for their effectiveness in the field.

Understanding the mechanisms of compatibility and synergistic interactions among microbial strains is essential to develop strategies for optimizing biofertilizer formulations and maximizing their efficacy [47].

Research endeavors need to be developed for effective formulations and delivery mechanisms for biofertilizers under varying environmental conditions, such as temperature, moisture, and soil pH, to enhance the shelf life, stability, and efficacy of biofertilizer products [48].

Research is needed to establish quality assurance protocols, including microbial purity, viability, and potency testing, for maintaining product consistency and efficacy [49].

Conducting large-scale field trials under diverse agroecological conditions is necessary to evaluate the performance, effectiveness, and economic viability of biofertilizer products [28].

Research is needed to assess the long-term impacts of biofertilizer application on soil fertility, crop yield, and ecosystem health, as well as to develop strategies for integrating biofertilizers into existing agricultural practices [30, 50].

Addressing knowledge gaps and promoting awareness about the benefits of biofertilizers among farmers, extension agents, and policymakers is essential for fostering their adoption. Collaboration among researchers, policymakers, industry stakeholders, and farmers is the key to promoting innovation in biofertilizer technology.

1.8 Regulatory Considerations

Addressing the regulatory considerations and challenges is essential for fostering the responsible development, registration, and adoption of biofertilizers, ensuring their safe and sustainable use in agriculture while facilitating market access and innovation in the biofertilizer industry [31, 51]. Some of these challenges include:

Developing harmonized regulations that define the criteria for biofertilizer registration, labeling, and marketing approval is crucial for streamlining the regulatory process and facilitating market access.

Unlike chemical fertilizers, which are subject to well-established registration procedures, biofertilizers may face additional regulatory hurdles due to their microbial nature and variability. Regulatory agencies may require comprehensive data on microbial characterization, safety, efficacy, and environmental impact assessment, posing challenges for biofertilizer manufacturers in meeting regulatory requirements and obtaining approval for commercialization. Establishing quality assurance protocols, Good Manufacturing Practices (GMP), and quality control measures is essential to meet regulatory standards and ensure product consistency and integrity throughout the manufacturing process. Conducting field trials, monitoring studies and ecological risk assessments can provide valuable data to inform regulatory decision-making and mitigate potential environmental risks associated with biofertilizer adoption. Compliance with labeling regulations, advertising standards, and claims substantiation requirements is essential to prevent misleading or false marketing practices and maintain consumer confidence in biofertilizer products. Promoting international collaboration, mutual recognition agreements, and harmonization efforts among regulatory agencies can facilitate cross-border trade and support the global adoption of biofertilizers.

1.9 Conclusions

As the world grapples with feeding a growing population while safeguarding our environment, traditional agriculture's reliance on polluting synthetic fertilizers poses a significant challenge. Bio-fertilizers derived from marine biomass offer a promising solution. These eco-friendly alternatives are rich in nutrients, readily absorbed by plants, and hold the potential to improve soil health, boost crop yields, and reduce environmental impact. Macroalgae and microalgae, abundant in our oceans, serve as a largely untapped source for this revolution. Harvesting them sustainably, followed by careful extraction of vital nutrients like nitrogen, phosphorus, and potassium, paves the way for bio-fertilizer formulation. These formulations not only contribute to a greener future by minimizing pollution and climate change but also enhance crop quality and performance through improved soil structure and microbial activity. Furthermore, the sustainable harvesting of marine biomass fosters a thriving blue bio-economy, benefiting coastal communities, and encouraging ocean conservation. However, challenges remain. Refining extraction and formulation methods, establishing standardized production processes, and ensuring safety and quality through regulations are key steps toward wider adoption. Additionally, raising awareness among farmers and stakeholders is crucial to overcome hesitations and encourage widespread use. The future of bio-fertilizers from marine biomass is bright, aligning with global sustainability goals for environmentally responsible and economically viable food production. By investing in research, development, and collaborative efforts, we can unlock the immense potential of this approach, paving the way for a more sustainable and bountiful agricultural future for generations to come.

References

1. Tilman D, Cassman K, Matson P et al (2002) Agricultural sustainability and intensive production practices. Nature 418:671–677. https://doi.org/10.1038/nature01014
2. Bashan A, Bartsch R, Kantelhardt J et al (2012) Network physiology reveals relations between network topology and physiological function. Nat Commun 3:702. https://doi.org/10.1038/ncomms1705
3. Khan S, Siddique R, Shereen MA, Ali A, Liu J, Bai Q, Bashir N, Xue M, Correction for Khan et al (2020) Emergence of a novel coronavirus, severe acute respiratory syndrome Coronavirus 2: biology and therapeutic options. J Clin Microbiol 58(8):e01297–e01220. https://doi.org/10.1128/JCM.01297-20. Erratum for: J Clin Microbiol. 2020 Apr 23;58(5): PMID: 32703895; PMCID: PMC7383521
4. Food and Agriculture Organization of the United Nations (FAO) (2016) ISBN: 978-92-5-109374-, http://www.fao.org/3/a-i6030e.pdf
5. Aghofack J, Schinzoumka P, Tatchago V (2015) Efets des extraits ou de la poudre de Spirulina platensis et Jatropha curcas sur la croissance et le développement de la tomateajol.info. J Appl Biosci. https://doi.org/10.4314/jab.v90i1.2
6. Egan S, Harder T, Burke C et al (2013) The seaweed holobiont: understanding seaweed-bacteria interactions. FEMS Microbiol Rev 37:462–476. https://doi.org/10.1111/1574-6976.12011
7. Nabti E, Jha B, Hartmann A (2017) Impact of seaweeds on agricultural crop production as biofertilizer. Int J Environ Sci Technol 14:1119–1134. https://doi.org/10.1007/S13762-016-1202-1
8. Khan W, Rayirath UP, Subramanian S et al (2009) Seaweed extracts as biostimulants of plant growth and development. J Plant Growth Regul 28:386–399
9. Selvam GG, Sivakumar K (2013) Efect of foliar spray from seaweed liquid fertilizer of Ulva reticulata (Forsk.) on Vigna mungo L. and their elemental composition using SEM–energy dispersive spectroscopic analysis. Asian Pac J Reprod 2:119–125. https://doi.org/10.1016/S2305-0500(13)60131-1
10. Coppens J, Grunert O, van den Hende S et al (2016) The use of microalgae as a high-value organic slow-release fertilizer results in tomatoes with increased carotenoid and sugar levels, vol 28. Springer, pp 2367–2377. https://doi.org/10.1007/s10811-015-0775-2
11. Mirparsa T, Ganjali HR, Dahmardeh M (2016) The efect of bio fertilizers on yield and yield components of sunfower oil seed and nut. Int J Agric Biosci 5:46–49
12. Hernández-Herrera RM, Santacruz-Ruvalcaba F, Ruiz-López MA et al (2014) Effect of liquid seaweed extracts on growth of tomato seedlings (Solanum lycopersicum L.). J Appl Phycol 26:619–628. https://doi.org/10.1007/S10811-013-0078-4
13. Eyras M, Rostagno C, Defossé G (2013) Biological evaluation of seaweed composting. Compost Sci Util 6:74–81. https://doi.org/10.1080/1065657X.1998.10701943
14. Arvanitoyannis IS (2010) Irradiation of food commodities: techniques, applications, detection legislation, safety and consumer opinion. Academic Press, Cambridge, MA, p 763
15. Mata-Gómez LC, Montañez JC, Méndez-Zavala A, Aguilar CN (2014) Biotechnological production of carotenoids by yeasts: an overview. Microb Cell Factories 13:12. https://doi.org/10.1186/1475-2859-13-12
16. Hillebrand H, Worm B, Lotze HK (2000) Marine microbenthic community structure regulated by nitrogen loading and grazing pressure. Mar Ecol Prog Ser 204:27–38
17. Leal MC, Pimentel T, Ricardo F, Rosa R, Calado R, Rosa R (2013) First insights into the gut microbiota of a deep-sea bone-eating snail (Protanthea simplex) using Illumina next-generation sequencing. PeerJ 1:e608
18. Dineshkumar R, Subramanian J, Gopalsamy J et al (2019) The impact of using microalgae as biofertilizer in maize (Zea mays L.). Waste Biomass Valor 10:1101–1110. https://doi.org/10.1007/s12649-017-0123-7
19. Bolton JJ, Lüning K (1982) Optimal growth and maximal survival temperatures of Atlantic Laminaria species (Phaeophyta) in culture. Mar Biol 66(1):89–94

20. Kris-Etherton PM, Harris WS, Appel LJ (2002) Fish consumption, fish oil, omega-3 fatty acids, and cardiovascular disease. Circulation 106(21):2747–2757
21. Capone DG, Zehr JP, Paerl HW, Bergman B, Carpenter EJ (1997) Trichodesmium, a globally significant marine cyanobacterium. Science 276(5316):1221–1229
22. Hilborn R, Orensanz J, Parma AM (2005) Institutions, incentives and the future of fisheries. Philos Trans Royal Soc B: Biol Sci 360(1453):47–57
23. Ammar EE, Rady HA, Khattab AM, Amer MH, Mohamed SA, Elodamy NI, AL-Farga A, Ahmed AA (2023) A comprehensive overview of eco-friendly bio-fertilizers extracted from living organisms. Environ Sci Pollut Res 30:113119
24. Scavo A, Fontanazza S, Restuccia A, Pesce GR, Abbate C, Mauromicale G (2022) The role of cover crops in improving soil fertility and plant nutritional status in temperate climates. A review. Agron Sustain Dev 42:93
25. Koudahe K, Allen SC, Djaman K (2022) Critical review of the impact of cover crops on soil properties. Int Soil Water Conserv Res 10:343–354
26. Kumar S, Diksha SS, Sindhu RK (2022) Biofertilizers: an ecofriendly technology for nutrient recycling and environmental sustainability. Curr Res Microb Sci 3:100094
27. Mahmud AA, Upadhyay SK, Srivastava AK, Bhojiya AA (2021) Biofertilizers: a Nexus between soil fertility and crop productivity under abiotic stress. Curr Res Environ Sustain 3:100063
28. Dineshkumar R, Ahamed Rasheeq A, Arumugam A, Nathiga Nambi KS, Sampathkumar P (2019) Marine microalgal extracts on cultivable crops as a considerable bio-fertilizer: a review. Indian J Tradit Knowl 18:849–854
29. Hashem HA, Mansour HA, El-Khawas SA, Hassanein RA (2019) The potentiality of marine macro-algae as bio-fertilizers to improve the productivity and salt stress tolerance of canola (Brassica napus L.) plants. Agronomy 9:146
30. Daniel AI, Fadaka AO, Gokul A, Bakare OO, Aina O, Fisher S, Burt AF, Mavumengwana V, Keyster M, Klein A (2022) Biofertilizer: the future of food security and food safety. Microorganisms 10:1220
31. Trujillo-Tapia MN, Ramírez-Fuentes E (2015) Bio-fertilizer: an alternative to reduce chemical fertilizer in agriculture. J Glob Agric Ecol 4:99–103. https://www.researchgate.net/publication/283014984
32. Kandpal G (2021) Review on impact of chemical fertilizers on environment. Int J Mod Agric 10:758–763
33. Khan MN, Mobin M, Abbas ZK, Alamri SA (2017) Fertilizers and their contaminants in soils, surface and groundwater. Encycl Anthr:225–240
34. Emadodin I, Reinsch T, Rotter A, Orlando-Bonaca M, Taube F, Javidpour J (2020) A perspective on the potential of using marine organic fertilizers for the sustainable management of coastal ecosystem services. Environ Sustain 3:105–115
35. Alexandraki V, Effie T, Konstantinos P, Wilhelm H (2013) Status and trends of the conservation and sustainable use of micro-organisms in food processes. Food and Agriculture Organization, pp 1–7
36. Mona S, Malyan SK, Saini N, Deepak B, Pugazhendhi A, Kumar SS (2021) Towards sustainable agriculture with carbon sequestration, and greenhouse gas mitigation using algal biochar. Chemosphere 275:129856
37. Zafar A, Ali I, Rahayu F (2022) Marine seaweeds (biofertilizer) significance in sustainable agricultural activities: a review. In: IOP Conference series: earth environmental science. p. 012080
38. Sharma B, Yadav L, Pandey M, Shrestha J (2022) Application of biofertilizers in crop production: a review. Peruvian J Agron 6:13–31
39. Badu Brempong M, Addo-Danso A (2022) Improving soil fertility with organic fertilizers. In: Turan M, Yildirim E (eds) New generation of organic fertilizers. IntechOpen, p 103944
40. Assefa S (2019) The principal role of organic fertilizer on soil properties and agricultural productivity: a review. Agric Res Technol Open Access J 22

41. Boschiero M, De Laurentiis V, Caldeira C, Sala S (2023) Comparison of organic and conventional cropping systems: a systematic review of life cycle assessment studies. Environ Impact Assess Rev 102:107187
42. Gamage A, Gangahagedara R, Gamage J, Jayasinghe N, Kodikara N, Suraweera P, Merah O (2023) Role of organic farming for achieving sustainability in agriculture. Farming Syst 1:100005
43. Sadafzadeh E, Javanmard A, Amani Machiani M, Sofo A (2023) Application of bio-fertilizers improves forage quantity and quality of Sorghum (Sorghum bicolor L.) intercropped with soybean (Glycine max L.). Plan Theory 12:2985
44. Moloşag A, Pârvulescu OC, Ion VA, Asănică AC, Soane R, Moţ A, Dobrin A, Frîncu M, Løes AK, Cabell J, Salifoglou A, Maroulis M, Matsia S, Bujor OC, Egri D, Dobre T, Bădulescu LA, Lagunosvchi-Luchian V (2023) Effects of marine residue-derived fertilizers on strawberry growth, nutrient content, fruit yield and quality. Agronomy 13:1221
45. Naseer I, Ahmad M, Nadeem SM, Ahmad I, Zahir ZA (2019) Biofertilizers for sustainable agriculture and environment. Springer, Berlin/Heidelberg. https://doi.org/10.1007/978-3-030-18933-4
46. Khan A, Singh AV, Shiv Shanker AA, Gautam SS, Punetha A, Upadhayay VK, Kukreti B, Bundela V, Jugran AK, Goel R (2023) Microbial bioformulation: a microbial assisted biostimulating fertilization technique for sustainable agriculture. Front Plant Sci 14:1270039
47. Garg D, Sridhar K, Stephen Inbaraj B, Chawla P, Tripathi M, Sharma M (2023) Nano-biofertilizer formulations for agriculture: a systematic review on recent advances and prospective applications. Bioengineering 10:1010
48. Bharti N, Suryavanshi M (2021) Chapter 10: Quality control and regulations of biofertilizers: current scenario and future prospects. In: Rakshit A, Meena VS, Parihar M, Singh HB, Singh AK (eds) Biofertilizers, vol I. Woodhead Publishing, pp 133–141
49. Kumawat A, Kumar D, Shivay YS, Bhatia A, Rashmi I, Yadav D, Kumar A (2023) Long-term impact of biofertilization on soil health and nutritional quality of organic basmati rice in a typic ustchrept soil of India. Front Environ Sci 11:1031844
50. Malusá E, Vassilev N (2014) A contribution to set a legal framework for biofertilisers. Appl Microbiol Biotechnol 98:6599–6607
51. Zambrano-Mendoza JL, Sangoquiza-Caiza CA, Campaña-Cruz DF, Yánez-Guzmán CF (2021) Use of biofertilizers in agricultural production. In: Ahmad F, Sultan M (eds) Technology in agriculture. IntechOpen

Marine-Derived Natural Cosmetics for a Green Future

2

Vivekananda Saha and Goutam Biswas

Abstract

In today's world, the demand for cosmetics and cosmeceuticals is drastically increasing day by day. With this highly growing market, a simultaneous increase of adverse consequences by the hazardous chemicals from cosmetics is also observed. To combat the adverse effects of artificial cosmetics, natural products are now used as active ingredients in cosmetics and cosmeceuticals. Although components derived from terrestrial plants are becoming more and more popular, the marine environment is also a more diverse source of both biological and chemical components. Starting from micro and macroalgae, seaweed, plants in coastal areas, marine fishes, bacteria, and the sea mud and water, all of these have minerals and nutrients that are very much essential for human skin health. These ingredients have selective skin care properties along with health benefits like antioxidant, anti-aging, immune protection, UV protection, stress protection, moisturizing agent, coloring agents, tanning cosmetics, anti-inflammatory (promote hyaluronan synthesis), skin repair activity, and so on.

This chapter focuses on studying the marine-based active components for making cosmetic products. The discussion in the whole chapter will be primarily based on the active ingredients within the sea and their beneficiary aspects in the cosmetic industry. Secondly, it will focus on the different marine resources of active natural components as alternatives for synthetic chemicals and emphasize the potential of these safe components to create a greener, cost-effective, and safer cosmetic industry.

V. Saha · G. Biswas (✉)
Department of Chemistry, Cooch Behar Panchanan Barma University, Cooch Behar, India

M. Rafatullah et al. (eds.), *Multidisciplinary Applications of Marine Resources*,
https://doi.org/10.1007/978-981-97-5057-3_2

Keywords

Cosmetics · Green technology · Heavy metals · Marine resources · Sustainable

2.1 Introduction

Cosmetics can be any substance or mixture intended for external application on the human body (the lips, nails, skin, hair, and other external parts of the body), as well as the teeth and within the mouth with the primary objective being to clean, odorize, protect, or modify the appearance of these areas (Regulation (EC) No 1223/2009). The word "cosmetic" comes from the Greek word 'kosm tikos' which means "capable of arrangement, skillful in decoration." In this sense, adornment is referred to as "kosmein," while order and harmony are referred to as kosmos [1]. In recent years, the requirements of cosmetics and cosmeceuticals have skyrocketed which creates a great business opportunity for all cosmetic companies. In the cosmetics industry, many new businesses have already been launched in this decade. One of the key reasons for such high demand of cosmetics and cosmeceuticals (cosmetics with pharmacological properties) is the detrimental consequences of our polluted environment and unhealthy food, which causes numerous skin illnesses in people and other health problems.

The global trend in business and society is towards creating a more sustainable and greener society in order to reduce health issues and the effects of climate change. Till the first decade of the twentieth century, consumers used conventional synthetic cosmetics, but in recent years a large portion of customers became aware of the cons of using synthetic cosmetics through the internet and other media sources. Nowadays, the cosmetics industry has been using more natural components in their products, partly due to consumer preferences. Still, the market share of natural cosmetics is lower than that of conventional cosmetics due to various reasons.

A major portion of our earth is composed of water. Five oceans and several seas have retained a vast biodiverse system containing many species and a wealth of possibly bioactive compounds beneath the water. Many of the compounds derived from the marine species exhibit different unique properties from the terrestrial plants as they grow in extreme environments. Scientists claim that marine-derived natural products have tremendous cosmeceutical and pharmaceutical properties, along with no toxicity to human life. Approximately 25,000 potentially active compounds from marine species were studied, representing only a small percentage of our complex marine ecosystem. Therefore, scientists are looking for effective research with the goal of finding new potentially bioactive compounds. There is already a lot of study being performed by scientists all around the world to investigate new species under the waters and try to explore the potent active compounds.

A wide range of potentially beneficial chemicals and nutrients may be found in many explored and unexplored organisms, starting with marine salt water, bacteria, seaweed, halophytes, fish, and sea mud. These all-important ingredients hydration, firmness, slimming, protection, and shine. This chapter reviews the main biological and functional properties of components that have been separated from marine sources, such as micro and macro-organisms, and how these properties relate to their use in cosmetics.

2.2 History Behind the Cosmetics

The history of the science of cosmetics is thought to have started in ancient Egypt and India, but the oldest known accounts of cosmetic ingredients and their use date to the Indus Valley Civilization between 2500 and 1550 B.C [2]. In Hindu culture, the application of kajal has a long tradition. Ancient Egyptian culture also served as the inspiration for many modern beauty procedures, including hair coloring, waxing, and exfoliating. While natural paints and dyes were used to color the face, mostly for ceremonial and religious occasions, oils containing essence and ointments were used to clean, soften, and disguise body odor. For example, a paste was made by powdering the rind of the Bel fruit (*Aegle marmelos* Corr.) and combining it with a woman's milk, which was then applied to the lips that are cracked. To remove unwanted extra hair, a mixture of dry fruits (Avalakatti and Pimpali) with the latex of Nivadunga was used on the desired place, and the hairs from that area were found to fall off [3]. A long time ago makeup on the eyes was considered an evil sign in religious occasions. So, if we look at our past there were evidence that many cosmetics used in that time originated from the depth of science thoughts and technology but were covered by spiritual events.

2.3 Broad Aspects of Cosmetics Formulation

The formulation of cosmetic products may have an impact on our health based on the chemical composition of the components and the quantity of each ingredient used within it. We apply 70–95 chemicals to our skin every day because the average cosmetic contains 15–20 ingredients and we use at least five cosmetics. Although cosmetics might correspond to a very wide range of dosage forms, all of them can be defined by a relatively general concept:

$$\begin{aligned}\text{One cosmetic} = \sum &\text{one or more of Active components}(\text{s}) \\ &+ \text{one or more Excipient}(\text{s}) + \text{one or more Additive}(\text{s}) \\ &+ \text{1st stage packaging} \pm \text{2nd stage of packaging}\end{aligned}$$

An additive is a component meant to enhance the product's organoleptic properties or preservation, whereas the excipient serves as the active ingredient's vector in cosmetics with specific functions like moisturizing, slimming, or anti-aging [4]. The main components of a cosmetic formulation include water, emulsifiers, thickeners, preservatives, color, emollients, and perfumes, though a product's formula may vary. Nearly all formulations, including those for creams, lotions, cosmetics, deodorants, shampoos, and conditioners, start with ultra-pure distilled water, which also serves as an emulsifier and solvent. Potassium cetyl phosphates and polysorbates (potent emulsifier) are added to the formulation to form a homogenous and uniform texture by decreasing the surface tensions between the water and the oil. To prolong the product's shelf life and prevent microbials from growing, water-soluble preservatives such as paraben, 2-phenoxyethanol, salicylic acid, chloro cresol, and chloro butanol are used. The preservative concentrations can range from 0.01% to 5% depending on the products [5].

The term "cosmeceutical" was initially introduced in 1984 by Dr. Albert Kligman of the University of Pennsylvania to refer a new class of goods that went beyond just medications and cosmetics. It is called cosmetic pharmaceuticals or cosmeceuticals when a pharmaceutical and cosmetics combine. Cosmetic and therapeutic benefits are shared by cosmeceuticals [6]. Today's cosmetic business must develop such skin-care formulations combined with consumer therapeutic benefits (Table 2.1). Conversely, cosmeceuticals are low-risk items with proven benefits for the skin according to guidelines for cosmetics. It is frequently employed in the cosmetics sector to highlight the unique mode of action of the components found in contemporary formulas for skin care. Generally, cosmeceuticals are categorized as skin whitening, anti-inflammatories, antioxidants, anti-photoaging, sunscreens, immune boosting, anti-inflammatory, etc. [7].

2.4 Emergency in Conventional Cosmetic Industries

Cosmetics are products that are applied to the body to enhance the attractive features, beautifying, cleansing, and improving the appearance of a person. A wide range of goods are included in the category of cosmetics, like toothpaste, shampoos, conditioners, mascara, after-shave lotion, styling gel, creams, lotions, powders,

Table 2.1 Common ingredients used in cosmeceuticals

Class	Examples	Functions
Sunscreens	Zinc oxide (ZnO), titanium dioxide (TiO_2), para-amino benzoic acid, etc.	Protection against tumors and increased reactive oxygen species (ROS)
Antioxidants	α-Lipoic acid, vitamin C, B_3, E, N-acetyl-glucosamine (NAG), and ubiquinone (CoQ10)	Reduce inflammation and pigmentation. Protect skin against sun damage due to collagen breakdown or skin cancers
Hydroxy acids	α-Hydroxy acids (AHAs), polyhydroxy acids (PHAs), and β-hydroxy acids (BHAs)	Improve skin texture and signs of aging
Retinoid	Retinol, retinoic acid, retinaldehyde	Reduce skin wrinkles, acts as anti-pigmenting agents and repairing agents
Whitening agents	Flavonoids, ascorbic acid, kojic acid, vitamin A, etc.	Inhibits melanin production
Botanicals	Plant extracts like soya, curcumin, and silymarin	Wound healer and damage repairing agents, anti-aging

perfumes, lipsticks, nail polish, eye and facial makeup, hair wavers, hair dyes, hair sprays, deodorants, and anti-perspirants. The size of the world market for cosmetics was estimated at $299.77 billion in 2022 and is expected to increase to $417.24 billion by 2030 from $313.22 billion in 2023. The global population's growing awareness of the importance of well-being and beauty is driving the rise of the cosmetics sector. These goods can improve people's appearances. People are utilizing more cosmetic items because of growing beauty consciousness, which leads to an increase in unfavorable incidents. Because cosmetics and beauty products contain over 10,000 substances that have been related to diseases, they are not particularly healthy. People who use cosmetics on a regular basis run the risk of developing side effects such as type IV hypersensitivity, allergies, contact dermatitis, and skin redness [8].

Beauty-enhancing and cosmetic products do not have very good effects on the physical health of users, and using them for extended periods of time can have negative effects on one's health. These substances have the potential to penetrate the skin and the other organs, producing carcinogenicity, as well as to have major adverse consequences on the skin [9]. Cosmetic compounds directly affect the human hormonal system, leading to the disruption of endocrine. One of the most prevalent substances in cosmetics, heavy metals irritate skin and harm epithelial cells. Mucous membranes suffer when heavy metals build up [10]. Of the population, 42% are unaware that the negative effects of cosmetics must be reported to the relevant authorities, whilst 57% are aware of this. Many stories on potentially hazardous substances included in cosmetic products, such as lead, aluminum, mercury, parabens, siloxanes, BHA, and BHT can be found on news websites or social media. A thorough investigation was reported about different adverse effects of the chemical ingredients present in cosmetics [11] (Table 2.2).

Factors like the ecology, sustainability, affordability, efficacy, and long-term health, with the toxicity of synthetic and fossil-derived and chemical compounds, necessitate an urgent transition from traditional to bio-based makeup. To address concerns about public safety, several nations, including Japan, China, Ethiopia, India, and Vietnam, are also prohibiting the use of chemical ingredients. Public awareness is also being raised by placing warnings on cosmetic labels [12, 13]. There is a growing market demand for natural products because of strict environmental regulations and a sharp drop in nonrenewable resource use. Productivity gains,recycling, reduced energy usage, safety, and well-being are all provided by bio-based personal care products [14]. To meet marketing requirements and to replace raw materials (having health-related issues) that are either outlawed or have lost customer trust, this business is constantly looking for new ingredients. As a result of its richness and diversity, the marine environment is a potential supply of intriguing materials. Marine-based cosmetic products are one of the most effective alternatives to bio cosmetics. In the past few decades, a multitude of marine-based compounds with intriguing medicinal properties has been discovered and utilized [15, 16].

Table 2.2 Conventional cosmetic formulations associated with health risks

Potent ingredients	Uses	Risks
Sodium lauryl sulfate (SLS)	Surfactants, emulsifier, and foam making	Carcinogenic agent, dryness of skin and eyes
Synthetic colors and petrochemicals	Pharmaceuticals and cosmetics, artificial dyes	PPD (*p*-phenylenediamine) in hair dye causes allergic reactions (e.g., weeping dermatitis, eczematous dermatitis, pruritic dermatitis)
Phthalates (DBP)	Smoothening and texturing agent	Miscarriage and infertility may be caused by damage to sperm DNA
Aluminum	Anti-perspirants	It is directly linked with breast cancer and Alzheimer's disease
Mercury	Both organic and inorganic forms are used in cosmetics (e.g., skin-lightening soaps, and lotions)	Damage to the kidneys is a frequent consequence of exposure
Lead	Mostly found as an impurity in lipsticks and eye powders such as kajal	High lead exposure in children and women is harmful to their health
Different derivatives of paraben	Preservative and anti-microbial agent	Skin aging through cosmetics and cancer risk
Paraffin wax	Moisturizing agents	Carcinogenic, causing allergy and asthma
Mineral oil	Keeps skin hydrated, helps to moisturize skin	Skin irritants such as rash and burning, causing shortness of breath
Cadmium	Mostly in body and hair products	Primary problems are vomiting and stomach pain. Prolonged exposure results in kidney damage
Siloxanes	Softening and smoothing agents in cosmetics	Acts as endocrine disruptors and have the potential to be reproductive toxins, which could harm fertility in humans
BHA	Moisturizing agents	A possible human carcinogen (International Agency for Research on Cancer)
BHT	Moisturizing agents and antioxidant	Cause reproductive issues by blocking the development of male sex hormones and mimicking hormones like estrogen
Triclosan	Anti-bacterial skin care and tooth care products	Helps in the formation of drug-resistant germs and colon cancer
EDTA	Stabilizer, preservative, improve the foaming abilities of body-cleansing products	Slight mutagenicity, inflammation, and harm to reproduction

2.5 Development of Marine-Derived Natural Products

The ocean is the genesis of life on earth. As a result, it is not surprising that the marine environment might be a source of numerous novel bioactive ingredients. The industry of marine bioprocessing has the aim of converting and utilizing most marine resources and marine food by-products as valuable components for cosmetic products. In today's world, the increasing demand for greener products in every field creates a worldwide natural product-based market. Cosmetic industries have already identified it as one of the important sectors and have started their journey in developing new greener cosmetics and cosmeceuticals. So, there is a growing interest in using marine products. Innovative bioprocessing techniques are continuously developing to separate certain bioactive compounds with cosmeceutical properties from marine food products for use in various applications.

The functional components of cosmetics are developed through a series of biotransformation processes. Recently, ultrafiltration membrane-equipped bioreactor technology has evolved for bioprocessing and the production of functional components, and it is being explored as a potential means of making efficient use of marine bioresources. This technology has lately gained attention in the creation of functional cosmeceutical components and is thought to be a viable way to use marine bioresources to create value-added cosmeceutical products that improve human skin health [17]. This method is great for synthesizing marine-based bioactive derivatives with specific molecular sizes as the specific functions of any components depend on their respective molecular weights. On the other hand, the development of functional properties in marine bioactive products (like peptides and chito-oligosaccharides) involves some biotransformation techniques which pass through selective enzyme hydrolysis at optimum temperature, pH, solvents, and other several conditions to maintain the activity of enzymes. A three-enzyme system is a good tool to employ sequential enzymatic digestion to produce peptides that are functionally active. Furthermore, serial enzymatic digestions can be obtained in a system that combines an ultrafiltration membrane system with a multistep recycling membrane reactor to bioprocess and create bioactive peptides and derivatives of chito-oligosaccharides derived from marine food [18].

2.6 Marine Resources in Cosmetics

2.6.1 Based on Types of Natural Products

Due to a global trend for products that are deemed healthy, environmentally sustainable, and ecologically friendly, the cosmetics industry is investing an increasing amount of money in the research and development (R&D) of new products incorporating chemicals or extracts derived from natural resources [19]. A large portion of informed consumers prefer natural-based products due to less adverse and numerous beneficial effects than the non-natural counterparts. Marine resources are becoming more and more popular as a rich supply of structurally diverse substances, even though cosmetics made from terrestrial plants have long been employed. To survive in harsh environments, marine organisms synthesize bioactive compounds through complex biochemical and physiological interactions. Since the marine sector emerged, there have been almost 25,000 novel bioactive compounds identified, providing several opportunities for the isolation of new active components. Many of these compounds have great potential as anti-inflammatory, anti-allergic, anti-aging, and anti-microbial effects [20].

2.6.1.1 Carotenoids

A family of colored substances known as carotenoids is only produced by plants, algae, fungus, and microorganisms. Carotenoids play a protective role in photosynthetic systems by efficiently dissipating surplus energy, inhibiting the production of ROS (reactive oxygen species), and stopping the synthesis of singlet oxygen that represents the antioxidant property. The efficient use of specific carotenoids as pigmenting agents in cosmeceuticals increases with increasing the conjugated double bond and presence of special groups and substituents at the ends of the carotenoid's structure. Seaweeds (e.g., marine algae), marine animals, marine bacteria, etc. are rich in a variety of carotenoids having cosmeceutical activity like β-carotene, fucoxanthin, astaxanthin, and so on. Besides cosmetic products are used in marine drugs also for various types of diseases (e.g., cardiovascular, neurological, cancer, obesity, and diabetes).

2.6.1.1.1 Fucoxanthin

Fucoxanthin (**3**) is one of the most bioactive xanthophyll carotenoids that was extracted from numerous types of algae, e.g., *Phaeodactylum tricornutum, Hijikiaa fusiformis*, and brown algae *Undariaa pinnatifida* and *Laminaria japonica.* Fucoxanthin from these organisms exhibits more antioxidant capacity than α tocopherol and α lipoic acid. Fucoxanthin has very potent anti-wrinkle and anti-photoaging activity by suppressing the effects of UV-B radiation. It is also used in melanoma and pigmentation problems due to its anti-tyrosinase and anti-melanogenesis activities. Recent studies showed that fucoxanthin is a potent inhibitor of H_2O_2-induced apoptosis, DNA damage, and intracellular ROS production. Anti-pigmenting (through suppression of prostaglandin E_2 synthesis), anti-inflammatory (by diminishing cytokines), anti-cancer activity are also observed by fucoxanthin-related cosmetics and pharmaceutical products [21].

2.6.1.1.2 Astaxanthin

Astaxanthin (**4**) is also a xanthophyll-type carotenoid extracted from microalgae *Haematococcus pluvialis,* having two extra oxygenated substituents than other active components (like α tocopherol (**36**) and β carotene (**8**), carotenoids) gives 100 times higher UV A protecting capacity and anti-radical activity. It also reduces moisture loss by scavenging ROS production. Furthermore, astaxanthin from marine creatures prevents lipoprotein oxidation when consumed. Astaxanthin, for instance, is 100 times more potent than α-tocopherol at shielding rat mitochondria from Fe2þ-catalyzed lipid peroxidation both in vivo and in vitro. It is particularly effective as α-tocopherol (**36**) in blocking lipid peroxidation induced by free radicals in rat liver microsomes [21]. Recent research on astaxanthin observed that it is one of the most potent bioactive compounds among the other carotenoids having two more oxygen in its structure and has other activities such as reducing inflammation by enhancing collagen production and strengthening the skin barrier very efficiently.

2.6.1.2 Collagen

Collagen serves as the primary structural component of all vertebrate organs, including blood vessels, teeth, corneas, cartilage, skin, tendons, and ligaments. Generally, collagen has been extracted from animals like cow and pig, but recently due to many health risk-related diseases and some religious issues the industrial production of collagen-derived nutricosmetics and pharmaceutical products from land animals has decreased. Safe and economic alternative companies are setting their focus on the extraction of collagen from marine fishes. Specifically, marine collagen is gaining popularity as a skin treatment and is thought to be superior to collagen derived from terrestrial sources. As a dietary supplement, fish collagen peptide has been shown to be effective in reducing UV-B-induced skin damage and photoaging. Collagen is useful for skin care because of its film-forming, protective, and moisturizing properties. A range of cosmeceuticals made from marine collagen have gained popularity as commercial goods recently. Because of its foaming and protecting qualities marine fish collagen (e.g., cod, tuna, and so on) is used as a conditioning agent and facial cleanser.

2.6.1.3 Polyphenolic Compounds

The secondary metabolism of algae produces phenolic or polyphenolic chemicals, which are a broad section of chemical components composed of a hydroxyl group (−OH) that is connected to a benzene or other arene rings. Due to environmental stressors including UV light, high salt concentrations, microbial stress, herbivore grazing, etc., algae frequently generate these chemicals, which are known for their antioxidant abilities [22]. The distinct chemical, physical, and biological features of a given class of polyphenols, such as phloroglucinols and phlorotannins, are determined by the kind and quantity of their phenol structural units. Phlorotannins can be further separated into other groupings, such as phloroglucinol, eckol (**1**), and dieckol (**6**) based on the inter-linkage. Due to their antioxidant function, phlorotannins have a variety of biological properties that make them useful in cosmetics. The scavenging capacity depends on the number of sites present in the benzene rings of phlorotannins. It can restrict the inflammatory symptoms brought on by dangerous radical elements through interacting with radicals and transforming them into stable molecules by donating electrons.

2.6.1.4 Polysaccharides

Polysaccharides are a major component of the cell walls of plants and microorganisms that have been employed as anti-inflammatory and anti-microbial treatment in numerous sectors of pharmaceutical and cosmetic industries. Various sea organisms, such as Phaeophyta, *Chondrus*, phyla Crustacea, *Gigartina*, *Kappaphycus*, and seaweeds like red or green, and even marine bacteria are rich in different types of polysaccharides, e.g., sulfated (fucoidan (**2**), exopolysaccharide (**5**), carrageenans, laminarans (**13**), galactans, and ulvan (**7**)) and nonsulfated polysaccharides like agars and alginates. Among all the organisms, algae is the major source of polysaccharides. Marine algae possess a multitude of bioactive polysaccharides found in their cell walls, including ulvans (**7**) in green algae, fucoidans in brown algae, and carrageenans in red algae. These polysaccharides have a variety of positive cosmetic impacts.

Fucoidans are present in brown algae consisting of high-fucose polymer backbone with a high degree of branching, sulfated, and acylated. Many biological studies conclude that fucoidans have anti-inflammatory effects through the inhibition of leucocyte requirement, anti-allergenic effect, and antioxidant potential by restricting some MMPs (matrix metalloproteinase) production and acting as a hydrating agent due to its hygroscopic nature [23].

According to a few studies algal polysaccharides, (e.g., chitin, chitosan, and hyaluronic acid) have moisturizing potential by restricting the TEWL (trans epidermal water loss). Red algae-derived polysaccharides also have additional advantageous cosmetic properties, such as skin repairing and anti-inflammatory activities [23].

2.6.1.5 Sea Water, Minerals, and Sea Mud

Sea water has long been recognized for its medicinal and beauty benefits. It also contains essential minerals including calcium, magnesium, potassium, sodium, sulfates, and chlorides that are required for the proper operation of the human body. Because of the health benefits of the minerals present in seawater, sea salts are effectively used in skin care products of prestigious cosmetic companies. Mineral additives, such as calcium derived from the waste of fish bones, can be added to a variety of food products, and utilized in cosmetics. Phoscalim, a mineral extract obtained from wild fish, offers bioavailable calcium and phosphorus to support good bone health in addition to calcium supplements [17]. The minerals and the nutrients present in sea mud are responsible for the numerous therapeutic activities. Sea mud is known for curing various skin-related issues, e.g., keeping the skin hydrated, maintaining the pH, acne repair and prevention, and acting as an anti-aging agent [24]. These days, due to pollution, ocean water is contaminated with a variety of dangerous metal ions that pose a serious risk to human health. Notable experiments have been conducted on dead sea mud to determine the presence of Ni and Cr. It has been shown that trace amounts (below safety level) of these metals in the form of solid particles were present in dead sea mud naturally. The minerals and salts from the Dead Sea were used by the ancient Egyptians to create cosmetics and balms for mummification. Some people think that adding Dead Sea ingredients to lotions and balms can help with joint pain, clear up acne and eczema, and minimize wrinkles. Dead Sea salt is used in various cosmetic products to detoxify skin, anti-aging, skin nourishing, etc. (Fig. 2.1).

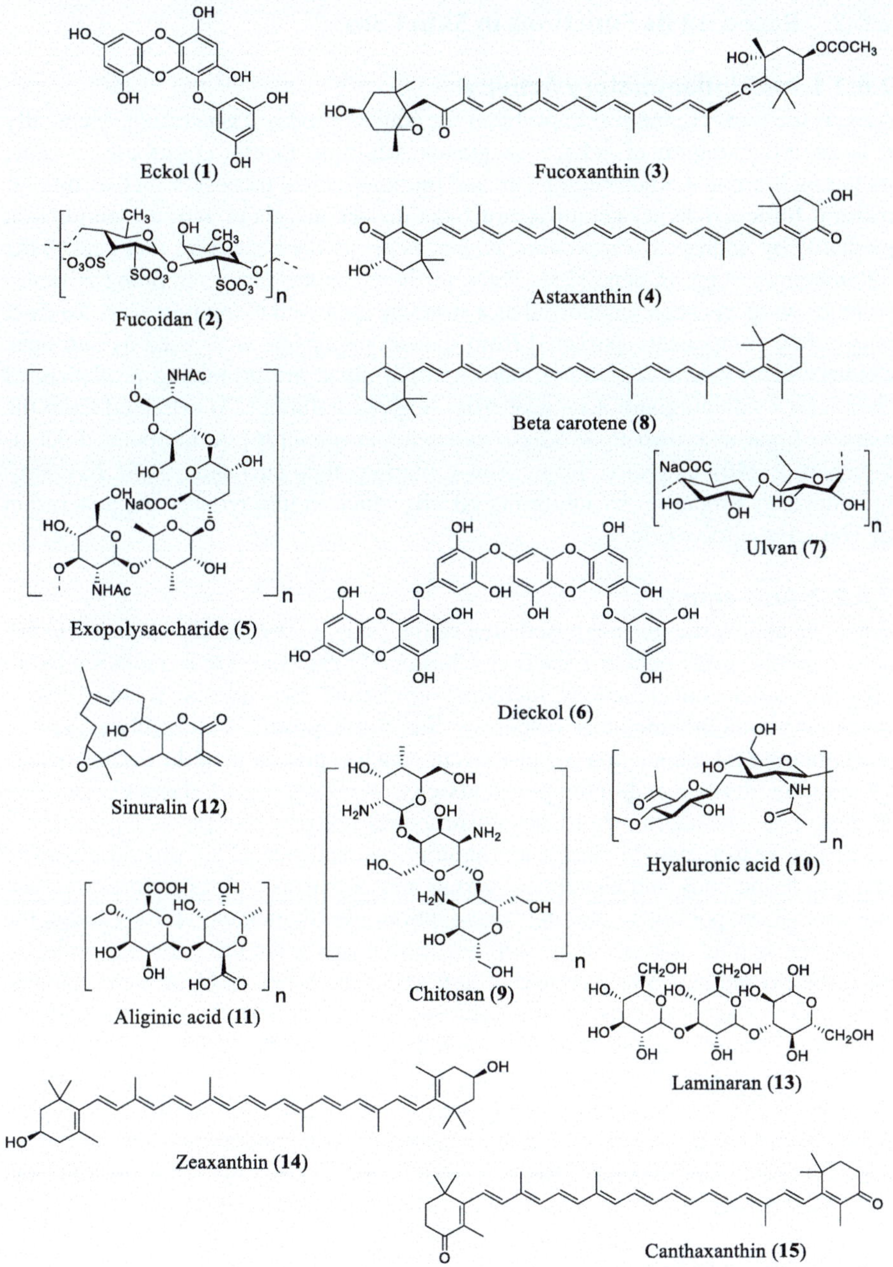

Fig. 2.1 Chemical structures of a few-marine based antioxidant compounds

2.6.2 Based on Its Functions in Skin Care

2.6.2.1 Anti-inflammatory Activity

Acne is the most common skin problem for adults of today's generation. Medically it is an inflammation of sebaceous glands. Multiple factors contribute to acne, including hormonal, microbiological, and immunological processes such as inflammation, follicular hyper-keratinization, and sebaceous gland activity stimulation mediated by androgen. *Propionibacterium acnes* is the bacterium that causes the inflammatory stage in acne [25]. Thus, in the cosmetic sector, stopping *P. acnes* from growing has been recognized as a strategic approach to treating acne. To date, only a few components extracted from marine organisms were used as anti-acne cosmeceutical products. Among them, marine algae are promising producers of chemicals containing anti-acne properties, e.g., sargafuran (**28**) extracted from the extract of marine brown algae *Sargassum macrocarpum* [26]. A diterpene, dolabelladienetriol (**40**) extracted from brown marine alga *Dictyota pfaffii* has anti-inflammatory properties by inhibiting NFkB, which in turn reduces the generation of TNF- and nitric oxide.

2.6.2.2 Anti-aging

Since the skin is the most exposed area of the body, excessive UV radiation exposure causes a severe sunburn reaction. Chronically exposed skin to radiation led to abnormal cutaneous reactions including epidermal hyperplasia, rapid collagen breakdown, and inflammatory responses. Sun overexposure is clinically expressed as erythema [27]. Skin aging is a process in which degradation of the dermis occurs by thinning, dryness, enlarging pores, laxicity, wrinkles, loss of elasticity, etc. Due to these phenomena, the skin loses its structural integrity and physiological function. The reason can be intrinsic (natural consequences like physicochemical changes in the body and anatomical variations) or extrinsic (temperature, weather, sun irradiation, pollution, nicotine consumption, overall diet, etc.) [28]. Though the mechanics behind skin aging are still not fully understood, the cosmetics industry nevertheless markets a large range of anti-aging therapies, most of them claim to accelerate collagen and glycosaminoglycan (GAG) synthesis by fibroblasts in the epidermis, improving the flexibility and hardness of the skin's corneal layer. Different protecting chemicals form in different types of living beings to protect them from UV irradiation of sun exposure, e.g., melanin for animals, for higher plants flavonoids do that work and in the case of marine microorganisms like algae and bacteria. Scytonemin (**16**), various carotenoids (e.g., β-carotene) and mycosporine-like amino acids behave as sunscreen agents [29] (Fig. 2.3).

2.6.2.3 Antioxidants

Oxidative stress is defined as an imbalance between the production of reactive oxygen species (ROS) and antioxidant defense activity. Overexposure to ROS can harm multiple cellular constituents, resulting in tissue damage. This is linked to aging and a range of chronic illnesses, including cancer, diabetes, neurodegenerative, cardiovascular disorders, and inflammatory diseases with severe tissue injuries among

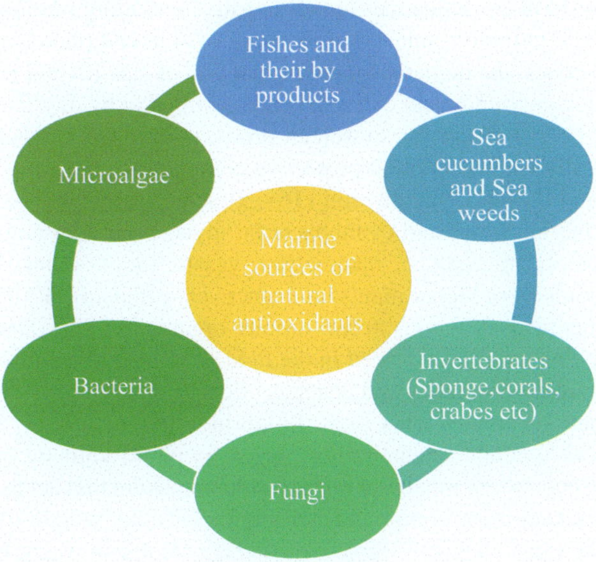

Fig. 2.2 Marine sources of natural antioxidants

others. The generation of ROS is remarkably inhibited by natural antioxidants [30]. Any substance that inhibits a free radical reaction to cease the oxidation process is an antioxidant. Antioxidants neutralize excessive amounts of highly reactive free radicals created during metabolism, protecting biological systems from their harmful effects. When human skin is exposed to UV-A radiation it causes ROS generation which results in lipid oxidation followed by damage to the skin cells. Because of that the youthful appearance of any person decreases. To inhibit ROS-induced oxidation, the cosmeceutical industries employed synthetic antioxidants such as propyl gallate (PG), tertbutyl hydroquinone (TBHQ), butylated hydroxytoluene (BHT), and butylated hydroxy anisole (BHA). However natural marine-based antioxidants are superior to synthetic antioxidants in terms of negative effects [31].

The marine environment is a reservoir of many antioxidants that are not fully explored and utilized. Sources of naturally occurring antioxidants include a wide range of marine macro and micro-organisms, seaweeds, microalgae, ascidians, bryozoans, sponges, lichen, bacteria, fungi, etc. Antioxidants from the same species have varying compositions worldwide because of the different climates under which the marine biota grows. Moreover, antioxidant activity depends on the procedure of extraction and purification technologies (Fig. 2.2).

2.6.2.4 Anti-Photoaging

Chronic exposure to sun UV radiations causes some temporary skin-related issues and, in some cases, severe permanent damage of skin like melanoma (typically mediated by ROS) also shown. Although short-term exposure can have positive effects on mood and vitamin D production, it can also result in skin burn, tanning, etc. Yet the long-term consequences are all negative including immunosuppression brought on by UV radiation-induced photoaging of skin and photo carcinogenesis. Based on these crises, the demand for the production of new natural products based anti-photoaging compounds and extracts is increasing day by day. To prevent photodamage, unique UV filters for sunscreen products have been developed using a variety of naturally occurring photoprotective compounds, including melanin (**19**) from animals and bacteria; scytonemin (**16**) extracted from cyanobacteria; mycosporines from marine fungi and cyanobacteria; mycosprine-like amino acids (**20**) from cyanobacteria, microalgae, macroalgae, yeasts, fungi, sponges, corals, and animals; flavonoids from higher plants; and other very efficient UV-filter substances of untapped chemical structures from various organisms [32]. The components from marine organisms suit very well in the field of producing ecofriendly and less toxic anti-photoaging cosmetics and cosmeceuticals.

Mycosporine-like amino acids (e.g., Shinorine (**18**), usujirene, and palythine (**17**)) are types of secondary metabolites present in photosynthetic organisms that are found in the ecosystem inhabited by high solar exposure that act as photoprotective agents. These colorless compounds with low molecular weight (often less than 400 Da) have the same chemical framework and are soluble in water. However, they differ in the kind and quantity of amino acids present as well as in the substituents they use. Their structures are composed of chromophore (cyclohexanone or cyclohexenimine which are conjugated to nitrogen of amino acids or its iminoalcohol). Because of these chromophoric linkages, it can adsorb UV light from the sun and radiate in the environment. Another most active component having radical scavenging capacity is scytonemin (**16**) a persistent yellow-brown pigment (soluble in lipids) found in the extracellular polysaccharide sheath of certain cyanobacteria. The dimeric chemical structure is made up of phenolic and indolic subunits. The two main forms of scytonemin are reduced (red) and oxidized (green). An olefinic carbon atom connects the two subunits of scytonemin. However, there are four distinct scytonemin derivatives: tetramethoxyscytonemin, dimethoxyscytonemin, and scytonin and they adsorb UV radiation between 250 and 320 nm [32]. It is also acts as a well-known antioxidant in various types of cosmetic products.

2.6.2.5 Anti-Wrinkle Activity

Skin aging occurs via the degradation of the dermis and epidermis layer of skin and producing outward symptoms like dryness in discrete portions, the color of the skin becoming dark by pigmentation, mild inflammations, premalignant lesions, laxness, and wrinkling. According to multiple studies, MMPs (matrix metalloproteinase), a family of zinc endopeptidases, are responsible for the decrease in the formation of collagen in photo-aged skin. A range of cells, including neutrophils, fibroblasts, keratinocytes, mast cells, and macrophages, create MMPs, which are thought to be important in the development of wrinkles. MMP expression is usually initiated by various extracellular stimulations, cytokines, and UV radiation. UV radiation influences ROS generation leading to gene expression of MMPs. The induction of MMPs results in collagen degradation, which simultaneously enhances wrinkle formation, aging, and skin thickening [33]. Marine algae *Corallina pilulifera*'s methanol extract decreased the expression of MMP-2 and -9, human MMP, in fibroblasts that had been exposed to UV light [34]. According to some researchers the Ecklonia species', Eckol (**1**) and Dieckol (**6**) suppressed the expression of MMP-1 in human dermal fibroblasts and were used as age preventing agents in cosmetics [35]. Generally, all the anti-wrinkle/anti-aging cosmetic formulations contain moisturizing ingredients since hydrated skin is necessary for skin functioning. Especially the compounds that form oil/water emulsion and create a layer to limit the TEWL (trans-epidermal water loss) are used as moisturizing agents. Traditionally γ-linolenic (**22**) and linolenic acids are used to keep the optimum hydration level of the skin. Recently in cosmetic industries to make green moisturizing cosmetics some marine microorganism-based bio surfactants mannosylerythritol (**21**), rhamnolipids, and sophorolipid (**23**) are used [28].

2.6.2.6 Coloring Agents

Natural pigments are a class of chemically diverse and biosynthetically unrelated molecules that are brought together by a chromophore present in their electronic structure, which gives these compounds their distinctive colors. Marine organisms are potential and sustainable sources of biopigments that support the development of novel cosmeceutical, pharmaceutical, and nutraceutical products. Species like micro- macro-organisms (algae), bacteria, and animals are rich in biopigmenting substances. Seaweeds like algae are one of the major resources of pigments such as carotenoids (fucoxanthin (**3**) from brown algae), phycobiliproteins, and chlorophylls. Pigments found in phytoplankton are employed as evidence for the biogeochemical processes that contribute to climate change. All derivatives of Chlorophylls (a, b, and c), α- and β-carotenes (**8**), α-cryptoxanthin, lutein, zeaxanthin (**14**), chlorophyllide (a), and lycopene are the primary pigments present in phytoplankton. Marine invertebrates such as sea urchins, mollusks, and crustaceans produce natural coloring agents such as quinone, carotenoid, and tetrapyrrole compounds. Some marine vertebrates, such as whales, dolphins, reptiles, and bony and cartilaginous fish, can create or accumulate pigments (containing more nitrogen), the main pigments in these animals are carotenoids. Mainly carotenoids are pigments of terpenoid class, like carotenes (unsubstituted carotenoids) and xanthophylls (oxygenated derivative), having hydroxyl groups like lutein and oxygen-containing like canthaxanthin (**15**) or astaxanthin. Different carotenoids have different colors due to their specific chromophoric groups like in marine fish-derived carotenoids astaxanthin (**4**) (pinkish-red), canthaxanthin (**15**) (orange-reddish), lutein (greenish-yellow), and zeaxanthin (**14**) (yellow and orange). Besides these marine microorganisms, bacteria, fungi, and yeasts also produce pigments like quinone, astaxanthin, and violacein (**39**). They are also responsible for the synthesis of violet, blue, and deep blue pigments like indigoidine and violacein. A violet pigment called violacein, derived from indole, is generated by a variety of marine bacteria, including *Pseudoalteromonas luteoviolacea*, *Collimonas* sp., *Serratia marcescens*, and *Chromobacterium violaceum*. These natural marine-based pigments have special characteristics like anti-microbial, anti-bacterial, antioxidant, and anti-radical properties. Moreover, pigments extracted from natural marine resources are biodegradable and biocompatible having fewer health issues than the conventional synthetic origin pigments, which raises their demand in biocosmetics as one of the major coloring components [24, 36, 37].

2.6.2.7 Depigmenting or Whitening

The skin is the largest organ in the body, therefore it is continually exposed to both internal and external factors. As a result, the pattern of constitutive pigmentation changes often in response to these factors. Whitening cosmetics are in high demand for the treatment of solar lengito, as well as for pregnancy masks, and residual hyperpigmentation.

Different skin tones have different quantities of melanin which leads to various skin colors from the darkest brown to the lightest pinkish-white. The main skin pigment is melanin, which can scavenge free radicals and operate as a broadband UV absorber [38]. In addition to excess or insufficiency of melanin, numerous physiological alterations in the human body as well as external stimuli are major contributors to pigmentation. Tyrosinase is the primary enzyme participating in the formation of melanin. A variety of naturally occurring molecules obtained from marine creatures have already been employed as tyrosinase suppressors, yet many of these chemicals (hydroquinone) have been demonstrated to be hazardous to human health. Recently, research has concentrated on finding novel skin-whitening substances produced by marine microorganisms. Zeaxanthin (**14**) from *Nannochloropsis oculata* extract and 7-phloroeckol, a phlorotannin derivative obtained from *E. cava* a brown seaweed, function as lightening agent due to its anti-tyrosinase capacity. *Pistacia lentiscus* is a typical medicinal halophyte plant that grows in salty environments across the Mediterranean area. Catechin (**24**), quercitrin (**26**), gallic acid, β-glucogallin (**25**), and epicatechin are some of the phenolic acids found in *Pistacia lentiscus*. These components are found to behave like tyrosinase inhibitors and therefore used in biocosmetic as whitening agents [24]. A marine fungi *Altenaria* sp. extract contains kjoic acid (**27**) that has been found to be a great anti-tyrosinase active agent. The marine brown alga *Ishige okamurae* was the source of diphlorethohydroxycarmalol, a phlorotannin that was found to have great potency for skin whitening as well as protective qualities against UVB-induced DNA damage via damaged tail and fibroblast morphological abnormalities [28].

2.6.2.8 Essential Oils

Perfumes are referred to as substances that emit and disperse a pleasant and aromatic odor. They are made up of synthetic blends of essential oils and aromatic compounds. Aromas go through the nasal canal and activate the olfactory cell, which is the olfactory center that sends the scent to the brain, producing smell sensations and eliciting happy memories. It has been used in a variety of fields, most notably aromatherapy. It has long been recognized that terrestrial plants can produce essential oils. Besides the synthetic perfume industries now greener industries based on plants from land to marine organisms have emerged as big opportunities for making less harmful perfume substances. The different chemo and bio-diversity with many extreme situations drive the production of many active aromatic compounds in marine organisms. Marine microbes could be exploited as a new source of perfume and many known aromatic ingredients can be created by biological transformation. *Synechococcus* PCC 6911 can convert Geranial to Geraniol and Menthone to Menthol, which are the most common ingredients used in perfumes [39]. On the other hand, seaweed-based essential oils have been in demand recently. Perfumers are finding fragrances with marine origins increasingly intriguing, two of the greatest examples are the species of *Dictyopteris* and *Dictyota*. *Crithmum maritimum* L. a halophyte found in coastal lines is unique for its essential oil with a fragrance of lemony (*p*-cymene) [40]. Many chemical classes, including terpenes, hydrocarbons, fatty acids, derivatives of esters, aldehydes, ketones, alcohols, polyphenols, C_{11}-hydrocarbons, halogenated chemicals, and sulfated polysaccharide compounds, are present in the volatile components of marine algae (*L. saccharina*, *Asparagopsis* sp., etc.). These components are not only recognized for aroma they are also well known for anti-fungal, anti-bacterial, anti-viral, and anti-cancer activities.

2.6.2.9 Moisturizing Activity

Water is an essential component in human skin to preserve its integrity. External influences accelerate skin aging by disrupting the ability to function as a barrier. The skin becomes more brittle as a result, and its inherent elasticity and moisturizing properties gradually disappear. Many compounds that have moisturizing qualities are produced by marine creatures, including proteins, polysaccharides, and fatty acids (mannosylerythritol, sophorolipids, and rhamnolipids) that are frequently employed in the human skin. Mostly chitooligosaccharide that contains N-acetyl D-glucosamine and linolenic acid γ linolenic acid (**22**) act as hydrating agents by forming oil/water emulsion. This emulsion helps to restrict trans-epidermal loss of water from the skin. Research on marine algae concludes that a variety of these organisms contain unsaturated fatty acids, e.g., EPA (*eicosapentaenoic* acid) (**29**) from *Vibrio cyclitrophicus* and *Nanochloropsis* sp.

Besides the seaweeds, marine bacteria also produce some efficient moisturizing ingredients like EPS (extracellular polysaccharides), which possess a higher protective capacity for dermal fibroblasts than chitosan, collagen, and glycerol due to their better water adsorption and retainability. Examples of few marine bacteria are *Phyllobacterium* sp.921F, *Polariacter* sp. SM1127, and *Peudoalteromonas* sp. [39].

Collagens are appropriate ingredients in food and cosmetic products because of their strong water-absorbing capacity. Several studies show that not only terrestrial animals, but marine animals also give an endless supply of collagen. Collagen peptides with low molecular weight have several cosmeceutical benefits with more hydrating ability. Collagen is found prominently in the bones and skins of marine vertebrates (finefish) as well as marine invertebrates (like sponge, corals, starfish, and sea cucumbers) [41]. Because of this behavior it is used by most cosmeceutical compounds as one of the main ingredients.

2.6.2.10 Preservatives

Cosmetic items need to have preservatives added to avoid microbial contamination and prevent modification. According to health experts for consumers besides the preserving character of the components it should not have any hazardous materials. For a long time, parabens (derivatives of *p*-hydroxy benzoic acid) was used which created a controversy due to its safety-related issues [42]. To make the cosmetic paraben-free, an alternative substitute methylisothiazolinone was used, which was later banned in terms of its skin tolerance. Though a lot of organisms from sea depth having anti-microbial effects were explored, a concrete application study of them is still in slow progress. In this greener world, there are research works on various activities of discovered marine organisms. Extracts from macro- and microalgae have several anti-microbial substances derived from marine environments. Some research demonstrates that extracts from the macroalgae *Himanthalia elongata* and *Synechocystis* sp. prevent the growth of *Staphylococcus aureus* and *Escherichia coli* [24]. Extracts from the microalgae (*Chlorella marina*, *Pavlova lutheri, Nannochloropsis oculata, Isochrysis galbana,* and *Dunaliella salina*) have some action against bacteria like *Klebsiella pneumoniae* and *Pseudomonas aeruginosa.* Terpenoids and phlorotannins are the most active components of macroalga extracts, which have demonstrated anti-bacterial and anti-fungal properties [16, 43]. These extracts can prevent the adverse effects and allergic reactions associated with artificial medications [43]. In vitro studies of extracts from brown and green seaweeds demonstrated efficacy against acne vulgaris [44]. In vitro studies of brown algal extracts demonstrated activity against oral bacteria, red algae are active against *Staphylococcus* and *Candida* sp. [45, 46], and extracts of brown algae against skin pathogenic bacteria, including methicillin-resistant *Staphylococcus aureus* [47].

Chitosan (**9**) (a polysaccharide found in marine arthropods) is used as a preservative in cosmetics due to its anti-microbial effect against fungi and viruses. Falcarindiol (**30**) was produced by extracting the leaves of the halophyte *Crithmum maritimum* using chloroform. The growth of certain microorganisms, including *Bacillus cereus* and *Micrococcus luteus*, is significantly inhibited by falcarindiol. As a result, *Crithmum maritimum* may find usage as a preservative in the cosmetics industry [24]. Anti-microbial activity has also been observed in extracts from the leaves and fruits of *P. lentiscus*, another halophyte [48] (Fig. 2.3; Table 2.3).

2.6.3 Based on Sources of Marine Natural Products

2.6.3.1 Seaweeds (Macroalgae)

Macroalgae generally grow in extreme sea environments which leads to the production of many active components that protect them from stress factors. Researchers mainly focus on the extraction of these potent components for extensive pharmaceutical treatments. Seaweed has long been utilized for skin healing treatment and its purifying qualities. Lots of active components like mineral salts, amino acids, and vitamins A, C, B_1, B_{12}, E, K, and D are present in seaweed abundantly. They were basically used in skin fostering and hydrating treatment because of the easily

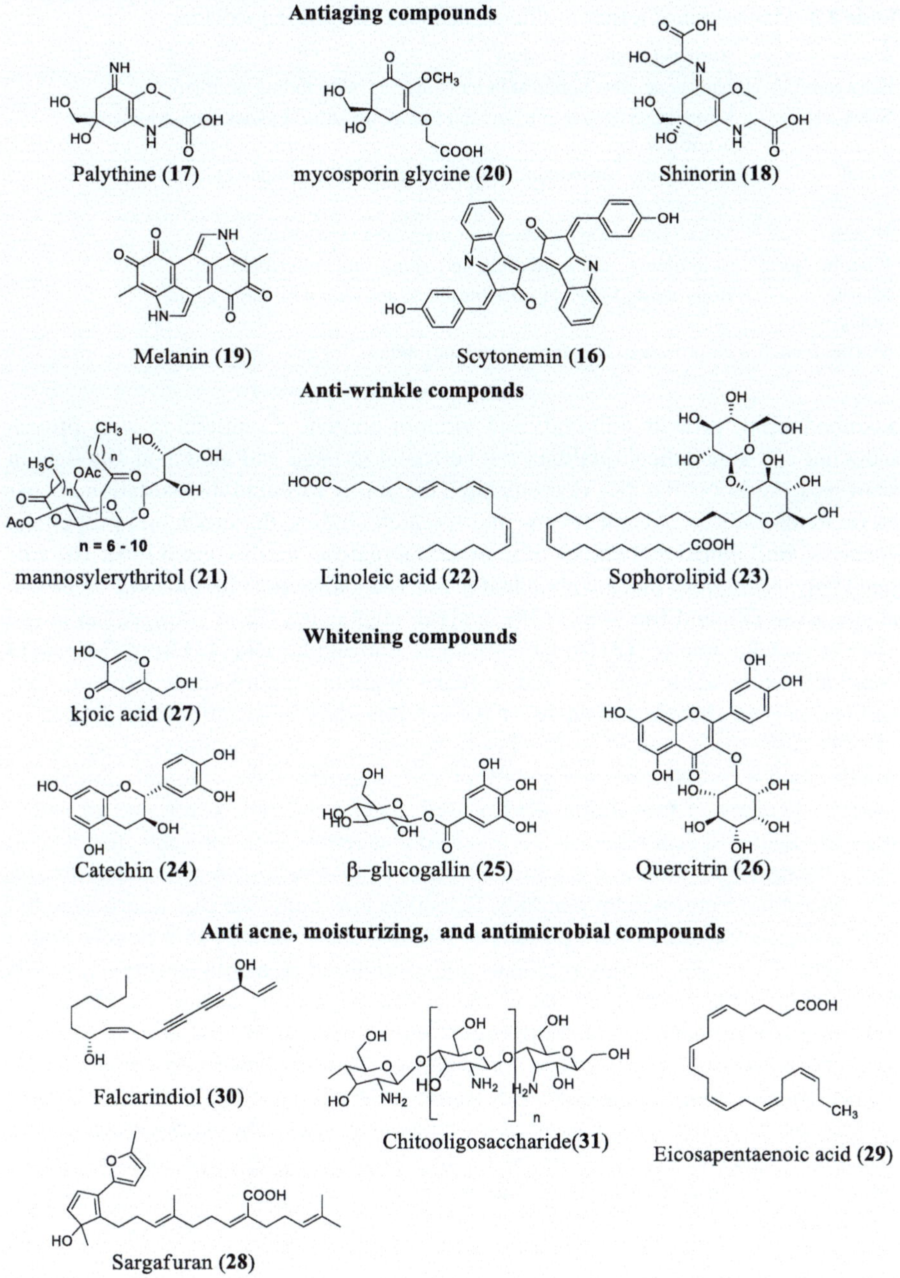

Fig. 2.3 Chemical structures of various marine soured active cosmeceuticals

Table 2.3 Marine sources related to different fields of cosmeceutical activities

Marine sources	Activity/cosmetic product
Seaweeds	Anti-acne, anti-aging, skin moisturizing, skin toning products
Shark fish	Moisturizer, deodorant, sun tan lotion, lip balm, lipstick, shampoo, and perfumes
Coral	Anti-aging, antioxidant, anti-acne, in lipstick, powder, etc.
Jellyfish	Anti-aging
Hydras	Sunscreen lotions, makeup remover, and moisturizer
Phytoplankton	Skin toning, anti-wrinkling, anti-aging, and skin whitening
Marine sponges	Body wash, baby skincare products, and skin whitening agents
Marine fennel	Antioxidant, skin toning, and anti-aging

adsorbed lipid, protein, mineral, and vitamin content. Compounds with proven reducing and anti-radical qualities can be found in large and abundant amounts in seaweeds. Additionally, they demonstrated the ability to shield the oxidation of cosmetic items [47]. In several review and research studies, the medicinal properties, cosmetic, and nutritional values of marine macroalgae are discussed [30]. The sulfated polysaccharides that are abundant in macroalgae include fucoidan (**2**) in brown algae, ulvan (**7**), and fucosterol (**35**) in green algal extract, and carrageenan in red algal extract, laminaran (**13**) from brown algae, and alginic acid (**11**) from *Turbinaria conoides*. Macroalgae can be used to make medicinal antioxidants because their sulfated polysaccharide derivatives in the cell boundary have antioxidant properties [49]. Besides these natural sulfated polysaccharides, the byproducts of above-sea weeds, such as edible brown algae by-product alginate, have a potent antioxidant property because of this alginic acid is used in cosmetic and cosmeceutical industries efficiently. Recent research has revealed that certain algae proteins, like the red algae *Palmaria palmate* and the protein waste of *Chlorella vulgaris* and *Chlorella ellipsoidea* have antioxidant peptides. Scientists now conclude that microalgae not only preserve the antioxidant properties but also show various skin health-related activities. There are various pigments for example scytonemin absorbs UV A/B radiation better than the synthetic elements that are used in cosmetics. Moreover, marine algae produce some tremendous anti-inflammatory compounds such as astaxanthin (**4**) (xanthophyll type carotenoid). There are other carotenoids as we previously mentioned (fucoxanthin **3**, canthaxanthin **15,** etc.) extracted from macroalgae that have anti-tyrosinase and anti-aging properties. The extract of *Laminaria saccharina*, another species with cosmetic potential, is rich in proteins, vitamins, minerals, and carbs. This algal extract has been claimed to have healing and anti-inflammatory qualities as well as to regulate sebaceous gland activity [28].

2.6.3.2 Marine Fishes and Invertebrates

Marine fishes are the major sources of antioxidant components like proteins, amino acids (Mycosporin) **20**, carotenoids, lipids, derived peptides, some polysaccharides (chondroitin sulfate, dermatan sulfate etc.), and chitooligosaccharides (**31**). It has a very high radical scavenging property than some synthetic antioxidants [50].

It has been observed that marine sponges, corals, and single bivalve also generate antioxidant chemicals as a key defense against toxicity mediated by free radicals. Some reports confirmed that secondary metabolites from marine sponges and soft corals showed potential antioxidant and anti-inflammatory, anti-fouling, anti-predator, and anti-microbial activities. Anti-melanogenic activity was observed by inhibiting tyrosinase expression in Gagunin D (**38**), a highly oxygenated diterpene that was isolated from the marine sponge *Phorbus* sp. Marine sponges' carotenoids are rich in photoaging and detoxification qualities. Methanolic extract of a marine sponge *Acanthella cavarnosa* is used in the production of anti-bacterial and anti-biofilm cosmetic products showing cell destroying properties against *P. acnes*. From soft coral, scientists found two compounds named sinularin (**12**) and dihydrosinularin that possess anti-cancer and antioxidant properties [39, 51].

2.6.3.3 Microorganisms

Microalgae, fungi, and marine bacteria are valuable producers of antioxidant enzymes like catalase and SOD (superoxide dismutase) and antioxidant components like carotenoids, exopolysaccharide (**5**), and bioactive peptides with a variety of biological features and applications. On the other hand, recently explored compounds, e.g., canthaxanthin (**15**), lutein, and lycopene with high immune-protected activity are extracted from sea microalgae. In terms of moisturizing properties microalgae-based compounds passed with good marks. Several studies observed that some proteins from *Spirulina* sp., and genus *Chlorella* have unique softening and smoothening effects on the skin and hair. Most of the natural compounds from marine fungi are used in cosmeceuticals because of their UV A adsorption capacity and anti-tyrosinase activity. A group of scientists observed that deep sea-derived fungus *Myrothecium* sp. could be a source of four new antioxidant and anti-inflammatory components: (−)-1-(S)-myrothecol (**34**), (+)-1-(R)-myrothecol (**33**), 1-methoxy-myrothecol (**32**) racemate, and myrothin (**37**) (an azaphilone derivative). Phytochemical analysis reveals that marine bacteria extracts contain steroids, saponins, tannins, flavonoids, anthocyanin, and β-cyanin which are used as antioxidants in various cosmetic, pharmaceuticals, and cosmeceuticals products. A high number of potent antioxidants like polysaccharides, scytonemin (**16**), phycobiliproteins, and phenolic compounds are observed in extracts of marine bacteria *Chroococcidiopsis* sp. [52]. Numerous bioactive secondary metabolites produced by bacteria are very valuable commercial products that are used in the cosmetics and pharmaceutical sectors. A wide range of substances including alkaloids, peptides, proteins, lipids, mycosporines, and glycosides that are derived from marine bacteria have anti-aging, moisturizing, antioxidant, and photoprotective properties [28].

2.6.3.4 Sea Cucumber

In addition, sea cucumbers are a great source of bioactive substances such as fatty acids, carotenoids, minerals, gelatin, chondroitin sulfate, saponins, collagen, bioactive peptides, vitamins, and amino acids [28]. A noteworthy feature of sea cucumbers is the abundance of unique sulfated polysaccharides, which hold immense promise for the creation of medicines and cosmeceuticals. The fucosylated chondroitin sulfate, a sulfated polysaccharide that is extracted from the body wall of sea cucumbers differs structurally from the sulfated polysaccharides that are extracted from algae, vertebrates, and other invertebrates. Research on sea cucumber collagen has mostly concentrated on the roles played by its hydrolytic bioactive peptides, which include antioxidant, anti-cancer, and angiotensin-converting enzyme inhibitory properties as well as the ability to repair injured tissue. Certain sea cucumber species, including *H. atra, H. fuscogilva*, *H. leucospilota*, *B. cousteaui*, and *H. nobilis*, exhibit strong anti-bacterial properties in their extracts. Strong anti-bacterial action against *S. aureus* was demonstrated in some research findings by β-cryptoxanthin and β-carotene (**8**), which were extracted from the Egyptian sea cucumber *H. scabra* [28]. Moreover, carotenoids derived from sea cucumber are used in various cosmetics to protect the elements from microbial contamination for a long period.

2.7 Uses of Marine-Derived Components on Therapeutic Treatment of Human Skin

The marine ecosystem is a remarkable source of bioactive natural products, many of which have unique structural and chemical characteristics not present in natural products on land. The number of new marine metabolites that are published in the literature each year is rising, which indicates that the marine environment will remain a major source of novel natural products for another several years. The potent compounds derived from different marine sources have not only cosmetic properties they also contain therapeutic properties like anti-allergic, anti-inflammatory, anti-obesity, anti-viral, anti-microbial, anti-cancer properties, and so on. There are numerous health benefits associated with substances like chitin, chitosan (**9**), omega-3 oils, algae, carotenoids, vitamins, minerals, calcium in fish bones, bioactive peptides, and fish protein hydrolysates. These benefits include a decrease in coronary heart disease (CHD) and anti-carcinogenic and anti-inflammatory properties. Among the 25,000 explored marine organisms only a few of them (fungi, cyanobacteria, some fishes, selective algae, etc.) possess such efficient clinical and medicinal properties. Some active anti-cancer compounds like Aplidine, Dideminin B, and Ecteinascidin-743 are extracted from marine invertebrate "Tunicate." Bryostatin 1 and Mycaperoxide B both are recognized for well-effective anti-cancer properties that are extracted from Bryozoan (marine invertebrate) and Sponges. Cyclodidemniserinol trisulfate and Lamellarin α 20 sulfate are anti-HIV components also found in tunicate (marine invertebrate) [53]. Bioactive proteins and peptides can be obtained in significant amounts from marine fish. Peptides produced from fish have a variety of pharmacological actions, such as immunomodulation, anti-hypertension, anti-bacterial, anti-cancer, and antioxidant properties. Dolastatin (from *Dolabella auricularia)* and Kahalalides (from *Elysia rufescens*) are the two peptides that are derived from

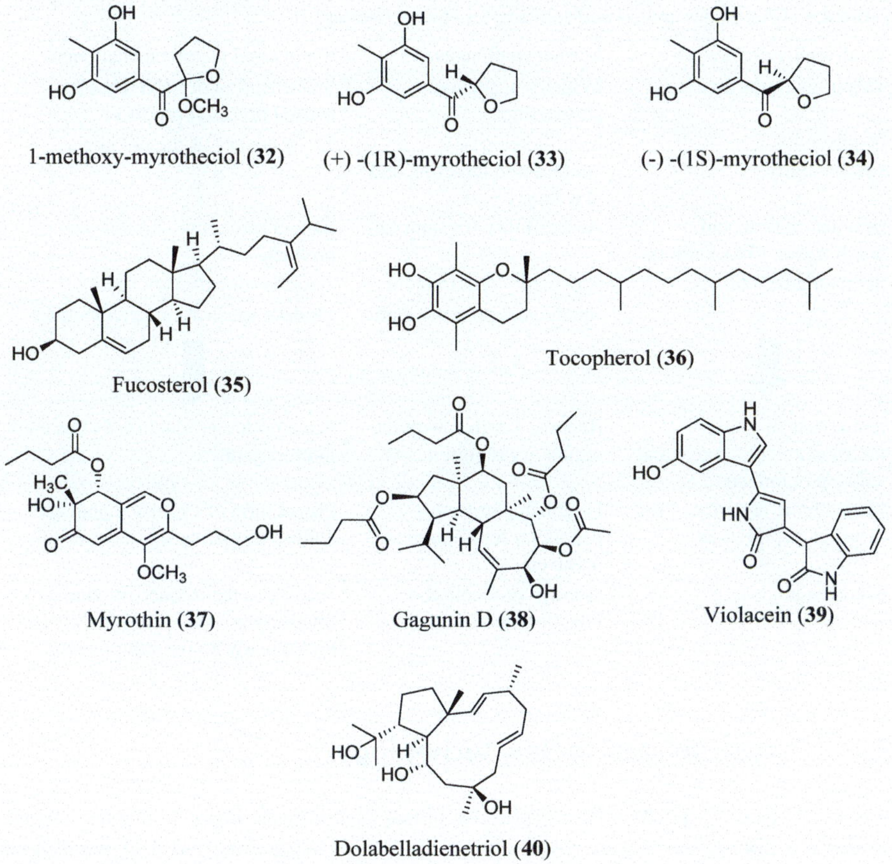

Fig. 2.4 Marine-derived compounds having cosmeceutical and pharmacological properties

marine fishes that have tremendous pharmacological activity in cancerous tumors. Moreover, anti-cancer peptides are employed as delivery systems in medication formulations to enhance and facilitate targeted drug transport across the proliferating cell membrane [54] (Fig. 2.4).

2.8 Challenges and Solutions Related to Marine-Based Cosmetic Industry

The market for natural product-based cosmetics is growing, which presents new difficulties for the research areas of bioactive marine resource exploration and development. Three categories emerged as the main obstacles in this ground: market, supply and technical issues, and biodiversity [55] (Table 2.4).

Table 2.4 Challenges and related solutions in new technology

Biodiversity challenges	Market challenges	Supply and technical challenges
Taxonomic expertise	Industry-academia partnerships	Identification of MNP (marine natural product) producer
Biodiversity access/ sustainability	Identification of competition targeting and new target	Re-discovery of the unknown
Genome mining and combinatorial biosynthesis	Accessibility for sampling	Virtual screening of natural products
Screening techniques	Regulatory requirement	Toxicity and efficacy studies
	Intellectual property status	Modern spectroscopic techniques
	Path of solutions	
Awareness of microorganism's usage and suitable organisms	Become familiar with the regulatory requirements	Scale up the production by cultured microorganisms
New robotic techniques for deep sea exploration	Establish partnerships beneficial for all parties involved	Preparation of exclusive plan for culture of marine microorganisms
Metagenomics	Aim for production of novel products	Advanced techniques for choosing microorganisms containing biologically active components

2.9 Issues Related to New Technology

It is found that, 25 percent of women use about 15 cosmetic items every day, while adults use about 9 cosmetic products per day worldwide. Around 1–3 percent of people have allergies to cosmetic chemicals [8]. The cosmetics industry, which produces both cosmetics and cosmeceuticals, is currently dealing with several issues. These include the need to accommodate certain personal beliefs (vegetarianism or veganism); environmental, cultural, or religious concerns; and, to name a few, health or safety-related issues. The market is currently seeing a rise in interest in "green products," or goods that buyers believe have less of an impact on the environment than conventional goods. The most conventional method for discovering bioactive compounds involves first removing a natural product from its source, screening it against a predetermined target, isolating it using a bioassay-guided isolation technique, fractionating it, and purifying it to produce essentially one biologically active compound. This conventional approach to discovering bioactive natural products is labor-intensive, sluggish, inefficient, and offers no assurance of success, despite being widely used. Biotechnological procedures have a significant impact on the cosmetic sector since they produce safe and effective active ingredients using inexpensive, noncontaminating technologies [7].

To confirm that raw materials are of high quality and in acceptable conditions for delivery to industrial processing facilities, the use of marine organism's by-products requires careful management in the areas of separation, classification, stabilization, and preservation. In addition, there are other factors that influence the viability of the by-product valorization techniques, such as seasonality, accumulation analysis, and raw material availability. It is common knowledge that humans have used marine habitats to produce various foods, decorations, tools, cosmetic elements, and even medical treatments since prehistoric times. Marine-based cosmetic formulations are in a better position to overcome the issues related to the conventional synthetic cosmetic industries. But for many years, the sustainability of their sources and manufacturing methods has not been considered as a crucial issue that needs to be addressed [56–59]. There is a widespread understanding of the necessity to manage marine resources to preserve their vast diversity and the health of the population for future generations. Industries need to plan how to make a sustainable ecosystem for this new natural product-based sector [56]. To get the best results in terms of revenue and product quality, businesses should enhance the management of their by-products. Additionally, as there are several parties involved in the cosmetics industry's actual adoption of these substances, it is necessary to coordinate and effectively manage some of the following issues:

- Accurate determination of the market for a specific ingredient and its applications in the industry.
- The proper taxonomic selection and classification of marine organisms must serve as the foundation for the selection of marine ingredients.
- Competitive price/kg of ingredients.
- A sustainable environment-friendly manufacturing process with sustainable ingredients supply [55].

2.10 Future Perspectives of Greener (Marine-Derived) Cosmetic Industries

Manufacturers and consumers are socially and environmentally conscious. Nonetheless, the absence of standardization and awareness regarding "eco-labels" permits producers, vendors, and marketers to employ this word freely without facing any pressure or repercussions. Businesses also use the tactic of selling natural components in cosmetics as a covert way to build their brands and increase consumer loyalty [7].

Social media's ability to enlighten the public and the efficient transmission of scientific research have increased public knowledge of the risks associated with the use of numerous chemicals in medications and cosmetics as well as the advantages of natural resource-based compounds for human health. Marine biotechnology, sometimes known as "blue biotechnology" is becoming an alternative to green technology as it provides a wide range of novel potent components with tremendous biological qualities that are not found in land ecosystems. In this emerging cosmetic market marine habitats attract the attention of researchers to conduct probing analysis [28]. However, more work must be done on the products' isolation and characterization to find their active sites, molecular modifications, assessment of their pharmacological qualities and safety aspects, enhancement of the products' quality, and most importantly, increased investment in R&D.

2.11 Discussion and Conclusion

The cosmetics industry is a profitable and gradually expanding sector of the world economy. Furthermore, natural products have a higher market value which can exceed the boundary of the synthetic cosmetic market. To make a sustainable, eco-friendly, and economically strong cosmetic market, the cosmetic industries are now inclined more toward the marine-based active sources. Moreover, cosmetics and beauty products contain over 10,000 substances that have been related to disease, therefore they are not particularly healthy. People who use cosmetics on a regular basis run the risk of developing side effects such type IV hypersensitivity, allergies, contact dermatitis, and skin redness. That is why people are becoming more aware about the products they are using. As a result, customer demand for greener products helps to focus research on novel undiscovered natural products and the optimization of their powerful ingredients. For the pharmaceutical and cosmetic sectors, the marine environment offers a comparatively unexplored reservoir of biologically active substances. So it becomes a great opportunity for marine researchers to developed the marine sourced ingredients that may help in not only in the production of cosmeceuticals but also in the improvement of human health. These marine-derived components have many benefits like anti-aging, anti-allergic, anti-inflammatory, anti-photo aging, antioxidant, and anti-bacterial devoid of any adverse effect which results in innovative success in the cosmetics sector. All opportunities with great good factors have some drawbacks. In the growing marine cosmetic market, there are lots of challenges as we mentioned previously. Because scientists have limited knowledge about the sea organisms that have been explored till now. But in the very depth of the sea, an ocean of potent active components is hidden from the world. To prepare a component for products, few steps should be followed: exploration, extraction, isolation, optimization, and finally the trial stage. So, to deal with these factors industries are looking into new technological developments that create a bright future for marine marine-based cosmetic market.

References

1. Liu J-K (2022) Natural products in cosmetics. Nat Prod Bioprospect 12:40
2. Schneider G et al (2001) Skin Cosmetics. In: Ullmann's Encyclopedia of industrial chemistry. Wiley. https://doi.org/10.1002/14356007.a24_219
3. Lucca JM, Joseph R, Hussain Al Kubaish Z, Mohammad Al-Maskeen S, Ali Alokaili Z (2020) An observational study on adverse reactions of cosmetics: the need of practice the Cosmetovigilance system. Saudi Pharm J 28:746–753
4. Pimentel FB, Alves RC, Rodrigues F, Oliveira MBPP (2017) Macroalgae-derived ingredients for cosmetic industry—an update. Cosmetics 5:2

5. Benson HAE, Roberts MS, Leite-Silva VR, Walters KA (2019) Cosmetic formulation. CRC Press. https://doi.org/10.1201/9780429190674
6. Pavlou P, Siamidi A, Varvaresou A, Vlachou M (2021) Skin care formulations and lipid carriers as skin moisturizing agents. Cosmetics 8:89
7. Goyal N, Jerold F (2021) Biocosmetics: technological advances and future outlook. Environ Sci Pollut Res 30:25148–25169
8. Nayak M, Sreedhar D, Prabhu SS, Ligade VS (2021) Global trends in cosmetics use-related adverse effects: a bibliometric analysis of literature published during 1957–2021. Cosmetics 8:75
9. Khan AD, Alam MN (2019) Cosmetics and their associated adverse effects: a review. J Appl Pharm Sci Res:1–6. https://doi.org/10.31069/japsr.v2i1.1
10. Kaličanin B, Velimirović D (2016) A study of the possible harmful effects of cosmetic beauty products on human health. Biol Trace Elem Res 170:476–484
11. Kanayochukwu Nduka J, Ijeoma Kelle H, Omoche Odiba I (2019) Review of health hazards and toxicological effects of constituents of cosmetics. In: Poisoning in the modern world—new tricks for an old dog? IntechOpen. https://doi.org/10.5772/intechopen.84590
12. Assali M, Zaid A-N (2022) Features, applications, and sustainability of lipid nanoparticles in cosmeceuticals. Saudi Pharm J 30:53–65
13. Pistollato F et al (2021) Current EU regulatory requirements for the assessment of chemicals and cosmetic products: challenges and opportunities for introducing new approach methodologies. Arch Toxicol 95:1867–1897
14. Secchi M, Castellani V, Collina E, Mirabella N, Sala S (2016) Assessing eco-innovations in green chemistry: Life Cycle Assessment (LCA) of a cosmetic product with a bio-based ingredient. J Clean Prod 129:269–281
15. Couteau C, Coiffard L (2016) Seaweed application in cosmetics. In: Seaweed in health and disease prevention. Elsevier, pp 423–441. https://doi.org/10.1016/B978-0-12-802772-1.00014-2
16. Wang H-MD, Chen C-C, Huynh P, Chang J-S (2015) Exploring the potential of using algae in cosmetics. Bioresour Technol 184:355–362
17. Kim S (2014) Marine cosmeceuticals. J Cosmet Dermatol 13:56–67
18. Ngo D-H, Vo T-S, Ngo D-N, Wijesekara I, Kim S-K (2012) Biological activities and potential health benefits of bioactive peptides derived from marine organisms. Int J Biol Macromol 51:378–383
19. Ariede MB et al (2017) Cosmetic attributes of algae—a review. Algal Res 25:483–487
20. Paradis S et al (2021) Persistence of biogeochemical alterations of deep-sea sediments by bottom trawling. Geophys Res Lett 48
21. Vo T, Kim S (2012) Cosmeceutical compounds from marine sources. In: Kirk-Othmer encyclopedia of chemical technology. Wiley, pp 1–17. https://doi.org/10.1002/0471238961.cosmkim.a01
22. Wang H-MD, Li X-C, Lee D-J, Chang J-S (2017) Potential biomedical applications of marine algae. Bioresour Technol 244:1407–1415
23. Fonseca S, Amaral MN, Reis CP, Custódio L (2023) Marine natural products as innovative cosmetic ingredients. Mar Drugs 21:170
24. Guillerme J-B, Couteau C, Coiffard L (2017) Applications for marine resources in cosmetics. Cosmetics 4:35
25. Kamei Y, Sueyoshi M, Hayashi K, Terada R, Nozaki H (2009) The novel anti-Propionibacterium acnes compound, Sargafuran, found in the marine brown alga Sargassum macrocarpum. J Antibiot (Tokyo) 62:259–263
26. Toyoda M, Morohashi M (2001) Pathogenesis of acne. Med Electron Microsc 34:29–40
27. Heo S-J et al (2009) Effect of phlorotannins isolated from Ecklonia cava on melanogenesis and their protective effect against photo-oxidative stress induced by UV-B radiation. Toxicol Vitr 23:1123–1130
28. Alves A, Sousa E, Kijjoa A, Pinto M (2020) Marine-derived compounds with potential use as cosmeceuticals and nutricosmetics. Molecules 25:2536

29. Cockell CS, Knowland J (1999) Ultraviolet radiation screening compounds. Biol Rev 74:311–345
30. El-Shafei R, Hegazy H, Acharya B (2021) A review of antiviral and antioxidant activity of bioactive metabolite of macroalgae within an optimized extraction method. Energies 14:3092
31. Kemp DC, Kwon JY (2003) Marine-derived antioxidant peptides: properties and applications. In: Marine antioxidants. Elsevier, pp 1–10. https://doi.org/10.1016/B978-0-323-95086-2.00034-5
32. Rastogi RP, Sonani RR, Madamwar D (2015) Cyanobacterial sunscreen scytonemin: role in photoprotection and biomedical research. Appl Biochem Biotechnol 176:1551–1563
33. Mukherjee PK, Maity N, Nema NK, Sarkar BK (2011) Bioactive compounds from natural resources against skin aging. Phytomedicine 19:64–73
34. Freitas-Rodríguez S, Folgueras AR, López-Otín C (2017) The role of matrix metalloproteinases in aging: Tissue remodeling and beyond. Biochim Biophys Acta, Mol Cell Res 1864:2015–2025
35. Joe M-J et al (2006) The inhibitory effects of eckol and dieckol from Ecklonia stolonifera on the expression of matrix Metalloproteinase-1 in human dermal fibroblasts. Biol Pharm Bull 29:1735–1739
36. Pereira DM, Valentão P, Andrade PB (2014) Marine natural pigments: chemistry, distribution and analysis. Dyes Pigments 111:124–134
37. Muñoz-Miranda LA, Iñiguez-Moreno M (2023) An extensive review of marine pigments: sources, biotechnological applications, and sustainability. Aquat Sci 85:68
38. Agrawal S, Adholeya A, Barrow CJ, Deshmukh SK (2018) Marine fungi: an untapped bioresource for future cosmeceuticals. Phytochem Lett 23:15–20
39. Ding J, Wu B, Chen L (2022) Application of marine microbial natural products in cosmetics. Front Microbiol:13
40. Coiffard L, Alliot A, Piron-Frenet M (1992) Seasonal and geographical adaptation of crithmum maritimum L.: variations in inorganic content. Ecol Food Nutr 28:261–269
41. Siahaan EA, Agusman Pangestuti R, Shin K-H, Kim S-K (2022) Potential cosmetic active ingredients derived from marine by-products. Mar Drugs 20:734
42. Routledge EJ, Parker J, Odum J, Ashby J, Sumpter JP (1998) Some alkyl hydroxy benzoate preservatives (Parabens) are estrogenic. Toxicol Appl Pharmacol 153:12–19
43. Pérez M, Falqué E, Domínguez H (2016) Antimicrobial action of compounds from marine seaweed. Mar Drugs 14:52
44. Lee J-H et al (2014) In vitro antibacterial and synergistic effect of phlorotannins isolated from edible brown seaweed Eisenia bicyclis against acne-related bacteria. Algae 29:47–55
45. Kim IH et al (2007) Antibacterial activity ofUlva lactuca against methicillin-resistantStaphylococcus aureus (MRSA). Biotechnol Bioprocess Eng 12:579–582
46. Wei Y et al (2015) Antibacterial mode of action of 1,8-dihydroxy-anthraquinone from Porphyra haitanensis against Staphylococcus aureus. Nat Prod Res 29:976–979
47. López-Hortas L et al (2021) Applying seaweed compounds in cosmetics, cosmeceuticals and nutricosmetics. Mar Drugs 19:552
48. Rodríguez-Pérez C et al (2013) A metabolite-profiling approach allows the identification of new compounds from Pistacia lentiscus leaves. J Pharm Biomed Anal 77:167–174
49. Rupérez P, Ahrazem O, Leal JA (2002) Potential antioxidant capacity of sulfated polysaccharides from the edible marine brown seaweed fucus vesiculosus. J Agric Food Chem 50:840–845
50. Zhou C et al (2020) Purification, characterisation and antioxidant activities of chondroitin sulphate extracted from Raja porosa cartilage. Carbohydr Polym 241:116306
51. Wang S-C et al (2021) Comparison of antioxidant and anticancer properties of soft coral-derived sinularin and dihydrosinularin. Molecules 26:3853
52. Assunção J et al (2021) Exploration of marine genus Chroococcidiopsis sp.: a valuable source for antioxidant industry? J Appl Phycol 33:2169–2187
53. Kijjoa A, Sawangwong P (2004) Drugs and cosmetics from the sea. Mar Drugs 2:73–82
54. Kang H, Choi M-C, Seo C, Park Y (2018) Therapeutic properties and biological benefits of marine-derived anticancer peptides. Int J Mol Sci 19:919

55. Martins A, Vieira H, Gaspar H, Santos S (2014) Marketed marine natural products in the pharmaceutical and cosmeceutical industries: tips for success. Mar Drugs 12:1066–1101
56. Sotelo CG, Blanco M, Ramos P, Vázquez JA, Perez-Martin RI (2021) Sustainable sources from aquatic organisms for cosmeceuticals ingredients. Cosmetics 8:48
57. McCullough JL, Kelly KM (2006) Prevention and treatment of skin aging. Ann N Y Acad Sci 1067:323–331
58. Pereira L (2018) Seaweeds as source of bioactive substances and skin care therapy—cosmeceuticals, algotheraphy, and thalassotherapy. Cosmetics 5:68
59. Boziaris IS (2017) Food ingredients from the marine environment. Marine biotechnol meets food science and technology. Front Mar Sci:1

Application of Diatoms in the Treatment of Industrial Effluent

3

John Tsado Mathew, Abel Inobeme, Elijah Yanda Shaba, Monday Musah, Yakubu Azeh, Muhammad Bini Etsuyankpa, Charles Oluwaseun Adetunji, Amos Mamman, and Musa Safiyanu Tanko

Abstract

Diatoms, a type of microalgae, have emerged as a promising solution for industrial effluent treatment. Their ability to efficiently remove heavy metals, organic pollutants, and nutrients from wastewater makes them a sustainable and eco-friendly option. This review explores the diverse applications of diatoms in bioremediation, emphasizing their cost-effectiveness, scalability, and low environmental impact. Leveraging their natural affinity for pollutant absorption and the potential for biomass production, diatoms offer a greener approach to mitigating water pollution in various industrial sectors, contributing to improved water quality and environmental sustainability.

Keywords

Biomass · Diatoms · Effluent · Environmental · Industrial · Pollutant

---

J. T. Mathew (✉) · M. Musah · Y. Azeh · A. Mamman
Department of Chemistry, Ibrahim Badamasi Babangida University, Lapai, Niger State, Nigeria

A. Inobeme
Department of Chemistry, Edo University Uzairue, Auchi, Edo State, Nigeria

E. Y. Shaba
Department of Chemistry, Federal University of Technology Minna, Minna, Nigeria

M. B. Etsuyankpa · M. S. Tanko
Department of Chemistry, Federal University of Lafia, Lafia, Nasarawa State, Nigeria

C. O. Adetunji
Applied Microbiology, Biotechnology and Nanotechnology Laboratory, Department of Microbiology, Edo University Uzairue, Auchi, Edo State, Nigeria

M. Rafatullah et al. (eds.), *Multidisciplinary Applications of Marine Resources*,
https://doi.org/10.1007/978-981-97-5057-3_3

3.1 Introduction

Industrial effluents, often contaminated with a myriad of pollutants, pose a significant threat to the environment and human health. The need for sustainable and efficient methods of treatment and remediation has never been more pressing. Diatoms, a group of microalgae encased in intricate silica shells, have emerged as a promising, eco-friendly, and cost-effective solution in the treatment of industrial effluents. This introduction explores the application of diatoms in the treatment of industrial effluents, highlighting their unique characteristics, mechanisms of action, and potential advantages in mitigating pollution and enhancing environmental sustainability [1].

Industrial activities across the globe generate vast volumes of wastewater, containing a wide range of contaminants, such as heavy metals, organic pollutants, and nutrients. Discharging these effluents into natural water bodies without proper treatment can have disastrous consequences for aquatic ecosystems, public health, and the broader environment. Therefore, it is imperative to develop innovative and sustainable technologies for their efficient removal and remediation [2].

Diatoms are a diverse group of unicellular microalgae characterized by their unique cell walls, composed of amorphous silica. These microorganisms are photosynthetic and play a pivotal role in aquatic ecosystems, where they form the base of the food chain. Their intricate, porous silica shells, called frustules, are the secret to their utility in effluent treatment. Diatoms come in a wide variety of shapes and sizes, making them adaptable to various environmental conditions [3].

Diatoms have a range of mechanisms that make them suitable for treating industrial effluents. Their high surface area, owing to the complex frustules, allows them to adsorb a multitude of contaminants effectively. This adsorption process can be enhanced by modifying diatom surfaces with specific functional groups, increasing their affinity for particular pollutants. In addition to physical adsorption, diatoms can also biologically degrade certain organic pollutants, transforming them into less harmful compounds through their metabolic activities [4].

Several case studies and applications have demonstrated the practical use of diatoms in industrial effluent treatment. For example, in the textile industry, diatoms have been employed to remove dyes and heavy metals from wastewater, significantly improving the quality of discharged effluents. In mining, diatom-based treatment processes have shown promise in reducing the environmental impact of mine tailings and acidic mine drainage [5].

This chapter presents the application of diatoms in the treatment of industrial effluents (Fig. 3.1). It discusses the characteristics and nature of diatoms. Reports on the use of diatom in this regard are also highlighted.

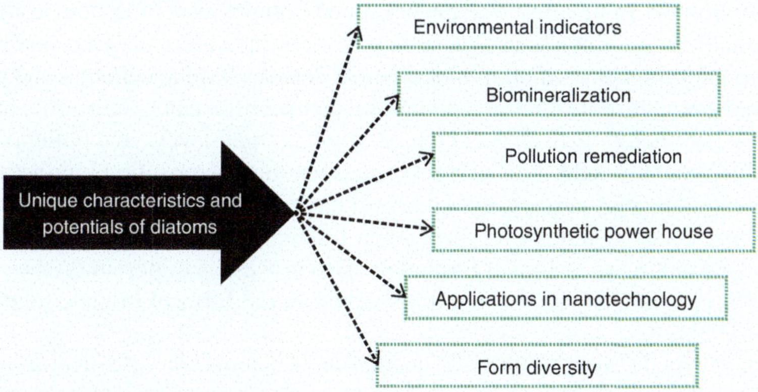

Fig. 3.1 Key characteristics of diatoms

3.2 Diatoms: Nature and Characteristics

Diatoms, a group of microalgae, are remarkable and diverse organisms that play a vital role in aquatic ecosystems. Their unique nature and characteristics make them a fascinating subject of study and have practical applications in various fields. This article provides an overview of diatoms, their nature, and their key characteristics (Fig. 3.1) [6].

Microscopic Marvels: Diatoms belong to the phylum Bacillariophyta and are a type of photosynthetic microalgae. They are unicellular organisms, and their size typically ranges from a few micrometers to a few hundred micrometers. This microscopic size is one of the defining features of diatoms, allowing them to be widely dispersed in aquatic environments, from oceans and lakes to rivers and even moist soils [7].

Intricate Silica Shells: One of the most distinctive features of diatoms is their cell walls, known as frustules, which are composed of silica. These frustules are like intricately designed glass cases that encase the diatom cell. The frustules have an outer and inner valve that fit together, much like a petri dish, and they are elaborately ornamented with pores, spines, and other structures. These frustules are what give diatoms their unique, often striking, and intricate shapes [8].

Diversity in Form: Diatoms exhibit remarkable diversity in their shapes and sizes. Some diatoms have a radial or centric symmetry, with circular or oval frustules, while others have a bilateral or pennate symmetry, with elongated, boat-shaped frustules. The incredible variety of shapes allows diatoms to occupy a wide range of ecological niches and adapt to different environmental conditions [9].

Photosynthetic Powerhouses: Diatoms are autotrophic organisms, meaning they are capable of photosynthesis. They contain chlorophyll and other pigments that enable them to convert light energy into chemical energy, primarily in the form of sugars. Diatoms are a crucial component of the primary production in aquatic ecosystems, serving as the base of the food chain and providing energy for a variety of aquatic organisms [10].

Environmental Indicators: The presence and composition of diatom communities in aquatic environments can serve as important indicators of water quality and environmental conditions. Different species of diatoms have specific habitat preferences, and changes in the diatom community composition can reflect shifts in environmental parameters such as water temperature, nutrient levels, and pollution [11].

Biomineralization and the Carbon Cycle: Diatoms are prolific biomineralizers, which means they play a role in the cycling of minerals, particularly silica, in aquatic ecosystems. When diatoms die, their silica frustules sink to the bottom of water bodies, contributing to sediment formation. This process is important in the regulation of the carbon cycle as it sequesters carbon in the form of organic matter and frustules [12].

Nutrient Uptake and Pollution Remediation: Diatoms are effective at nutrient uptake, particularly for silica and nitrate. In some applications, they are used to remediate water polluted with excess nutrients. They can act as natural filters, reducing the availability of nutrients for harmful algal blooms and restoring the ecological balance in affected water bodies [13].

Applications in Nanotechnology: The intricate and porous nature of diatom frustules has garnered interest in the field of nanotechnology. Diatom-derived silica nanostructures have shown promise in various applications, including drug delivery, biosensors, and as templates for the synthesis of nanoparticles [14].

In addition, diatoms are fascinating microorganisms with unique characteristics, such as their silica frustules, diverse shapes, and ecological significance. Their importance in aquatic ecosystems, as well as their potential applications in environmental monitoring and nanotechnology, make them a subject of ongoing research and appreciation in the scientific community. Understanding the nature and characteristics of diatoms is essential not only for ecological studies but also for harnessing their potential in diverse fields [15].

3.3 Industrial Effluent and Its Impact

Industrial effluent refers to the wastewater generated by industrial processes and activities. This effluent typically contains a mixture of chemicals, contaminants, and pollutants, making it a significant environmental concern [16]. In this article, we will explore what industrial effluent is, its sources, and the environmental impact it can have (Table 3.1).

3.4 Industrial Effluent Treatment Using Diatoms

Diatoms are a group of single-celled algae with unique characteristics that make them valuable for various applications, including industrial effluent treatment. Industrial effluent, which includes wastewater generated by industrial processes, can contain a wide range of pollutants and contaminants that must be treated before

Table 3.1 Industrial effluent and its impact

Aspect	Description	Examples of sources	Environmental impact	References
Sources	Various industries, such as chemical manufacturing, mining, and agriculture, generate industrial effluent.	Chemical manufacturing, mining, textile, agriculture	Water pollution, eutrophication, habitat destruction, toxicity, bioaccumulation, acidification, soil contamination, atmospheric pollution	[17]
Composition	The composition of industrial effluent varies depending on the industry, containing pollutants like heavy metals, chemicals, organic matter, and nutrients.	Heavy metals, dyes, pesticides, organic waste, thermal pollution	Contamination of water bodies, disruption of ecosystems, harm to aquatic life, changes in water quality, and formation of dead zones	[18]
Regulatory measures	Governments and international bodies have implemented regulations to control the quality of effluents, limit the discharge of harmful substances, and ensure responsible treatment of industrial wastewater	Water quality standards, discharge permits, environmental protection agencies	Legal frameworks to enforce responsible effluent management and reduce environmental harm	[19]
Mitigation and treatment	Advanced treatment technologies like biological treatment, chemical precipitation, and filtration are used to remove or reduce pollutants from industrial effluent before discharge. Sustainable practices are also adopted to minimize waste generation and environmental impact	Wastewater treatment plants, sustainable production techniques, pollution prevention measures	Reduction of pollutants, improved effluent quality, preservation of ecosystems, and protection of human health and the environment	[20]

discharge to protect the environment and public health. Diatoms can play a role in this treatment process through a method known as phytoremediation [21].

How diatoms can be used in industrial effluent treatment:

Biological Filtration: Diatoms are photosynthetic organisms that require light, carbon dioxide, and nutrients to grow. In an effluent treatment system, diatoms can be cultivated in a controlled environment such as a photobioreactor. The effluent is passed through this reactor, and the diatoms capture and metabolize contaminants. They can remove various pollutants, including heavy metals, organic compounds, and nutrients like nitrogen and phosphorus [22].

Bioaccumulation: Diatoms can accumulate contaminants within their cells. This process, known as bioaccumulation, helps in reducing the concentration of pollutants in the effluent. Over time, the diatoms can be harvested and disposed of or treated to recover valuable materials if the contaminants are valuable metals, for example [1].

Oxygen Production: Diatoms produce oxygen as a byproduct of photosynthesis. This can be advantageous in effluent treatment as it helps maintain aerobic conditions in the treatment system, which is beneficial for the growth of other beneficial microorganisms that can further degrade contaminants [23].

Nutrient Uptake: Diatoms are efficient in taking up nutrients like nitrogen and phosphorus. In industrial effluent treatment, these nutrients can be a major concern as they can lead to eutrophication if released into natural water bodies. Diatoms can help in removing these nutrients, improving the quality of treated effluent [6, 24].

Sustainable Approach: Diatoms offer a sustainable approach to effluent treatment as they are part of the natural aquatic ecosystem and can be maintained and cultured relatively easily.

However, it is important to note that the effectiveness of using diatoms for industrial effluent treatment may vary depending on the specific contaminants in the effluent and the environmental conditions. Also, it is often used as a part of a broader treatment strategy, which may include other physical and biological processes [25].

The use of diatoms in industrial effluent treatment is an area of ongoing research, and the specific methods and techniques may continue to evolve as our understanding of these organisms and their capabilities improves [26].

3.5 Diatoms in Bioremediation

Diatoms, a group of single-celled algae with a silica shell, are increasingly being recognized for their potential in bioremediation, which is the use of biological organisms to remove or mitigate pollutants from various environments [27]. Diatoms have several characteristics that make them valuable for bioremediation processes (Table 3.2).

High Surface Area: The intricate, porous silica shells of diatoms provide a high surface area, making them effective adsorbents for a wide range of pollutants. This includes heavy metals, organic compounds, and even radionuclides [28].

Selective Uptake: Diatoms can be tailored to selectively accumulate specific pollutants. Researchers have explored the genetic modification of diatoms to enhance their ability to absorb and concentrate specific contaminants [29].

Nutrient Uptake: Diatoms can help in removing excess nutrients such as nitrogen and phosphorus from aquatic environments, which can cause issues like eutrophication. By reducing nutrient levels, they can indirectly mitigate water pollution [30].

Table 3.2 The main aspects of diatoms in bioremediation

Aspect	Description	Example	Benefits	References
High surface area	Intricate, porous silica shells provide a large surface area for adsorption of various pollutants	Adsorption of heavy metals, organic compounds	Efficient pollutant removal	[36, 37]
Selective uptake	Diatoms can be genetically modified for selective accumulation of specific contaminants.	Enhanced uptake of heavy metals, radionuclides	Tailored to target pollutants	[38]
Nutrient uptake	Diatoms help reduce excess nutrients, such as nitrogen and phosphorus, mitigating eutrophication.	Removal of nutrients in polluted aquatic systems	Indirect pollution control	[39]
Oxygen production	Diatoms generate oxygen during photosynthesis, maintaining aerobic conditions in polluted environments.	Supports growth of beneficial microorganisms	Enhanced microbial bioremediation	[40]
Sustainable approach	Diatoms are part of natural ecosystems and can be easily cultivated, providing an eco-friendly solution	Low environmental impact, easy maintenance	Environmentally sustainable	[41]
Biomineralization	Some diatom species facilitate mineral formation, aiding in the precipitation of heavy metals	Removal of heavy metals through biomineralization	Additional method for pollutant removal	[42]
Biosensors	Engineered diatoms can serve as biosensors, responding to specific pollutants by producing visible signals	Real-time monitoring of pollution levels	Pollution detection and measurement	[43]

Oxygen Production: Diatoms produce oxygen as a byproduct of photosynthesis, which can help maintain aerobic conditions in polluted environments. This oxygen can also support the growth of other beneficial microorganisms that further aid in bioremediation [31].

Sustainable Approach: Diatoms are part of natural aquatic ecosystems, and their use in bioremediation is relatively eco-friendly. They can be cultivated and maintained easily, offering a sustainable approach to pollution control [6].

Biomineralization: Some diatom species have the ability to facilitate the formation of minerals, such as iron, manganese, and silica. This can be harnessed in bioremediation efforts, such as the precipitation of heavy metals [32].

Biosensors: Diatoms can be engineered to function as biosensors, responding to the presence of specific pollutants by producing visible changes or signals. These biosensors can be used to monitor pollution levels in real time [33].

Diatoms can be applied in various bioremediation scenarios, such as the treatment of industrial wastewater, contaminated groundwater, and polluted surface waters. They are often used in combination with other bioremediation techniques, like constructed wetlands, biofilters, and biostimulation, to enhance the overall efficacy of the process [34].

However, the practical application of diatoms in bioremediation is an area of ongoing research, and the specific methods may vary depending on the contaminants, environmental conditions, and the desired outcome. It is important to conduct site-specific studies and assessments to determine the suitability of diatoms in any given bioremediation project [35].

3.6 Effluent Treatment Technologies

Effluent treatment technologies are methods and processes used to treat and manage various types of wastewater or effluent, which may contain contaminants, pollutants, or other undesirable components. The choice of treatment technology depends on the nature of the effluent and the specific contaminants present [16]. Table 3.3 provides an overview of various effluent treatment technologies, their applications, and the advantages associated with each technology.

3.6.1 Future Prospects and Research Directions

The future prospects and research directions in various fields are highly dynamic and can vary significantly depending on technological advancements, societal needs, and scientific breakthroughs. Here are some general trends and research directions across different domains.

Table 3.3 Various effluent treatment technologies

Treatment technology	Description	Typical applications	Advantages	References
Physical treatment	Involves physical processes to remove contaminants	Municipal wastewater, industrial effluents	Simple and cost-effective	[44]
– Screening	Removal of large solids through screens	Preliminary treatment	Removes coarse debris	[45]
– Sedimentation	Gravity-based separation of suspended solids	Primary treatment	Effective for settling solids	[46]
– Filtration	Use of filters to remove fine particles	Tertiary treatment, polishing	Effective for smaller particles	[22]
Chemical treatment	Chemicals are added to facilitate contaminant removal	Industrial wastewater, metal finishing	Precipitates and removes pollutants	[47]
– Coagulation and flocculation	Destabilizes particles for easier removal	Treatment of colloidal suspensions	Effective for removing fine particles	[48]
– Chemical precipitation	Chemicals are added to form insoluble precipitates	Heavy metal removal	Targeted removal of specific ions	[49]
– Chemical oxidation	Chemicals (e.g., chlorine, ozone) break down organic matter	Removal of organic contaminants	Effective for organic pollutant degradation	[50]
Biological treatment	Utilizes microorganisms to break down organic matter	Municipal sewage, industrial wastewater.	Effective for organic pollutant removal	[51]
– Activated sludge process	Aerobic microbial treatment with suspended growth	Secondary treatment	Efficient organic matter degradation	[52]
– Trickling filters	Effluent passes over microbial biofilms on media	Secondary or tertiary treatment	Effective for organic and BOD removal	[53]
– Constructed wetlands	Natural plants and microbes treat effluent	Tertiary treatment, ecological systems	Effective for nutrient and BOD removal	[54]

(continued)

Table 3.3 (continued)

Treatment technology	Description	Typical applications	Advantages	References
– Anaerobic digestion	Bacteria break down organic matter in the absence of oxygen	Industrial wastewater	Effective for sludge treatment	[55]
Advanced treatment	Utilizes advanced processes for specific contaminant removal	Pharmaceutical wastewater, complex effluents	High removal efficiency	
– Membrane filtration	Separation using ultrafiltration or reverse osmosis membranes	Water purification, desalination	Efficient removal of small particles	
– Adsorption	Pollutants adhere to adsorbents (e.g., activated carbon)	Removal of organic contaminants	Effective for diverse pollutants	[56]
– Ion exchange	Ions are exchanged with specific resins or materials	Removal of heavy metals, softening	Targeted removal of specific ions	
– Advanced Oxidation Processes (AOPs)	Combines UV, ozone, and/or hydrogen peroxide	Degradation of complex organics	Effective for recalcitrant contaminants	
Disinfection	Inactivation of microorganisms to ensure safe discharge	Final treatment step for wastewater	Ensures treated effluent is safe	
– Chlorination	Addition of chlorine for disinfection	Municipal wastewater, swimming pools	Proven disinfection method	[57]
– UV Disinfection	Use of ultraviolet (UV) light to inactivate pathogens	Drinking water treatment	Chemical-free disinfection	
– Ozonation	Ozone gas is used for disinfection	Municipal and industrial wastewater	Effective against various pathogens	
Desalination	Removal of salts and dissolved solids from water	Seawater desalination, brackish water treatment	Provides fresh water from saline sources	

(continued)

Table 3.3 (continued)

Treatment technology	Description	Typical applications	Advantages	References
– Reverse Osmosis	High-pressure membrane separation for desalination	Seawater and brackish water desalination	Effective salt and contaminant removal	[58]
Zero liquid discharge	System designed to recover and recycle all wastewater	Industries with water scarcity concerns	Minimizes effluent discharge	
Solid waste management	Techniques for handling and disposing of the generated sludge or solid residues	Sewage sludge management, industrial processes	Safe disposal of waste products	
Bioremediation	Microorganisms used to break down specific contaminants	Soil and groundwater remediation	Effective for organic pollutant removal	[59]
– Electrocoagulation	Electrochemical process to destabilize and remove solids	Industrial wastewater	Efficient solids removal	

3.6.1.1 Biotechnology and Healthcare

Gene editing and CRISPR technology for personalized medicine. Development of vaccines, therapies, and diagnostics for emerging diseases. Artificial organs, regenerative medicine, and 3D bioprinting.

3.6.1.2 Climate Change and Sustainability

Renewable energy sources, energy storage, and carbon capture technologies. Climate modeling and prediction to mitigate the impact of climate change. Sustainable agriculture and eco-friendly materials.

3.6.1.3 Space Exploration

Mars exploration and potential human colonization, advanced propulsion systems and asteroid mining, and search for extraterrestrial life.

3.6.1.4 Information Technology

Quantum computing and quantum cryptography, 6G and beyond for faster and more reliable communication, cybersecurity to counter evolving threats.

3.6.1.5 Neuroscience and Brain–Computer Interfaces

Understanding the brain's functions and mechanisms, developing brain–computer interfaces for medical and non-medical applications, mental health research and treatment.

3.6.1.6 Environmental Science

Biodiversity conservation and ecosystem restoration, monitoring and combating pollution and environmental degradation, climate adaptation and resilience strategies.

3.6.1.7 Robotics and Automation

Collaborative robots (cobots) for industrial and service applications, autonomous vehicles and drones for various industries, Human-robot interaction, and social robotics.

3.6.1.8 Blockchain and Cryptocurrency

Decentralized finance (DeFi) and digital currencies, Blockchain applications in supply chain, voting, and identity management, Scalability and security improvements.

3.6.1.9 Social Sciences and Humanities

Behavioral economics and the study of human behavior, Research on the impact of technology on society. Addressing global challenges like inequality and migration.

3.6.1.10 Renewable Energy and Environmental Engineering

Advanced materials for solar and wind energy. Energy storage solutions for a more reliable grid. Sustainable urban planning and transportation.

3.6.1.11 Education and E-Learning

Remote and online learning tools and platforms. Personalized learning and adaptive education. Bridging the digital divide for equitable access to education.

3.6.1.12 Biomedical Engineering

Implantable medical devices and wearable health technology. Tissue engineering and organ transplantation. Biomaterials and drug delivery systems.

3.6.1.13 Materials Science and Nanotechnology

Advanced materials with unique properties. Nanomedicine for diagnostics and treatment. Sustainable and environmentally friendly materials.

It is important to note that these trends and research directions may have evolved or new areas may have emerged since my last update. Researchers and institutions often adapt their focus based on emerging challenges and opportunities. If you are interested in a specific field or have a particular research question, it's advisable to consult the latest academic journals, conference proceedings, and experts in that field for the most up-to-date information and insights.

3.7 Conclusion

Diatoms, a diverse group of microalgae with intricate silica shells, hold substantial promise for the treatment of industrial effluents. Their unique characteristics, including high adsorption capacity, rapid growth rates, and adaptability to various environmental conditions, make them an environmentally friendly and effective solution for mitigating the environmental impact of industrial discharges. The application of diatoms in effluent treatment offers several advantages. Their ability to sequester heavy metals, nutrients, and organic pollutants through biosorption and biomineralization processes not only helps in reducing the toxic load in wastewater but also supports the recovery of valuable metals and minerals. This dual benefit promotes a sustainable approach to effluent management. Furthermore, diatoms can be cultivated in wastewater, utilizing the available nutrients, thereby contributing to the remediation process and reducing the overall treatment cost. Their role in oxygen production during photosynthesis further enhances water quality. Moreover, the biomass generated from diatom cultivation can be repurposed for biofuel production, thus adding economic value to the treatment process. Nevertheless, some challenges, such as optimizing cultivation methods, scaling up production, and ensuring the long-term stability of diatom-based treatment systems, need to be addressed. Collaborative efforts between researchers, industries, and environmental regulatory agencies are essential to overcome these challenges and advance the implementation of diatoms in industrial effluent treatment. Overall, diatoms represent a promising and sustainable solution for the treatment of industrial effluents, aligning with the global goal of reducing pollution and conserving natural resources. Their potential to enhance water quality, reduce environmental impact, and create economic value underscores the significance of continued research and development in this field. As we move towards more eco-conscious industrial practices, diatoms offer a ray of hope for a cleaner, greener future.

References

1. Khan MJ, Rai A, Ahirwar A, Sirotiya V, Mourya M, Mishra S, Schoefs B, Marchand J, Bhatia SK, Varjani S, Vinayak V (2021) Diatom microalgae as smart nanocontainers for biosensing wastewater pollutants: recent trends and innovations. Bioengineered 12(2):9531–9549
2. Afrad MSI, Monir MB, Haque ME, Barau AA, Haque MM (2020) Impact of industrial effluent on water, soil and Rice production in Bangladesh: a case of Turag River Bank. J Environ Health Sci Eng 18(2):825–834
3. Sardo A, Orefice I, Balzano S, Barra L, Romano G (2021) Mini-review: potential of diatom-derived silica for biomedical applications. Appl Sci 11(10):4533
4. Sen TK (2023) Agricultural solid wastes based adsorbent materials in the remediation of heavy metal ions from water and wastewater by adsorption: a review. Molecules 28(14):5575
5. Raji Z, Karim A, Karam A, Khalloufi S (2023) Adsorption of heavy metals: mechanisms, kinetics, and applications of various adsorbents in wastewater remediation—a review. Waste 1(3):775–805

6. Saxena A, Tiwari A, Kaushik R, Iqbal HMN, Parra-Saldívar R (2021) Diatoms recovery from wastewater: overview from an ecological and economic perspective. J Water Process Eng 39:101705
7. Hachicha R, Elleuch F, Ben Hlima H, Dubessay P, de Baynast H, Delattre C, Pierre G, Hachicha R, Abdelkafi S, Michaud P, Fendri I (2022, 1924) Biomolecules from microalgae and cyanobacteria: applications and market survey. Appl Sci 12(4)
8. Rogato A, De Tommasi E (2020) Physical, chemical, and genetic techniques for diatom frustule modification: applications in nanotechnology. Appl Sci 10(23):8738
9. Inobeme A, Nayak V, Mathew JT, Okonkwo S, Ekwoba L, Ajai AI, Bernard E, Inobeme J, Agbugui MM, Singh KRB (2022a) Chemometric approach in environmental pollution analysis: a critical review. J Environ Manag 309(114653):1–18
10. Sethi D, Butler TO, Shuhaili F, Vaidyanathan S (2020) Diatoms for carbon sequestration and bio-based manufacturing. Biology 9(8):217
11. Shibabaw T, Beyene A, Awoke A, Tirfie M, Azage M, Triest L (2021) Diatom community structure in relation to environmental factors in human influenced rivers and streams in tropical Africa. PLoS One 16(2):e0246043
12. Berges JA, Driskill AM, Guinn EJ, Pokrzywinski K, Quinlan J, von Korff B, Young EB (2021) Role of nearshore benthic algae in the Lake Michigan silica cycle. PLoS One 16(8):e0256838
13. Rahman A, Agrawal S, Nawaz T, Pan S, Selvaratnam T (2020) A review of algae-based produced water treatment for biomass and biofuel production. Water 12(9):2351
14. Inobeme A, Adetunji CO, Ajai AI, Inobeme J, Mathew JT, Obar A, Maliki M, Nwakife N, Eziukwu C (2023a) Chemical nanosensors for monitoring environmental pollution. In: Advanced application of nanotechnology to industrial wastewater. Springer Nature, Singapore, pp 93–103
15. Abimbola OF, Adetunji CO, Dare BJ, Mathew JT, Inobeme A, Ghazanfar S, Olaniyan OT, Ajiboye MD, Dauda W (2023) Packaging of animal-based food products: the need for intelligent freshness sensors. In: Sensing and artificial intelligence solutions for food manufacturing. CRC Press, pp 89–102
16. de Medeiros ADM, da Silva Junior CJG, de Amorim JDP, Durval IJB, de Costa AFS, Sarubbo LA (2022) Oily wastewater treatment: methods, challenges, and trends. Processes 10(4):743
17. Adetunji CO, Olaniyan OT, Inobeme A, Adeyomoye O, Mathew JT, Akinbo O (2023a) Recent advances in the characterization and application of graphene in the food industry. In: Sensing and artificial intelligence solutions for food manufacturing. CRC Press, pp 153–164
18. Tripathi M, Singh P, Singh R, Bala S, Pathak N, Singh S, Chauhan RS, Singh PK (2023) Microbial biosorbent for remediation of dyes and heavy metals pollution: a green strategy for sustainable environment. Front Microbiol 14:1168954
19. Mathew JT, Adetunji CO, Inobeme A, Musah M, Azeh Y, Otori AA, Shaba EY, Mamman A, Tanko E (2023) Removal of heavy metals using bio-remedial techniques. In: Shah MP (ed) Modern approaches in waste bioremediation. Springer Nature, Switzerland AG, pp 117–130
20. Inobeme A, Adetunji CO, Mathew JT, Okonkwo S, Bamigboye MO, Ajai AI, Afoso E, Inobeme J (2023b) Advanced nanotechnology for the degradation of persistent organic pollutants. In: Shah M (ed) Microbial degradation and detoxification of pollutants. De Gruyter, Berlin, Boston, pp 51–72
21. Hashmi Z, Bilad MR, Fahrurrozi Zaini J, Lim JW, Wibisono Y (2023) Recent progress in microalgae-based technologies for industrial wastewater treatment. Fermentation 9(3):311
22. Adetunji CO, Olaniyan OT, Singh KRB, Bodunrinde RE, Inobeme A, Mathew JT, Ogundolie AF, Akinbo O, Singh J, Nayak V, Singh RP (2023b) Applications of nanofiltration for wastewater treatment. In: Bio-nano filtration in industrial effluent treatment, vol 1. CRC-Taylor and Francis, pp 139–152
23. Oruganti RK, Katam K, Show PL, Gadhamshetty V, Upadhyayula VKK, Bhattacharyya D (2022) A comprehensive review on the use of algal-bacterial systems for wastewater treatment with emphasis on nutrient and micropollutant removal. Bioengineered 13(4):10412–10453

24. Mathew JT, Adetunji CO, Inobeme A, Musah M, Shaba EY, Azeh Y, Abulude FO, Mamman A (2022) Organic compounds in atmospheric aerosols: atmospheric aerosols properties, sources and detection. Nova Science Publishers, Inc., pp 43–59
25. Gururani P, Bhatnagar P, Kumar V, Vlaskin MS, Grigorenko AV (2022) Algal consortiums: a novel and integrated approach for wastewater treatment. Water 14:3784
26. Nieri P, Carpi S, Esposito R, Costantini M, Zupo V (2023) Bioactive molecules from marine diatoms and their value for the nutraceutical industry. Nutrients 15:464
27. Singh S, Acharyya T, Raulo S, Sudatta BP, Sahoo C, Srichandan S et al (2023) A review of phytoplankton primary production in relation to environmental forcing in Indian Estuaries. In: Dynamics of planktonic primary productivity in the Indian Ocean, pp 1–27
28. Damiri F, Andra S, Kommineni N, Balu SK, Bulusu R, Boseila AA, Akamo DO, Ahmad Z, Khan FS, Rahman MH, Berrada M, Cavalu S (2022) Recent advances in adsorptive nanocomposite membranes for heavy metals ion removal from contaminated water: a comprehensive review. Materials (Basel, Switzerland) 15(15):5392
29. Tramontano C, Chianese G, Terracciano M, de Stefano L, Rea I (2020) Nanostructured biosilica of diatoms: from water world to biomedical applications. Appl Sci 10(19):6811
30. Adetunji CO, Mathew JT, Singh KRB, Inobeme A, Olaniyan O, Singh J (2022) Antibiofilm potential of zero-dimensional carbon nanomaterials. In: Zero-dimensional carbon nanomaterials: fundamentals and applications. IOP Publishing
31. Ankit BK, Korstad J (2022) Phycoremediation: use of algae to sequester heavy metals. Hydrobiology 1(3):288–303
32. Newsome L, Falagán C (2021) The microbiology of metal mine waste: bioremediation applications and implications for planetary health. GeoHealth 5(10):e2020GH000380
33. Adetunji CO, Inobeme A, Mathew JT, Inobeme J, Ajai AI, Adetunji JB, Maliki M, Olayinka AS, Akinbo O (2023c) The role of biosensors in inspection of food and agricultural products: current advances. In: Sensing and artificial intelligence solutions for food manufacturing. CRC Press, pp 71–88
34. Gupta PK, Gandhi M (2023) Bioremediation of organic pollutants in soil-water system: a review. Biotech (Basel, Switzerland) 12(2):36
35. Stepanova AY, Gladkov EA, Osipova ES, Gladkova OV, Tereshonok DV (2022) Bioremediation of soil from petroleum contamination. Processes 10(6):1224
36. Grozdov D, Zinicovscaia I (2023) Mesoporous materials for metal-laden wastewater treatment. Materials (Basel, Switzerland) 16(17):5864
37. Wang B, Lan J, Bo C, Gong B, Ou J (2023) Adsorption of heavy metal onto biomass-derived activated carbon: review. RSC Adv 13(7):4275–4302
38. Jach ME, Sajnaga E, Ziaja M (2022) Utilization of legume-nodule bacterial symbiosis in phytoremediation of heavy metal-contaminated soils. Biology 11(5):676
39. Inobeme A, Adetunji CO, Mathew JT, Ajai AI, Mann A, Inobeme J, Oyedolapo B, Adekoya MA, Onyeaku S (2023c) Diversity, biodegradation and bioremediation of polycyclic aromatic hydrocarbons. In: Microbial technologies in industrial wastewater treatment. Springer Nature Singapore, pp 31–50
40. Sepúlveda-Muñoz CA, de Godos I, Muñoz R (2023) Wastewater treatment using photosynthetic microorganisms. Symmetry 15(2):525
41. Forio MAE, Goethals PLM (2020) An integrated approach of multi-community monitoring and assessment of aquatic ecosystems to support sustainable development. Sustain For 12(14):5603
42. Disi ZA, Attia E, Ahmad MI, Zouari N (2022) Immobilization of heavy metals by microbially induced carbonate precipitation using hydrocarbon-degrading ureolytic bacteria. Biotechnol Rep (Amst) 35:e00747
43. Huang H, Xu S, Li S, Wang X, Guo K, Yan R, Xie W, Yin K, Hou S, Jiang H (2023) Diversity and distribution of harmful algal bloom species from seamount to coastal waters in the South China Sea. Microbiol Spectrum 11(2):e0416922
44. Younas F, Mustafa A, Farooqi ZUR, Wang X, Younas S, Mohy-Ud-Din W, Ashir Hameed M, Mohsin Abrar M, Maitlo AA, Noreen S, Hussain MM (2021) Current and emerging adsorbent

technologies for wastewater treatment: trends, limitations, and environmental implications. Water 13(2):215

45. Adetunji CO, Shariati MA, Olugbenga SM, Osarenkhoe OO, Odoh UE, Olaniyan OT, Rebezov M, Ajayi OO, Baibalinova G, Bodunrinde RE, Adetunji JB, Adeniyi MJ, Ugwu PN, Inobeme A, Mathew JT, Elufisan TO, Omotayo OO (2023) Pineapple [*Ananas comosus* (L.) Merr.]: a biological and pharmacological active medicinal plant. In: Pullaiah T (ed) Phytochemical composition and pharmacy of medicinal plants, vol 1. Apple Academic Press, Inc. Co-published with CRC Press (Taylor & Francis), pp 537–567
46. Mansour AT, Ashour M, Alprol AE, Alsaqufi AS (2022) Aquatic plants and aquatic animals in the context of sustainability: cultivation techniques, integration, and blue revolution. Sustain For 14(6):3257
47. Alsukaibi AKD (2022, 1968) Various approaches for the detoxification of toxic dyes in wastewater. Processes 10(10)
48. Badawi AK, Salama RS, Mostafa MMM (2023) Natural-based coagulants/flocculants as sustainable market-valued products for industrial wastewater treatment: a review of recent developments. RSC Adv 13(28):19335–19355
49. Vidu R, Matei E, Predescu AM, Alhalaili B, Pantilimon C, Tarcea C, Predescu C (2020) Removal of heavy metals from wastewaters: a challenge from current treatment methods to nanotechnology applications. Toxics 8(4):101
50. Lu H, Li Q, Feng W (2022) Application progress of O3/UV advanced oxidation technology in the treatment of organic pollutants in water. Sustain For 14:1556
51. Sharma M, Agarwal S, Agarwal Malik R, Kumar G, Pal DB, Mandal M, Sarkar A, Bantun F, Haque S, Singh P, Srivastava N, Gupta VK (2023) Recent advances in microbial engineering approaches for wastewater treatment: a review. Bioengineered 14(1):2184518
52. Waqas S, Harun NY, Sambudi NS, Abioye KJ, Zeeshan MH, Ali A, Abdulrahman A, Alkhattabi L, Alsaadi AS (2023) Effect of operating parameters on the performance of integrated fixed-film activated sludge for wastewater treatment. Membranes 13(8):704
53. Sathya K, Nagarajan K, Carlin Geor Malar G, Rajalakshmi S, Raja Lakshmi P (2022) A comprehensive review on comparison among effluent treatment methods and modern methods of treatment of industrial wastewater effluent from different sources. Appl Water Sci 12(4):70
54. Hassan I, Chowdhury SR, Prihartato PK, Razzak SA (2021, 1917) Wastewater treatment using constructed wetland: current trends and future potential. Processes 9(11)
55. Vítězová M, Kohoutová A, Vítěz T, Hanišáková N, Kushkevych I (2020) Methanogenic microorganisms in industrial wastewater anaerobic treatment. Processes 8(12):1546
56. Belete B, Desye B, Ambelu A, Yenew C (2023) Micropollutant removal efficiency of advanced wastewater treatment plants: a systematic review. Environ Health Insights 17:11786302231195158
57. Othman NH, Alias NH, Fuzil NS, Marpani F, Shahruddin MZ, Chew CM, David Ng KM, Lau WJ, Ismail AF (2021) A review on the use of membrane technology systems in developing countries. Membranes 12(1):30
58. Vievard J, Alem A, Pantet A, Ahfir ND, Arellano-Sánchez MG, Devouge-Boyer C, Mignot M (2023) Bio-based adsorption as ecofriendly method for wastewater decontamination: a review. Toxics 11(5):404
59. Czupryński P, Płotka M, Glamowski P, Żukowski W, Bajda T (2022) An assessment of an ion exchange resin system for the removal and recovery of Ni, Hg, and Cr from wet flue gas desulphurization wastewater-a pilot study. RSC Adv 12(9):5145–5156

Diatoms as Tools for Aquaculture and Food Production

4

John Tsado Mathew, Abel Inobeme, Monday Musah, Yakubu Azeh, Elijah Yanda Shaba, Muhammad Bini Etsuyankpa, Charles Oluwaseun Adetunji, Amos Mamman, and Musa Safiyanu Tanko

Abstract

Diatoms, microscopic photosynthetic algae, are increasingly recognized as valuable tools in aquaculture and food production. Their role in promoting water quality, serving as a natural feed source for various aquatic organisms, and their potential for high-value bioactive compounds make them essential in sustainable aquaculture practices. This review highlights the multifaceted contributions of diatoms to the aquaculture industry, emphasizing their capacity to enhance growth, nutritional value, and overall environmental sustainability, thus revolutionizing the production of seafood and contributing to global food security.

Keywords

Aquaculture · Diatoms · Food · Production · Tools

J. T. Mathew (✉) · M. Musah · Y. Azeh · A. Mamman
Department of Chemistry, Ibrahim Badamasi Babangida University, Lapai, Niger State, Nigeria

A. Inobeme
Department of Chemistry, Edo University Uzairue, Auchi, Edo State, Nigeria

E. Y. Shaba
Department of Chemistry, Federal University of Technology Minna, Minna, Nigeria

M. B. Etsuyankpa · M. S. Tanko
Department of Chemistry, Federal University of Lafia, Lafia, Nasarawa State, Nigeria

C. O. Adetunji
Department of Chemistry, Federal University of Lafia, Lafia, Nasarawa State, Nigeria

Applied Microbiology, Biotechnology and Nanotechnology Laboratory, Department of Microbiology, Edo University Uzairue, Auchi, Edo State, Nigeria

M. Rafatullah et al. (eds.), *Multidisciplinary Applications of Marine Resources*,
https://doi.org/10.1007/978-981-97-5057-3_4

4.1 Introduction

Diatoms, a diverse group of microalgae, have gained increasing attention in recent years for their potential as valuable tools in the fields of aquaculture and food production. These microscopic photosynthetic organisms are an essential component of aquatic ecosystems and are known for their unique characteristics and wide-ranging applications. Diatoms have been harnessed for various purposes due to their nutritional value, rapid growth, and eco-friendly attributes [1].

In this introductory overview, we will explore the pivotal role that diatoms play in aquaculture and food production, shedding light on their versatile applications and the benefits they bring to these vital sectors. The significance of diatoms lies in their capacity to provide solutions to some of the most pressing issues facing the global food industry. These issues include the need for sustainable and nutritious food sources, environmentally responsible production methods, and innovative technologies that enhance efficiency. By harnessing the potential of diatoms, we can unlock a world of possibilities to meet the ever-growing demand for food in a manner that is both economically viable and ecologically responsible [2].

Diatoms belong to a diverse group of photosynthetic microalgae characterized by their silica cell walls, which give them a distinctive appearance under the microscope. They are found in a wide range of aquatic environments, from oceans and rivers to lakes and even puddles. Their ecological importance cannot be overstated, as diatoms are responsible for a significant portion of the Earth's primary production, contributing to the global carbon and oxygen cycles. Beyond their ecological role, diatoms have garnered attention for their potential in aquaculture and food production [1].

Diatoms are often praised for their nutritional value, as they are rich in essential nutrients, including proteins, lipids, vitamins, and minerals. These microalgae can serve as a source of highly nutritious food for various aquatic organisms, including fish, shellfish, and shrimp, which are integral to the aquaculture industry. By incorporating diatoms into the diets of farmed aquatic species, aquaculturists can enhance the growth and overall health of their livestock, resulting in higher yields and improved product quality [3].

Furthermore, diatoms exhibit rapid growth rates, a characteristic that is particularly advantageous in aquaculture. Unlike some traditional feed sources that rely on land-based agriculture, diatoms can be cultivated in controlled aquatic environments, offering a more efficient and sustainable approach to food production. Their ability to multiply quickly mean that diatom-based aquaculture feeds can be produced on a shorter timescale, reducing the pressure on land and freshwater resources. This not only supports the expansion of the aquaculture industry but also lessens its impact on the environment [4].

In addition to their nutritional value and rapid growth, diatoms have another remarkable feature that makes them valuable in aquaculture: the ability to produce bioactive compounds. Diatoms are known to synthesize a wide range of bioactive molecules, including antioxidants, polyunsaturated fatty acids (PUFAs), and antimicrobial substances. These compounds can contribute to the overall health and

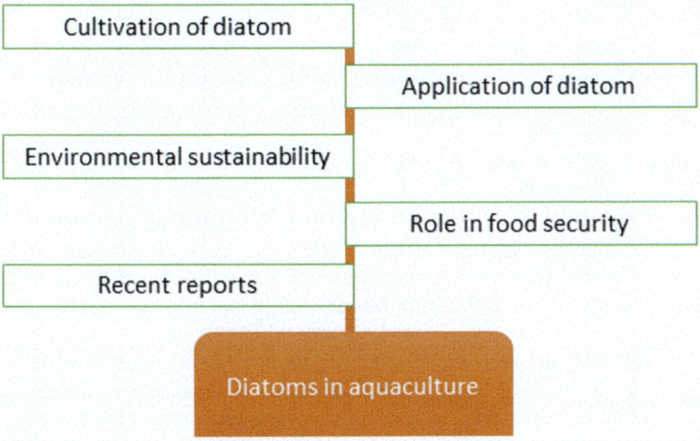

Fig. 4.1 Schematic representation of chapter

disease resistance of aquaculture species. By including diatoms in aquaculture diets, we can potentially enhance the quality and safety of seafood products while reducing the need for chemical additives [3].

Diatoms are not limited to their role as a feed source in aquaculture; they also find applications in wastewater treatment. As industries and populations grow, wastewater management becomes an increasingly critical concern. Diatoms, with their natural ability to absorb and assimilate nutrients, can be employed in wastewater treatment processes. These microalgae can help in the removal of excess nutrients, such as nitrogen and phosphorus, from wastewater, reducing pollution and supporting the restoration of water bodies. In this way, diatoms contribute to both environmental protection and the sustainable management of resources [5].

This article will delve into the multiple facets of diatoms as tools for aquaculture and food production, from their nutritional value and growth characteristics to their role in wastewater treatment, biofuel production, and beyond (Fig. 4.1). It will also explore the current research, developments, and potential future applications that make diatoms a compelling choice for shaping the future of these critical industries. Ultimately, diatoms offer a promising avenue for advancing sustainable practices and ensuring a resilient food supply chain, addressing the challenges of our time [6].

4.2 Diatoms: Nature and Characteristics

Diatoms are a type of microscopic, photosynthetic algae that belong to the group of microorganisms known as phytoplankton. They are found in various aquatic environments, including freshwater and marine habitats. Diatoms are known for their unique and intricate cell walls, which are made of silica (essentially glass), giving them distinctive and beautiful geometric shapes (Table 4.1).

Table 4.1 The nature and characteristics of diatoms

Nature	Characteristics	Habitat	Significance	Reference
Type of organism	Microscopic, single-celled algae with a hard, silica-based cell wall (frustule)	Aquatic environments	Key role in the food chain, oxygen production, and indicators of water quality	[7]
Biodiversity	Estimated to comprise around 100,000 known species	Found in freshwater, marine, and terrestrial environments	Important in ecology, carbon cycling, and paleoclimate research	[8]
Ecological importance	Primary producers in aquatic food chains, responsible for a significant portion of global photosynthesis	Planktonic and benthic forms	Major contributors to global carbon fixation and removal of carbon dioxide	[9]
Morphological diversity	Diverse shapes and sizes, including pennate (elongated) and centric (circular) forms	Planktonic diatoms are often responsible for harmful algal blooms	Used in various scientific disciplines, including paleoceanography and biostratigraphy	[10]

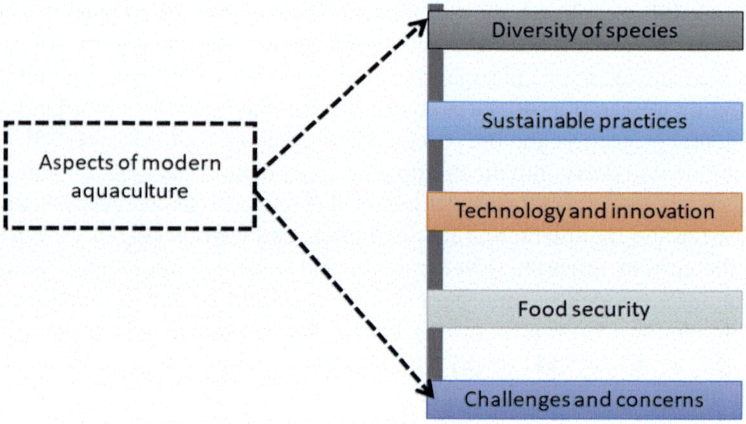

Fig. 4.2 Aspects of modern aquaculture

4.3 Aquaculture in the Modern Context

Aquaculture, often referred to as aquafarming, is the practice of cultivating aquatic organisms, such as fish, shellfish, and aquatic plants, in controlled environments. In the modern context, aquaculture has become a vital and rapidly growing sector of the global food industry. Some main aspects of modern aquaculture (Fig. 4.2):

Global Growth: Over the past few decades, aquaculture has experienced remarkable growth, meeting the increasing demand for seafood. According to the Food and Agriculture Organization (FAO) of the United Nations, aquaculture accounts for over 50% of the world's seafood production, surpassing wild-caught fisheries [11].

Diverse Species: Modern aquaculture encompasses a wide range of species. While fish, including species like tilapia, salmon, and catfish, are the most common, aquaculturists also rear shrimp, oysters, mussels, and even seaweeds. The diversity of species allows for adaptation to specific ecological and market demands [12].

Technology and Innovation: Advances in technology and innovation have transformed aquaculture. Modern facilities often use recirculating aquaculture systems (RAS) that carefully control water quality, temperature, and other environmental factors. These closed systems reduce environmental impact and disease risks. Additionally, selective breeding programs have improved the growth rates and disease resistance of farmed species [13].

Sustainable Practices: As concerns about the environmental and ethical aspects of aquaculture have risen, sustainable practices have gained prominence. Sustainable aquaculture seeks to minimize environmental impact, reduce the need for wild fish in feed, and adhere to responsible practices in waste management. Certification programs like the Aquaculture Stewardship Council (ASC) and Best Aquaculture Practices (BAP) have been established to encourage sustainability and responsible farming [14].

Economic Impact: Aquaculture has significant economic importance in many regions, providing livelihoods for millions of people, from farmers and processors to those involved in the distribution and marketing of aquaculture products. It has become a driving force for economic development and job creation in both developed and developing countries [15].

Food Security: Aquaculture plays a crucial role in global food security. With the world's population expected to reach nearly 10 billion by 2050, aquaculture helps meet the growing demand for protein-rich seafood. Moreover, it provides a reliable source of nutrition and income in areas where traditional fisheries may be unreliable due to overfishing or changing environmental conditions [16].

Challenges and Concerns: Despite its successes, aquaculture faces various challenges. Disease outbreaks, overuse of antibiotics, and habitat degradation are concerns. Responsible management and regulation are necessary to address these issues. Additionally, the sustainability of aquaculture feed, which often includes wild-caught fish, remains a topic of discussion [17].

Future Prospects: The future of aquaculture is likely to involve continued innovation and expansion. Technologies such as aquaponics, which combines aquaculture with hydroponics (the cultivation of plants without soil), offer the potential for integrated and sustainable food production. The development of land-based RAS facilities can help address environmental concerns [18].

In addition, modern aquaculture is a dynamic and vital component of the global food industry. Its rapid growth, technological advancements, and focus on sustainability have made it a key player in addressing the world's food security challenges. As the industry continues to evolve, responsible practices and innovative solutions will be critical to meeting the increasing demand for seafood while minimizing its environmental impact [19].

4.4 Diatoms in Aquaculture

Diatoms, which are microalgae known for their unique silica-based cell walls, have several potential applications in aquaculture. While they are not commonly used as a primary feed source for farmed aquatic organisms, diatoms can play important roles in improving water quality, supporting the growth of beneficial microorganisms, and contributing to the overall health of the aquaculture system [20]. Here are some ways diatoms are relevant in aquaculture:

Water Filtration and Purification: Diatoms are effective natural filters. In recirculating aquaculture systems (RAS) and other closed-loop systems, diatoms can help improve water quality by removing suspended particles and excess nutrients. They act as a biological filter, reducing the levels of ammonia, nitrites, and nitrates in the water, which is crucial for maintaining a healthy aquatic environment for the farmed species [21].

Biofloc Systems: In biofloc aquaculture systems, diatoms can become a part of the complex microbial community within the floc. The microbial flocs contain a diverse population of microorganisms, including diatoms, which help in nutrient cycling and converting organic matter into a usable form for the farmed species. These systems are particularly beneficial in intensive shrimp farming, where they can enhance water quality and provide additional food sources for the cultured organisms [22].

Live Food for Larval Stages: Diatoms, in their planktonic form, can be used as live food for the early larval stages of some aquaculture species. Many marine species, such as bivalves and certain species of fish, require small, nutritious particles for their initial growth. Diatoms can serve as a suitable source of microalgae for these larvae [23].

Diatom Cultivation: Diatom cultures can be established and maintained in separate tanks or containers within an aquaculture system. These cultures can be used to inoculate the water with diatoms, promoting their growth as a natural food source for filter-feeding organisms like oysters, clams, and some larval fish species [24].

Diatom Biofilms: The diatom biofilms that form on various surfaces, such as substrates and tank walls, can be beneficial for aquaculture systems. These biofilms can serve as a habitat for beneficial microorganisms and provide a natural food source for some aquatic organisms. They help create a more stable and diverse ecosystem within the aquaculture environment [25].

Biological Balance: Diatoms contribute to the overall balance of the aquatic ecosystem in aquaculture systems. By creating diverse and stable microenvironments within the water column, diatoms can help control the growth of potentially harmful microorganisms and contribute to the health and well-being of the farmed species [26].

It is important to note that while diatoms have several beneficial roles in aquaculture, they are not typically the primary food source for most commercially farmed fish species. Instead, they are often used as a supplement to other types of feed or as a natural part of the aquatic ecosystem. Additionally, the specific use of diatoms in aquaculture may vary depending on the species being cultured and the environmental conditions of the aquaculture system. Careful management and monitoring are necessary to ensure that diatoms contribute positively to the overall success of the aquaculture operation [4].

4.5 Cultivation and Utilization of Diatoms

Diatoms are a type of microalgae that are found in various aquatic environments, such as freshwater and marine ecosystems. They are known for their unique, silica-based cell walls and are an important component of the phytoplankton community. Diatoms play a crucial role in aquatic food chains and are also of interest for various practical applications (Table 4.2).

4.6 Diatoms for Improved Aquaculture Nutrition

Diatoms can play a significant role in improving aquaculture nutrition and the overall health of farmed aquatic species. Aquaculture is an important industry that provides a significant portion of the world's seafood, and diatoms can be beneficial in several ways [1]. Table 4.3 provides a concise overview of the benefits of incorporating diatoms into aquaculture nutrition, along with the challenges and considerations associated with their use in aquafeeds.

4.7 Biotechnological Applications of Diatoms

Diatoms are a group of microalgae that are known for their unique and intricate silica cell walls, which give them a distinctive glass-like appearance. These single-celled organisms have a wide range of biotechnological applications due to their abundance, diversity, and various unique features [26]. Some of the notable biotechnological applications of diatoms include:

Biofuel production: Diatoms can be used to produce biofuels, such as biodiesel and bioethanol. They are rich in lipids, making them a potential source of renewable energy. Researchers are exploring ways to optimize diatom growth and lipid extraction techniques for biofuel production [36].

Bioremediation: Diatoms are effective in water and soil remediation due to their ability to absorb heavy metals, toxins, and nutrients. They can be used to clean up polluted environments and mitigate the effects of eutrophication, a common problem in bodies of water [37].

Table 4.2 Cultivation and utilization of diatoms

Aspect	Cultivation	Utilization	Benefits	Reference
Cultivation methods	– Diatoms can be cultivated in culture flasks or photobioreactors under controlled conditions – They require a light source for photosynthesis, suitable temperature, and nutrient-rich water – Nutrients like nitrogen and phosphorus are essential for their growth – Batch and continuous culture systems are commonly used	– Diatoms are used as live food in aquaculture for larval stages of marine species – They can be cultured to form natural biofilms on various surfaces within aquaculture systems – Diatoms are employed in recirculating aquaculture systems (RAS) for water purification – They are an important component of biofloc systems in shrimp farming	– Provides a sustainable source of live food for aquaculture – Contributes to the natural biofilter and nutrient cycling within aquaculture systems – Enhances water quality by reducing ammonia and other pollutants in RAS – Supports the microbial community in biofloc systems, promoting nutrient conversion	[27]
Suitable species	– Diatoms are a diverse group with numerous species, making them adaptable to a wide range of aquaculture systems – Some species may be more suitable for specific aquaculture applications	– Larval stages of marine species like bivalves and some fish – Filter-feeding organisms, such as oysters and clams – Beneficial microorganisms in biofloc systems – Microbial communities in aquaculture systems	– The diversity of diatom species allows for versatility in different aquaculture contexts – Effective for supporting filter-feeding organisms and larval stages – Contributes to the ecological balance of the aquaculture environment – Provides a natural food source for beneficial microorganisms	[28]

(continued)

Table 4.2 (continued)

Aspect	Cultivation	Utilization	Benefits	Reference
Environmental impact	– Diatom cultivation can help reduce nutrient levels and improve water quality in aquaculture systems – They contribute to maintaining a stable and healthy aquatic ecosystem – Their cultivation reduces the need for chemical water treatment	– Diatoms help remove excess nutrients and particles from the water, reducing the risk of algal blooms – Their presence contributes to overall water quality improvement.	– Reduces the environmental impact of aquaculture by promoting sustainability – Minimizes the use of chemical water treatments, enhancing the ecological balance – Supports responsible and eco-friendly aquaculture practices	[29]
Challenges and considerations	– Diatom cultures may be susceptible to contamination by unwanted algae or pathogens, requiring careful management – Monitoring of nutrient levels and light conditions is necessary for optimal growth – Diatom biofilms and microbial communities should be maintained in a balanced state to avoid overgrowth	– Diatom cultures should be monitored to ensure they are serving their intended purpose as live food or biofilm contributors – In biofloc systems, the balance between diatoms and other microorganisms must be carefully maintained	– Effective management and monitoring are crucial for success in diatom cultivation – A proactive approach to controlling contamination and nutrient balance is essential – The benefits of diatoms in aquaculture can be maximized with proper care and attention	[30]

Table 4.3 The main points about diatoms for improved aquaculture nutrition

Aspect	Benefits	Challenges and considerations	Reference
Nutrient content	– High in proteins, lipids, vitamins, minerals, and omega-3 fatty acids	– Quality control for consistent composition	[31]
Sustainability	– Can be sustainably cultivated, reducing reliance on wild-caught fish	– Efficient large-scale cultivation techniques	[32]
Omega-3 fatty acids	– Natural source of EPA and DHA, vital for fish health	– Cost-effectiveness compared to traditional feeds	[33]
Improved growth and quality	– Enhances growth and overall quality of farmed fish	– Regulatory approval for novel feed ingredients	[34]
Reduced environmental impact	– Lowers the carbon footprint and preserves marine ecosystems	– Species selection for specific aquaculture needs	[35]

Bioplastic production: Diatom biomass can be used as a feedstock for bioplastic production. The silica shells of diatoms can also be utilized in the creation of bio-based materials, enhancing their sustainability [38].

Drug delivery systems: Researchers have explored the use of diatoms as drug delivery vehicles. The porous silica shells of diatoms can be loaded with drugs or other therapeutic agents and then targeted to specific cells or tissues [39].

Nanotechnology: Diatom silica shells have unique nanostructures that are of interest in nanotechnology. These shells can be used as templates for the creation of nanoscale materials, such as nanoparticles and nanowires [40].

Aquaculture: Diatoms are a natural source of essential nutrients for aquatic organisms. They are used as a food source in aquaculture to promote the growth of fish and shellfish larvae [3].

Pharmaceuticals: Diatoms produce bioactive compounds, including polyunsaturated fatty acids, antioxidants, and antimicrobial substances, which have potential pharmaceutical applications. Research is ongoing to explore their therapeutic properties [41].

Food and dietary supplements: Diatoms can be used in the production of food and dietary supplements. They are a source of essential nutrients, such as omega-3 fatty acids and various vitamins, which can be added to human and animal diets [42].

Environmental monitoring: Diatoms are commonly used as bioindicators for water quality assessment. Changes in diatom species composition and abundance can provide insights into environmental conditions and water quality [43].

Carbon sequestration: Diatoms play a role in the global carbon cycle by sequestering carbon dioxide from the atmosphere. Harnessing their ability to capture and store carbon could be beneficial for mitigating climate change [44].

Overall, diatoms offer a wide range of biotechnological applications, from environmental remediation to advanced materials and pharmaceuticals. Their unique characteristics and abundance in various ecosystems make them a valuable resource for numerous industries and research fields [45].

4.8 Environmental and Economic Sustainability

Using diatoms in aquaculture can contribute to both environmental and economic sustainability in several ways:

Reduced Pressure on Wild Fisheries: Incorporating diatoms into aquaculture feeds reduces the need for fishmeal and fish oil, which are typically sourced from wild-caught fish. By relying less on these finite and sometimes overexploited resources, aquaculture can alleviate the pressure on wild fisheries, helping to maintain their ecological balance and biodiversity [34].

Energy Efficiency: Diatoms are photosynthetic microorganisms, meaning they use sunlight to produce their own food. This photosynthetic process can be harnessed in closed-loop aquaculture systems to recycle nutrients and maintain water quality. It reduces the need for energy-intensive processes, such as aeration and water treatment, which can have a positive impact on the overall energy efficiency of aquaculture operations [46].

Nutrient Recycling: The waste generated by aquaculture can negatively impact water quality and surrounding ecosystems. Diatoms can be employed to help recycle excess nutrients, as they can take up nitrogen and phosphorus compounds from aquaculture effluents. This reduces nutrient pollution and minimizes the environmental impact of aquaculture operations [1].

Enhanced Growth and Disease Resistance: By improving the nutritional quality of aquaculture feeds with diatoms, the health and growth of farmed species are enhanced. This can lead to reduced mortality rates and the need for medications, contributing to economic sustainability by lowering operational costs [47].

High-Quality Product: Diatom-enhanced feeds can result in higher-quality seafood products. Enhanced coloration, improved flavor, and higher nutritional content can command premium prices in the market, improving the economic viability of aquaculture ventures [48].

Market Demand: Consumers are increasingly concerned about the sustainability and environmental impact of their food choices. Aquaculture operations that can demonstrate sustainable practices, such as incorporating diatoms into their feed, may attract more environmentally conscious consumers, leading to increased market demand and profitability [48].

Strengthened Regulatory Compliance: Many regions have implemented stricter regulations on aquaculture practices, with an emphasis on sustainability. Incorporating sustainable and environmentally friendly techniques, such as diatom-based nutrition, can help aquaculture operations comply with these regulations and avoid fines or sanctions [32].

Long-Term Viability: Diatom-based approaches promote a more balanced and resilient aquaculture ecosystem, reducing the risk of environmental crises and disease outbreaks. This contributes to the long-term viability and stability of aquaculture operations [32].

In addition, the use of diatoms in aquaculture contributes to both environmental and economic sustainability by reducing environmental impacts, improving the quality of farmed products, and meeting consumer demand for more sustainable seafood choices. It also helps aquaculture operations adhere to stricter regulations and adapt to changing market dynamics, ultimately enhancing their long-term viability [49].

4.9 Future Directions and Research Opportunities

The use of diatoms in various biotechnological applications, including aquaculture, is an evolving field with several promising directions and research opportunities for the future. Here are some areas where further research can make a significant impact:

Diatom Strain Selection: Identifying and optimizing diatom strains that are best suited for specific aquaculture species is a priority. Research can focus on selecting diatom species or strains with the most beneficial nutritional profiles and growth characteristics for different aquatic organisms [24].

Nutrient Manipulation: Investigate the optimal conditions for diatom cultivation, including nutrient composition and light intensity, to maximize their growth and nutritional value. This research can lead to more efficient and sustainable diatom production.

Diatom-Based Bioremediation: Further, explore the use of diatoms in water and soil remediation, especially in aquaculture systems, to reduce the impact of excess nutrients and pollutants. Developing cost-effective and scalable methods for implementing diatom-based bioremediation is an important research avenue [50].

Bioplastic and Material Development: Investigate the use of diatom silica shells and biomass in bioplastic production and the creation of eco-friendly materials. Research can focus on developing innovative and sustainable products that can replace traditional plastics and materials [51].

Diatom-Derived Pharmaceuticals: Explore the potential pharmaceutical applications of diatoms, including their bioactive compounds and therapeutic properties. This research can lead to the development of new drugs and health products [51].

Carbon Sequestration: Investigate the role of diatoms in carbon sequestration and their contribution to mitigating climate change. Understanding how diatoms capture and store carbon can lead to strategies for enhancing this natural process [52].

Diatom Genetics and Genetic Engineering: Investigate the genetics of diatoms to better understand their growth and metabolic processes. Genetic engineering can be used to optimize diatom strains for specific applications, such as aquaculture.

Integrated Aquaculture Systems: Research the integration of diatoms into multifunctional aquaculture systems where diatoms can provide both nutrition for farmed species and water treatment, creating more sustainable and resilient aquaculture operations [53].

Market Acceptance and Consumer Perception: Examine consumer attitudes and perceptions of diatom-enhanced aquaculture products. Understanding consumer preferences and willingness to pay for sustainably produced seafood can guide market strategies.

Economic Viability: Analyze the economic aspects of diatom-based aquaculture, including production costs, market competitiveness, and return on investment. Identifying the economic advantages and challenges is crucial for the widespread adoption of these techniques [1].

Regulatory Frameworks: Collaborate with regulatory bodies to establish clear guidelines and standards for the use of diatoms in aquaculture. Regulatory support is essential for the integration of innovative technologies into established practices.

Education and Outreach: Develop educational programs and outreach initiatives to raise awareness about the benefits and potential of diatom-based aquaculture. This can include knowledge dissemination to farmers, researchers, and policymakers.

Future research in these areas can help unlock the full potential of diatoms in aquaculture and other biotechnological applications, promoting environmental sustainability and economic growth in the aquaculture industry.

4.10 Conclusion

Diatoms offer promising potential as tools for aquaculture and food production. These microscopic, photosynthetic algae play multifaceted roles in enhancing sustainability, improving nutrition, and mitigating environmental challenges within these industries. First and foremost, diatoms are valuable sources of omega-3 fatty acids, which are crucial for both aquatic organisms and human consumers. These essential nutrients can be efficiently transferred up the food chain, benefiting the health and growth of farmed fish and shellfish, ultimately improving the quality of seafood products. As the demand for seafood continues to rise, diatoms can contribute to meeting this demand sustainably.

Moreover, diatoms possess unique qualities that make them excellent candidates for bioremediation. They have the ability to absorb excess nutrients, such as nitrogen and phosphorus, from aquaculture wastewater, reducing environmental pollution and eutrophication. This not only promotes cleaner water quality but also benefits the surrounding ecosystems. Diatoms are also being explored as feed additives for aquaculture. Their natural production of bioactive compounds can enhance the immune systems of aquatic organisms, reducing the need for antibiotics and improving overall disease resistance. This, in turn, contributes to more sustainable and responsible aquaculture practices. Furthermore, the cultivation of diatoms can serve as an additional income stream for aquaculture operations. The mass production of diatoms can provide a source of revenue through the sale of diatom-based products, such as feed additives or omega-3 supplements, diversifying income sources for aquaculture farmers. In addition, diatoms offer a range of benefits to both aquaculture and food production industries. Their contributions to nutrition, sustainability, and environmental protection make them invaluable tools for addressing the challenges faced by these sectors. Harnessing the potential of diatoms can lead to a more sustainable, resilient, and nutritious future for both aquatic ecosystems and the growing human population.

References

1. Saxena A, Tiwari A, Kaushik R, Iqbal HMN, Parra-Saldívar R (2021) Diatoms recovery from wastewater: overview from an ecological and economic perspective. J Water Process Eng 39:101705
2. Ruocco N, Albarano L, Esposito R, Zupo V, Costantini M, Ianora A (2020) Multiple roles of diatom-derived oxylipins within marine environments and their potential biotechnological applications. Mar Drugs 18:342
3. Nieri P, Carpi S, Esposito R, Costantini M, Zupo V (2023) Bioactive molecules from marine diatoms and their value for the nutraceutical industry. Nutrients 15:464
4. Araujo GS, da Silva JWA, Cotas J, Pereira L (2022) Fish farming techniques: current situation and trends. J Mar Sci Eng 10:1598
5. Marella TK, López-Pacheco IY, Parra-Saldívar R, Dixit S, Tiwari A (2020) Wealth from waste: diatoms as tools for phycoremediation of wastewater and for obtaining value from the biomass. Sci Total Environ 724:137960
6. Fung KCL, Dornelles HS, Varesche MBA, Gutierrez T (2023) From wastewater treatment plants to the oceans: a review on synthetic chemical surfactants (SCSs) and perspectives on marine-safe biosurfactants. Sustain For 15:11436
7. Sangeetha J, Codreanu S, Thangadurai D (eds) (2023) Microalgal biotechnology: bioprospecting microalgae for functional metabolites towards commercial and sustainable applications. CRC Press
8. Alçiçek H, Gross M, Bouchal JM, Wesselingh FP, Neubauer TA, Meijer T et al (2023) Paleobiodiversity and paleoenvironments of the eastern Paratethys Pleistocene lacustrine-palustrine sequence in the Baklan Basin (SW Anatolia, Turkey). Palaeogeogr Palaeoclimatol Palaeoecol 626:111649
9. Singh S, Acharyya T, Raulo S, Sudatta BP, Sahoo C, Srichandan S et al (2023) A review of phytoplankton primary production in relation to environmental forcing in Indian Estuaries. In: Dynamics of planktonic primary productivity in the Indian Ocean, pp 1–27
10. Huang H, Xu S, Li S, Wang X, Guo K, Yan R, Xie W, Yin K, Hou S, Jiang H (2023) Diversity and distribution of harmful algal bloom species from seamount to coastal waters in the South China Sea. Microbiol Spectrum 11(2):e0416922
11. Rocha CP, Cabral HN, Marques JC, Gonçalves AMM (2022) A global overview of aquaculture food production with a focus on the activity's development in transitional systems—the case study of a South European Country (Portugal). J Mar Sci Eng 10(3):417
12. Kim DY, Shinde SK, Kadam AA, Saratale RG, Saratale GD, Kumar M, Syed A, Bahkali AH, Ghodake GS (2022) Advantage of species diversification to facilitate sustainable development of aquaculture sector. Biology 11(3):368
13. Benjamin EO, Ola O, Buchenrieder GR (2022) Feasibility study of a small-scale recirculating aquaculture system for sustainable (peri-)urban farming in sub-Saharan Africa: a Nigerian perspective. Land 11(11):2063
14. da Silva MÚ, Rotta MA, Fornari DC, Streit DP (2022) Aquaculture sustainability assessed by emergy synthesis: the importance of water accounting. Agriculture 12(11):1947
15. Parrao GC, Moratti M, Shisler S, Snilstveit B, Eyers J (2021) Protocol: aquaculture for improving productivity, income, nutrition and women's empowerment in low- and middle-income countries: a systematic review and meta-analysis. Campbell Syst Rev 17(3):e1188

16. Boyd CE, McNevin AA, Davis RP (2022) The contribution of fisheries and aquaculture to the global protein supply. Food Secur 14(3):805–827
17. Inobeme A, Nayak V, Mathew JT, Okonkwo S, Ekwoba L, Ajai AI, Bernard E, Inobeme J, Agbugui MM, Singh KRB (2022a) Chemometric approach in environmental pollution analysis: a critical review. J Environ Manage 309(114653):1–18
18. Vasdravanidis C, Alvanou MV, Lattos A, Papadopoulos DK, Chatzigeorgiou I, Ravani M, Liantas G, Georgoulis I, Feidantsis K, Ntinas GK, Giantsis IA (2022) Aquaponics as a promising strategy to mitigate impacts of climate change on rainbow trout culture. Animals (Basel) 12(19):2523
19. Pawlak K, Kołodziejczak M (2020) The role of agriculture in ensuring food security in developing countries: considerations in the context of the problem of sustainable food production. Sustain For 12(13):5488
20. Khan MJ, Rai A, Ahirwar A, Sirotiya V, Mourya M, Mishra S, Schoefs B, Marchand J, Bhatia SK, Varjani S, Vinayak V (2021) Diatom microalgae as smart nanocontainers for biosensing wastewater pollutants: recent trends and innovations. Bioengineered 12(2):9531–9549
21. Ramli NM, Verreth JAJ, Yusoff FM, Nurulhuda K, Nagao N, Verdegem MCJ (2020) Integration of algae to improve nitrogenous waste management in recirculating aquaculture systems: a review. Front Bioeng Biotechnol 8:1004
22. Mathew JT, Adetunji CO, Inobeme A, Musah M, Shaba EY, Azeh Y, Abulude FO, Mamman A (2022) Organic compounds in atmospheric aerosols: atmospheric aerosols properties, sources and detection. Nova Science Publishers Inc, pp 43–59
23. Abimbola OF, Adetunji CO, Dare BJ, Mathew JT, Inobeme A, Ghazanfar S, Olaniyan OT, Ajiboye MD, Dauda W (2023) Packaging of animal-based food products: the need for intelligent freshness sensors. In: Sensing and artificial intelligence solutions for food manufacturing. CRC Press, pp 89–102
24. Mansour AT, Ashour M, Alprol AE, Alsaqufi AS (2022) Aquatic plants and aquatic animals in the context of sustainability: cultivation techniques, integration, and blue revolution. Sustain For 14(6):3257
25. Sousa J, Félix PM, Brito AC, Venâncio E, Azevedo e Silva F, Simões T, Amorim A, Dâmaso-Rodrigues ML, Pombo A (2023) Optimizing growth and rearing techniques for Larvae and Juveniles of the sea cucumber Holothuria arguinensis. Diversity 15(6):722
26. Sardo A, Orefice I, Balzano S, Barra L, Romano G (2021) Mini-review: potential of diatom-derived silica for biomedical applications. Appl Sci 11(10):4533
27. Benner P, Meier L, Pfeffer A, Krüger K, Oropeza Vargas JE, Weuster-Botz D (2022) Lab-scale photobioreactor systems: principles, applications, and scalability. Bioprocess Biosyst Eng 45(5):791–813
28. Adetunji CO, Shariati MA, Olugbenga SM, Osarenkhoe OO, Odoh UE, Olaniyan OT, Rebezov M, Ajayi OO, Baibalinova G, Bodunrinde RE, Adetunji JB, Adeniyi MJ, Ugwu PN, Inobeme A, Mathew JT, Elufisan TO, Omotayo OO (2023a) Pineapple [*Ananascomosus* (L.) Merr.]: a biological and pharmacological active medicinal plant. In: Pullaiah T (ed) Phytochemical composition and pharmacy of medicinal plants, vol 1. Apple Academic Press, Inc. Co-published with CRC Press (Taylor & Francis), pp 537–567
29. Adetunji CO, Olaniyan OT, Inobeme A, Adeyomoye O, Mathew JT, Akinbo O (2023b) Recent advances in the characterization and application of graphene in the food industry. In: Sensing and artificial intelligence solutions for food manufacturing. CRC Press, pp 153–164
30. Koedooder C, Stock W, Willems A, Mangelinckx S, De Troch M, Vyverman W, Sabbe K (2019) Diatom-bacteria interactions modulate the composition and productivity of benthic diatom biofilms. Front Microbiol 10:1255
31. Aouji M, Imtara H, Rkhaila A, Bouhaddioui B, Alahdab A, Parvez MK, Saleh Alzahrani M, AichaLrhorfi L, Bengueddour R (2023) Nutritional composition, fatty acids profile, mineral content, antioxidant activity and acute toxicity of the flesh of *Helix aspesra* Müller. Molecules (Basel, Switzerland) 28(17):6323
32. Engle CR, van Senten J (2022) Resilience of communities and sustainable aquaculture: governance and regulatory effects. Aust Fish 7(5):268

33. Mathew JT, Adetunji CO, Inobeme A, Musah M, Azeh Y, Otori AA, Shaba EY, Mamman A, Tanko E (2023) Removal of heavy metals using bio-remedial techniques. In: Shah MP (ed) Modern approaches in waste bioremediation. Springer Nature Switzerland AG, pp 117–130
34. Sarker PK (2023) Microorganisms in fish feeds, technological innovations, and key strategies for sustainable aquaculture. Microorganisms 11(2):439
35. Inobeme A, Adetunji CO, Ajai AI, Inobeme J, Mathew JT, Obar A, Maliki M, Nwakife N, Eziukwu C (2023a) Chemical nanosensors for monitoring environmental pollution. In: Advanced application of nanotechnology to industrial wastewater. Springer Nature Singapore, pp 93–103
36. Govindan N, Maniam GP, Ab. Rahim M, Sulaiman A, Ajit A, Chatsungnoen T, Chisti Y (2021) Production of renewable lipids by the diatom Amphora copulata. Fermentation 7(1):37
37. Rahman A, Agrawal S, Nawaz T, Pan S, Selvaratnam T (2020) A review of algae-based produced water treatment for biomass and biofuel production. Water 12(9):2351
38. Kong U, Mohammad Rawi NF, Tay GS (2023) The potential applications of reinforced bioplastics in various industries: a review. Polymers 15(10):2399
39. Zobi F (2022) Diatom biosilica in targeted drug delivery and biosensing applications: recent studies. Micro 2(2):342–360
40. Inobeme A, Adetunji CO, Mathew JT, Okonkwo S, Bamigboye MO, Ajai AI, Afoso E, Inobeme J (2023b) Advanced nanotechnology for the degradation of persistent organic pollutants. In: Shah M (ed) Microbial degradation and detoxification of pollutants. De Gruyter, Berlin, Boston, pp 51–72
41. Inobeme A, Adetunji CO, Maliki M, Onyeachu BI, Kelani T, Eziukwu CA, Olori E, Mathew JT, Bamigboye MO (2023c) Strategies to synthesize, advantages, and disadvantages of pharmaceutical nanoparticles. In: Nanotechnology for drug delivery and pharmaceuticals. Academic Press is an imprint of Elsevier, pp 371–402
42. Świderska-Kołacz G, Jefimow M, Klusek J, Rączka N, Zmorzyński S, Wojciechowska A, Stanisławska I, Łyp M, Czerwik-Marcinkowska J (2021) Influence of algae supplementation on the concentration of glutathione and the activity of glutathione enzymes in the mice liver and kidney. Nutrients 13(6):1996
43. Barinova S, Gabyshev V, Genkal S, Gabysheva O (2023) Diatoms of small water bodies as bioindicators in the assessment of climatic and anthropogenic impacts on the coast of Tiksi Bay. Russian Arctic Water 15(8):1533
44. Viswanaathan S, Perumal PK, Sundaram S (2022) Integrated approach for carbon sequestration and wastewater treatment using algal–bacterial consortia: opportunities and challenges. Sustain For 14(3):1075
45. Balycheva D, Anufriieva E, Lee R, Prazukin A, Shadrin N (2023) Salinity-dependent species richness of Bacillariophyta in hypersaline environments. Water 15(12):2252
46. Diaz CJ, Douglas KJ, Kang K, Kolarik AL, Malinovski R, Torres-Tiji Y, Molino JV, Badary A, Mayfield SP (2023) Developing algae as a sustainable food source. Front Nutr 9:1029841
47. Nasr-Eldahan S, Nabil-Adam A, Shreadah MA, Maher AM, El-Sayed Ali T (2021) A review article on nanotechnology in aquaculture sustainability as a novel tool in fish disease control. Aquac Int: J Eur Aquac Soc 29(4):1459–1480
48. Zlaugotne B, Pubule J, Blumberga D (2022) Advantages and disadvantages of using more sustainable ingredients in fish feed. Heliyon 8(9):e10527
49. Inobeme A, Mathew JT, Adetunji CO, Ajai IA, Okonkwo S, Inobeme J, Adekoya MA, Oyedolapo BM (2022b) Trace elements and rare earth elements in aerosols: atmospheric aerosols properties, sources and detection. Nova Science Publishers, Inc., pp 1–22
50. Inobeme A, Adetunji CO, Mathew JT, Ajai AI, Mann A, Inobeme J, Oyedolapo B, Adekoya MA, Onyeaku S (2023d) Diversity, biodegradation and bioremediation of polycyclic aromatic hydrocarbons. In: Microbial technologies in industrial wastewater treatment. Springer Nature Singapore, pp 31–50
51. Sethi D, Butler TO, Shuhaili F, Vaidyanathan S (2020) Diatoms for carbon sequestration and bio-based manufacturing. Biology 9(8):217

52. Adetunji CO, Mathew JT, Singh KRB, Inobeme A, Olaniyan O, Singh J (2022) Antibiofilm potential of zero-dimensional carbon nanomaterials. In: Zero-dimensional carbon nanomaterials: fundamentals and applications. IOP Publishing
53. Tramontano C, Chianese G, Terracciano M, de Stefano L, Rea I (2020) Nanostructured biosilica of diatoms: from water world to biomedical applications. Appl Sci 10(19):6811

Diatoms for Value-Added Products: Challenges and Opportunities

5

Krishna Neeti, Reena Singh, Shaz Ahmad, Sakshi, and Abhishek Kumar

Abstract

Diatoms are a promising feedstock for developing value-added products like carotenoids, functional foods, bioactive pharmaceuticals, cosmetics, and biofuel. However, developing diatoms for these applications is not without its challenges. This chapter discusses the challenges and opportunities in developing diatoms for value-added products. The impact of abiotic environmental stresses, such as intensity and quality of incident light, nutrient deficiency, and silicon depletion, on diatoms has been examined to determine key factors that affect the growth performance and productivity of valuable products. The nutraceutical value of the myriad compounds like omega 3 fatty acids-DHA, polysaccharides, and phenols in diatoms has also been elaborated. Finally, the challenges in sustainable diatom biorefinery have been discussed. This chapter provides insights into the biotechnology potential of diatoms for value-added products and highlights the need for further research in this area.

Keywords

Diatoms · Value-added products · Biotechnology · Nutraceuticals · Sustainable biorefinery

K. Neeti (✉) · R. Singh · A. Kumar · Sakshi
Department of Civil Engineering, National Institute of Technology Patna, Patna, India
e-mail: krishnan.phd19.ce@nitp.ac.in

S. Ahmad
Government Engineering College, Kaimur, Jaitpur Kalan Bhagwanpur Kaimur, Bhabua, Bihar, India

M. Rafatullah et al. (eds.), *Multidisciplinary Applications of Marine Resources*,
https://doi.org/10.1007/978-981-97-5057-3_5

5.1 Introduction

Diatoms are a group of unicellular microalgae that have diverse and fascinating shapes, sizes, and structures. They are found in almost all aquatic habitats, from oceans to lakes, rivers, and ponds [1]. They are not only important components of the aquatic food web and biogeochemical cycles, but also valuable sources of various bioactive compounds that have applications in nutrition, industrial commodities, ecological studies, and biotechnology [2]. They produce and store various types of lipids, carbohydrates, proteins, pigments, polysaccharides, and silica, which can be used for producing biofuels, bioplastics, biosensors, biocatalysts, nutraceuticals, pharmaceuticals, cosmetics, biomimetics, and nanomaterials [3]. Moreover, they have unique and adaptable metabolic pathways that enable them to synthesize novel and rare compounds, such as fucoxanthin, diatoxanthin, and diadinoxanthin, which have potential uses in various fields [4]. However, developing diatoms for value-added products are not without its challenges. Diatom biotechnology faces several technical, economic, social, and environmental challenges that limit its potential and viability [5, 6]. Some challenges include maintaining the desired diatom species and preventing contamination. Additionally, there is the need to enhance the growth rate, biomass yield, and productivity of valuable compounds by diatoms. There is also the need for efficient and cost-effective methods for separating and extracting diatoms and their products. Moreover, we need to improve the quality and stability of diatom-derived products. We also need to reduce the production and processing costs, and increase the market competitiveness and profitability of diatom biotechnology. Additionally, we need to comply with the safety and quality standards and regulations of diatom-based products. We have to increase the public awareness and acceptance of diatom-based products, and minimize the environmental impact and risk of diatom biotechnology.

To overcome these challenges, diatom biotechnology needs to exploit the opportunities and create value from the scientific and technological advances and discoveries. The growing demand and preference for natural, renewable, and biodegradable products, the political and regulatory support and incentives, and the environmental and social benefits and impacts of diatom biotechnology would motivate research in this field [3]. Some of these opportunities include exploring the diversity and functionality of diatom species and strains. Understanding and manipulating the molecular mechanisms and regulation of diatom biosynthesis and metabolism. Improving the genetic engineering and synthetic biology tools and methods for diatom modification and improvement. Developing novel and high-quality products based on diatom-derived compounds and materials, conducting market research and analysis would help in development in diatom biotechnology. Developing effective marketing and branding strategies for diatom-based products. Engaging and influencing the policy makers and regulators. Participating and contributing in the policy making and regulation processes of diatom biotechnology. Assessing and reporting the environmental and social impacts and benefits of diatom biotechnology. Communicating and demonstrating the environmental and social value and significance of diatom biotechnology.

In this chapter, we aim to provide a comprehensive and updated overview of diatom biotechnology and its products. We will discuss the challenges and opportunities in developing diatoms for value-added products. Lastly, we will suggest some recommendations and future directions for diatom biotechnology research and innovation.

5.2 Diatoms: An Overview

5.2.1 Diversity and Characteristics of Diatoms

Diatoms are estimated to have more than 200,000 species, belonging to two major classes: centric and pennate diatoms [7]. Centric diatoms have radial symmetry and are mostly planktonic, while pennate diatoms have bilateral symmetry and are mostly benthic. They have a unique life cycle that involves alternation of generations and size reduction [8]. They reproduce asexually by mitosis, resulting in two daughter cells that inherit one of the two halves of the parent cell wall. The daughter cell that inherits the smaller half has to secrete a new wall inside the old one, leading to a decrease in cell size over generations [9]. When the cell size reaches a critical limit, diatoms switch to sexual reproduction by meiosis, resulting in the formation of auxospores that restore the original cell size [10]. They have a distinctive cell wall composed of hydrated silica embedded in an organic matrix. The cell wall, also known as the frustule, consists of two overlapping valves that fit together like a box and a lid [11]. The frustule has intricate patterns of pores, spines, and ridges that are species-specific and can be used for identification. They provide protection, support, and buoyancy for the diatom cell, they also play a role in light harvesting, nutrient uptake, and gas exchange [12]. Additionally, diatoms are photosynthetic organisms that use chlorophylls a and c, as well as accessory pigments such as fucoxanthin, diadinoxanthin, and diatoxanthin, to capture light energy [13]. They can store carbon in the form of lipids or polysaccharides, depending on the environmental conditions. Lipids are mainly accumulated in oil bodies, which are spherical organelles surrounded by a membrane. Polysaccharides are mainly stored in the cytoplasm or extracellularly as chrysolaminarin or fucoidan, respectively. Moreover, diatoms can also synthesize various secondary metabolites, such as polyunsaturated fatty acids, terpenes, sterols, and bioactive compounds, that have potential applications in nutrition, health, and industry [7].

5.2.2 Diatom Biotechnology and Bioprocessing

Diatom biotechnology is the use of diatoms or their products for industrial or commercial purposes. This biotechnology has been developing rapidly in the past few decades, thanks to the advances in genome sequencing, metabolic modeling, genetic engineering, and bioreactor design [3]. Several diatom species, such as Phaeodactylum tricornutum and Thalassiosira pseudonana, have been established as model organisms for diatom research and biotechnology. Their genomes have been sequenced and annotated, revealing the complexity and diversity of their metabolic pathways and gene families [7, 14]. Genome-scale metabolic models have been constructed to simulate and optimize the metabolic fluxes and yields of diatoms under different conditions. Genetic engineering tools, such as CRISPR-Cas9, have been developed and used to modify the diatom genome. This allows diatoms to produce new compounds or improve their natural traits by inserting foreign genes [15]. Bioreactor systems, such as photobioreactors and fermenters, have been designed and optimized to cultivate diatoms at large scale and high efficiency. They use various modes of operation and control parameters. Diatom bioprocessing is the extraction and purification of diatom products for downstream applications [3]. This involves several steps, such as harvesting, cell disruption, product separation, and product refinement. Harvesting is the process of concentrating the diatom biomass from the culture medium, which can be done by centrifugation, filtration, flocculation, or flotation. Cell disruption is the process of breaking the diatom cell wall and releasing the intracellular contents, which can be done by mechanical, chemical, or enzymatic methods. Product separation is the process of isolating the desired product from the cell lysate, which can be done by solvent extraction, chromatography, electrophoresis, or precipitation. Product refinement is the process of improving the quality and purity of the product, which can be done by crystallization, distillation, or recrystallization.

5.3 Nutraceutical Value of Diatoms

Diatoms have been found to have various nutraceutical properties that make them a valuable source of bioactive compounds [16, 17]. Here are some of the nutraceutical compounds found in diatoms:

5.3.1 Omega 3 Fatty Acids-DHA

Omega-3 fatty acids are essential for human health, as they play important roles in various physiological functions, such as anti-inflammatory, antithrombotic, antiarrhythmic, and neuroprotective effects. Among the omega-3 fatty acids, docosahexaenoic acid (DHA, 22:6ω-3) is especially recognized for its nutritional value and ability to alleviate many diseases, such as cardiovascular, neurodegenerative, and metabolic disorders [18]. However, the human body cannot synthesize DHA efficiently, and therefore it needs to be obtained from dietary sources. Diatoms are excellent producers of DHA, as well as other polyunsaturated fatty acids (PUFAs), such as eicosapentaenoic acid (EPA, 20:5ω-3) and arachidonic acid (ARA; 20:4ω-6). They can accumulate up to 40% of their dry weight as lipids, and their fatty acid composition varies depending on the species, strain, growth conditions, and environmental factors [19]. They are also rich in other bioactive compounds, such as pigments, antioxidants, vitamins, and minerals, that can enhance the nutraceutical value of DHA [17]. Additionally, they have been considered as a potential source of DHA for human consumption, as they offer several advantages over other sources, such as fish oil, krill oil, or other microalgae [20]. Moreover, they can also be genetically modified to enhance their DHA production and quality, as well as to introduce other desirable traits, such as increased growth rate, stress tolerance, and lipid accumulation.

5.3.2 Polysaccharides

Polysaccharides are complex carbohydrates that consist of repeating units of monosaccharides or their derivatives. They have various biological functions, such as energy storage, structural support, cell recognition, and signalling. Diatoms produce and secrete different types of polysaccharides, such as chrysolaminarin, fucoidan, alginic acid, and mannitol. These polysaccharides have diverse chemical structures and properties, depending on the diatom species, strain, growth phase, and environmental conditions [21]. Diatom polysaccharides have been explored for their potential applications in various fields, such as biotechnology, medicine, and nutraceuticals. For example, chrysolaminarin, a β-1,3 glucan, is a major intracellular reserve polysaccharide that can be used as a source of bioethanol or bioplastics [21]. Fucoidan, a sulfated polysaccharide, is an extracellular polymer that has anti-inflammatory, antiviral, anticoagulant, and anticancer activities. Alginic acid, a uronic acid polysaccharide, is another extracellular polymer that has gel-forming, emulsifying, and stabilizing properties and can be used as a thickener, binder or drug delivery agent. Mannitol, a sugar alcohol, is a trace intracellular polysaccharide that has antioxidant, osmoprotective, and sweetening effects, and can be used as a food additive, a pharmaceutical excipient, or a cosmetic ingredient.

5.3.3 Phenols

Phenols are aromatic compounds that contain one or more hydroxyl groups attached to a benzene ring. They have various biological functions, such as antioxidant, antimicrobial, anti-inflammatory, and signalling effects [22]. Diatoms produce and secrete different types of phenols, such as phlorotannins, phenolic acids, and flavonoids. These phenols have diverse chemical structures and properties, depending on the diatom species, strain, growth phase, and environmental conditions [23]. Diatom phenols have been explored for their potential applications in various fields, such as biotechnology, medicine, and nutraceuticals. For example, phlorotannins, a group of polyphenols derived from phloroglucinol, are extracellular polymers that have antioxidant, antiviral, antifungal, antidiabetic, and neuroprotective activities. Phenolic acids, such as caffeic acid, ferulic acid, and p-coumaric acid, are intracellular metabolites that have antioxidant, anti-inflammatory, and anticancer activities. Flavonoids, such as quercetin, kaempferol, and apigenin, are intracellular metabolites that have antioxidant, anti-inflammatory, and antiallergic activities.

5.4 Potential Value-Added Products from Diatoms

5.4.1 Carotenoids

Carotenoids are a group of pigments that are responsible for the yellow, orange, and red colours of many plants, animals, and microorganisms. They have various biological functions, such as antioxidant, photoprotective, anti-inflammatory, and immunomodulatory activities and are also important for human health, as they can prevent or reduce the risk of certain chronic diseases, such as cancer, cardiovascular disease, and age-related dermal degeneration [7].

5.4.2 Carotenoids Types and Its Applications

Diatoms are rich sources of carotenoids, especially fucoxanthin, which is the most abundant carotenoid in marine diatoms, accounting for up to 67% of the total carotenoid content. Fucoxanthin has a unique molecular structure, with an allenic bond and a 5,6-monoepoxide group, that confers it distinctive biological properties, such as anti-obesity, anti-diabetic, anti-cancer, anti-angiogenic, and cancer confining effects [7]. It can also be converted into fucoxanthinol and amarouciaxanthin A, which have higher bioavailability and potency than fucoxanthin. Additionally, Fucoxanthin and its metabolites have potential applications in various fields, such as nutraceuticals, functional foods, cosmetics, and pharmaceuticals. Other carotenoids that can be produced by diatoms include β-carotene, lutein, zeaxanthin, violaxanthin, and diadinoxanthin. These carotenoids have different roles in the photosynthetic process of diatoms, such as light harvesting, photoprotection, and xanthophyll cycle. These carotenoids also have various industrial applications, such as fish food (aquaculture), food (nutritional supplements), and cosmetics (skin care and anti-aging) [7].

5.4.3 Functional Foods

Functional foods are foods that have beneficial effects on health beyond their basic nutritional value. They may contain bioactive components, such as nutrients or non-nutrients, that can prevent or reduce the risk of certain diseases, or enhance physiological functions [24]. They can be classified into two types: conventional and modified. Conventional functional foods are natural, whole-food ingredients that are rich in important nutrients, such as vitamins, minerals, antioxidants, and fibre whereas modified functional foods are foods that have been fortified or enriched with additional ingredients, such as probiotics, prebiotics, or phytochemicals, to increase their health benefits [25]. Figure 5.1 shows a comparison of conventional and modified functional foods highlighting their key components.

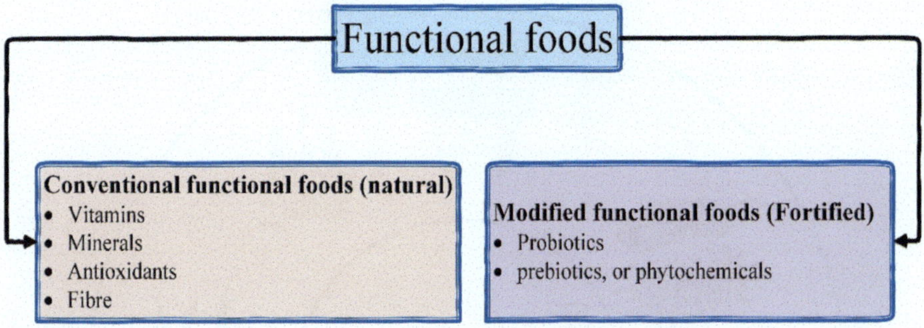

Fig. 5.1 Key component of conventional and modified functional food

5.4.4 Diatoms as Functional Food Ingredients

Diatoms can be used as functional food ingredients, either as whole biomass or as extracts, to enhance the nutritional and functional quality of food products. It can be incorporated into various food products, such as bread, pasta, cookies, cakes, yogurt, cheese, beverages, and snacks, to improve their texture, colour, flavour, and shelf life [26, 27]. Additionally, it can also be used as natural food additives, such as emulsifiers, stabilizers, thickeners, gelling agents, and colorants, to modify the physicochemical and palatability properties of food products. Moreover, Diatoms can also be used as carriers or delivery systems for other functional ingredients, such as vitamins, minerals, probiotics, and phytochemicals, to enhance their stability, bioavailability, and efficacy.

5.4.5 Bioactive Pharmaceuticals

Bioactive pharmaceuticals are drugs or drug candidates that are derived from natural sources, such as plants, animals, or microorganisms, and have specific biological activities against various diseases or disorders [28]. They have several advantages over synthetic drugs, such as higher diversity, complexity, and selectivity, lower toxicity and side effects, and higher compatibility with human physiology. Figure 5.2 illustrates the diversity and complexity of these natural sources in providing beneficial compounds for drug development.

5.4.6 Compounds with Medicinal Properties

Diatoms are a rich source of bioactive compounds with medicinal properties [29, 30], such as:

• Polyunsaturated fatty acids (PUFAs), especially eicosapentaenoic acid (EPA) and docosahexaenoic acid (DHA), which have anti-inflammatory, anti-

Fig. 5.2 Natural sources for bioactive pharmaceuticals

thrombotic, anti-cholesterol, and neuroprotective effects, and can be used for the prevention and treatment of cardiovascular diseases, diabetes, arthritis, Alzheimer's disease, and depression.

- Fucoxanthin and its metabolites, which have anti-obesity, anti-diabetic, anti-cancer, anti-angiogenic, and anti-metastatic effects, and can modulate various signalling pathways, such as AMPK, PPAR, NF-κB, MAPK, and PI3K/Akt.
- Polysaccharides, such as fucoidan, laminarin, and alginate, which have immunomodulatory, anticoagulant, antiviral, and anticancer effects, and can activate various immune cells, such as macrophages, natural killer cells, and dendritic cells.
- Silica, which has osteogenic, wound healing, and drug delivery properties, and can be used for the fabrication of various biomaterials, such as scaffolds, nanoparticles, and microneedles.

5.4.7 Cosmetics

Cosmetics are products that are applied to the skin, hair, or nails to enhance their appearance, health, or function. Cosmetics can be classified into different categories, such as skin care, hair care, nail care, makeup, fragrance, and personal hygiene. Cosmetics can contain various ingredients, such as water, oils, waxes, emulsifiers, preservatives, colorants, fragrances, and active agents [31].

5.4.8 Diatom-Based Cosmetic Products

Diatoms can be used as cosmetic ingredients, either as whole biomass or as extracts, to formulate various cosmetic products [31], such as:

- Diatom-based skin care products, such as creams, lotions, gels, masks, and serums, that can improve skin hydration, elasticity, smoothness, and brightness, and can also prevent or reduce skin aging, wrinkles, spots, and inflammation.
- Diatom-based hair care products, such as shampoos, conditioners, sprays, and oils, that can improve hair strength, shine, volume, and colour, and can also prevent or treat hair loss, dandruff, and scalp irritation.
- Diatom-based nail care products, such as polishes, removers, and cuticle oils, that can improve nail growth, hardness, and appearance, and can also prevent or treat nail infections, brittleness, and splitting.
- Diatom-based makeup products, such as foundations, powders, blushes, eyeshadows, and lipsticks, that can provide natural and long-lasting colour, coverage, and protection, and can also enhance skin health and appearance.
- Diatom-based fragrance products, such as perfumes, deodorants, and candles, that can provide pleasant and distinctive aromas, and can also have therapeutic and relaxing effects.

Figure 5.3 illustrate some examples of cosmetic products that can be formulated with diatom-based ingredients.

5.4.9 Biofuel

Biofuel is a type of fuel that is derived from biomass, such as plants, animals, or microorganisms, and can be used as a renewable and environmentally friendly alternative to fossil fuels. It can be classified into different generations, depending on the feedstock and the conversion technology. The first generation of biofuel uses edible crops, such as corn, sugarcane, and soybean, and converts them into ethanol or biodiesel. The second generation of biofuel uses non-edible crops, such as switchgrass, poplar, and jatropha, and converts them into cellulosic ethanol or synthetic diesel. The third generation of biofuel uses microalgae, such as diatoms, and converts them into biocrude or bio-oil [32, 33].

Fig. 5.3 Diatom-based cosmetic products

5.4.10 Diatoms as a Source of Biofuel

Diatoms have several advantages as a source of biofuel [17, 32], such as:

- Diatoms have a high biomass productivity and can be cultivated in various environments, such as freshwater, seawater, brackish water, and wastewater, using different technologies, such as open ponds, photobioreactors, and fermenters.
- Diatoms have a high lipid content and can produce different types of lipids, such as triglycerides, phospholipids, and glycolipids, that can be converted into biodiesel, biogas, or bio-oil.
- Diatoms have a diverse and adaptable metabolism and can produce novel compounds, such as fucoxanthin, diatoxanthin, and diadinoxanthin, that have unique properties and potential applications in various fields.

Figure 5.4 illustrates different types of diatoms, for biofuel production including biomass of seaweed, algae, and plants.

Fig. 5.4 Diatom-based biofuel

5.5 Environmental Stresses on Diatoms

Developing diatoms for value-added products is not without its challenges. Here are some of the challenges that researchers have identified:

5.5.1 Abiotic Environmental Stresses

Diatoms are sensitive to various abiotic environmental stresses, such as the intensity and quality of incident light, nutrient deficiency, and silicon depletion [34]. These stresses can have a significant impact on the growth and productivity of diatoms.

5.5.1.1 Intensity and Quality of Incident Light

Diatoms are photosynthetic microalgae that live in various aquatic environments, where they are exposed to changing light conditions. Light is essential for diatoms, as it provides the energy for their growth and metabolism. However, too much light can also be harmful. Therefore, diatoms have developed various mechanisms to regulate the amount of light they receive and use for photosynthesis. These mechanisms can be classified into two categories: behavioural and physiological [34].

Behavioural mechanisms involve the movement of diatoms in response to light. They can move vertically within the water column or the sediment, or horizontally along the substrate, to adjust their position to the optimal light level. They can also orient their cells toward or away from the light source, depending on their light preference. These movements are mediated by various factors, such as phototaxis, buoyancy, mucilage secretion, and biofilm formation [34]. Physiological mechanisms involve the modulation of the photosynthetic activity and pigment composition of diatoms in response to light. Diatoms can adjust the efficiency of their photosystems by changing the distribution and connectivity of their light-harvesting complexes and reaction centres. They can also dissipate excess light energy as heat or fluorescence, by activating non-photochemical quenching (NPQ). Diatoms can also synthesize or degrade different types of pigments, such as chlorophylls, carotenoids, and fucoxanthin, to optimize their light absorption and protection [13]. Both behavioural and physiological mechanisms are influenced by the intensity and quality of the incident light. Light intensity refers to the number of photons per unit area and time, while light quality refers to the spectral composition or wavelength of the light. Different light intensities and qualities can have different effects on the photosynthesis and photoprotection of diatoms. For example, low light intensity can limit the photosynthetic rate and induce the synthesis of more light-harvesting pigments, while high light intensity can saturate or inhibit the photosynthetic rate and induce the activation of NPQ and the degradation of light-harvesting pigments. Similarly, different light qualities can have different effects on the photosynthesis and photoprotection of diatoms. For example, blue light can penetrate deeper into the water and stimulate the photosynthesis and migration of diatoms, while red light can be absorbed more by the water and induce the synthesis of more light-harvesting pigments and the activation of NPQ.

The intensity and quality of the incident light can also interact with each other and with other environmental factors, such as temperature, salinity, nutrients, and biotic interactions, to affect the photosynthesis and photoprotection of diatoms.

5.5.1.2 Nutrient Deficiency

Nutrient deficiency can lead to reduced growth and biomass production in diatoms, as well as changes in their lipid accumulation and fatty acid composition. Therefore, understanding the mechanisms involved in the responses of diatoms to nutrient deficiency is important for optimizing their cultivation and biotechnological applications [35]. Nutrients, such as nitrogen (N), phosphorus (P), and silicon (Si), are essential for diatom growth and are closely linked to their photosynthetic and physiological status. It is widely recognized that diatoms exhibit lower sinking rates under nutrient-replete conditions, while nutrient deficiency can induce the formation of resting spores or increase the cell density and sinking rate. Nutrient deficiency can also affect the lipid accumulation and fatty acid composition of diatoms, which are important factors for biodiesel production. For example, N deficiency can increase the lipid content and the ratio of saturated and monounsaturated fatty acids to polyunsaturated fatty acids, while P deficiency can decrease the lipid content and increase the proportion of polyunsaturated fatty acids [36].

5.5.1.3 Silicon Depletion

Diatoms produce silica-based cell walls (frustules) that have unique structures and functions. Silicon is an essential nutrient for diatoms, as it is required for frustule synthesis and cell division [37]. However, they often face silicon limitation in their natural habitats, due to the high demand and low availability of this element. Silicon depletion can have various effects on diatom physiology, ecology, and evolution. It can affect the growth rate, cell size, and shape of diatoms, as well as their ability to form resting stages or colonies [38]. Additionally, it can also affect the expression of genes and proteins involved in silicon metabolism, such as silicon transporters (SITs), silicic acid synthases (SASs), and silaffins. Moreover, silicon depletion can affect the biogeochemical cycling of carbon and other nutrients, as diatoms play a major role in the export of organic matter to the deep ocean. Diatoms have developed various strategies to cope with silicon depletion, such as increasing their silicon uptake efficiency, recycling their internal silicon pools, modifying their frustule morphology, and switching to alternative sources of silicon. They can also adapt to silicon depletion by evolving different silicon requirements, frustule compositions, and gene expression patterns. Furthermore, it can interact with other organisms, such as bacteria and viruses, to modulate their silicon metabolism and response.

5.5.1.4 Factors Affecting Growth Performance and Productivity of Diatom

Several factors affect the growth performance and productivity of diatoms. These include the choice of cultivation mode, the type of nutrient source, the light regime, and the temperature. Researchers have explored different nutrition modes, namely, phototrophy, mixotrophy, and heterotrophy, to optimize the growth conditions and scale-up the production of diatoms in a cost-effective and sustainable way. Growth on organic carbon substrates promotes higher biomass, lipid, and carbohydrate productivity, which further triggers the yield of various biomolecules. The choice of cultivation mode is an important factor affecting the growth performance and productivity of diatoms. Phototrophic cultivation is the most common method used for diatom cultivation, where diatoms are grown under light conditions [2]. Mixotrophic cultivation is another method that involves the use of organic carbon sources in addition to light for diatom growth. Heterotrophic cultivation is a third method that involves the use of organic carbon sources as the sole energy source for diatom growth. The type of nutrient source is another key factor affecting the growth performance and productivity of diatoms. It requires various nutrients, including nitrogen, phosphorus, and silicon, for their growth and metabolism [39]. Researchers have explored different nutrient sources, such as inorganic salts, organic compounds, and wastewater, for diatom cultivation. The light regime is also an important factor affecting the growth performance and productivity of diatoms. Light intensity, quality, and duration are key parameters that affect growth and metabolism. Researchers have explored different light regimes, such as continuous light, intermittent light, and light–dark cycles, for diatom cultivation.

Temperature is another key factor affecting the growth performance and productivity of diatoms. They have a wide temperature range for growth, with optimal

Table 5.1 Effect of factors on diatom growth and productivity

Factor	Effect on diatom growth and productivity
Cultivation mode	Phototrophy: common but low productivity; Mixotrophy: higher productivity and biomolecule yield; Heterotrophy: highest productivity but requires organic substrates
Nutrient source	Inorganic salts: standard but costly and polluting; Organic compounds: alternative but variable and complex; Wastewater: cheap and eco-friendly but may contain contaminants
SLight regime	Continuous light: high growth rate but low lipid content; Intermittent light: lower growth rate but higher lipid content; Light–dark cycles: natural but variable and dependent on diurnal rhythms
Temperature	15 °C–25 °C: optimal range for most diatoms; Higher or lower temperatures: may affect growth rate, lipid content, and composition

growth occurring at temperatures between 15 °C and 25 °C [40]. Researchers have explored the effect of temperature on diatom growth and metabolism and found that temperature affects the lipid content and composition of diatoms. Table 5.1a shows the effect of different factors, such as cultivation mode, nutrient source, light regime, and temperature, on the growth performance and productivity of diatoms.

5.6 Challenges and Opportunities of Diatom Value-Added Products

Diatoms are remarkable microalgae encased in intricate silica shells, holding immense promise. As we explore their potential for value-added products, we encounter challenges and seek possible solutions or alternatives to overcome these hurdles. Simultaneously, we identify opportunities and consider ways to exploit them. Figure 5.5 visually illustrates challenges and opportunities in the realm of diatom value-added products.

5.6.1 Challenges

5.6.1.1 Technical Challenges

The main technical challenges in developing diatom-based products are related to the optimization of diatom cultivation, harvesting, and extraction processes [41]. These include:

- Maintaining the desired diatom species and preventing contamination by invasive species or predators.
- Enhancing the growth rate, biomass yield, and productivity of valuable compounds by diatoms under different environmental conditions.
- Developing efficient and cost-effective methods for separating diatoms from the culture medium and breaking their silica cell walls.

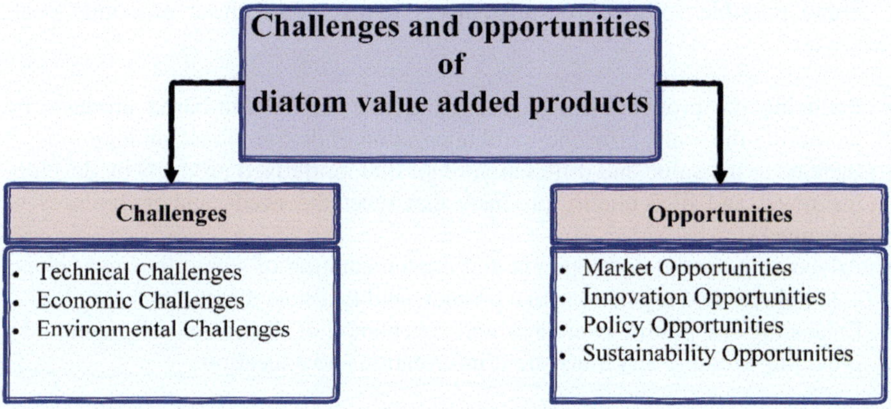

Fig. 5.5 Challenges and opportunities diatom value-added products

- Improving the quality and stability of diatom-derived products and reducing the loss of valuable compounds during extraction and purification.

Some possible solutions or alternatives to overcome these technical challenges are:

- Using genetic engineering and synthetic biology tools to modify diatom strains for improved traits and performance.
- Applying systems biology and metabolic modelling approaches to understand and optimize diatom metabolism and biosynthesis pathways.
- Developing novel and integrated biorefinery processes that utilize the whole diatom biomass and produce multiple products.
- Exploring new and innovative applications of diatom-derived products that can create value and demand.

5.6.1.2 Economic Challenges

The main economic challenges in developing diatom-based products are related to the market competitiveness and profitability of diatom biotechnology [3, 13]. These include:

- The high capital and operational costs of diatom production and processing facilities [42].
- The low and variable prices of diatom-derived products and the competition from other sources of similar products.
- The lack of clear and consistent policies and regulations that support and incentivize diatom biotechnology.
- The limited public awareness and acceptance of diatom-based products and their benefits.

Some possible solutions or alternatives to overcome these economic challenges are:

- Reducing the production and processing costs of diatom-based products by increasing the scale, efficiency, and integration of diatom biotechnology.
- Increasing the value and differentiation of diatom-derived products by developing novel and high-quality products that meet the needs and preferences of consumers.
- Advocating for the development and implementation of supportive and favourable policies and regulations that promote and facilitate diatom biotechnology.
- Enhancing the public awareness and acceptance of diatom-based products by providing credible and transparent information and education.

5.6.1.3 Environmental Challenges

The main environmental challenges in developing diatom-based products are related to the environmental sustainability and impact of diatom biotechnology [3]. These include:

- The high water and energy consumption and greenhouse gas emissions of diatom production and processing.
- The potential environmental pollution and contamination by diatom cultivation and harvesting waste and residues.
- The possible ecological disruption and invasion by genetically modified or exotic diatom strains.
- The uncertain environmental fate and effects of diatom-derived products and their degradation products.

Some possible solutions or alternatives to overcome these environmental challenges are:

- Improving the water and energy efficiency and reducing the greenhouse gas emissions of diatom production and processing.
- Minimizing the environmental pollution and contamination by diatom cultivation and harvesting waste and residues.
- Preventing the ecological disruption and invasion by genetically modified or exotic diatom strains.
- Evaluating the environmental fate and effects of diatom-derived products and their degradation products.

5.6.2 Opportunities

5.6.2.1 Market Opportunities

The main market opportunities of developing diatom-based products are related to the growing demand and preference for natural, renewable, and biodegradable products [3, 7, 13]. These include:

- The increasing consumer awareness and interest in the health benefits and potential uses of diatom-derived nutraceuticals, such as omega-3 fatty acids, carotenoids, polysaccharides, and phenols.
- The expanding market potential and value of diatom-derived industrial commodities, such as biofuels, bioplastics, biosensors, and biocatalysts.
- The emerging market opportunities and niches of diatom-derived novel and high-quality products, such as cosmetics, pharmaceuticals, biomimetics, and nanomaterials.

Some possible ways to exploit these market opportunities and create value are:

- Conducting market research and analysis to identify and target the most profitable and promising segments and customers of diatom-based products.
- Developing and implementing effective marketing and branding strategies to promote and differentiate diatom-based products from other competitors.
- Establishing and maintaining strong and reliable partnerships and collaborations with other stakeholders and actors in the diatom value chain, such as suppliers, distributors, retailers, and consumers.

5.6.2.2 Innovation Opportunities

The main innovation opportunities of developing diatom-based products are related to the scientific and technological advances and discoveries in diatom biotechnology [3, 7, 13]. These include:

- The availability and accessibility of various tools and methods for diatom genomics, transcriptomics, proteomics, and metabolomics, which enable the exploration and manipulation of diatom biology and metabolism.
- The development and improvement of various techniques and platforms for diatom cultivation, harvesting, and extraction, which enhance the efficiency and quality of diatom production and processing.
- The creation and invention of various products and applications based on diatom-derived compounds and materials, which expand the diversity and functionality of diatom-based products.

Some possible ways to exploit these innovation opportunities and create value are:

- Investing and supporting in research and development activities and projects that focus on diatom biotechnology and its applications.
- Encouraging and facilitating the exchange and transfer of knowledge and technology among diatom researchers and practitioners.
- Protecting and commercializing the intellectual property and innovation outputs of diatom biotechnology.

5.6.2.3 Policy Opportunities

The main policy opportunities of developing diatom-based products are related to the political and regulatory support and incentives for diatom biotechnology [7]. These include:

- The alignment and integration of diatom biotechnology with the national and international goals and agendas for sustainable development and green economy.
- The development and implementation of supportive and favourable policies and regulations that promote and facilitate diatom biotechnology and its products.
- The provision and allocation of adequate and appropriate resources and funds for diatom biotechnology and its products.

Some possible ways to exploit these policy opportunities and create value are:

- Engaging and influencing the policy makers and regulators to recognize and appreciate the benefits and potential of diatom biotechnology and its products.
- Participating and contributing in the policy making and regulation processes of diatom biotechnology and its products.
- Monitoring and evaluating the policy impacts and outcomes of diatom biotechnology and its products.

5.6.2.4 Sustainability Opportunities

The main sustainability opportunities of developing diatom-based products are related to the environmental and social benefits and impacts of diatom biotechnology [13, 43]. These include:

- The contribution and role of diatom biotechnology in mitigating and adapting to climate change and its effects, such as reducing greenhouse gas emissions, enhancing carbon sequestration, and increasing resilience.
- The contribution and role of diatom biotechnology in restoring and improving the quality and health of aquatic ecosystems, such as removing excess nutrients, pollutants, and toxins, and providing food and habitat for other organisms.
- The contribution and role of diatom biotechnology in enhancing and ensuring the well-being and livelihood of human society, such as providing food security, income generation, and employment opportunities.

Some possible ways to exploit these sustainability opportunities and create value are:

- Assessing and reporting the environmental and social impacts and benefits of diatom biotechnology and its products.
- Adopting and implementing the best practices and standards for environmental and social responsibility and accountability of diatom biotechnology and its products.

- Communicating and demonstrating the environmental and social value and significance of diatom biotechnology and its products.

5.7 Recommendations

5.7.1 Improving the Current State of Diatom-Based Products

Some recommendations or suggestions for improving the current state of diatom-based products are:

- Developing and applying standardized and harmonized protocols and methods for diatom cultivation, harvesting, and extraction.
- Enhancing the quality control and assurance of diatom-based products and ensuring their safety and efficacy.
- Increasing the collaboration and coordination among diatom researchers, practitioners, and stakeholders from different disciplines and sectors.
- Establishing and joining diatom networks and platforms that facilitate the sharing and dissemination of diatom-related information and resources.

5.7.2 Identifying the Gaps and Needs for Further Research and Development

Some gaps and needs for further research and development in diatom biotechnology are:

- Exploring the diversity and functionality of diatom species and strains and discovering new and valuable compounds and molecules from diatoms.
- Understanding the molecular mechanisms and regulation of diatom biosynthesis and metabolism and identifying the key genes and enzymes involved in diatom production and processing.
- Improving the genetic engineering and synthetic biology tools and methods for diatom manipulation and modification and creating new and improved diatom strains and products.
- Evaluating the techno-economic and life cycle analysis of diatom biotechnology and its products and assessing their feasibility and viability.

5.7.2.1 Highlighting the Potential Impact and Implications of Future Research and Innovation

Some potential impact and implications of future research and innovation in diatom biotechnology are:

- Contributing to the development and advancement of a bio-based and circular economy that relies on renewable and biodegradable resources and products.

- Addressing the global challenges and issues related to climate change, energy security, food security, and human health and well-being.
- Creating new and diverse opportunities and values for diatom biotechnology and its products in various sectors and industries.
- Enhancing the public awareness and appreciation of diatom biotechnology and its products and fostering a culture of innovation and sustainability.

5.7.3 Conclusion

In this chapter, we have explored the potential of diatoms as a source of value-added products for various applications in nutrition, industrial commodities, ecological studies, and biotechnology. We have shown that diatoms are a valuable and versatile resource that can offer a wide range of benefits and solutions for human society and the environment. However, we have also discussed the challenges and opportunities in developing diatoms for value-added products, covering the technical, economic, social, and environmental aspects of diatom biotechnology. We have suggested some recommendations and future directions for diatom biotechnology research and innovation, highlighting the potential impact and implications of diatom biotechnology for a greener and healthier future.

References

1. Ragni R, Cicco SR, Vona D, Farinola GM (2018) Multiple routes to smart nanostructured materials from diatom microalgae: a chemical perspective. Adv Mater 30:1704289. https://doi.org/10.1002/ADMA.201704289
2. Marella TK, Bhattacharjya R, Tiwari A (2021) Impact of organic carbon acquisition on growth and functional biomolecule production in diatoms. Microb Cell Factories 20:1–13. https://doi.org/10.1186/S12934-021-01627-X/FIGURES/3
3. Sharma N, Simon DP, Diaz-Garza AM et al (2021) Diatoms biotechnology: various industrial applications for a greener tomorrow. Front Mar Sci 8:636613. https://doi.org/10.3389/FMARS.2021.636613/BIBTEX
4. Kuczynska P, Jemiola-Rzeminska M, Strzalka K (2015) Photosynthetic pigments in diatoms. Mar Drugs 2015(13):5847–5881. https://doi.org/10.3390/MD13095847
5. Fabris M, Abbriano RM, Pernice M et al (2020) Emerging technologies in algal biotechnology: toward the establishment of a sustainable, algae-based bioeconomy. Front Plant Sci 11:515802. https://doi.org/10.3389/FPLS.2020.00279/BIBTEX
6. Rizwan M, Mujtaba G, Memon SA et al (2018) Exploring the potential of microalgae for new biotechnology applications and beyond: a review. Renew Sust Energ Rev 92:394–404. https://doi.org/10.1016/J.RSER.2018.04.034
7. Mal N, Srivastava K, Sharma Y et al (2021) Facets of diatom biology and their potential applications. Biomass Convers Biorefinery 12(5):1959–1975. https://doi.org/10.1007/S13399-020-01155-5
8. Seckbach J, Gordon R (2019) Diatoms: fundamentals and applications. John Wiley & Sons. https://doi.org/10.1002/9781119370741
9. Davidovich NA, Davidovich OI, Podunai YA et al (2015) Reproductive properties of diatoms significant for their cultivation and biotechnology. Russ J Plant Physiol 62:153–160. https://doi.org/10.1134/S1021443715020041/METRICS

10. Chen J, Huang Y, Shu Y et al (2022) Recent progress on systems and synthetic biology of diatoms for improving algal productivity. Front Bioeng Biotechnol 10:908804. https://doi.org/10.3389/FBIOE.2022.908804/BIBTEX
11. Tramontano C, Chianese G, Terracciano M et al (2020) (2020) Nanostructured biosilica of diatoms: from water world to biomedical applications. Appl Sci 10:6811. https://doi.org/10.3390/APP10196811
12. Tiwari A, Marella TK (2019) Potential and application of diatoms for industry-specific wastewater treatment. In: Application of microalgae in wastewater treatment, pp 321–339. https://doi.org/10.1007/978-3-030-13913-1_15
13. Fu W, Wichuk K, Brynjólfsson S (2015) Developing diatoms for value-added products: challenges and opportunities. New Biotechnol 32:547–551. https://doi.org/10.1016/J.NBT.2015.03.016
14. Dhaouadi F, Awwad F, Diamond A et al (2020) Diatoms' breakthroughs in biotechnology: Phaeodactylum tricornutum as a model for producing high-added value molecules. Am J Plant Sci 11:1632–1670. https://doi.org/10.4236/AJPS.2020.1110118
15. Moosburner M, Allen AE, Daboussi F (2022) Genetic engineering in marine diatoms: current practices and emerging technologies. In: The molecular life of diatoms, pp 743–773. https://doi.org/10.1007/978-3-030-92499-7_25/COVER
16. Tiwari A, Melchor-Martínez EM, Saxena A et al (2021) Therapeutic attributes and applied aspects of biological macromolecules (polypeptides, fucoxanthin, sterols, fatty acids, polysaccharides, and polyphenols) from diatoms—a review. Int J Biol Macromol 171:398–413. https://doi.org/10.1016/J.IJBIOMAC.2020.12.219
17. Nieri P, Carpi S, Esposito R et al (2023) Bioactive molecules from marine diatoms and their value for the nutraceutical industry. Nutrients 15:464. https://doi.org/10.3390/NU15020464
18. Djuricic I, Calder PC (2021) Beneficial Outcomes of Omega-6 and Omega-3 polyunsaturated fatty acids on human health: an update for 2021. Nutrients 13:2421. https://doi.org/10.3390/NU13072421
19. von Schacky C, Kuipers RS, Pijl H et al (2023) Omega-3 fatty acids in heart disease—why accurately measured levels matter. Netherlands Heart J 31:415–423. https://doi.org/10.1007/S12471-023-01759-2/FIGURES/1
20. Barta DG, Coman V, Vodnar DC (2021) Microalgae as sources of omega-3 polyunsaturated fatty acids: Biotechnological aspects. Algal Res 58:102410. https://doi.org/10.1016/J.ALGAL.2021.102410
21. Kamalanathan M, Chiu MH, Bacosa H et al (2019) Role of polysaccharides in diatom Thalassiosira pseudonana and its associated bacteria in hydrocarbon presence. Plant Physiol 180:1898–1911. https://doi.org/10.1104/PP.19.00301
22. Dhyani S, Bramhanwade K (2023) Diatoms in forensics: adding new dimension to the growing relevance of diatoms in improving lives. Springer Nature Singapore, Singapore, pp 165–174. https://doi.org/10.1007/978-981-19-5920-2_10
23. Del Mondo A, Sansone C, Brunet C (2022) Insights into the biosynthesis pathway of phenolic compounds in microalgae. Comput Struct Biotechnol J 20:1901–1913. https://doi.org/10.1016/J.CSBJ.2022.04.019
24. Essa MM, Bishir M, Bhat A et al (2023) Functional foods and their impact on health. J Food Sci Technol 60:820–834. https://doi.org/10.1007/S13197-021-05193-3/FIGURES/3
25. Bhaskarachary K, Vemula SR, Gavaravarapu SRM, Joshi AKR (2016) Traditional foods, functional foods and nutraceuticals. In: Proceedings of the Indian National Science Academy. Indian National Science Academy, pp 1565–1577
26. Ji J, Shi L, Ampofo J, Abbey L (2022) Microalgae: bioactive composition, health benefits, safety and prospects as potential high-value ingredients for the functional food industry. Food Secur 11:1744. https://doi.org/10.3390/FOODS11121744
27. Batista AP, Niccolai A, Bursic I et al (2019) Microalgae as functional ingredients in savory food products: application to wheat crackers. Food Secur 8:611. https://doi.org/10.3390/FOODS8120611

28. Akhtar MS, Swamy MK, Sinniah UR (2019) Natural bio-active compounds. In: Natural bio-active compounds: Volume 1: production and applications, pp 1–608. https://doi.org/10.1007/978-981-13-7154-7/COVER
29. Song C, Shieh CH, Wu YS et al (2016) The role of omega-3 polyunsaturated fatty acids eicosapentaenoic and docosahexaenoic acids in the treatment of major depression and Alzheimer's disease: acting separately or synergistically? Prog Lipid Res 62:41–54. https://doi.org/10.1016/J.PLIPRES.2015.12.003
30. Xiao H, Zhao J, Fang C et al (2020) Advances in studies on the pharmacological activities of fucoxanthin. Mar Drugs 18:634. https://doi.org/10.3390/MD18120634
31. Guerra E, Llompart M, Garcia-Jares C (2018) Analysis of dyes in cosmetics: challenges and recent developments. Cosmetics 5:47. https://doi.org/10.3390/COSMETICS5030047
32. Neeti K, Gaurav K, Singh R (2023) The potential of algae biofuel as a renewable and sustainable bioresource. Eng Proc 37:22. https://doi.org/10.3390/ECP2023-14716
33. Gaurav K, Neeti K, Singh R (2024) Microalgae-based biodiesel production and its challenges and future opportunities: a review. Green Technol Sustain 2:100060. https://doi.org/10.1016/J.GRETS.2023.100060
34. Prins A, Deleris P, Hubas C, Jesus B (2020) Effect of light intensity and light quality on diatom behavioral and physiological photoprotection. Front Mar Sci 7:516428. https://doi.org/10.3389/FMARS.2020.00203/BIBTEX
35. Marella TK, López-Pacheco IY, Parra-Saldívar R et al (2020) Wealth from waste: diatoms as tools for phycoremediation of wastewater and for obtaining value from the biomass. Sci Total Environ 724:137960. https://doi.org/10.1016/J.SCITOTENV.2020.137960
36. Lin Q, Zhuo WH, Wang XW et al (2018) Effects of fundamental nutrient stresses on the lipid accumulation profiles in two diatom species Thalassiosira weissflogii and Chaetoceros muelleri. Bioprocess Biosyst Eng 41:1213–1224. https://doi.org/10.1007/S00449-018-1950-Z/TABLES/4
37. Sardo A, Orefice I, Balzano S et al (2021) Mini-review: potential of diatom-derived silica for biomedical applications. Appl Sci 11:4533. https://doi.org/10.3390/APP11104533
38. McNair HM, Brzezinski MA, Krause JW (2018) Diatom populations in an upwelling environment decrease silica content to avoid growth limitation. Environ Microbiol 20:4184–4193. https://doi.org/10.1111/1462-2920.14431
39. Bhattacharjya R, Singh PK, Tiwari A (2021) Aquaculture water as a source of sustainable growth medium for diatom cultivation and its nutritive suitability as a potential aqua feed. Environ Technol Innov 24:101987. https://doi.org/10.1016/J.ETI.2021.101987
40. Zhang Y, Peng C, Wang Z et al (2018) The species-specific responses of freshwater diatoms to elevated temperatures are affected by interspecific interactions. Microorganisms 6. https://doi.org/10.3390/MICROORGANISMS6030082
41. Wang JK, Seibert M (2017) Prospects for commercial production of diatoms. Biotechnol Biofuels 10:1–13. https://doi.org/10.1186/S13068-017-0699-Y/TABLES/2
42. Vinayak V, Manoylov KM, Gateau H et al (2015) Diatom milking: a review and new approaches. Mar Drugs 13:2629–2665. https://doi.org/10.3390/MD13052629
43. Dhanker R, Saxena A, Tiwari A et al (2024) Towards sustainable diatom biorefinery: recent trends in cultivation and applications. Bioresour Technol 391:129905. https://doi.org/10.1016/J.BIORTECH.2023.129905

Marine Resources: A Sustainable and Promising Source for Cosmetic Industries

6

W. V. Celcia Gnana Rathinam, S. Bragadeeswaran, S. Kumaresan, K. Gunamathy, B. Visnu, J. Mohamed Asarudeen, T. Sasidharan, and S. Srikavibharathi

Abstract

Marine-derived substances or compounds are unique and have many benefits when compared to terrestrial sources. These marine-derived products also play an important role in cosmetic industries. Cosmetic industries also known as beauty industries formulate chemical compounds to create or produce many products that will serve the purpose of humans in skincare, hair care, makeup, etc. These cosmetic industries have a desirability toward the natural products that are derived from marine resources. Marine resources produce many useful active components, excipients, and bioactive substances that might act as natural skincare agents. This chapter gives a detailed explanation of how and what kind of bioactive substances from different marine resources have anti-aging, antioxidant, photo-protection, anti-acne, moisturizing, and skin whitening properties. It includes the uses and availability of bio-pigments extracted from marine resources as colorants in cosmetic industries. It also gives a brief idea about how marine by-products are used in cosmetics.

Keywords

Marine resources · Cosmetics · Drug delivery system · Skin health · Marine by-products

W. V. Celcia Gnana Rathinam · S. Bragadeeswaran (✉) · S. Kumaresan · K. Gunamathy · B. Visnu · J. Mohamed Asarudeen · T. Sasidharan · S. Srikavibharathi
Faculty of Marine Sciences, CAS in Marine Biology, Annamalai University, Parangipettai, Tamil Nadu, India

M. Rafatullah et al. (eds.), *Multidisciplinary Applications of Marine Resources*, https://doi.org/10.1007/978-981-97-5057-3_6

6.1 Introduction

One of the most dynamic and ever-evolving sectors is the cosmetic industry, which has continuous challenges for innovating and formulating the ingredients to meet the diverse and evolving preferences of consumers. As the term "organic" has become a trend nowadays, consumers have gained more interest toward natural and sustainable beauty products. In search of these natural ingredients, the cosmetic industries are turning their attention toward the marine world, which has a vast wealth of marine ingredients ranging from seaweeds and algae to deep-sea minerals. Seventy percent of the world is covered by water. Many macro- and microorganisms from marine environments have improved unique metabolic abilities to survive in diverse and hostile habitats, which result in the biosynthesis of an array of secondary metabolites with specific activities. As the pharmaceutical industry is highly interested in high-value ingredients from marine sources, the cosmetic industry is also progressively turning to sea in search of new ingredients. Cosmeceuticals have grabbed more attention nowadays than cosmetics since they have both cosmetics and pharmaceuticals used as bioactive ingredients, which are combined with creams, lotions, and ointments. The suppliers of cosmetic industries are being pushed to add extracts from coastal plants, seaweeds, sea minerals, marine invertebrates, and fishes into cosmetic ingredients nowadays. Vitamins and minerals from these extracts will have high ultraviolet and antioxidant protection and general anti-aging benefits. The usage of marine resources in cosmetics not only helps in providing a rich source of bioactive compounds with potential skincare benefits but also ranges with the growing global focus toward sustainable sourcing. To preserve marine ecosystems, the extraction of marine elements is frequently done using ecologically friendly techniques including aquaculture and wild harvesting. The cosmetic business has made a deliberate change toward using ingredients that are ethically obtained and environmentally benign, as evidenced by the usage of these marine resources. This movement is part of a larger trend toward sustainability in the cosmetics sector. As this trend picks up steam, it highlights the industry's dedication to producing cutting-edge and successful cosmetics while also acting responsibly and with consideration for the environment.

6.2 Cosmetics and Cosmeceuticals

Cosmetic product definition mostly depends on the intended use of cosmetics or drugs. Most of the time, they are established based on claims made for the product. As an exception, neither the cosmetic product nor its ingredients are subject to FDA premarket clearance authority. Instead, color, additives, and cosmetic corporations are responsible for proving the safety of cosmetic products and ingredients before marketing. Cosmetic product refers to any substance or mixture that is intended to be placed in contact with the external parts of the human body, i.e., epidermis, hair system, nails, lips, and external genital organs, or with the teeth and the mucous membranes of the oral cavity with an exclusively or mainly purpose of cleaning perfuming, changing their appearance, protecting, keeping them in good condition, or correcting body odors. This regulation includes skin-whitening products, anti-wrinkle products, and products with nanomaterials [1].

There is a huge difference between cosmetics and cosmeceuticals. Similar to cosmetics, cosmeceuticals are also mostly applied externally. Cosmeceuticals differ from cosmetics in the way that cosmeceuticals have potent ingredients such as enzymes, antioxidants, essential oils, phytochemicals, and vitamins. The main purpose of these cosmeceuticals is to promote healthy skin as they deliver nutrients and influence biological functions. The words cosmetic and pharmaceutical form the term "cosmeceutical," which refers to cosmetic products that have drug-like benefits, and help enhance or protect the external appearance of the human body [2]. The grouping of products placed between formulated and non-formulated products or prescription and nonprescription products is represented as cosmeceuticals. Cosmeceuticals have four major categories, and they are non-bleaching agents, antioxidants, peptides, and growth factors [3].

6.2.1 Role of Cosmetics and Cosmetic Formulations

To know about the role of cosmetics, one should know the basics of skin. Skin being the physiological barrier of our immune system plays an important role in protecting the internal organs against harmful microbes, restricting inflammation, etc. Our skin is made up of three layers: The first layer is the epidermis *stratum corneum*, the second layer is the dermis, and the innermost layer is the hypodermis. The first outermost layer the epidermis, i.e., *stratum corneum,* has protein protein-rich cornified layer of dead cells. These cells that are embedded in lipid matrix are known as corneocytes. The lipid matrix contains more ceramides, free fatty acids, cholesterols, and a small amount of water. Beneath the stratum corneum lies the viable epidermis that has melanocytes, Merkel cells, and Langerhans cells. This viable epidermis also consists of laminin, type 4 collagen, nidogen, and proteoglycan perlecan. A high amount of elastin is present to provide elastic properties to skin seen in the dermal matrix. The fibroblasts pervaded by blood and lymph vessels together form the dermis. The innermost hypodermis consists of adipocytes that help insulation and conserve the body's heat being a shock absorber [4].

Being not a modern invention, cosmetics are used by humans at least 10,000 years before and also very long before that. People nowadays are also concerned about their appearance, health, etc. To meet the demand to overcome skin problems and health issues, cosmetic industries are bringing up many new innovative products. Also, demand for natural products as active ingredients has also become a more concerned thought nowadays. Sources from plants and marine sources have paved many new ways to obtain active ingredients from natural sources in cosmetic industries.

The main role of the cosmetic industry is to create products that are designed in such a way that helps in cleansing, protecting, and changing the appearance of external parts of our body. The active molecule or active ingredient present in a

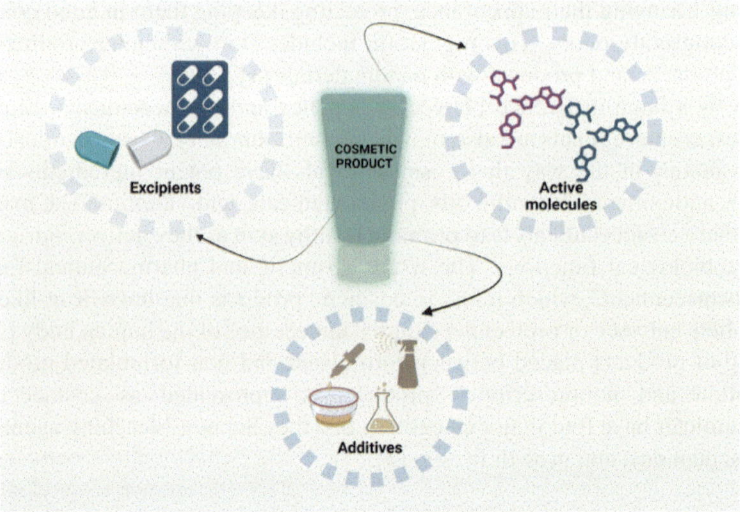

Fig. 6.1 Cosmetic formulations

particular cosmetic product plays an important role in serving its purpose. The sweat gland and hair follicles are different options, and the active molecules present in cosmetic products penetrate the skin in several ways such as transepidermal or transfollicular. It is said that the small soluble molecules that are lipophilic and amphipathic have greater ability to cross the outermost layer of skin, i.e., *stratum corneum* through intracellular way [5].

A cosmetic product contains an active ingredient (i.e., the active molecule), excipients, and additives. The sum of all these together forms the composition in a cosmetic product. Active ingredients play a major role in providing the desired cosmetic activity. The excipient ensures the desired pharmaceutical dosage is obtained. The additives are substances that help preserve the products and also improve organoleptic property [6] (Fig. 6.1).

For skin hydration, the basic formulation for cosmetic cream has simple ingredients such as humectants (1–5%), emollients (10–40%), emulsifiers (1–6%), stabilizers (0.01–0.2%), thickeners (0.1–0.5%), and neutralizers (0.01–0.05%). The antioxidants, anti-acne substances, and sunscreens are biologically active compounds that can be added to the product to increase its value and also depend on the property of the product. Also, to attract and meet the needs of the consumers, some cosmetic products contain fragrances and dye [7].

6.2.2 Cosmetic Industry Turning Towards Natural Products for Sustainable and Organic Way of Production

The cosmetics industry has changed dramatically in the last several years, moving toward a new paradigm where organic, ecological, and natural production methods are given priority. The primary driving force behind this transformation is customers' growing knowledge of the possible health dangers and environmental effect of conventional beauty products. Consumers are looking for plant-based substitutes and a better understanding of the components used in cosmetics; therefore, there is an increasing desire for transparency in formulas. This shift has been mostly attributed to environmental concerns, which have prompted cosmetic companies to embrace eco-friendly cultivation techniques and sustainable sourcing strategies. To reduce plastic waste, this calls for the ethical collection of natural ingredients and a switch to recyclable and biodegradable packaging materials. In addition, the pursuit of organic certifications from respectable agencies such as USDA organic or COSMOS highlights the dedication of cosmetic companies to strict organic production guidelines. Innovations in sustainable formulations, with a focus on plant-derived actives such as botanical extracts and essential oils, are being driven by research and development initiatives in green chemistry. Initiatives for corporate social responsibility, which include community involvement and ethical sourcing methods to promote sustainable farming and improve nearby communities, are also essential. Larger cosmetic companies are progressively adding terrestrial, marine natural sources, and sustainable product lines, indicating a wider industry recognition and embrace of these eco-conscious trends, even if niche businesses and start-ups have started this movement. This transition not only represents a fundamental shift in industry standards and customer expectations toward a more organic and sustainable future, but it also reshapes the cosmetics landscape.

6.3 Use of Marine Sources in Drug Delivery Systems of Cosmetics

The main target organ in cosmetic product is skin, and others include nails and hair. It says that the outermost layer of the skin, the epidermis, acts as physical, biochemical, and immunological barrier representative. Since our epidermis has a continuous layer of protein-enriched cells, which are embedded in an intercellular matrix, they are enriched with non-polar lipids, which thus help in preventing the skin from water and electrolytes loss, have immune defense mechanism, also protect the body against ultraviolet radiation, and thus prevent oxidative damage.

The ingredients that are used for specific purpose in cosmetics are active ingredients. These active ingredients need delivery vehicles to reach the target organs. Based on the function of active ingredients in combination with innovative delivery system, the skincare methods have been targeted on aging mechanisms by utilizing this combination. This increases the permeability of *stratum corneum* for cosmetic formulations. The active ingredients carrying delivery systems reach the target through transdermal delivery that enters the bloodstream by penetrating the skin. The most critical aspect of developing cosmetic product is incorporating the cosmetic formulations to delivery system for the product to work effectively. The delivery system in cosmetics can be incorporated directly into the final product using any aqueous phase or in the oily phase. The product is blended during manufacturing after the product has been incorporated into gels or they are dissolved in solutions or slurries. A cosmetic product must be effective, must give guarantee for uniform product, and also must have stability for long period. For each cosmetic product to have effective outcome, lot of new experiments are needed. The novel delivery system, such as lipid systems (liposomes), niosomes, microemulsion, and nanoparticles (nanoemulsion), is used on cosmetic formulations (Fig. 6.2).

Liposomes are used as delivery system in cosmetics due to their capacity of self-assembly, biocompatibility, and their ability to hold huge amount of drugs. Microemulsions are the potential carrier for active ingredients that serve as delivery vehicles in cosmeceutical applications. These microemulsions are optically clear liquids or gels that are isotropous. They have improved capacity to solubilize drug and also increase bio-availability. The nanoemulsion (containing nanoparticles) drug delivery system is transparent and semi-transparent and has high solubilization capability including leophilic drug. These nanoemulsion systems are used as formulations in foams, liquids, creams, and sprays of cosmeceuticals. Lipid-based nanostructures are used to increase the efficacy of cosmetics. The nanoemulsion is one of the delivery system using nanoparticals including nanocapsules, nanopigments, and nanoliposomes. Microparticles are used as delivery system and are applied as scrub and exfoliating agents. They help to free the pores and also exfoliate dead cells. They also increase cosmetic product's tactile proficiency. They help in solubilizing poor-soluble drugs and also help in decreasing dosage and toxicity. The microspheres such as microparticle delivery systems are found to be the best for cosmetic applications since they have the ability to protect the drug from catalyst and photolytic cleavage. There is another type of delivery system known as

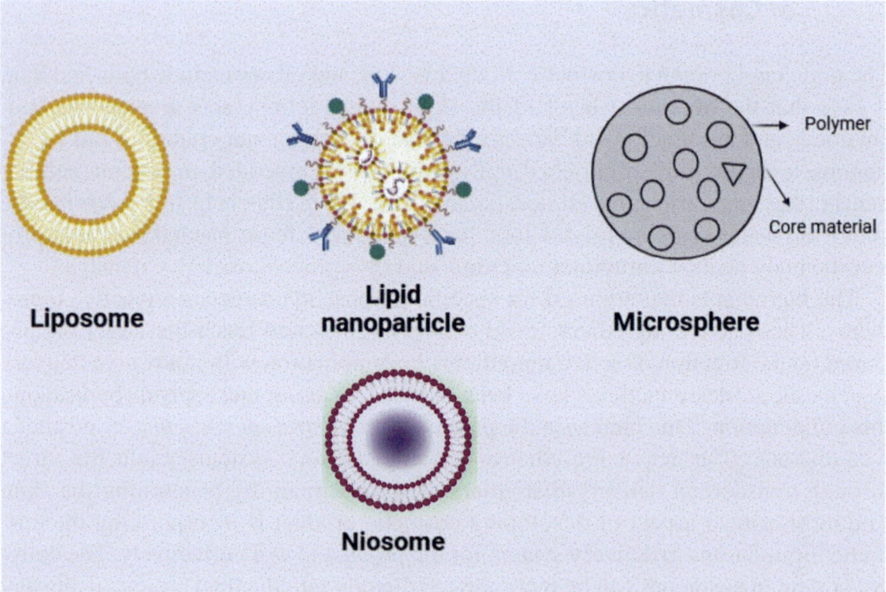

Fig. 6.2 Example of few drug delivery systems used in cosmetics

microsponge delivery system that contains porous microspheres. They are considered to have enhanced product performance. The hydrophobic drug and hydrophilic drugs are usually delivered using these microsponges. They also help to increase stability, enhance solubility, also improve pharmacological activity, and have sustained delivery with physical and chemical degradation protection [8]

Some of the marine bioactive compounds or components are also used as drug delivery vehicles in the cosmetic industries (Table 6.1).

1. The alginates from the marine sources such as brown seaweeds (such as *Laminaria digitata, Laminaria japonica, Macrocystis pyrifera, and Ascophyllum nodosum*) are used for its low toxicity, high bioavailability, lower extraction. The alginates are not only used as delivery vehicles but they are also used as excipients in tablets as they provide greater protection and stabilization for the drug. Alginate is also utilized in the synthesis of microparticles that can include various bioactive substances, including proteins [9].
2. The red algae contains the sulfated polysaccharide, and carrageenan that is potential delivery vehicle also can be used as potent emulsifier and stabilizer in cosmetic industries. As the oral delivery of drugs mainly depends on the physiochemical properties such as water stability and jellification, the carrageenan is the potent molecule that can be used as excipient [10].
3. The sulfated polysaccharide, fucoidans, found in the brown algae is a polymer chain of (1 -> 3)-linked α-L-fucopyranosyl residues. The fucoidans extracted from the seaweed *Fucus vesiculosus* are rich in fucose and sulfate, and the fucoidans containing the galactose, glucuronic acid, glucose, xylose, and mannose are obtained from *Sargassum stenophyllum*. They are delivered as nanoparticle delivery system and have the ability to synthesize hydrogels [11].
4. The construction of nanocarriers for biomolecules can be done using the ulvan, a sulfated polysaccharide extracts from green algae of *Ulva sp.* and *Enteromorpha sp*. In cosmetic industries, the ulvan is used for its potential in the development of new drug delivery systems. Ulvan in wide range of concentrations possesses low cytotoxicity levels [12].
5. The sources from marine animals also have the ability of being drug delivery systems as they are the excellent sources of polysaccharides, e.g., chitin-derived polymers and glycosaminoglycans. Because of its antibacterial action, chitosan can be used to create films that shield wounds from infection. It also demonstrates anti-inflammatory and anticancer properties. Because of all these biological characteristics, chitosan is a great option for excipients in DDSs and for building devices that need to come into contact with biological settings. Electrostatic interactions between a bioactive agent and a polysaccharide are essential for drug stability, protection, and release acceleration (or slowing) in chitosan-based DDSs [13].
6. The chondroitin sulfates are extracted from the marine animal sources such as cartilage of sharks and rays, salmon fish sea cucumbers, and also mollusks and cnidarians. They are suitable material to build drug delivery system as they have

Table 6.1 Some of the marine bioactive compounds or components that are used as drug delivery vehicles in the cosmetic industries

Bioactive compounds	Sources
Alginates 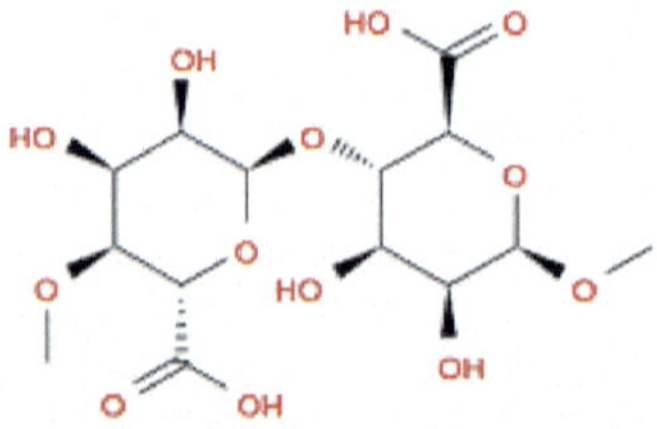	Brown seaweeds (e.g., *Laminaria digitata, Laminaria japonica, Macrocystis pyrifera, and Ascophyllum nodosum*) [9]
Carrageenan (sulfated polysaccharides) 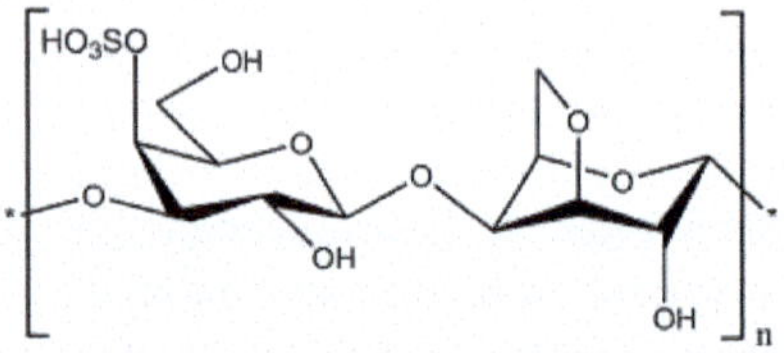	Red algae [10]
Fucoidans (sulfated polysaccharides) 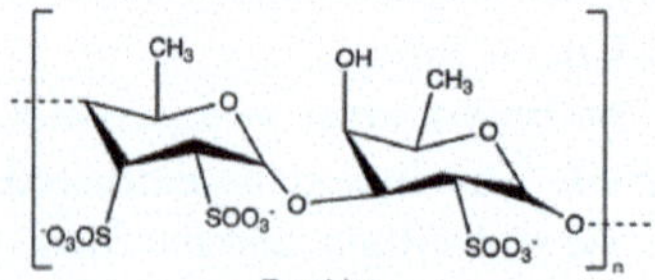 Fucoidan	*Fucus vesiculosus* and *Sargassum stenophyllum* [11]
Ulvan 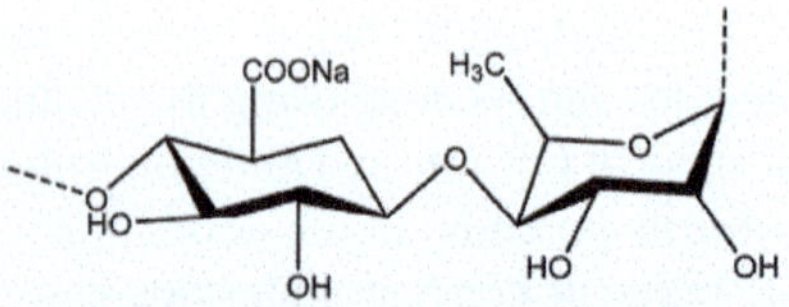	*Ulva sp.* and *Enteromorpha sp.* [12]
Chitin-derived polymers and glycosaminoglycans 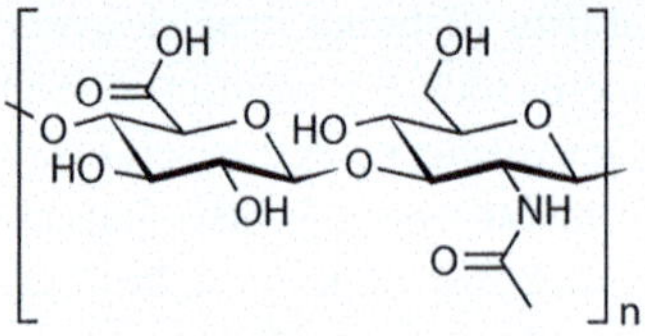 **Glycosaminoglycans structure**	Marine animals [13]
Chondroitin sulfates 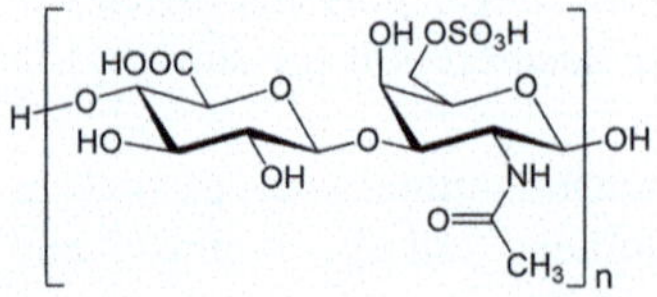	Sea cucumber, mollusks, Cnidarians, salmon fish,and from the cartilage of shark and rays [14]

(continued)

Table 6.1 (continued)

Bioactive compounds	Sources
Omega-3 lipids 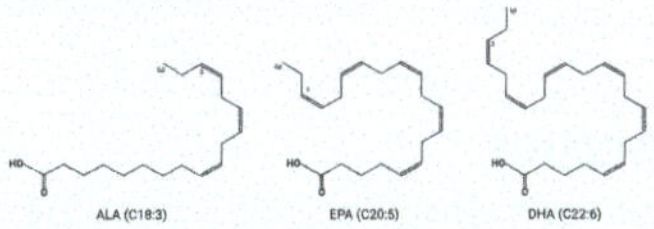 ALA (C18:3) EPA (C20:5) DHA (C22:6)	Fishes [14]

anticoagulant properties and also have water retention capacity. These are considered as a natural substitute for heparin [14].

7. Using encapsulation technology, the omega-3 lipids from marine sources such as fishes are used as delivering systems to incorporate into skin [14].

6.4 Marine Sources as Cosmetic Ingredients

6.4.1 Anti-Aging

All living beings have an irreversible physical growth and also aging metabolism. By seeing and observing the outer dermal texture of any individual, one can find their age, but because of many external factors in many people, their skin's texture changes. Skin aging is a slow process that happens because of external and internal factors. Some of the major signs that happen during skin aging are fine lines, wrinkles, skin thinning, and also skin sensitivity. For many people nowadays at very early age, their skin becomes more fragile and lose, and also it loses some of the protective layer of fat which ultimately makes the skin look more old and wrinkled. In order to overcome these wrinkling problems, to make skin look healthier, the cosmetic industry has brought many anti-aging creams. The radical theory and anti-radical substances are some main theories of aging, which would be a relevant choice when formulating anti-aging products [15]. By using the active compounds present in these anti-aging creams, the skin becomes healthier that makes an individual look young. Marine resources are also the potential source for producing active compounds that are present in these anti-aging creams. In our body, we have special cells known as fibroblasts. These fibroblasts together form extracellular matrix, which in turn together forms the dermis, i.e., our skin. When the aging process starts, the dermis undergoes many changes. During aging process, the collagen in the extracellular matrix of our skin breaks down and the total amount of collagen gets decreased. This process leads and links to the increased activity of matrix metalloproteinases (MMPs) that make our skin look thinner and weak, and wrinkling happens [15].The matrix metalloproteinases (MMPs) come under the family of endopeptidases that are capable of breaking down the constituents of the extracellular matrix and play a key role in biological processes [16]. These roles are linked to the fact that MMPs intervene in the regulation of the synthesis and secretion of cytokines and growth factors. In the skin, they apparently play a substantial role in wrinkle formation [17, 18].

6.4.1.1 Marine Resources with Antiaging Properties

Formation of wrinkle on the skin is caused by matrix metalloproteinase enzymes (MMPs). MMPs are available in several types such as MMP-2 and MMP-9. These matrix metalloproteinase enzymes are closely associated with wrinkle formation and play an important role in the digestion of extracellular matrix components on the skin. To avoid wrinkle formation, the activity of MMPs can be inhibited [18]. The inhibition effects in in vitro of phlorotannins that are derived from marine brown algae *Ecklonia cava* were studied by Kim et al. [19] for the first time. Marine-derived chitooligosaccharides and fatty acids are also studied against the activity of MMPs. Also, studies show that the extracts of *Ecklonia cava* induce the activity of MMP-2 and MMP-9 in the human dermal fibroblast. Few more studies have also demonstrated the prevention of skin aging by using compounds such as eckol and dieckol from *Ecklonia stolonifera* also show inhibitory effects on MMP-1 expression in human dermal fibroblasts [20].

Some of the algae such as Rhodophyceae (red algae), Phaeophyceae (brown algae), and Chlorophyceae (green algae) contain substances such as sulfated polysaccharides and polyphenols derived from phloroglucinol that has properties and capability to inhibit the activity of these MMPs; thus, these marine-derived substances can also be used in antiaging products [21].

Some of microalgae species such as *Arthrospira sp.* and *Chlorella sp.* were also established on skincare market as main ones for anti-aging property. Extracts of microalgae are also incorporated as main ingredients in anti-aging creams. A powerful novel microalgae compound, alguronic acid, plays an important role in regenerating and protecting the microalgae cell in harsh environments and conditions. This alguronic acid when scientifically tested helped in rejuvenating the skin, which significantly demonstrated the anti-aging property [22].

Traditionally to treat wounds and burns on skin, some red algae such as *Porphyra atropurpurea* and *Chondrus crispus* were used as they had regeneration properties on skin. Brown algae such as *Eisenia arborea*, *Ecklonia cava*, *Laminaria japonica*, or *Fucus* contain interesting raw materials that can be used in radical and antiradical theories. Some studies show that anti-aging effects are also seen in Sea muds [15].

Progerin, being physiological biomarkers of aging process, begins to increase at the age of 30 years. Sometimes in young adults, progerin accumulation in skin is seen when the skin starts to age, so the expression of the progerin is higher in young fibroblasts [23]. Some research works show that the extracts of an edible seaweed, *Alaria esculenta* (winged kelp), at low concentrations have significant effects in decreasing the progerin in older fibroblasts [24]. The ability of stimulating collagen synthesis in skin by the extracts of *Chlorella vulgaris* makes it suitable agent for anti-aging and wound-healing properties [25].

The sea cucumber *Protankyra bidentata* has anti-skin aging capacity and anti-elastase activity because of its high hyaluronidase inhibitory effects [26]. In case of UV-A and UV-B radiation exposure, the extracts of marine planktonic cyanobacteria *Chlorogloeopsis sp.* protect the keratinous tissue, thus prevent the free radical production, and lead to the prevention of photo-aging. To repair premature skin aging and to treat stretch marks, the skin tightening agents of *Spirulina* extracts are also used [27].

Terpenoids, or isoprenoids, are the largest and most widely distributed class of secondary metabolites found in significant numbers in marine algae. Diverse rare terpenoids are found in marine organisms, including algal terpenoids that replace bromo- and chloro-atoms. They find specific applications as antiaging agents due to their capacity to scavenge free radicals, or reactive oxygen species (ROS), which cause aging [28].

6.4.2 Anti-Acne

Acne or atopic eczema is the common problem that happens in young individuals. It happens when there is excess sebum secretion and when hair follicles get clogged under skin. This leads to outbreak in skin causing pimples or zits. Acne is of many types such as nodules, cysts, and truncal and fungal acne, which after getting reduced may leave pores and scars on the skin. This acne is caused by many factors such as improper diet, excess sebum production, by infection through microbes such as fungi and bacteria, inflammation, hair follicles closed by oil and dead cells, increased stress, and hormonal changes.

A common kind of acne is the acne vulgaris. In the formation of acne vulgaris (pimple), the major inhibiting bacteria in human adult skin is a gram-positive, non-spore-forming, and anaerobic rods *Propionibacterium acnes,* which usually resides within the sebaceous follicles. The bacterium *Propionibacterium acnes* releases lipases not only digests more amount skin oil and sebum that has been produced but also stimulates an intense of local inflammation that bursts of hair follicle, which results in acne formation [15]. Thus, to control the acne, at first controlling or inhibition of this *Propionibacterium acnes* is needed. To do, so many antimicrobial activities have been tested using different extracts from different organisms. Marine organisms also contribute in the production of anti-acne products. Marine brown algae *Sargassum macrocarpum* has Sargafuran, which showed possible anti-acne activity against *Propionibacterium acnes* [29].

Chitooligosaccharide derivatives from marine resources also have antibacterial compounds against these acne-causing bacteria [15]. Sea mud not only has various nutrients and minerals but also has antibacterial components. Mud therapy to skin is common practice in many places, and sea mud is used since it helps in maintaining pH, blood circulation, and mainly for acne repair. Fish oil contains alpha-linolenic acid (ALA), one of the important omega-3 fatty acids. Derivatives of ALA from the epidermis act on Toll-like receptors and, through influencing T cells, trigger caspase cascades, which relieve inflammatory dermatoses such as acne [30].

Extracts from five species of sea cucumbers such as H. impatiens from Lampung waters, *Holothuria scabra* from Makassar waters, *Pseudocolochirus sp.* from Makassar waters, *Stichopus vastus* from Lampung waters, and H. atra from Lampung waters showed antibacterial activity against three species of acne-causing bacteria such as *P. acnes, S. epidermidis, and S. aureus* [31].

Extracts of seaweed *Gelidium spinosum* are used in cosmetic industry known as Cream F3, an melanin-incorporated cream that has great wound-healing property and also has antibacterial activity against skin pathogens such as *Streptococcus pyogenes* and *S. aureus* sp. [32]. Also, some other marine resources such as seaweeds, microalgae, and sea minerals and also some of the coastal plants contain extracts such as vitamins and minerals that help in controlling acne.

6.4.3 Antioxidant

Free radicals are unstable molecules produced by various factors such as UV radiation, pollution, and also normal metabolic processes within the body. These free radicals can cause premature aging, fine lines, and wrinkles by causing damage to skin cells. To protect the skin health against these free radicals, the antioxidants plays a crucial role. These antioxidants work by neutralizing these skin-damaging free radicals and thus prevent free radicals from causing oxidative stress. The common antioxidants used in skincare products are vitamin C (for skin brightening), vitamin E (for skin moisturizing and healing), polyphenols (protection against UV damage), resveratrol (prevents skin damage), coenzyme Q10 (helps in reducing fine lines), and so on.

Some of the benefits of these antioxidants are:

- The antioxidants such as vitamin C, vitamin E, and flavonoids neutralize the free radicals by donating electrons to unstable free radicals and make them stable and thus prevent skin damage.
- The essential proteins that maintain skin elasticity and firmness are collagen and elastin. Due to oxidative stress, the collagen and the elastin break down. The antioxidants reduce oxidative stress in the skin and thus preserve these proteins, which help in reducing the appearance of fine lines and wrinkles.
- Antioxidants can also inhibit the melanin production, thus leading to more even skin tone, and help in reducing the hyperpigmentation or dark spots, e.g., vitamin C.
- The free radicals generated by the UV radiation can be neutralized by using antioxidants. The antioxidants cannot replace the sunscreen but can be used as complement along with sunscreens and help the skin against UV damage.
- Anti-inflammatory properties are possessed by many antioxidants, and they help in soothing and calming the skin of people with sensitive or reactive skin conditions (rosacea or eczema).

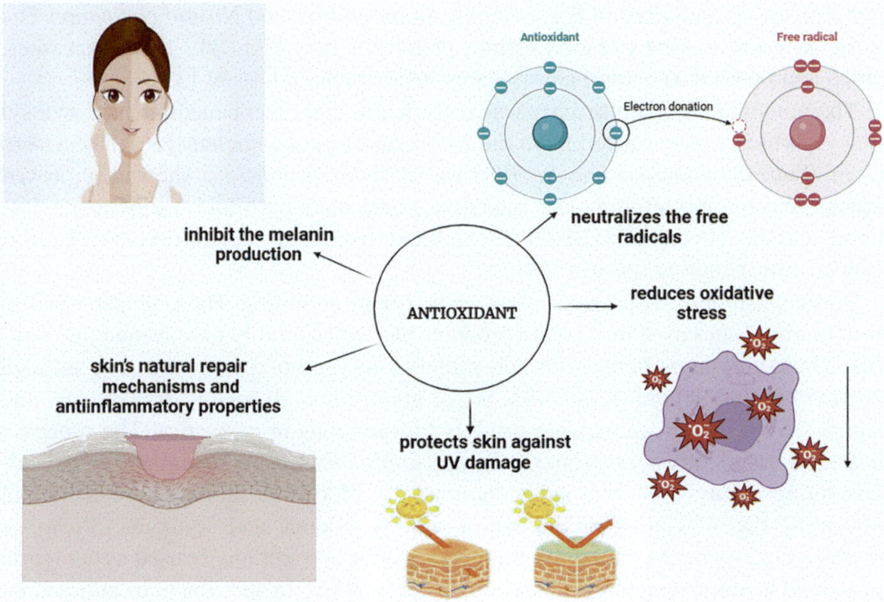

Fig. 6.3 Properties of antioxidants on skin

- Antioxidants can support the skin's natural repair mechanisms, by repairing the damaged cells, thus promoting overall skin health (Fig. 6.3).

Antioxidants can be formulated in natural or synthetic way. The synthetic antioxidants are inexpensive that contains butylated hydroxyanisole (BHA), propyl gallate, and butylated hydroxytoluene (BHT), but high consumption of the synthetic antioxidant may lead to health issues. In recent years, people prefer natural ones over synthetic products. Thus, the demand for natural antioxidants over synthetic ones is increasing. In cosmetic industries, the extracts and substances are from plants, grains, fruits, and marine resources.

A study shows that the bioactive compounds from the extracts of the *Nannochloropsis sp.* have anti-inflammatory and antioxidant activity [33]. To find out the anti-inflammation activity, the anti-collagenase and anti-elastase activities were done for nearly 38 marine animals, and among them, the extracts of *Patiria pectinifera* (blue bat star) showed the highest elastase inhibitory effects and the *Spirastrella insignis* showed both collagenase and elastase inhibition [26].

Many cyanobacteria have natural antioxidant sources such as vitamins A, B1, B2, and B12; thus, when they are exposed to oxidative stress naturally, they produce multiple efficient defense system against reactive oxygen species and free radicals. As a result, the exopolysaccharides of marine cyanobacteria *Arthrospira platensis* contains MAAs, carotenoids (beta-carotene, lycopene), scytonemin, vitamins (A, C, D, and E), and enzymes (catalase, superoxide dismutase, and glutathione peroxidase) that has antioxidant properties. The antioxidant compounds such as B complex group and vitamin E are rich in *Anabaena sp.* and *Nostoc commune*. The nitrogen-fixing marine cyanobacterium possesses very high B12 extraction rates, which can be used as natural antioxidants in cosmetic industries [33].

The potent antioxidant activity was recorded in diapolycopenedioic acid xylosyl ester A, which was extracted from the *Rubritalea squalenifaciens* [34]. The *Vibrio sp.* of *Phorbas tenacious* and *Bacillus sp.* of *Tedania anhelans* showed improved radical scavenging activity. The bioactive compound pyrrolo[1,2-a]pyrazine-1,4-dione and hexahydro-C7H10N2O2 extracted from the sponge-derived *Bacillus* showed antioxidant properties [34].

Marine fish-derived peptides show antioxidant activities. The collagen derived from marine fishes used in skincare products has the capability to scavenge free radicals. The active ingredients with anti-inflammatory properties comprising sulfated polysaccharides, saponins, lactones, sterol glycosides, glycogens, protamines, and peptides from sea cucumbers are used as formulations in cosmetics. The omega 3 polyunsaturated fatty acids from the *Perna canaliculus* having potential free radical-scavenging ability show that anti-inflammatory effects are useful for cosmetics in preventing the alleviating inflammation, allergy, edema, and erythema. *Stichopus japonicas* belonging to the family Stichopodidae was traditionally used as analgesic for wound healing, which is also used in cosmetics [35]. In addition to its antioxidant properties, "carrageenan"-derived sulfated galactans from red algae, including *Gigartina sp., Eucheuma sp., Kappaphycus sp., and Hypnea sp.,* is a significant ingredient in the cosmetics industry due to its anti-aging and anti-carcinogenic properties derived from its physical and functional abilities. It also has strong thickening and gelling qualities that help produce cosmetics such as toothpaste, skin lotions, binders, and shaving foams with a high texture and consistency [28].

According to Je et al. [36], alginates made from brown algae, such as *Fucus vesiculosus* and *Turbinaria conoides*, have antioxidant qualities and help fend off cutaneous diseases and skin aging. Furthermore, by blocking lipid oxidation, antioxidants can help preserve the organoleptic qualities of cosmetic goods by averting changes in flavor, texture, and odor. Lipid-soluble substances known as tocopherols are the body's primary source of vitamin E (a-tocopherol), with the maximum antioxidant activity found in the RRR-a-tocopherol stereoisomer (natural or D-tocopherol). "Vitamin E" refers to a class of fat-soluble compounds that include four main tocopherols and four main tocotrienols. They are composed of an aliphatic hydrophobic side chain (C-12) with two methyl groups at the terminal position and two more in the middle, and a polar chromanol ring, which are where the antioxidant capabilities are located. Tocopherols' four side chains are saturated, but tocotrienols' four side chains have three double bonds that are all next to methyl groups. Furthermore, it has been proposed that vitamin E could help prevent the deterioration of collagen in the skin. Numerous studies have demonstrated that when vitamin E is used as an excipient in conjunction with sun-blocking drugs, it can function as a protective agent to lessen the physical look of aging caused by sun exposure. A-tocopherol becomes UV-absorbing dimers and trimers when exposed to radiation; these compounds function similarly to sunscreen agents. Microalgal biotechnology produces tocopherols, which have been found in marine microalgae such as Nannochloropsis *oculata* and *Tetraselmis suecica* in varying compositions and quantities [28].

Phlorotannins have inhibitory effects on melanogenesis and provide protection against UVB-induced photooxidative stress. Skin is continually exposed to UV radiation.

Exposure to various environmental toxins and sun radiation, leads to oxidative stress. This may initiate a sunburn reaction that manifests in a clinical condition such as erythema. Thus, it is thought that regularly using products containing antioxidants will help prevent or minimize damage caused by UV radiation [28].

6.4.4 Topical Photoprotection

Our skin, i.e., the dermis, protects our internal organ from many external factors that cause damage, and one of the main external factors that damage our protective layer (skin) is the UV rays of sunlight. Ultraviolet radiation from the sunlight is the major factor that brings adverse effects to skin. Our skin has a pigment known as melanin. When the UV radiation from sun reaches the body, it affects this melanin that in turn causes burning of skin, thinning and aging of skin, tanning, and many other effects that spoil our skin's health. Naturally, our skin has repair mechanism against UV radiation-induced damages such as DNA repair enzymes, stress signaling, and antioxidants.

Increased expression of matrix metalloproteinases (MMPs) is by the activation of protein kinase C (PKC) by reactive oxygen species (ROS) causing skin damage, which also leads to the degradation of collagen. The expressions of MMP1, MMP3, and MMP9 can also be used by UVB in normal human epidermis in vivo, and among them, the two main photoaging contributors are considered to be MMP1 and MMP3 [37, 38]. To avoid such problems, many sunscreen products are introduced and commercialized by cosmetic industries. The main component in the formulation of sunscreen is its Sun Protection Factor (SPF), which has great potential to protect our skin against ultraviolet radiation B. The use of photoprotective agents mainly determines the delay of the effects of photoaging by decreasing the adverse effects of free radicals [1]. The photoprotective sunscreens are available as both gels and creams. Sunscreens are considered as cosmetics in Europe and as drugs in other parts of the world like in the USA. The sunscreens protect the human skin from short-term UV radiation exposures such as sunscreen and also adverse effects such as photodermatoses, photoaging, and skin cancer by absorbing UV radiation using UV filters. The UV filters present in the sunscreens are available in market as inorganic UV filters and organic UV filters. The inorganic UV filters such as titanium dioxide (TiO2) and zinc oxide (ZnO) are used as nanoparticles in the sunscreens, which might bring harmful toxicity to the human dermal cells. In order to prevent from the skin from toxic inorganic UV filters, organic UV filters are used extensively. Some of the organic particles used as UV filters such as butyl methoxydibenzoylmethane (BM-DBM), ethylhexyl methoxycinnamate (EHMC), octocrylene (OCR), and benzophenone-4 (BP-4) are used in sunscreens [39].

The mycosporine-like amino acid (MAAs) also known as palythine has the ability to protect the cells from UV-induced cell death and and also has protective effects on human cell, which can be isolated from marine sources that have great potential applications in cosmetic industries as anti-photoaging or photoprotective agents. Because *Porphyra umbilicalis* (red algae) contains MAAs that may absorb UV rays, these compounds can be utilized in sunscreen formulations. The genuine sunscreen is thought to be scytonemin, which is generated by cyanobacteria [40].

The exopolysaccharides from bacterium *Alteromonas macleodii* have been put forward as a potential photo protection candidate since they reduce ultraviolet B radiation aggression. Some UV-absorbing algal compounds are considered as new candidates for natural sunscreens and photoprotective and antiphotoaging agents as many studies in recent years have been focused on marine organisms as a source of natural bioactive molecules [41]. Sargaquinoic acid and sargachromenol from *Sargassum sagamianum* can also be considered as an effective therapeutic agents against UV-B since they clinically help in treating hyperproliferative skin diseases by using UV-B and topical reagents as combination of phototherapy [42]. Few of the prokaryotes such as Cyanobacteria, eukaryotes such as microalgae, yeasts, fungi, variety of macroalgae, corals, and other organisms have accumulated mycosporines and mycosporines such as amino acids, which can be used components for sunscreen. The presence of high phenolic content in methanol extracts of algae Corallina pilulifera (CPM) helps in reducing the expression of UV-induced MMP-2 and MMP-9 in human dermal fibroblast (HDF) cells, and based on dosage, it helps in inhibiting free radicals. A skincare product created from peptides found in sponges acts as an antiphlogistic and/or UV radiation protective agent. A natural, safe substance that is nontoxic and safe that can bind heavy metals and absorb UV-B radiation is sporopollenin. Several algae species have reported experiencing it. Cosmetics containing sporopollenin are linked to protecting skin from the sun and the onset of age-related wrinkles [28].

6.4.5 Moisturizers

Our body needs a natural moisture balance to keep our skin hydrated and rejuvenated. Many moisturizing skincare products are designed to nourish and hydrate in order to maintain a healthy skin. For normal physiological skin functioning, the water homeostasis in skin is required. Our skin's outermost layer, epidermis, i.e., stratum corneum, has natural moisturizing factor (NMF), which comprises lactic acid, urea pyrrolidine carboxylic acid, and many amino acids such as serine. Moisturizers work by confining the existing moisture present in the skin, and thus, they help in preventing the moisture from evaporating and keep the skin soft, smooth, and supple. For different skin types and concerns, moisturizers are available in market as creams, lotions, gels, and ointments. The most commonly and long-used moisturizing cosmetic ingredient was glycerol.

Long used as a cosmetic ingredient, glycerol also plays a crucial physiological role in stratum corneum hydration. Glycerol is naturally present in the epidermis and stratum corneum, and it can originate from the general circulation, the metabolism of fat in sebaceous glands, the conversion of phospholipids to free fatty acids (FFAs), and so on. In the latter case, glycerol is then transported through the epidermis through specific aquaporin water/glycerol channels. Aquaporin 3 is abundant in the epidermis, and its expression is stimulated by a number of molecules, including retinoic acid, along with an increase in glycerol transport activity. For dry or high dehydrated skin, it is important to formulate an ingredient that has the capability to withstand the dehydrating skin texture. Based on two principles, it is easy to formulate active cosmetic products for dry or dehydrated skin. The ingredients with film-forming properties and the ingredients with occlusive properties, for example, silicones, fatty acids, vegetable oils, hydrocarbons (petroleum, paraffin oil, and phytosqualane), waxes, and butters, can be used. To increase the capacity to capture water in stratum corneum, humectant agents are also used. The humectant agents are constituents of natural moisturizing factors (glycerin and moisturizers), and they retain the water within the outer epidermal cells by binding and penetrating the corneous layer [6].

Marine-derived ingredients have gained popularity in skincare due to their unique properties that offer hydration, rejuvenation, and overall skin health. Some of the common marine sources used in cosmetic industries for formulating moisturizers are seaweed, marine collagen, sea salt, and some marine exopolysaccharides. For skin to be healthy and to be moist, the water retention capacity of the skin layers must be in normal condition. The naturally occurring hydraulic acid helps to retain moisture in skin. The cyanobacteria can be used in cosmetics as they are potential moisturizing agents since these organisms have their own protective mechanisms against dehydration. Microalgae such as *Thalassiosira sp., Monodus subterraneus, and Nannochloropsis sp.* are rich in compounds such as serin, polyunsaturated fatty acids, and linolenic acid content, respectively. They help in restoring the transepidermal water [43]. The unsaturated and saturated fatty acids of the filamentous green microalgae, Cladophora glomerata, can act as emollient and helps in protecting our skin from excess water loss [6]. The brown algal cell wall is formed by the structural components polysaccharides, alginic acids, and alginate salts. The alginates are used as emulsifies in creams, lotions, soaps, shampoos, foam, and gels. They are also used as stabilizers and used on cosmetics as foundation for body lotion, facemasks, and body wash. They play a role in preserving water, thus repairing, and maintaining the skin structure [28].

The collagen from the skin, scales, and bones of some marine fishes (such as *Lagocephalus gloveri, Sardinella longiceps, and Priacanthus tayenus*) has skin-hydrating and skin-firming effects of cosmetic formulations [44]. Moisturizing and whitening cosmetic composition uses fermented extracts from *Thelenota ananas, Cucumaria frondosa, Apostichopus nigropunctatus, and Stichopus japonicus*. In addition, it improves suppleness and has anti-inflammatory, anti-wrinkle, and anti-atopy properties [35].

6.4.5.1 Marine Collagen

Collagen is a fibrous protein and main structural material of extracellular matrix (ECM) of all connective tissues in our body that has biocompatibility in which it can be used as ingredients in skincare products. Collagen provides mechanical strength and physiological functions to our tissues. The high water absorbing capacity makes the collagen a good component for texturizing, thickening, and gel formation of skincare products. The collagen is usually extracted from bovine and pigs, which can spread diseases such as bovine spongiform encephalopathy (BSE), transmissible spongiform encephalopathy (TSE), and foot and mouth disease (FMD). In order to avoid these problems, the cosmetic industries went for alternate sources such as marine organisms which contain collagen that has high biocompatibility, weak immunogenicity, and low risk of disease transmission. They are also suitable sources as they give high yield at low cost. Depending on the sources, collagen from marine can be categorized into two: (i) from marine invertebrate animals such as sponges, sea urchin, jellyfish, sea anemone, cuttlefish, squid, and octopus; and (ii) marine collagen from marine vertebrate sources that includes fish and marine mammals. The type 1 collagen is the most extracted protein from marine sources and has suitable factors that can be used in cosmetics.

The climatic changes during winter with low relative humidity cause dry skin or xeroderma, which can be treated using emollients and moisturizers. Emollients and moisturizers are sophisticated concoctions of cosmetic components that work to hydrate and soften skin by boosting its moisture levels and occlusively retaining water in the skin after application. Because of its various advantages as a natural humectant and moisturizer, collagen is a common constituent in creams and lotions used in cosmetic formulations. Collagen is a big molecular weight humectant that is not absorbed by the skin, in contrast to low molecular weight humectants such as urea and glycine. It stays on the stratum corneum, where it creates a thin layer that binds water vapor from the air to keep the skin hydrated while also obstructing water in the skin from evaporating (preventing moisture loss). This enhances the plumpness and softness of the skin.

Physical, chemical, and cosmetic qualities of *Chondrosia reniformis* (leather sponge) when compared to those of ordinary collagen, and it was found to be similar to conventional collagen. Furthermore, sponge collagen has the ability to raise skin lipids after topical application, suggesting that in addition to its humectant qualities, collagen can also increase sebum lipids and impart a glossy look to the skin. This collagen is acceptable for use in cosmetics and may be substituted for traditional collagen [45].

Good water retention potential was found in the collagen from salmon skin (*Salmo salar*) and codfish skins (*Gadus morhua*). This collagen ingredient from these marine fishes is suitable for dermal applications as it has no irritant or non-inflammation potential unlike conventional collagen. Along with moisturizing effects, the collagen-containing cosmetic products also help to reduce winkles in skin [45]. The collagen also helps to increase the production of hyaluronic acid, which simultaneously helps in boosting up the skin hydration, skin suppleness, and also reduction in wrinkle formation. Collagen peptides also used to treat UV-induced damage in skin and also help in the anti-aging property [45].

6.4.6 Skin-Whitening Anti-Tyrosinase Inhibitors

The amount of melanin produced by melanocytes in the skin determines the level of pigmentation in skin. The melanin helps the skin against photo-carcinogenesis. By the action of specific oxidative enzyme, the amino acid dihydroxyphenylalanine (DOPA) helps in the formation of melanin pigment. The enzyme dopa oxidase is responsible for the reaction of melanin formation. When there is overproduction of melanin, dark patches or spots appear on the skin, and this condition is known as hyperpigmentation often causes sun exposure, hormonal changes, or skin injuries. There are many other factors that regulate the production of skin pigmentation. Of these factors, the skin melanin content mainly depends on the level of tyrosinase present in human melanocytes. In order to overcome the hyperpigmentation condition various treatments, such as topical creams, laser therapies or chemical peels are under practice. Due to abnormal accumulation of melanin, the pigmentation disorders such as melisma, freckles, and lentigo senilis occur, and to treat this, skin whitening is needed. The inhibition of melanogenesis happens only when the regulation of tyrosinase is limited. The whitening of skin has become the most demand in cosmetic formulations. Since the enzyme tyrosinase plays a key role in melanin production, many formulations include extracts or compounds that have anti-tyrosinase activity. The compounds with anti-tyrosinase activity help in regulating melanin production and help in reducing hyperpigmentation by interfering with the function of this tyrosinase enzyme. These substances or extracts with anti-tyrosinase activity are often used in skincare products targeting skin-lightening or skin-brightening effects.

6.4.6.1 Role of Tyrosinase in Melanin Biosynthetic Pathway

The most important rate-limiting enzyme in the biosynthetic pathway of melanin production is the enzyme tyrosinase. This enzyme converts the tyrosine to dihydroxyphenylalanine (DOPA), which is later on converted to dopaquinone. Then, the dopaquinone by autoxidation gets converted into a brown-black pigment (eumelanin) in the presence of dopachrome tautomerase. On the other hand, the dopaquinone in the presence of glutathione or cysteine can also be converted to cysteinyl DOPA, which forms yellow-red pigment (pheomelanin). By the mRNA expression, this tyrosinase and tyrosinase-related proteins are regulated by the signal transduction pathways that are responsible for increasing melanin production. Thus, the regulators of mRNA expression of tyrosinase are also used as natural whitening agents along with the inhibitors regulated of these enzymes. The MITF (microphthalmia-associated transcription factors) regulates the complex process of melanogenesis; thus, compounds that help in inhibiting the MITF are also used in cosmetic as skin-whitening agents [46] (Fig. 6.4).

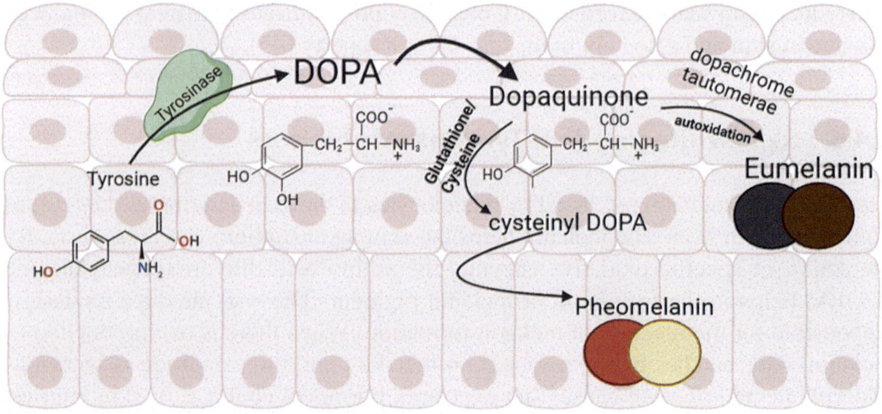

Fig. 6.4 Melanin biosynthesis pathway

6.4.6.2 Marine Sources with De-Pigmenting Properties

Marine resources being rich in many compounds also have the compounds or extracts that have anti-tyrosinase activity. Cyanobacteria from marine source have potential compounds with tyrosinase inhibitory activity that helps in the treatment of pigmentation disorders. A study shows that *Oscillatoria agardhii* was found to induce tyrosinase inhibition in 55% of the control [47]. Spirulina also can negatively regulate tyrosinase gene expression resulting in surpassing melanin synthesis. The extracts from the G1-5 strain of *Nannochloropsis* sp. isolated from the southern West Sea of Korea contained bioactive compounds such as PUFAs (including EPA), and carotenoids such as astaxanthin, canthaxanthin, β-carotene, zeaxanthin, violaxanthin, and phenolic compounds have anti-melanogenic activities. This NG15 extract has a skin-whitening effect as it influences tyrosinase activity and intercellular melanin synthesis [33]. The pure extracts of the Nannochloropsis *oculata* contain an tyrosinase substance, zeaxanthin. The zeaxanthin from these marine microalgae was patented to be used in creams. The bioactive compounds such as canthaxanthin extracted from marine *Nannochloropsis salina, Nannochloropsis oculata, and Nannochloropsis gaditana* have potential activities against skin tanning (potential use of marine microalgae). The potential skin-whitening agent phlorotannin and 7-phloroeckol extracts from *E. cava* (marine brown seaweed) help in antityrosinase activity [48].

The penis fish, *Urechis unicinctus,* has the strongest tyrosinase inhibitory effect among the 38 examined marine animals [26]. The next highest marine organism that has tyrosinase inhibition was the Asian murex (sea snail) *Chicoreus asianus,* and the extracts of this marine species showed 42 ± 5.10% tyrosinase inhibition activity. In the same 38 marine animals when they were tested for DOPA oxidation inhibitory activity, the marine sponge *Petrosia corticata* showed the highest activity. From this experiment, they concluded that *Urechis unicinctus* inhibits melanin synthesis since it has both tyrosinase inhibition and DOPA oxidation inhibitory activity [26]. Algae develop certain efficient protective systems against reactive oxygen species and free radicals under oxidative stress conditions producing compounds that can act in cosmetics. These compounds produced by algae have similar actions to organic and inorganic filters used in the market that act against the harmful effects of UV radiation. The extracts of brown seaweed, *Fucus vesiculosus,* help in reducing the dark circle appearance under the eye. This works by stimulating the expression of heme oxygenase-1 (HO-1). Heme oxygenase is a molecule that eliminates the heme production on the skin by the removal of heme catabolites [49].

The extracts of marine algae such as *Endarachne binghamiae, Sargassum siliquastrum, Ecklonia cava, Ishige okamurae, Sargassum polycystum, and Padina tenuis* are known to have more effects for anti-melanogenesis activities [50]. The *Sargassum polycystum* contains the phlorotannins which showed potent anti-melanogenesis effects in both cell-free mushroom and cellular tyrosinase assays. Marine plants and algae have polyphenolic tyrosinase inhibitors that showed moderate success for skin-whitening properties.

The potential anti-pigment treatment was analyzed using the secondary metabolite fucoidan of brown seaweed *Fucus vesiculosus*. The downregulation of melanin synthesis by activation of the extracellular signal-related kinase (ERK) pathway was reported in the sulfated polysaccharides of *Fucus vesiculosus.* The most effective and widely used depigmenting cosmetic ingredient is kojic acid (5-hydroxyl-2 hydroxymethyl-4-pyrone) for skin-whitening properties. It is a naturally occurring hydrophilic product, which is extracted from marine fungal species such as *Acetobacter sp., Aspergillus sp.*, and *Penicillium sp.* isolated from acetone extracts. This kojic acid reduces hyperpigmentation usually used at the concentration from 1% to 4%. The marine sponge, *Geodia japonica,* has antimelanogenic compounds such as fucoxanthin, geoditin A, and isomalabarcane triterpene [1].

6.5 Cosmetic Ingredients for Sensitive Skin from Marine Sources

Sensitive skin is a condition characterized by heightened reactivity and vulnerability to various environmental factors, skincare products, or substances. Individuals with sensitive skin may experience redness, itching, burning sensations, or dryness in response to stimuli that typically do not affect others. Common triggers include harsh chemicals, fragrances, weather extremes, certain fabrics, and certain skincare ingredients. Managing sensitive skin often involves using hypoallergenic and fragrance-free products, avoiding potential irritants, and adopting a gentle skincare routine. Individuals with sensitive skin should consult with dermatologists to determine specific triggers and develop a personalized skincare regimen.

Sensitive skin toward cosmetic products is a common concern, as many skincare and makeup items contain ingredients that can trigger adverse reactions in individuals with heightened skin sensitivity. A sensitive skin person must consider fragrance-free and hypo-allergic cosmetic products, must do patch testing before the cosmetic product application, and can avoid the usage of harsh chemicals to avoid skin irritation, rashes, and redness.

The marine resources can offer various benefits for sensitive skin due to their natural, soothing, and hydrating properties. Some of the marine-derived sources that are found in skincare which are beneficial for sensitive skin are [25]:

- The combination of extracts from *Ascophyllum nodosum* was found to reduce tingling sensation, resiliency, and skin comfort for sensitive skin types.
- The neurogenic inflammation was reduced using the extracts of *Laminaria ochroleuca* by decreasing the pro-inflammatory cytokine release by mast cells simultaneously by increasing MT-1 and SOD antioxidant enzyme production. This extract of *Laminaria* also can improve the skin barrier function.
- *Chlorella vulgaris* contains sulfated polysaccharides that can prevent free radical accumulation and activity, and it also acts as a protective agent against oxidative and radical stress systems.
- Cosmetic formulations contain ingredients extracted from the brown macroalgae such as Fucales (*Ascophyllum Nodosum*) and Laminariales (*Laminaria ochroleuca*) for sensitive skin application. *Ascophyllum nodosum* and *Laminaria ochroleuca* also serve as active ingredients for cosmetic applications (Figs. 6.5 and 6.6).
- The spirulina due to their high advantages as anti-inflammatory and antioxidant effects toward skin health is used in cosmetic formulations, especially for sensitive skin.

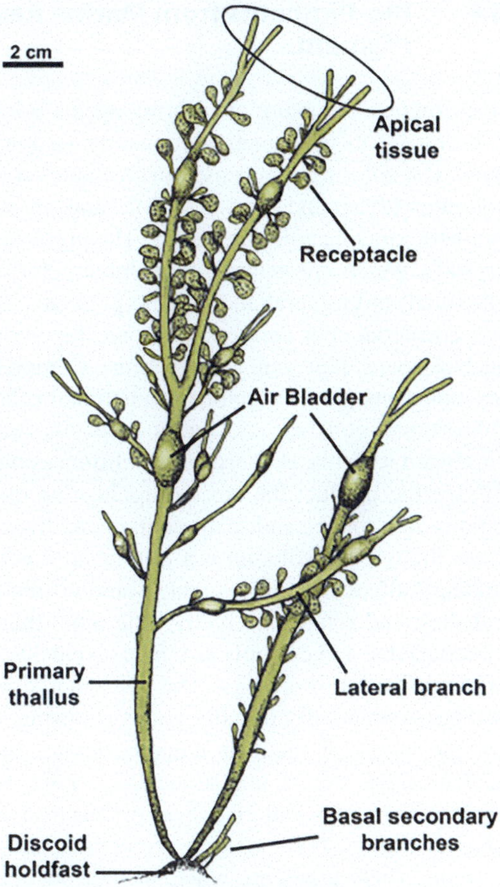

Fig. 6.5 *Ascophyllum nodosum* [51]

Fig. 6.6 *Laminaria ochroleuca* [52]

6.6 Bio-Pigments from Marine Resources as Colorants/Pigments

In cosmetics, the using of coloring agents is common for accessory purposes. The coloring agents in cosmetics are mainly used in lipsticks, nail polishes, and also hair dyes. The coloring agents used in cosmetics are categorized into two groups such as colorants (dyes) and pigments. Colorants are water and oil-soluble and are usually synthetic organic coloring agents. The pigments are in the form of fine powder and crystals, which are generally insoluble coloring agents. The pigments are further classified into mineral and organic pigments. Colorants are added to foods, drugs, and cosmetics. For aesthetic purpose, the colorants are added in the skin products such as soaps, hair tonics, and lotions. Comprising half the total weight of cosmetics, colorants play a significant functional role. Certain cosmetics, such as rouges (blush), nail polishes, and eyebrow pencils, are really nothing more than colorants combined with one or more other substances that act as binders, carriers, or diluents to give the product the desired application qualities. Products such as aftershave lotions, hair tonics, and soaps contain additives purely for esthetic reasons. In many cases though, the colorant is a significant functional part of a cosmetic, often comprising half of its total weight. Some cosmetics including eyebrow pencils, nail polishes, and rouges are really little more than colorants mixed with one or more materials that serve simply as binders, vehicles, or diluents to give the product desirable application properties. The consumer's demand for cosmetics toward natural and environmentally friendly goods is raising; thus, the cosmetic industries mostly use the natural pigments or natural colorants as marketing strategy. Marine species such as algae, corals, and some bacteria are the source of biopigments, which are organic chemicals that give these organisms their vivid hues. They are essential for photosynthesis, UV radiation protection, and even as defense or communication systems. These pigments, which include phycobilins, carotenoids, and chlorophylls, have drawn interest because of their natural, environmentally beneficial, and frequently antioxidant qualities and because they may find use in food, cosmetics, and pharmaceuticals. Because of these chemicals' sensitivity to environmental changes, researchers are investigating their possible applications as natural colorants, antioxidants, and even environmental health indicators.

6.6.1 In Lipsticks, Nail Polish, Eyeshadows, and Dyes

Lipsticks are beauty cosmetics that were evolved centuries ago. It is available in different colors based on different skin tones and preferences. The characteristics of lipsticks are as follows: Under room temperature, they remain solid, must be hard as well as soft enough to make layer on lips, must be even-toned, and must long last after applying on the lips. Lipsticks are generally a composition of wax (for texture), oil (for moisture), and pigments (for color) and also contain preservatives, antioxidants, fragrance, and alcohol (as solvents). Pigments are the main ingredient in lipsticks since they fulfill the purpose of this cosmetic. The colorants or the pigments used in the lipsticks can be classified as two: organic and inorganic. The inorganic pigments or colorants are ZnO, TiO_2, and pearl powder. These inorganic or synthetic pigments offering wide range of colors and effects are produced in laboratories. Though they produce varied desired colors, the synthetic pigments have some disadvantages as they have allergic reaction, toxicity concerns, and also environmental impacts. Microplastics are indeed a concerning component found in some synthetic pigments used in cosmetics, including lipsticks. To avoid such harmful health risks by synthetic pigments, cosmetic industries are shifting towards natural pigments as lipstick ingredients due to demand for safer, eco-friendly products. These natural pigments might have certain limitations in color range or stability when compared to synthetic pigments.

Marine organisms have been historically utilized as sources of natural dyes and colorants for cosmetic products. The vibrant and diverse hues found in certain marine species have made them desirable for enhancing the aesthetic appeal of various beauty and cosmetic formulations. In lipsticks, the DHA and EPA of PUFA from fishes are used as cosmetic formulation.

Cosmetic industries use natural pigments derived from various sources such as plants, minerals, and insects. The tyrian purple color extracted from marine mollusks such as *Murex sp.* can be used in cosmetics for different shades ranging from purple to blue [10].

Phycobiliproteins are found in the phycobilisomes of marine algae such as Rhodophyta, Cryptophyta, and cyanobacteria. These phycobiliproteins contain fluorescent protein pigments like red or pink pigment (phycoerythrin), bluish-green pigment (allophycocyanin) from *Gracilaria gracilis,* and blue pigment (phycocyanin) that has anti-oxidant and anti-free radical properties. These phycobiliproteins have intrinsic brilliant color with high fluorescence. Since the phycobiliproteins are nontoxic and noncarcinogenic with antioxidant properties, they can be used as natural dyes and also can be used as colorants in lipsticks, nail polish, eyeliners, and eyeshadows. It also acts as photoprotective agents [53].

The blue pigment phycocyanin is marketed in Japan as cosmetic colorant, which can be used in lipsticks, eye shadows, and nail polishes. The fat-soluble accessory pigments such as yellow, orange, or red in color produced by carotenoids from cyanobacteria. Along with producing colors for dyes and colorants in cosmetic beauty products, they are also used as food color.

In the formulation of colored cream and cosmetic products, the biotechnologically derived R-phycoerythrin is used as natural dye [28].

The majority of dyes used in the cosmetics business nowadays are synthetic; the remaining hues are derived from oxides and minerals and to a lesser extent from plants due to stability concerns. Since some of the dyes already in use cause allergies and some consumers prefer cosmetics produced with more naturally occurring ingredients, the industry is looking for alternative colors. The cosmetics sector receives phycobilins and carotenoids from the marine environment, which implies a broad spectrum of blue, yellow, orange, and red colors.

The primary photosynthetic accessory pigments of some cyanobacteria and eukaryotic algae that are members of the Glaucophyte, Crystophyte, and Rhodophyte families are tetrapyrroles known as phycobilins. These pigments are proteins known as phycobiliproteins that are covalently bonded to one another. Phycobilisomes are present in thylakoid membranes of cyanobacteria. Phycoerythrin, phycocyanins, phycoerythrocyanins, and allophycocyanins are the four groups into which they are divided. Since they are all soluble in water, their application in cosmetology will be restricted to altering organoleptic characteristics; they cannot be used for makeup. Of all the phycobiliproteins, R-phycoerythrin (R-PE) has received the greatest amount of research. Red microalgae are used to extract R-phycoerythrin, which has a wide range of potential uses, most notably in immunodiagnostics. *Corallina officinalis* is a fascinating alga that can be used in beauty products due to the presence of red pigment phycoerythrin. Moreover, *Gracilaria longa*, *Gracilaria fisheri*, and *Palmaria palmata* are frequently used to extract R-PE (R-phycoerythrin). Because of their rich composition, cyanobacteria may be a more intriguing source of pigments on an industrial scale. Many cyanobacteria species, including *Porphyridium sp., Synechocystis, Arthrospira platensis* (Spirulina), have been identified to be sources of phycocyanin. Carotenoids, the second major family of colors helpful in cosmetics, are likewise derived from cyanobacteria. Carotene, astaxanthin, lutein, lycopene, canthaxanthin, and fucoxanthin are the primary carotenoids. As the best natural source of astaxanthin, the microalga strain *Haematococcus pluvialis* (Chlorophyceae, Volvocales) is thought to have. Microalgae are used to extract pigments such as carotenoids, phycobiliproteins, and chlorophylls. The phycoerythrin and phycocyanin are produced by the red algae *Porphyridium cruentum* and cyanobacterium *Spirulina platensis,* respectively. These pigments are used in cosmetic industries as dyes [6] (Figs. 6.7, 6.8, 6.9 and 6.10; Table 6.2).

Fig. 6.7 *Corallina officinalis* [54]

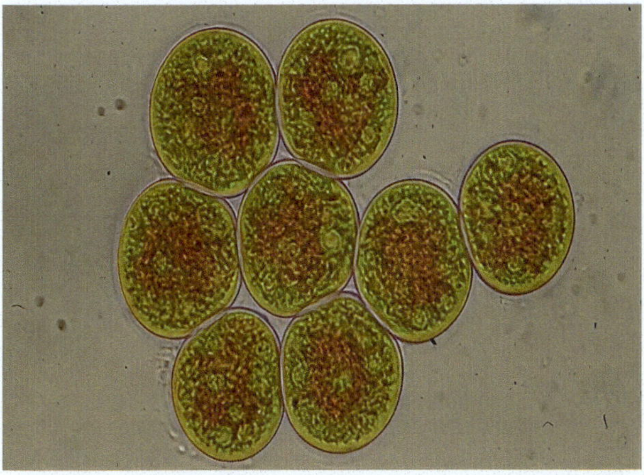

Fig. 6.8 *Haematococcus pluvialis* [55]

Fig. 6.9 *Gracilaria gracilis* [56]

6.6.2 Perfumes

The marine sources have odor molecules that can be used in perfumes. Many different algae species produce varied odors based on their taxa. The algal species and cyanobacteria produce various odorous components such as carotenoids, terpenoids, fatty acid derivatives, and sulfur compounds. Some species of cyanobacteria such as *Lyngbya, Anabaena, Symploca,* and *Oscillatoria* have odor-producing compounds like geosmin. It has been reported that the green algae *Ulothrix fimbriata* can produce β-cyclocitral (non-carotenoid group) and β-ionone that have tobacco odor and woody, fruity odor, respectively. The PUFA derivatives of cyanobacteria have odorous substances that have fishy, cucumber, and rancid odors. These odorous compounds can be used in cosmetic industries [6].

Fig. 6.10 *Arthrospira platensis* [57]

Table 6.2 Natural pigments extracted from marine algae used in cosmetics [6]

Pigments	Producing color	Marine source
Phycoerythrin	Red or Pink Pigment	*Corallina officinalis, Gracilaria longa, Gracilaria fisheri*, and *Palmaria palmata*
Phycocyanins	Bluish-green pigment	*Porphyridium cruentum, Synechocystis, Arthrospira platensis*
Allophycocyanins	Blue pigment	*Gracilaria gracilis*
Carotenoids (carotene, astaxanthin, lutein, lycopene, canthaxanthin, and fucoxanthin)	Yellow, orange, red, and purple	*Haematococcus pluvialis*

6.7 Cosmetic Active Ingredients from Marine By-Products

Marine resources though being promising sources of novel bioactive compounds, concern on sustainability must be considered in cosmetic industries. The bioactive substances can be extracted not only directly from marine organisms but also from marine by-products, which could potential source for formulating cosmetic ingredients. Marine by-products contain valuable protein, lipids, vitamins, minerals, and also bioactive compounds that will be useful for human health. By-products are the raw materials or unused parts, i.e., marine discards, which can be turned into useful or valuable cosmetic products. The marine-by-products such as viscera, heads, trimmings, bones, cartilage, tails, skin, scales, blood, shells, carcasses, or damaged fish have unique functionality and marked biological activities and thus can be used to convert the raw materials into fine chemicals.

Finfishes by products are the head (9–12%), intestine (12–18%), skin (1–3%), bones (9–15%), and scales (12–18%) [This composition varies based on the size, category process, species, and technology used], are removed during fish processing, and can be used for further use. The finfish by-products contain valuable ingredients such as collagen, gelatin, fish oil, and calcium phosphates, which will be useful for cosmeceutical applications such as moisturizing agents and anti-aging agents.

Collagen type I, including collagen recovered from fish scale, bone, scale, and swim bladder, is more desirable for cosmeceutical as human skin is predominantly by the same type of collagen, so it is compatible with human skin. Since collagen has water absorption activity, it is used as moisturizers in cosmetic industries [19]. Type I collagen is the most desirable type of protein as human skin has the same type of collagen. This type of collagen can be derived from the fin-fish by-products such as swim bladder, bones, and scales of fishes that have great demand as an ingredient in cosmetic products as they are rich in a collagen matrix.

Gelatin is a derivative obtained when the collagen is partially hydrolyzed with acid. Gelatin is a useful bioactive compound, can be used in cosmetic products such as body lotions, shampoos, hair sprays, face creams, sunscreens, bath salts, and bubble baths because of its gel-forming abilities and thickening activities, and also helps to promote skin health. Gelatins can be derived from finfish by-products such as skin of yellowfin tuna, bones of grouper and red snappers, swim bladder of tuna, and shark cartilage. Gelatin also has antioxidant properties, e.g., potential antioxidant hydrolysate extracted from thornback rays (*Raja clavata*) and substantial antioxidants from the skin gelatin of skatefish (Okamejei kenojei). The hydrolyzed collagen from the milkfish scale can act as moisturizers, anti-aging agents, and skin-whitening agents [19].

In formulating cosmetic products, oils are needed for moisture content. Fish oil (from marine by-products) can be used as ingredients. Salmon by-products being rich in oil can be used to obtain fish oil from the head, viscera, skin, and trimmings. The fish oil is not only as combining agents but also is used to promote skin health by reducing acne included in cosmetic products. The omega-3 fatty acids (in small dosage) from fish oil have effects on the skin reaction due to UV radiation. Because of the MMP-1 suppressing activity, the skin and liver oil of Antarctic fish (marbled rock cod) can be used as bioactive nutrients to improve skin health. Natural calcium phosphates (CaPs) are inorganic components that are present in the bones and teeth of marine vertebrates. CaPs are used in cosmetic products such as skin care, hair care oral care, and deodorant care products. More studies can be done on these CaPs from marine by-products as they may contain variety of trace materials, which can be used in cosmetics.

By-products such as carapace and head of crustaceans such as crabs, prawns, shrimps, and lobsters during seafood processing may possess cosmetic ingredients. The cosmetic ingredients are chitin, chitosan, and astaxanthin. Chitin is used as good delivery systems and active ingredients in cosmetic industries. Chitosan and its derivatives have the potential to function as a vehicle for the transport of active ingredients, and they combine well with those ingredients to produce the best possible effects on skin health in people. Chitosan is occasionally fashioned to be smaller or more conventionally shaped as the delivery mechanism for topical treatment. In order to create UV-protective cosmetic emulsions, natural and pure annatto, and saffron—sunscreen agents—have been encapsulated using chitosan nanoparticles as an active carrier and stabilizer. At room temperature, such emulsions demonstrated good color stability and storage for up to 90 days. The astaxanthin is derived from carapaces of crustaceans by various methods such as green extraction technique, organic solvent, and oil extraction techniques. Astaxanthin's topical treatment and oral administration provide brief benefits for skin health. It may have anti-aging, anti-wrinkle, moisturizing, anti-cancer, anti-ray, anti-eczema, and wound-healing properties. Astaxanthin functions as an antioxidant and stimulates the growth of skin cells, according to an in vitro experiment. The antioxidant action of astaxanthin may function by promoting the cellular synthesis of moderate amounts of reactive oxygen species (ROS), as an alternative to directly scavenging free radicals. An in vivo experiment showed that astaxanthin could still be beneficially applied topically even when taken orally. Using astaxanthin-liposomal topically before exposure to UV light can help prevent skin thickening [14].

Astaxanthin-liposomal pre-administration also decreased UV-induced collagen loss. In skin exposed to UV light, topical administration of astaxanthin-liposomal cationic lipid inhibited melanin production. Topical astaxanthin-liposomal prevented UV-induced skin damage, according to an in vivo experiment. For skin beauty applications, astaxanthin may also be administered as an emulsion or film.

Astaxanthin-carboxymethyl chitosan nanoemulsion, which was generated by the low-energy emulsion phase, demonstrated improved stability and good skin permeability without having any harmful effects on skin cells. Furthermore, squid skin astaxanthin-collagen film was demonstrated by an in vivo experiment. Shrimp *Litopenaeus vannamei* shell astaxanthin at doses of 3–50 μg/mL has dramatically suppressed TYR in a dose-dependent manner. Two oxygenated groups that can chelate TYR's copper may be the cause of the TYR suppression of astaxanthin from shrimp feces. The inhibition of tyrosinase was recorded using protein hydrolysate of *Metapenaeus monoceros* shrimp by-product. Thus, these compounds extracted from by-products of shrimps have skin-whitening effects. The chitin nanofibrils, oligochitosan, tetracycline, and erythromycin from the crab shell help to prevent skin dryness and also have anti-inflammatory and antioxidant activities. The commercially important crustaceans such as *P. monodon, P. indicus, and M. monoceros* have carotenoids in head wastes [14].

Followed by fishes and crustaceans the most important delicacy from the marine world is mollusks. The major bivalve mollusks for international trade are oysters, scallops, mussels, and clams. The outer shell by-product of mollusks can be used as valuable ingredients in various industries. Studies show that scallop shell extract can be used to improve skin health as it has high collagen content as they have collagen-promoting metabolism. Powdered pearl shells or powdered nacreous shell layers are becoming a popular cosmetic component.

In pigs with deeply burned skin, the water-soluble nacre extracted from the finely ground pearl oysters (*Pteria martensii*) aided in wound healing. The skin was returned to its normal state by the topical (spraying) administration of water-soluble nacre directly on the burned area, which encouraged the synthesis of collagen. *P. margaritifera* (black-lip pearl oyster) shell powder was reported by Agarwal et al.

Table 6.3 List of few Marine-by-product extracts used in cosmetics

Marine by-product extracts	Marine sources	Cosmetic uses
Hydrolyzed collagen	Scales of milkfish	Moisturizers, antiaging agents, and skin-whitening agents [19]
Gelatin	Skin of yellow fish tuna, swim bladder of tuna, shark cartilage, bones of groupers and snappers, etc.	Antioxidant properties [19]
Fish oil	By-products of salmon (viscera, head, trimmings, etc.)	Reduces acne, promotes skin health [19]
Chitin and chitosan	Head and carapace of crustaceans	As drug delivery systems and active ingredients [14]
Astaxanthin	Carapace of crustaceans	Antiaging, anti-wrinkle, moisturizing, anti-cancer, anti-ray, anti-eczema, and wound-healing properties [14]
Water-soluble nacre	Shell of pearl oysters	It helps in the synthesis of collagen and has wound-healing properties [58]
Hydrogel-trapped nacre pearl powder	Shell of black lip pearl oyster	Anti-inflammatory and anti-apoptotic properties [58]
N-(2-hydroxyl) propyl-3-trimethyl ammonium chitosan chloride (HTCC)	Squid ink	Good moisture retention capacity, prevents oxidative stress [19]

[58] to have minimal cytotoxicity at high doses and to show no signs of oxidative stress in primary skin fibroblast cells and epidermal skin cells. The anti-inflammatory and anti-apoptotic properties of the hydrogel-trapped nacre pearl powder were studied [23].

The by-products from the squids such as squid pens and squid ink have functional products like N-(2-hydroxyl) propyl-3-trimethyl ammonium chitosan chloride (HTCC) and squid ink polysaccharides that help in cosmeceutical functions such as good moisture retention capacity and help in preventing oxidative stress in human dermal fibroblasts, respectively [19] (Table 6.3).

6.8 Use of Bioactive Compounds from Marine Resources for Hair Care

Humans at young age face many hair problems due to stress, environmental conditions, food habits, and hereditary. Common hair problems faced by individuals are hair loss, decreased hair growth, hair thinning, dry scalp, premature graying, etc. In order to avoid these problems, hair care products containing needful ingredients can be used to prevent hair damage. Hair growth in humans not only depends on external factors by mainly internal factors such as hormones, nutrients, minerals, and vitamins. Our body needs hair for external protection and also as an accessory need. It also protects our skin from damage and maintains body temperature. The hair has hair root and shaft. Beneath the layers of skin, the hair root is present surrounded by hair follicle (sheath of skin and connective tissue). This follicle is also connected to sebaceous glands. The tiny muscle near the hair follicle and the nerves sense the hair movement. The hair has the growth phase also known as anagen phase. The growth phase is further classified into two phases as transitional phase and resting phase. We regularly lose hair because hairs keep going into the resting period and then falling out. A healthy adult's daily hair loss may range from 70 to 100 hairs. However, because new hair constantly grows to replace the old, this normal hair loss goes unnoticed. If the hair roots are harmed during the growth phase or if many hairs enter the resting phase simultaneously, the rate of hair loss may dramatically rise. That area of skin goes bald if no new hair grows to replace the lost hair. Alopecia is the term for this kind of hair loss, regardless of the size of the bald spot or whether it affects the body or scalp hair. Alopecia can sometimes result in hair growth. However, baldness can also be permanent; males who have progressive hair loss are a common example.

Cosmetic industry uses natural ingredients for hair care since they have high potential and safety over synthetic ones. Since marine sources are rich in nutrients (vitamins, minerals, fatty acids, etc.) and may be beneficial to hair health, they are often used in the development of hair care products. Because of their nourishing and renewing qualities, a number of marine components are included in shampoos, conditioners, and other hair care products.

The premature graying in hair can be prevented using the extracts of *Monodus sp., Thalassiosira sp., Chaetoceros sp., and Chlorococcum sp.* which can be used as formulations in cosmetic hair products as they have the ability to modulate the melanogenesis in human hair and skin. These extracts also help in stimulating and improving keratinocyte differentiation, melanocyte proliferation and help in human hair and hair follicle growth. The microalgae Chlorella has the ability to flexibilize and soften the hair and can be used as cosmetic formulation [6].

There are cosmeceutical products that contain marine sources such as marine algae. Marine algae such as *Palmaria palmate, Ecklonia cava, Undaria pinnatifida, Laminaria japonica, and Sargassum fulvellum* have the secondary metabolites and extracts that have effects on hair growth. The edible red algae *P. palmate* have the ability to regenerate hair because of the presence of vitamin B group. The Dieckol (sea polyphenol) extracts isolated from *Ecklonia cava* have potential hair growth-promoting effects with the ability to reduce the essential enzyme, 5α-reductase, that helps transform the testosterone into DHT. Moreover, the *Ecklonia cava* contains various metabolites such as Astaxanthin, fucoxanthin, and other phenolic compounds. The synergistic action of folic acid, iodine, and fucoidan present in the brown algae *U. pinnatifida* makes it beneficial for attaining healthy hair [59]. Hair growth-promoting effects are shown by the ethanolic extracts of *E .cottonii* because of the presence of various compounds such as phycoerythrin, phycocyanin, sulfated polysaccharides, and flavonoids. *C. crispus*, the red algae, is enriched in nutrients containing calcium, iodine, potassium, and magnesium that promote hair follicle health; this can be used in hair care products due to its soothing action by specific binding to keratin fibers.

The cartilaginous fishes are said to have hair growth-promoting effects, especially in females. Fishes usually contain essential nutrients and also proteins, lipids, vitamins, and minerals that are good sources for human skin and hair health. Reports say that the extracts from the cartilage of sharks can be used as ingredients for hair growth promoters such as viviscal (helps in the treatment of androgenic alopecia) and nourkrin (an alternative for pharmacotherapy that helps in treating hair loss) [60]. Sometimes, our hair might have omega-3 and omega-6 (essential fatty acids) deficiency that might cause abnormality in hair. Omega-3 and omega-6 cannot be synthesized by our body but are an accessory need that can be taken through the foods we are consuming. Marine resources such as fish and shellfish containing omega-3 or omega-6, docosahexaenoic acid, and docosahexaenoic acids are potential sources that can be used in hair care products. Some hair care products are available commercially that contain fish protein and oil as ingredients.

In hair care cosmetics such as shampoos and conditioners, the PUFA is used as an ingredient. Omega-3 fatty acids are used in cosmeceutical formulations incorporated as encapsulated or nanoencapsulated form. The encapsulation of fatty acids is done so that the fatty acids can be fresh with optimized antioxidant mixture [14].

6.9 Commercially Available Marine Natural Products from Cosmetic Industries

Several marine natural products are commercially available for use in cosmetics, offering various benefits for skincare and beauty formulations. Some of the commonly used marine-derived ingredients in the cosmetic industries are [61]:

- The Algowhite from CODIF Recherche & Nature contains ingredients extracted from *Ascophyllum nodosum* that can inhibit melanin synthesis and thus help in the whitening of skin.
- The product LIGHTOCEANE from the company Gelyma controls the production of melanin by decreasing the tyrosinase activity and also reduces the melanosome by transferring it into keratinocytes. This product also claims to have the property of protecting the skin against UVA/UVB radiation extracted from Halidrys siliquosa [50].
- The company SEPPIC markets a product known as SeaShine that has ingredients from *Undaria pinnatifida* that are highly effective in reducing melanin synthesis.
- The deepsane (EPS HYD657), extracted from the marine *Alteromonas sp.,* can protect keratinocytes from inflammatory agents. As the deepsane is an effective agent in soothing and reducing the irritation caused by chemical, physical, and UVB factors on sensitive skin types, it is used as the ingredient in the cosmetic product Abyssine® by Unipex. Since the deepsane bioactive ingredient is marketed as ferment extract, the product Abyssine® is also rich in proteins, minerals, amino acids, and organic and inorganic compounds.

- RefirMAR®, a product containing potential hydrating and anti-wrinkle properties, is marketed by the BIOALVO Company. The product containing these properties is due to the presence of *Pseudoalteromonas* sp. as its ingredient. Due to its defensive function in natural vents, the extracts of this *Pseudoalteromonas* sp. are used in cosmetics as topical application.
- The skincare product Resilience® by Estée Lauder contains extracts from the extracellular portions of Caribbean sea whip *Pseudopterogorgia elisabethae*. The extract is composed of pseudopterosins that help in inhibiting the phospholipase and 5-lipoxygenase of eicosanoid biosynthesis and thus have a potential effect as an anti-inflammatory and analgesic agent.
- The extracellular polysaccharides along with other glucidic exopolymers are produced from *Pseudoalteromonas* sp. (biotechnologically fermented). This species was isolated from the intertidal coasts of Antarctic waters. The Lipotec company launched SeaCode®, which contains these extracts from *Pseudoalteromonas* sp. By significantly increasing the in vitro synthesis of vital dermal proteins (like type I collagen) in adult human dermal fibroblast (HDFa) cells and preserving the cutaneous tonicity and firmness of those fibroblasts, *Pseudoalteromonas sp.* have been shown to improve skin characteristics. SeaCode®'s replenishing impact has been shown in in vivo studies on a panel of volunteers to improve skin roughness by 16.5% and 25.1% after 1 and 4 weeks, respectively. These improvements are predominantly seen at the upper lip creases.
- The microalgae compound Alguronic acid helps in protecting and regenerating the cell of microalgae during harsh environment and conditions. This alguronic acid can act as anti-aging agent, help rejuvenating the skin, and make the skin to look younger. The company Algenist (San Francisco, CA, USA) markets this alguronic acid® because of its potent cosmeceutical properties.
- The natural sulfated polysaccharide complex Alguard® was extracted from single red microalgae (*Porphyridium* sp.). It functions as a shield, enveloping the cell in a thick layer of protection and shielding it from the harsh intertidal environment. Numerous biological activities have been discovered through research, proving that Alguard® is more than just a physical barrier—rather, it actively protects the skin from photodamage, aging, and microabrasion.
- Dermochlorella DG®, a Chlorella sp. extract with oligopeptides from CODIF Recherche & Nature (Britany, France), functions as a restructuring agent on the dermal-epidermal junction to improve skin tone and firmness. As it increases collagen production in blood vessels, the active also acts on the epidermis to eliminate vascular imperfections (dark circles and minor vascular imperfections).
- Squalene and other extracts from pink marine algae are commercially available as Biossance Marine Algae Eye Cream, used in under eye treatment by firming and lifting.
- Hydra-marine serum, by the company Thalgo, helps reactivate the skin's natural detoxification process to deliver pure radiance and the original luminosity of the complexion. This product has marine bacteria Thermus thermophilus as one of its ingredients since it has self-protection ability in harsh environmental conditions.

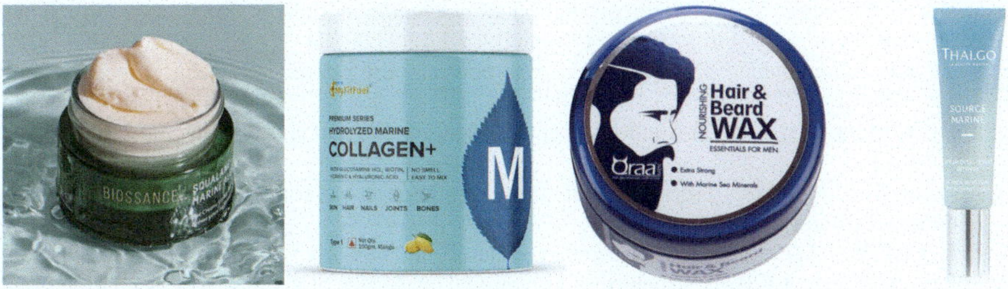

Fig. 6.11 Some of the commercially available cosmetic products that contain marine ingredients

- Dead sea minerals are used in the product Nourishing Styling Hair and Beard Wax With Marine Sea Minerals by the company Qraa that helps in maintaining hair health (Fig. 6.11).

6.10 Summary

The cosmetic industry is witnessing a shift toward sustainability and natural sourcing, with a notable trend involving the incorporation of marine resources in cosmetic formulations. This move is driven by consumer demand for eco-friendly products and the industry's commitment to sustainable practices. Cosmetic companies are increasingly utilizing marine ingredients such as seaweed and algae, sourced through environmentally responsible methods such as aquaculture and wild harvesting. This not only taps into the rich bioactive compounds found in marine resources but also ensures the preservation of marine ecosystems. The use of these ingredients reflects the industry's dedication to ethical and sustainable sourcing, contributing to a broader global movement toward environmentally conscious beauty products. As companies embrace marine resources for their formulations, they align with consumer expectations for effective and innovative cosmetics that are not only beneficial for the skin but also support responsible and eco-friendly practices.

References

1. Balboa EM, Conde E, Soto ML, Pérez-Armada L, Domínguez H (2015) Cosmetics from marine sources. In: Springer handbook of marine biotechnology, pp 1015–1042
2. Kligman D (2000) Cosmeceuticals. Dermatol Clin 18(4):609–615
3. Amer M, Maged M (2009) Cosmeceuticals versus pharmaceuticals. Clin Dermatol 27(5):428–430
4. Rodrigues F, de la Luz Cádiz-Gurrea M, Nunes MA, Pinto D, Vinha AF, Linares IB, Carretero AS (2018) Cosmetics. In: Polyphenols: properties, recovery, and applications. Woodhead Publishing, pp 393–427
5. Rawlings AV, Matts PJ (2005) Stratum corneum moisturization at the molecular level: an update in relation to the dry skin cycle. J Invest Dermatol 124(6):1099–1110
6. Couteau C, Coiffard L (2020) Phycocosmetics and other marine cosmetics, specific cosmetics formulated using marine resources. Mar Drugs 18(6):322

7. Scalia S, Mezzena M (2009) Incorporation of quercetin in lipid microparticles: effect on photo-and chemical-stability. J Pharm Biomed Anal 49(1):90–94
8. Pagar KR, Khandbahale SV (2019) A review on novel drug delivery system: a recent trend. Asian J Pharmacy Technol 9(2):135–140
9. Lee KY, Mooney DJ (2012) Alginate: properties and biomedical applications. Prog Polym Sci 37:106–126
10. Rinaudo M (2008) Main properties and current applications of some polysaccharides as biomaterials. Polym Int 57:397–430
11. Li B, Lu F, Wei X, Zhao R (2008) Fucoidan: structure and bioactivity. Molecules 13:1671–1695
12. Lahaye M, Robic A (2007) Structure and functional properties of ulvan, a polysaccharide from green seaweeds. Biomacromolecules 8:1765–1774
13. Cardoso MJ, Costa RR, Mano JF (2016) Marine origin polysaccharides in drug delivery systems. Mar Drugs 14(2):34
14. Kim SK, Bak SS (2011) 16 Hair biology and care product ingredients from marine organisms. Mar Cosmeceut Trends Prospects 201
15. Kim SK (2014) Marine cosmeceuticals. J Cosmet Dermatol 13(1):56–67
16. Sternlicht MD, Werb Z (2001) How matrix metalloproteinases regulate cell behavior. Annu Rev Cell Dev Biol 17(1):463–516
17. Hu J, Van den Steen PE, Sang QXA, Opdenakker G (2007) Matrix metalloproteinase inhibitors as therapy for inflammatory and vascular diseases. Nat Rev Drug Discov 6(6):480–498
18. Sang QXA, Jin Y, Newcomer RG, Monroe SC, Fang X, Hurst DR, Schwartz MA (2001) Matrix metalloproteinase inhibitors as prospective agents for the prevention and treatment of cardiovascular and neoplastic diseases. Curr Top Med Chem 6(4):289–316
19. Kim SK, Mendis E (2006) Bioactive compounds from marine processing byproducts—a review. Food Res Int 39(4):383–393
20. Joe MJ, Kim SN, Choi HY, Shin WS, Park GM, Kang DW, Kim YK (2006) The inhibitory effects of eckol and dieckol from Ecklonia stolonifera on the expression of matrix metalloproteinase-1 in human dermal fibroblasts. Biol Pharm Bull 29(8):1735–1739
21. Matsukawa R, Dubinsky Z, Masaki K, Takeuchi T, Karube I (1997) Enzymatic screening of microalgae as a potential source of natural antioxidants. Appl Biochem Biotechnol 66:239–247
22. Patented Breakthrough Ingredient. Available online: http://www.algenist.com/why-algenist/patented-breakthrough-ingredient (accessed on 25 November 2013)
23. Senaratne LS, Park PJ, Kim SK (2006) Isolation and characterization of collagen from brown backed toadfish (Lagocephalus gloveri) skin. Bioresour Technol 97(2):191–197
24. Xu Y, Han X, Li Y (2010) Effect of marine collagen peptides on long bone development in growing rats. J Sci Food Agric 90(9):1485–1491
25. Ferreira MS, Resende DI, Lobo JMS, Sousa E, Almeida IF (2021) Marine ingredients for sensitive skin: market overview. Mar Drugs 19(8):464
26. Im ST, Jang Y, Park S, Mun H, Kim DS, Lee DS, Lee SH (2022) In vitro screening of extracts from 38 marine animal resources for novel cosmeceutical activities. Fish Aquat Sci 25(6):326–334
27. Nowruzi B, Sarvari G, Blanco S (2020) The cosmetic application of cyanobacterial secondary metabolites. Algal Res 49:101959
28. Jahan A, Ahmad IZ, Fatima N, Ansari VA, Akhtar J (2017) Algal bioactive compounds in the cosmeceutical industry: a review. Phycologia 56(4):410–422
29. Kamei Y, Sueyoshi M, Hayashi K et al (2009) The novel anti-propionibacterium acnes compound, Sargafuran, found in the marine brown alga Sargassum macrocarpum. J Antibiot 62:259–263
30. Huang TH, Wang PW, Yang SC, Chou WL, Fang JY (2018) Cosmetic and therapeutic applications of fish oil's fatty acids on the skin. Mar Drugs 16(8):256
31. Ardiansyah A, Nugroho A, Rasyid A, Putra MY (2021) Screening of antioxidant and anti-acne activities in 16 sea cucumber in Indonesia. In: IOP conference series: earth and environmental science, vol 695, No. 1. IOP Publishing, p 012048

32. Poulose N, Sajayan A, Ravindran A, Sreechitra T, Vardhan V, Selvin J, Kiran GS (2020) Photoprotective effect of nanomelanin- seaweed concentrate in formulated cosmetic cream: with improved antioxidant and wound healing properties. J Photochem Photobiol B Bio 205
33. Kim SY, Kwon YM, Kim KW, Kim JYH (2021) Exploring the potential of Nannochloropsis sp. extract for cosmeceutical applications. Mar Drugs 19(12):690
34. Esposito R, Ruocco N, Viel T, Federico S, Zupo V, Costantini M (2021) Sponges and their symbionts as a source of valuable compounds in cosmeceutical field. Mar Drugs 19(8):444
35. Kapoor S, Nailwal N, Kumar M, Barve K (2019) Recent patents and discovery of anti-inflammatory agents from marine source. Recent Patents Inflamm Allergy Drug Discov 13(2):105–114
36. Je JY, Park PJ, Kim EK, Park JS, Yoon HD, Kim KR, Ahn CB (2009) Antioxidant activity of enzymatic extracts from the brown seaweed Undaria pinnatifida by electron spin resonance spectroscopy. LWT-Food Sci Technol 42(4):874–878
37. Brenneisen P, Oh J, Wlaschek M, Wenk J, Briviba K, Hommel C, Scharffetter-Kochanek K (1996) Ultraviolet B wavelength dependence for the regulation of two major matrix-metalloproteinases and their inhibitor TIMP-1 in human dermal fibroblasts. Photochem Photobiol 64(4):649–657
38. Jeffrey SW, MacTavish HS, Dunlap WC, Vesk M, Groenewoud K (1999) Occurrence of UVA- and UVB-absorbing compounds in 152 species (206 strains) of marine microalgae. Mar Ecol Prog Ser 189:35–51
39. Morone J, Alfeus A, Vasconcelos V, Martins R (2019) Revealing the potential of cyanobacteria in cosmetics and cosmeceuticals—a new bioactive approach. Algal Res 41:101541
40. Derikvand P, Llewellyn CA, Purton S (2017) Cyanobacterial metabolites as a source of sunscreens and moisturizers: a comparison with current synthetic compounds. Eur J Phycol 52(1):43–56
41. De la Coba F, Aguilera J, De Galvez MV, Alvarez M, Gallego E, Figueroa FL, Herrera E (2009) Prevention of the ultraviolet effects on clinical and histopathological changes, as well as the heat shock protein-70 expression in mouse skin by topical application of algal UV-absorbing compounds. J Dermatol Sci 55(3):161–169
42. Hur S, Lee H, Kim Y, Lee BH, Shin J, Kim TY (2008) Sargaquinoic acid and sargachromenol, extracts of Sargassum sagamianum, induce apoptosis in HaCaT cells and mice skin: its potentiation of UVB-induced apoptosis. Eur J Pharmacol 582(1–3):1–11
43. Couteau C, Coiffard L (2018) Microalgal application in cosmetics. In: Microalgae in health and disease prevention. Academic Press, pp 317–323
44. Venkatesan J, Anil S, Kim SK, Shim MS (2017) Marine fish proteins and peptides for cosmeceuticals: a review. Mar Drugs 15(5):143
45. Felician FF, Xia C, Qi W, Xu H (2018) Collagen from marine biological sources and medical applications. Chem Biodivers 15(5):e1700557
46. Liu JK (2022) Natural products in cosmetics. Nat Prod Bioprospect 12(1):40
47. Sano T, Kaya K, Oscillapeptin G (1996) A tyrosinase inhibitor from toxic Oscillatoria agardhii. J Nat Prod 59(1):90–92
48. Yoon NY, Eom TK, Kim MM, Kim SK (2009) Inhibitory effect of phlorotannins isolated from Ecklonia cava on mushroom tyrosinase activity and melanin formation in mouse B16F10 melanoma cells. J Agric Food Chem 57(10):4124–4129
49. Sun Y, Chavan M (2017) U.S. Patent No. 9,603,790. Washington, DC: U.S. Patent and Trademark Office
50. Kim JH, Lee JE, Kim KH, Kang NJ (2018) Beneficial effects of marine algae-derived carbohydrates for skin health. Mar Drugs 16(11):459
51. Pereira L, Morrison L, Shukla PS, Critchley AT (2020) A concise review of the brown macroalga Ascophyllum nodosum (Linnaeus) Le Jolis. J Appl Phycol 32:3561–3584
52. Smirthwaite J (2007) *Laminaria ochroleuca* Golden kelp. In: Tyler-Walters H, Hiscock K (eds) Marine life information network: biology and sensitivity key information reviews. Marine Biological Association of the United Kingdom, Plymouth

53. Freitas MV, Pacheco D, Cotas J, Mouga T, Afonso C, Pereira L (2021) Red seaweed pigments from a biotechnological perspective. Phycology 2(1):1–29
54. Bernard Spragg (2014) NZ. Calcareous red seaweed. https://www.flickr.com/photos/volvob12b/1431682468
55. Andrepotter (2017) Haematococcuspluvialiscultivation. http://algaeanalytics.com/blog/2017/6/15/haematocccus-pluvialis-cultivation
56. Lamare Véronique (2022) VERLAQUEMarc. In: Steentoft M, Irvine LM, Farnham WF (eds) DORIS, 07/29/2022: *Gracilaria gracilis* (Stackhouse). https://doris.ffessm.fr/ref/specie/3425
57. Fedor. Karali. *Arthrospira platensis.* Bio203 University of Wisconsin-La Crosse. Web (30 Oct. 2011)
58. Agarwal V, Tjandra ES, Iyer KS, Humfrey B, Fear M, Wood FM, Raston CL (2014) Evaluating the effects of nacre on human skin and scar cells in culture. Toxicol Res 3(4):223–227
59. Dupont E (2002) U.S. Patent No. 6,383,522. Washington, DC: U.S. Patent and Trademark Office
60. Couteau C, Coiffard L (2016) Seaweed application in cosmetics. In: Seaweed in health and disease prevention. Academic Press, pp 423–441
61. Martins A, Vieira H, Gaspar H, Santos S (2014) Marketed marine natural products in the pharmaceutical and cosmeceutical industries: tips for success. Mar Drugs 12(2):1066–1101

Seaweeds are Potential Source for Production of Sustainable Bioethanol for the Imminent Future

7

J. Immanuel Suresh and S. Divyeswari

Abstract

The pursuit of Sustainable and renewable energy sources has gained immense importance in the face of Global energy and environmental challenges. The current fossil fuel reserves are not sufficient and very soon will become exhausted. Seaweeds, commonly known as marine algae, have emerged as a promising candidate for Bioethanol production due to their abundance, rapid growth, and minimal competition with food crops. Bioethanol has attracted attention as an alternative to petroleum-derived fuel; the production process involves the extraction of polysaccharides, primarily cellulose and alginate, from seaweed biomass. These polysaccharides are subjected to enzymatic hydrolysis and fermentation to convert them into bioethanol. Some seaweed species that have been studied and explored for bioethanol production includes *Saccharina japonica* (Kelp) which is known for its high cellulose content, making it a valuable source of biomass for bioethanol production. Similarly, *Gracilaria verrucosa*, a red seaweed, can be utilized in bioethanol production. These approach not only mitigates the demand for traditional feedstock like corn or sugarcane but also reduces greenhouse gas emissions and land use conflicts. This review delves into the intricacies of bioethanol production tailored to these seaweed species, and their associated microorganisms involving enzymatic hydrolysis with cellulolytic and xylanolytic bacteria like *Clostridium* sp., *Cellulomonas* sp., *Bacillus* sp.; fungi such as *Trichoderma* sp., *Fusarium* sp.; and fermentation with *Saccharomyces cerevisiae*. Species-specific optimizations are critical to ensure efficient and high-yield bioethanol production. Challenges such as seasonal variability, logistical issues, and the need for large-scale seaweed cultivation infrastructure are

J. Immanuel Suresh (✉) · S. Divyeswari
Department of Microbiology, The American College, Madurai, Tamil Nadu, India
e-mail: immanuelsuresh@americancollege.edu.in

M. Rafatullah et al. (eds.), *Multidisciplinary Applications of Marine Resources*,
https://doi.org/10.1007/978-981-97-5057-3_7

also explored, along with potential solutions. Overall, this chapter emphasizes the production of bioethanol from multiple seaweed species presenting a sustainable and eco-friendly alternative in the quest for renewable energy sources, offering the potential to reduce our reliance on fossil fuels and mitigate the impacts of climate change.

Keywords

Seaweed · Sustainable energy · Cellulose · Fossil fuel · Renewable energy source

7.1 Introduction

Globally, there is a strong push for biofuels like bioethanol, biodiesel, biogas, and bio-butanol due to their perceived eco-friendliness and sustainability. In particular, bioethanol leads the pack, with a staggering production volume of 106 billion litres in 2017. Its primary use lies in fuelling engines and serving as a fuel additive. However, a significant drawback is the heavy reliance on edible feedstock such as sugarcane, corn, sweet sorghum, and sugar beet, raising concerns about potential food security issues, especially in developing nations. In response, researchers are now turning their attention to alternative feedstock options, including crop residues and algae. Seaweed, or macroalgae, is emerging as a promising candidate for third generation biofuels [1].

Marine macroalgae encompass a diverse group of large, photoautotrophic, multicellular, sessile, benthic plants, collectively known as seaweeds. With as many as 25,000 species worldwide, seaweeds offer considerable morphological and functional diversity. The ongoing exploration of seaweeds for various chemical applications has led to a significant increase in seaweed production, from 3.8 million tons in 1990 to a staggering 15.8 million tons fresh in 2008, positioning them as one of the major mariculture crops with a market value exceeding US$7.4 billion [2].

Seaweeds have a wide range of industrial applications, spanning from conventional food, feed, and hydrocolloids to more recent uses as a source of fertilizer, bioactive compounds, and energy molecules. Their rich composition, characterized by easily depolymerizable polysaccharides and minimal lignin content, makes them an attractive renewable biomass for bioethanol production, surpassing terrestrial resources in several aspects. Notably, seaweeds require no agricultural land usage and minimal resource inputs such as fertilizer, pesticides, and water. Furthermore, their low contents of recalcitrant compounds like lignin enable enzymatic hydrolysis with little or no pre-treatment, contributing to their potential as a sustainable bioethanol source. Moreover, seaweed bioethanol production offers carbon neutrality, thanks to the favourable net CO_2 emissions between production and combustion, coupled with seaweed CO_2 uptake during photosynthesis [3].

Seaweeds exhibit a higher photosynthetic efficiency (6–8%) compared to terrestrial biomass (2%), highlighting their potential as an eco-friendly and sustainable feedstock for bioethanol production. The carbohydrate content of seaweed, constituting 40–77% dry matter (DM), serves as the primary source for bioethanol production. Notably, 50–70% of seaweed carbohydrates can often be converted into fermentable sugars. Bioethanol has successfully been produced from various seaweed species, including red species such as *Kappaphycus* sp., *Gelidium* sp., and *Gracilaria* sp.; brown species such as *Laminaria* sp. and *Sargassum* sp.; and green species such as *Ulva* sp., *Chaetomorpha* sp., and *Rhizoclonium* sp., through stages like biomass pre-treatment, hydrolysis, fermentation, and ethanol recovery. However, the long-term commercial success of seaweed bioethanol production hinges on the optimization of critical stages like hydrolysis and fermentation, ensuring efficiency, cost-effectiveness, and sustainability, especially at commercial scales. Achieving this optimization remains a major challenge globally in bioethanol production [1].

7.2 Marine Macroalgae (Seaweeds)

Seaweeds, acting as primary producers, sustain multiple benthic animal communities and serve as the cornerstone of the marine food web. They actively vie for vital resources such as illumination, nutrients, and territory, while depending on carbon dioxide and water for growth. Interestingly, both algae and flora share resemblances in their storage substances and defensive tactics against predators and parasites. They have devised effective mechanisms to combat various living threats, including bacterial, viral, and fungal infections. As immobile organisms, seaweeds have evolved to flourish in varied and often severe non-living environmental conditions. They exhibit impressive adaptability to factors like temperature oscillations, shifts in salinity, exposure to pollutants, and UV radiation. This adaptability underscores the resilience and significance of seaweeds in upholding the equilibrium of marine ecosystems [4].

7.3 Macroalgae in Biofuel Production, Its Cultivation Methods and Environmental Impact

Macroalgae, a diverse, non-phylogenetic macroscopic aquatic eukaryote, belongs to Rhodophyta (red algae), phaeophyta (green algae), and Phaeophyceae (brown algae). Algae cultivation is possible in various water bodies, including wastewater. Additionally, the growth rate of algae is approximately 20–30 times faster than that of traditional fodder crops, and the oil content in macroalgae surpasses that of conventional feedstocks by around 30 times. The biodegradability and sulfur-free nature of algal sources contribute to their superior quality. Moreover, the absence of lignin facilitates microbial digestion during the biorefinery process, making macroalgae conversion into biofuel easier than land-based plants. Residues from the conversion processes find utility in heating, fertilizer production, and other fuel types. Macroalgae boast water content rich in carbohydrates (25–50%), protein (7–15%), and lipid (1–5%), positioning them as a promising feedstock for biodiesel, bioethanol, and biohydrogen production. Similarly, macroalgae serve as food supplements, hydrocolloids, healing materials, fertilizer, and animal feed. In the food industry, macroalgae contribute significantly to annual revenues, comprising 83–90% of the total seaweed industry. Biofuel production from macroalgae encompasses biodiesel, bioethanol, biohydrogen, biomethane, and bio-oil. Most findings express optimism regarding biofuel production from macroalgae [5].

The macroalgae biomass market is witnessing expansion in market capitalization on an annualized basis. According to statistical data from the Food and Agriculture Organization of the United Nations (FAO), global macroalgal biomass production in 2016 amounted to approximately 30 million tonnes, valued at USD$ 11.6 billion. Asia emerges as the world's leading producer of macroalgae, with China at the forefront, producing 14 million tonnes valued at USD$ 8.6 billion, followed by Africa with approximately 140,000 tonnes and the Americas with 15,634 tonnes. Various macroalgal species, including *Laminaria japonica, Eucheuma spp., Kappaphycus alverezii, Pyropiayezoensis, Undaria pinnatifida,* and *Graciliaria verrucosea*, have already undergone mass culture in Asia [6].

In contrast, Europe has a nascent aquaculture industry for macroalgae, with cultivation methods still lagging. Nevertheless, efforts are underway to stimulate the European macroalgae market and aquaculture industry, propelled by both academic and commercial interests. Macroalgae cultivation can occur both offshore and onshore using various methods. Offshore cultivation methods include kelp growth, raft cultivation, and floating cultivation. Cultivation using ropes or nets is prevalent due to lower installation and maintenance costs. Lagoons provide nutrients for macroalgae cultivation, while fixed off-bottom, long lines, and rock-based farming are alternative methods. Transplantation involves growing species saplings indoors, then culturing them in tanks before transplanting them into the sea using ropes [7].

Onshore cultivation techniques utilize seawater extensively, offering strict control over safety and high product yield. This method boasts adaptability to a wide range of macroalgae and is more sustainable than offshore cultivation. Proper agitation or circulation promotes better algal growth in onshore cultivation. However, this method lacks a sustainable low-cost innovative approach for large-scale implementation. Open farming may intrude upon biotic and abiotic ecosystems, diminishing the algae quality for use in pharmaceutical, chemical, and cosmetic industries. A ring-shaped culture technique for algal cultivation on land has been developed, significantly reducing cultivation medium requirements. This method also absorbs heat during agitation, enhancing growth efficiency [8].

The growth rate of cultivated algal biomass is crucial, standard formula for calculating growth rate as follows [9]:

$$\text{Growth rate}(\text{G}) = \left[\left(Wf \,/\, Wi \right) 1 / T - 1 \right] \times 100\% \tag{7.1}$$

where,

Wi—Initial weight of algal biomass, *Wf*—final weight of algal biomass, and *T*—number of days in culture.

Macroalgae create a unique environment for marine organisms, sustaining ecosystems. Factors like light intensity, turbidity, water temperature, nutrient concentrations, pH, and salinity influence algae growth. However, algal harvesting may disrupt ecosystems, and the use of inorganic fertilizers to stimulate macroalgae growth can induce algal blooming, disrupting coastal ecosystems and causing hypoxia. Seaweed farming serves as a promising nutrient removal technology. Additionally, reduced irradiance under macroalgae culture sites may adversely affect marine organisms in shallow areas. Integrated multitrophic aquaculture utilizing macroalgae can employ nutrients for fish farming in both land-based and offshore marine culture systems [5].

7.4 Biofuel from Macroalgae

Biofuels encompass a range of solid, liquid, or gaseous fuels derived from biological sources, suitable for use in vehicles and industrial applications (Fig. 7.1). Depending on the biomass used, biofuels are categorized into first, second, and third-generation types. Macroalgae can serve as a source for biofuels, with extraction methods including biochemical and thermochemical processes. Common production techniques involve transesterification, liquefaction, fermentation, anaerobic digestion, and pyrolysis. However, the intricate structure of algal biomass can hinder the hydrolysis process, slowing down biofuel production. To address this, various pre-treatment methods, such as physical, chemical, biological, mechanical, or combinations thereof, are employed to break down complex molecules and enhance solubilization. This, in turn, facilitates the conversion process and boosts biofuel yields [10].

Methods such as physical, chemical, biological, mechanical, and combinative methods are used for the solubilization of complex substrate in macroalgae which enhance biofuel production [5].

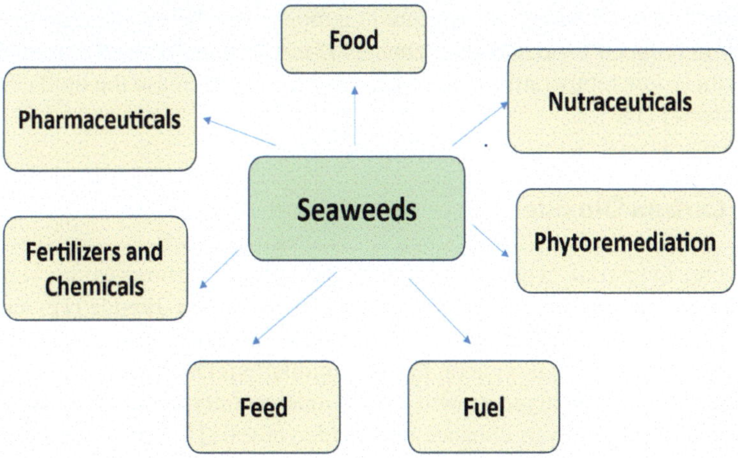

Fig. 7.1 Potential application of seaweeds

7.5 Factors Impacting Algal Cultivation and Biofuel Production

Various factors can affect the cultivation of algae and the production of biofuels. These factors include both living (biotic) and non-living (abiotic) elements. Abiotic factors such as light, temperature, pH, nutrients, and carbon dioxide levels play a crucial role, while the choice of algae species utilized for cultivation is a biotic factor that significantly influences yield and productivity [11].

7.5.1 Light

The growth and biomass accumulation of algae during cultivation heavily depend on both the wavelength and intensity of light. Research indicates that increased light intensity often leads to higher lipid content in some algae species. Algae are typically chosen for cultivation based on their photosynthetic capabilities, as light is essential for photosynthesis and growth. Consequently, light plays a crucial role in controlling lipid accumulation and enhancement. Some studies have noted a shading effect of light, where shading inhibits the growth of certain algal strains, but growth resumes quickly once shading is removed. Additionally, fluorescent light sources have been found to enhance growth compared to other light sources [11, 12].

7.5.2 Temperature

Temperature is a crucial factor influencing the growth of algae, lipid accumulation, and biofuel production. Most algae species thrive optimally within the temperature range of 20–35 °C. However, certain species categorized as mesophilic can tolerate temperatures of up to approximately 40 °C. Deviating from the optimal temperature range, either by overheating or exposure to temperatures below the required range, can result in reduced yield and cell damage [13]. Additionally, both extremely high and extremely low temperatures have been observed to decrease the lipid content in algal biomass [11].

7.5.3 Carbon Dioxide

Atmosphere, industrial exhaust emissions, and soluble carbonates are significant sources of carbon dioxide. A higher concentration of carbon dioxide is essential for achieving greater algal lipid content. Various studies offer differing evidence regarding the impact of carbon dioxide on lipid accumulation. Decreasing carbon dioxide concentration enhances algae growth but diminishes fatty acid synthesis, whereas higher carbon dioxide levels enhance fatty acid synthesis [11]. However, increasing carbon dioxide concentration significantly affects the carbon chain. Research on *Chlorella pyrenoidosa* SJTU-2 and *Scenedesmus obliquus* SJTU-3 demonstrated enhanced growth at 10% carbon dioxide, while lipid accumulation and fatty acid synthesis increased at 30–50% CO_2 levels [14].

7.5.4 pH

pH plays a crucial role in lipid accumulation, oil synthesis, and enzymatic activity, all of which are vital for algae growth. An environment with slightly acidic or neutral pH typically enhances algae growth. However, the presence of carbonic acids in the nutrient medium can lower the pH, creating conditions unfavourable for algal growth. Additionally, the lower concentration of bicarbonate at low pH negatively impacts carbon integration for lipid formation. Changes in pH affect microalgal biological reactions differently. Higher pH levels have been observed to suppress cell division in *Chlorella*, trigger auto-spore release, and ultimately lead to TAG exploitation [11].

7.5.5 Nutrients

Assuming various nutrient limitations, algae can exhibit diversification in biochemical compounds. The extent of this diversification depends on the type and degree of nutrient limitation. When pH and temperature are maintained within optimal ranges, the rate of algae growth is directly correlated with the uptake rate of the most limiting nutrient [11]. Phosphate and nitrogen are among the most crucial macronutrients essential for algae growth and development. Additionally, carbon, oxygen, and hydrogen, as non-mineral nutrients, play essential roles in algal growth, with abundance of oxygen and hydrogen not posing challenges to algae metabolic pathways [15]. Some researchers have demonstrated that nutrients released from algal biomass can be recycled and reused in nutrient media. These nutrients are initially converted into a liquid co-product and then made available for reuse. For example, in a study involving a bi-culture of two algal strains, nutrients released during carbonization were recycled and utilized as feedstock for biodiesel production [16].

7.5.6 Algal Species

The biochemical composition of algal communities significantly influences its potential for biofuel production at lower costs. Colonial species often dominated in high-rate algal ponds and easily harvested through gravity settling were understudied. Five wastewater colonial algae species—*Coelastrum* sp., *Desmodesmus* sp., *Pediastrum boryanum*, *Micractinium pusillum*, *Mucidosphaerium pulchellum*—were evaluated for wastewater treatment and biofuel production. During summer, *Mucidosphaerium pulchellum* and *Micractinium pusillum* exhibited higher biomass, lipid, and energy content, with *Micractinium pusillum* showing superior settling characteristics. This suggests *Micractinium pusillum* holds a promise for wastewater treatment at low-cost biofuel production [17]. While no polyculture outperformed the most prolific monoculture, diverse species in polycultures demonstrated increased output compared to average monocultures, with potential benefits in avoiding undesirable crop function trade-offs [11].

7.6 Bioethanol

Bioethanol, also known as ethyl alcohol or grain alcohol (CH_3–CH_2–OH or ETOH), is a liquid biofuel derived from various biomass feedstocks and conversion methods. It offers an appealing alternative to traditional fuels due to its renewable nature and oxygenation properties, which have the potential to decrease particulate emissions in compression–ignition engines [18].

7.7 Properties of Seaweeds as the Raw Materials for Bioethanol Production

Various types of seaweeds are utilized in industrial applications, such as alginate, agar, and carrageenan production, as well as in foods like nori, wakame, and kombu. Globally, seaweed has significant industrial and food-related consumption, with total 1.54 and two million wet tonnes per year, respectively. However, seaweeds not utilized for these purposes could serve as raw materials for bioethanol production. While some unused seaweeds provide feed for shellfish and habitats for marine fish, cultivated seaweeds and those not utilized in the marine ecosystem are suitable for bioethanol production. Producing large quantities of bioethanol from cultivated biomass requires highly productive seaweeds. Small-scale cultivation experiments with green seaweeds like *Ulva*, *Chaetomorpha*, and *Cladophora* suggest that their productivity in terms of volatile solids (organic compounds) is around 50 tons/(ha y) [19]. *Ulva* spp. have volatile-solid contents of approximately 70% on a dry-weight basis, with a calculated productivity of around 7100 dry g/(m^2 y) [20].

Brown and red seaweeds such as *Macrocystis*, *Laminaria*, *Ecklonia*, and *Sargassum* exhibit productivities ranging from 3300 to 11,300 dry g/(m^2 y). Seaweeds, with higher productivities compared to lignocellulosic biomass, offer promise for biomass cultivation for bioethanol production. Moreover, seaweed wastes, which contribute to issues like green tide, can be repurposed for bioethanol production, addressing disposal challenges. In regions like Japan and France, the accumulation of Ulva and other seaweeds poses environmental challenges, while invasive species like *Gracilaria salicornia* in Hawaii require proper disposal, suggesting bioethanol production as a potentially effective and economically advantageous solution [19].

7.8 Composition of Seaweeds

Seaweeds display a wide range of chemical compositions, particularly in their primary metabolic constituents like polysaccharides, which are essential for bioethanol production. Important polysaccharides include carrageenan and agar from red seaweeds, alginate and laminarin from brown seaweeds, and cellulose and ulvan from green seaweeds [3].

Carrageenan, found in red seaweeds, is a linear sulfated polysaccharide consisting of alternating D-galactose and anhydrogalactose units. Agar, another polysaccharide from red seaweeds, is made up of β-D-galactose and α-L-galactose derivatives [21].

Brown seaweeds are abundant in alginate, a major structural polysaccharide containing β-D-mannuronic acid and α-L-guluronic acid units. Additionally, brown seaweeds contain laminarin, a linear β-1,3-linked glucose residue with some β-1,6 linkages, and fucoidan, a sulfated complex polysaccharide. Green seaweeds, like *Ulva* species, are rich in cellulose and ulvan, a water-soluble sulfated polysaccharide comprising D-glucuronic acid, D-xylose, L-rhamnose, and sulfate. These hydrocolloids have become commercially significant, with market values rising from US$644 million to US$1018 million over the past decade [21].

The carbohydrate content varies among seaweed divisions, with green, red, and brown seaweeds containing approximately 33–58%, 31–61%, and 21–33% of dry weight, respectively. Seaweed proteins have also been investigated as dietary supplements due to their digestibility [22]. Although seaweed's nutritional value has long been acknowledged, recent studies emphasize its proximate composition, which is comparable to terrestrial vegetables and other food ingredients [3]. The carbohydrate content and proximate compositions of seaweeds are summarized in Table 7.1.

Table 7.1 Proximate composition of seaweeds

Component (% DW)	Green Seaweeds	Red Seaweeds	Brown Seaweeds
Carbohydrates	30–60	30–50	20–30
Polysaccharides	Ulvan, starch, Cellulose, mannan	Agar, carrageenan, Cellulose, lignin	Laminarin, alginate, mannitol, fucoidan, Cellulose
Monosaccharides	Glucose, mannose, xylose, rhamnose, galactose, uronic acid, glucuronic acid	Glucose, galactose	Glucose, galactose, xylose, uronic acid, glucuronic acid, fucose, mannuronic acid, guluronic acid
Proteins	10–20	6–15	10–15
Lipids	1–3	0.5–1.5	1–2
Ash	13–22	5–15	14–28

7.8.1 Green Seaweeds

Green seaweeds like *C. lentillifera* (with 45% carbohydrate) and *C. fragile* (with 59% carbohydrate) are characterized by the presence of ulvan, starch, and cellulose as primary polysaccharides. Ulvan, prevalent in *Ulva* species, is composed of oligosaccharide units containing L-rhamnose-3-sulfate, D-xylose-2-sulfate, and various uronic acids. It is soluble in water and finds applications across food, pharmaceutical, and chemical industries [23]. Starch and cellulose in green seaweeds are both made up of glucose units, differing in their configuration at the anomeric carbon (β-in cellulose, α-in starch) [24]. Cellulose forms robust crystalline structures resistant to enzymatic and physical degradation, while starch adopts a more open, loosely bonded helical structure, aiding in its solubility [1].

7.8.2 Red Seaweeds

Red seaweeds, exemplified by *E. cottonii* (with 27% carbohydrate) and *G. amansii* (with 66% carbohydrate), feature polysaccharides like carrageenan and agar. Agar, sourced from *Gelidium* and *Gracilaria*, is composed of hydrophilic galactans characterized by alternating α-1,3 and β-1,4 linkages. Carrageenan derived from *Kappaphycus*, *Chondrus*, and *Eucheuma* consists of linear sulfated galactans, each with distinctive structures: iota (yielding elastic gels), kappa (forming rigid gels), and lambda (resulting in viscous solutions). Enzymes such as cellulase can be employed to selectively extract hybrid carrageenans with specific properties, proving invaluable in the food industry, particularly in dairy products and enzyme immobilization [1, 24, 25].

7.8.3 Brown Seaweeds

In brown seaweeds, the carbohydrate content varies between 40% (e.g., *S. fulvellum*) and 64% (L. *digitata*). Components include laminarin (major storage polysaccharide), mannitol (osmoregulatory function), cellulose, alginate (50% of carbohydrate fraction), and fucoidan (mainly L-fucose and sulphated ester groups). Laminarin and mannitol concentrations peak in June and July in *L. digitata*, declining in the later part of the year. Alginate enhances bioethanol recovery during yeast fermentation. Fucoidans in *Sargassum* and *Laminaria* exhibit antioxidant, anticoagulant, antiviral, and therapeutic properties. *Laminaria hyperborea* analysis shows 0–30% laminarin, 4–25% mannitol, and 17–34% alginate in dry matter [1].

7.9 Seaweed Biomass Handling and Pre-treatment

The management of seaweed biomass post-harvest is critical to prevent decay during transit and storage, as well as to avoid contamination by impurities like sand and aquatic organisms, which can alter its composition. Key processes include washing, drying, and grinding. Washing seaweed after harvest, preferably with seawater to conserve freshwater, removes undesirable impurities. However, rapid washing under tap water can lead to significant loss of carbohydrates, making it less favourable for large-scale bioethanol production [26].

Drying seaweed, which has high moisture content initially, is a major energy-intensive step in bioethanol production and can be done through methods like freeze-drying, sun drying, or oven drying. Among these, freeze-drying shows promise but may not be economically feasible for seaweed farmers from poor coastal communities due to high energy costs. Sun drying, though less controlled than oven drying, remains popular due to its low cost, despite potential effects on seaweed composition. Commercial bioethanol production typically requires dried seaweed to minimize decay and transportation costs. However, if a bioethanol plant is nearby and can match the cultivation site's output, skipping the drying step might be feasible. Size reduction, achieved through milling, is crucial as it increases biomass surface area, aiding subsequent hydrolysis and fermentation stages, while also improving transportation and storage efficiency [27]. Milling has been observed to have some hydrolytic effect on seaweed biomass [1].

7.10 Hydrolysis of Seaweeds

The hydrolysis phase in bioethanol production involves breaking down complex sugars found in seaweeds, such as laminarin, cellulose, mannitol, alginate, ulvan, carrageenan, and agar, into simpler sugars like glucose, galactose, mannose, fucose, xylose, and arabinose, which can then be fermented into ethanol [28]. Various methods are employed for seaweed hydrolysis in bioethanol production, including dilute acid thermal, dilute alkaline thermal, enzymatic, thermal, ball milling, hydrothermal (HTT), and ultrasound treatments, which are categorized as physical, chemical, thermal, and enzymatic. Combinations of these treatments are often used to optimize yield. However, there hasn't been a study yet to determine the most suitable hydrolysis method for specific groups of seaweed. The following sections delve into these methods in detail, focusing on their efficiency and sustainable utilization [1].

7.10.1 Dilute Acid Thermal Hydrolysis

Dilute acid thermal hydrolysis is extensively utilized in research on seaweed bioethanol, primarily due to its cost-effectiveness and shorter reaction duration. Common catalysts like sulfuric acid (H_2SO_4) and hydrochloric acid (HCl) are employed, yet they produce inhibitors (5-HMF, furfural, levulinic acid) that affect fermentation. The use of activated carbon filtration effectively reduces the concentration of inhibitors. The yeast strain *Pichia stipitis* KCTC 7228 exhibits promise in converting furfural and 5-HMF without impeding glucose fermentation. Dilute acid hydrolysis proves cost-effective, benefiting from a well-established market for strong acid catalysts used in low concentrations (0.01–0.90 M). Parameters for optimization include substrate concentration, reaction time, temperature, and catalyst concentration, which vary due to factors such as seaweed species, carbohydrate types, and seasonal variations. Challenges in catalyst recovery affect sustainability. Despite being efficient and economical, concerns about environmental impact, toxicity, and salt disposal render acid-catalysed seaweed hydrolysis less sustainable [1, 29–31].

7.10.2 Dilute Alkaline Thermal Hydrolysis

Dilute alkaline thermal hydrolysis offers a direct alternative to dilute acid hydrolysis for seaweed processing, employing a base as a catalyst. Limited attention has been given to base-catalysed seaweed hydrolysis due to the higher concentrations of base required for efficiency and the challenges related to gel formation in red seaweeds. To address viscosity issues, the biorefinery approach involves extracting hydrocolloids before applying diluted alkaline hydrolysis [1]. In a study on *E. cottonii* carrageenan residue, acid treatment (1% H_2SO_4) resulted in a glucose yield of 91%, while base treatment (1% NaOH) yielded 80% DM glucose. However, the impact of acid by-products, like 5-HMF, on fermentation was not reported. Dilute alkaline hydrolysis produces degradation products acting as fermentation inhibitors, but their effects remain uncertain [32, 33]. Hydrolysis of *Ulva rigida* by base catalysts showed a 79% w/w total carbohydrate yield, with inhibitors only at very high catalyst concentrations. Like acid hydrolysates, base hydrolysates require pH adjustment before fermentation. Currently, no significant advantage of base over acid catalysts in seaweed hydrolysis has been identified; a techno-economic analysis may provide further insights [1].

7.10.3 Enzymatic Hydrolysis

Enzymatic hydrolysis is considered highly efficient for converting complex seaweed polysaccharides into sugars with increased efficiency and fewer harmful by-products. Cellulases, produced by cellulolytic bacteria (such as *Clostridium*, *Cellulomonas*, *Bacillus*, etc.) and fungi (like *Trichoderma*, *Penicillium*, etc.), consist of endoglucanases, exoglucanases, and β-glucosidases [34]. Endoglucanases break down cellulose randomly, producing oligosaccharides, while exoglucanases act on remaining ends, creating cellobiose. β-Glucosidases further break down cellobiose into glucose. Pentose-specific enzymes, such as endo-1,4-β-D-xylanases and α-L-arabinofuranosidases, are also involved. Given the diversity of polysaccharides in seaweed, enzyme mixtures are beneficial for seaweed hydrolysis. However, the sensitivity of enzymes to pH (ideal range: 4–6) requires pH adjustment, leading to additional costs. Despite the high production costs, lengthy reaction times, and challenges in enzyme recovery, enzymatic hydrolysis offers advantages in maximizing sugar yields compared to alternative methods. The development of seaweed-specific enzymes could help reduce costs [1].

In a study [35], cellulase from *Trichoderma reesei* and β-glucosidase from *Aspergillus niger* were utilized on *G. verrucosa*, resulting in a yield of 0.87 g of reducing sugars per gram of cellulose. In many seaweed hydrolysis experiments, enzymatic hydrolysis is combined with acid or thermal treatment to enhance the yield of reducing sugars. For instance, *Sargassum* sp. was treated with 4% H_2SO_4 followed by enzymatic hydrolysis, yielding 17.7 g/L of reducing sugars [36]. Other studies have also combined acid and enzymatic hydrolysis on *G. amansii* and *K. alvarezii*, achieving higher sugar yields. However, the combined effects of dilute acid and enzymatic hydrolysis raise concerns about sustainability due to enzyme costs and the formation of inhibitors [1].

7.10.4 Other Methods of Hydrolysis

Various unconventional techniques have been utilized for the pre-treatment and hydrolysis of seaweeds, such as high-temperature treatment (HTT), ultrasound treatment, hot water washing, and ball milling. These methods are often employed either independently to address concerns related to traditional chemical and enzymatic approaches or in conjunction with them to enhance the yield of reducing sugars. For instance, autoclave (thermal) treatment, as described by [37], involved subjecting *G. Amansii* species to autoclaving at 121 °C for durations ranging from 20 to 80 min, followed by enzymatic hydrolysis using cellulase and β-glucosidase enzymes. The total reducing sugar yields between the untreated and autoclave-treated biomass were comparable, but the glucose yield was significantly higher with autoclave treatment [37]. Similarly, gamma irradiation combined with dilute acid hydrolysis in *Undaria* sp. achieved a notable reducing sugar concentration. Integrated hydroxyl radicals and hot water pre-treatment (IHRHW) resulted in significant sugar recovery for *Macrocystis pyrifera* [1].

A comparative analysis on *U. rigida* demonstrated differing reducing sugar concentrations across various methods, with enzymatic hydrolysis yielding the highest concentration [33]. Hydrothermal treatment with subcritical water on *S. japonica* proved to be an efficient alternative to traditional acid and enzymatic hydrolysis, achieving a maximum glucose concentration under specific temperature and pressure conditions [1].

7.11 Seaweed Fermentation Techniques

Fermentation plays a crucial role in the bioethanol production process, as it is during this phase that organisms convert reducing sugars, obtained from hydrolysis, into ethanol. While not always the bottleneck in bioethanol production, the efficiency of fermentation significantly impacts the final ethanol yield. The common techniques employed for seaweed fermentation include Separate Hydrolysis and Fermentation (SHF) and Simultaneous Saccharification and Fermentation (SSF) [1].

7.11.1 Separate Hydrolysis and Fermentation (SHF)

Separate Hydrolysis and Fermentation (SHF) is widely adopted due to its flexibility in hydrolysis process selection and the ability to optimize conditions for catalysts and fermenting organisms [37]. In SHF, hydrolysis is completed first, with optional recovery of reducing sugars through centrifugation or filtration, before fermentation begins. This approach was employed in a study on *G. Tenuistipitata*, where acid hydrolysis yielded reducing sugars subsequently fermented with *Saccharomyces cerevisiae*, resulting in an ethanol yield of 0.042 g/g reducing sugars. Similarly, SHF was applied to *U. pinnatifida*, using *Pichia angophorae* as the fermenting organism, achieving an ethanol yield of 0.33 g/g reducing sugars. Another study utilized SHF for bioethanol production from agar extraction pulp of *G. Verrucosa* [35], with enzymatic hydrolysis and fermentation by *S. cerevisiae* yielding an ethanol yield of 0.43 g/g reducing sugars [1].

7.11.2 Simultaneous Saccharification and Fermentation (SSF)

In Simultaneous Saccharification and Fermentation (SSF), both enzyme saccharification and fermentation with organisms like yeast occur concurrently in the same reactor, typically under similar operational conditions. While SSF offers cost-effectiveness by combining hydrolysis and fermentation in one step, achieving optimal conditions for both processes simultaneously can be challenging. This is due to variations in temperature requirements, with enzymatic hydrolysis typically occurring at 45–55 °C and fermentation at 28–40 °C [1]. For example, SSF was utilized in the production of bioethanol from brown algae, *L. Japonica,* after pre-treatment with 0.06% H_2SO_4 at 170 °C for 15 min. Cellulase and *S. cerevisiae* were employed as the enzyme and fermenting organism, respectively, at temperatures ranging from 37 to 46 °C for 48 h at 130 rpm, resulting in an ethanol concentration of 6.65 g/L. [24]. SSF has been shown to yield higher concentrations compared to SHF, as evidenced in a study on *G. amansii*, where SSF achieved maximum ethanol concentrations of 3.78 g/L compared to 3.33 g/L for SHF [37]. Conversion efficiencies of 84.9% and 74.7% were achieved for SSF and SHF, respectively. These findings suggest that SSF not only offers higher efficiency but also saves time. However, the limited application of SSF in bioethanol processes, particularly due to enzyme catalyst usage, remains a significant challenge [1].

7.11.3 Other Fermentation Methods

In addition to SSF and SHF, there are alternative fermentation pathways such as Simultaneous Saccharification and Co-fermentation (SSCF) and Consolidated Bioprocessing (CBP). CBP processes occur in a single reactor, combining cellulose production, biomass hydrolysis, and fermentation of hydrolysates into ethanol.

These processes utilize single or combined consortia of microorganisms, reducing capital costs. Microorganisms capable of CBP include *Clostridium thermocellum, Neurospora crassa, Fusarium oxysporum,* and *Paecilomyces sp.* SSCF involves simultaneous fermentation of multiple substrates with one or more inoculum in a single system [1]. It may also encompass fermentation of sugars from different process streams. For instance, SSCF resulted in 64.3 g/L ethanol yield using *K. alvarezii* as feedstock, with separate fermentation of galactose-rich liquid fraction and cellulose-rich solid fraction. In comparison, the same study utilized SSF separately for liquid and solid fractions, yielding ethanol concentrations of 38 g/L and 53 g/L, respectively [31]. These findings suggest that while SSCF is valuable, separate application of SSF may lead to higher combined ethanol yields from various fractions of the same biomass. Co-fermenting microorganisms for SSCF include strains of *S. cerevisiae, E. coli, Z. mobilis, P. tannophilus, C. shehatae,* and *P. Stipitis* [1].

7.11.4 Organisms Used in Fermentation

Different fungal and bacterial species have been utilized for fermenting reducing sugars into ethanol, particularly from seaweed sources. Species like *S. cerevisiae, P. Angophorae, P. Stipitis,* and *Escherichia coli* are commonly employed. *S. cerevisiae* is favoured due to its affordability and accessibility, with strains like baker's and brewer's yeast readily available. S. *cerevisiae* selectively ferments hexose reducing sugars such as glucose, galactose, and mannose, leaving pentose reducing sugars like xylose and arabinose unfermented [38]. *P. stipitis*, a pentose fermenting organism, has been identified and used in studies, yielding ethanol from hydrolysates of various seaweed species [39].

For instance, ethanol yields of 0.08, 0.07, and 0.05 g/g reducing sugars were obtained from *Ulva* sp., *Gracilaria* sp., and *Costaria costata* hydrolysates, respectively [1]. Additionally, the yeast strain *P. angophorae* (CBS 5830) has been noted for its ability to ferment mannitol, a sugar alcohol present in brown seaweeds, while producing the enzyme laminarinase, which hydrolyzes laminarin. This yeast strain achieved an ethanol concentration of 5.1 g/L with a conversion efficiency of 73% when fermenting the brown seaweed *S. Latissima* [40]. The study also showed varied ethanol concentrations when mannitol or glucose was used as substrates, indicating the potential of different microorganisms for efficient glucose and mannitol conversion in seaweed fermentation [41].

Furthermore, acclimating yeasts to higher concentrations of specific substrates has been explored to improve ethanol yield. A two-stage fermentation approach was employed on hydrolysates from *S. Japonica,* with *S. cerevisiae* targeting glucose conversion to ethanol in the first stage and *P. angophorae* targeting mannitol conversion in the second stage [30]. Acclimated *P. angophorae* showed significantly improved ethanol production compared to non-acclimated strains. Despite these advancements, a significant challenge in seaweed hydrolysate fermentation remains the identification of organisms capable of efficiently converting the diverse sugars found in seaweeds, including both hexoses, pentoses, and sugar alcohols. Additionally, there is potential for further exploration regarding the effect of different yeast strains on ethanol yield when acclimated to pentoses like xylose and arabinose, an area currently lacking in literature [30].

7.12 Bioethanol Production

Bioethanol production from *Sargassum* spp. achieved an impressive 89% conversion yield. *Gracilaria verrucosa*, a type of red seaweed, was utilized for bioethanol production with yield of 0.43 g/g sugars. An experiment with *Ulva reticulate* macroalgae resulted in a bioethanol concentration of 0.37% v/v, equivalent to approximately 90 litres of ethanol per dry tonne of macroalgae [42]. From *Ulva intestinalis,* a bioethanol yield of 0.081 g/g dry weight was recovered. Saccharification and fermentation methods achieved a remarkable 90.9% bioethanol conversion yield from treated seaweed waste. In anaerobic fermentation using *B. Custersii,* a batch reactor produced 11.8 g/L ethanol from 90 g/L sugar, while a continuous reactor yielded approximately 27.6 g/L ethanol from 72.2 g/L sugar. *E. Cottonii* was identified as a promising feedstock for bioethanol production [1]. Red algae, *Palmira palmata,* rich in carrageenan, released glucose, galactose, and sugars through acid hydrolysis, subsequently fermented to ethanol. Saccharification of *Kappaphycus alvarezii* at 100 °C in 0.9 M H_2SO_4 yielded 26.2% and 30.6% (w/w) at laboratory and bench scales, respectively. Enhanced laminarin hydrolysis was achieved by washing macroalgae in acidic water at 65 °C. Various macroalgal species' bioethanol production data are summarized in Table 7.2 [5].

Table 7.2 Summaraisation of Various macroalgal species' bioethanol production data

Species type	Pre-treatment methods of conversion techniques	Pre-treatment or conversion technique conditions	Bioethanol yield or production potential
Chaetomorpha linum	Wet oxidation method	Temperature-200 °C	44 g ethanol/100 g glucan
Saccharina japonica	Low-acid pre-treatment	Acid-0.06% (w/w) sulphuric acid, temperature-170 °C, time-15 min	6.65 g/L
Saccharina japonica	Thermal acid hydrolysis	Acid-40 mM H_2SO_4, temperature-121 °C, time-60 min	7.7 g/L
Laminaria digitata	Oven drying	Temperature-70 °C, time-72 h	13.6 ± 0.2 μL/g
Ulva linza	Mild acid hydrolysis	Acid condition-3% H_2SO_4	12.01%

7.13 Bioethanol Recovery Processes

Bioethanol finds its most extensive application as a fuel for engines and as a fuel additive. Following fermentation, it undergoes refining through distillation and dehydration processes. Distillation, being the primary method, extracts ethanol from the fermentation broth—a mixture of ethanol, solid biomass residue, residual sugars, enzymes (if used), and fermenting organisms. This stage is energy-intensive mainly due to the heat required to vaporize ethanol–water mixtures. Distillation units and rotary evaporators are employed for analytical and small-scale research purposes, while larger distillation columns are utilized in commercial processes. Commercial methods combine distillation columns with ethanol dewatering techniques like molecular sieves to surpass the ethanol–water azeotrope. Distillation columns can refine fermentation broths to 95.6% ethanol concentration (the ethanol–water azeotrope), followed by dehydration processes such as vacuum distillation, pressure swing, membranes, or molecular sieves to achieve >99% ethanol purity. Molecular sieves, such as zeolite, exploit the difference in size between water and ethanol molecules to selectively adsorb water. They offer an advantage in dehydration as they can be continuously regenerated for reuse through drying [1].

7.14 Limitations and Benefits of Algal Biofuels

Algal biofuels, unlike first- and second-generation counterparts, face challenges in competing with fossil fuels. The considerable energy needs of equipment and capital for augmented agronomy rotations result in a lower net energy return. Photo bioreactors (PBRs) incur higher cultivation costs, mainly due to energy-intensive construction and culture circulation. Financial risks in algae cultivation include weather and pricing changes, with the US 2018 Farm Bill recognizing algae for crop insurance. Despite uncertainties, advancements in algae strains and manufacturing tools could enhance net energy balance. Algae-derived biofuels show potential for air travel petroleum due to their dense energy characteristics [11].

Algae offer numerous by-products for commercial use, impacting market feasibility. Comparatively, PBRs exhibit more efficient energy ratios than open pond systems. Biomass production via algae photosynthesis is more effective, but the conversion process remains economically challenging. Commercial-scale microalgae production is expected to yield positive net fossil fuel by-products. Recycling outlet gas during cultivation and introducing fuel gas as carbon dioxide input can decrease carbon emissions. Microalgae, requiring inorganic compounds like nitrogen, present opportunities for wastewater treatment. The advantage of microalgae not competing for food resources addresses concerns about biofuel impacts on food security and agronomical resources. Algae cultivation, especially in wastewater, reduces water competition. The cultivation of algae minimizes land use compared to land-dwelling biomass, reducing the fuel versus food debate. Algal biofuel production has socio-economic benefits, contributing to public sustainability, economic growth, and job opportunities. Industries based on algae provide chances for economic enhancement, particularly in rural and topical areas, overcoming limitations of first- and second-generation biofuels [11].

7.15 Conclusion

Bioethanol production from seaweeds holds significant promise as a sustainable and renewable energy source. With its abundant availability, minimal competition for arable land, and potential for carbon neutrality, seaweed-based bioethanol presents a viable alternative to traditional feedstocks. Ongoing research and innovation are likely to refine cultivation techniques, optimize conversion processes, and enhance overall efficiency. As global energy demands continue to rise, seaweed bioethanol stands poised to play a pivotal role in meeting these needs sustainably. Moreover, the versatility of seaweed cultivation, which does not compete with traditional agriculture for land and freshwater resources, positions it as a resilient and scalable solution. Collaborative efforts among researchers, industries, and policymakers are crucial to creating an enabling environment for the widespread adoption of seaweed-based bioethanol production.

References

1. Offei F, Mensah M, Thygesen A, Kemausuor F (2018) Seaweed bioethanol production: a process selection review on hydrolysis and fermentation. Fermentation 4(4):99. https://doi.org/10.3390/fermentation4040099
2. Food and Agriculture Organisation of the United Nations (2010) The state of world fisheries and aquaculture. FAO, Rome, Italy
3. Trivedi N, Gupta V, Reddy CR, Jha B (2015) Marine macroalgal biomass as a renewable source of bioethanol. In: Marine bioenergy: trends and developments. CRC Publisher, Taylor and Francis Group, USA. https://doi.org/10.1201/b18494-16
4. Suraiya S, Park HD, Jang WJ, Choi YB, Rafiquzzaman SM, Sarker MK, Kong IS (2020) Utilization of Saccharina japonica as a solid-fermented substrate for the production of bioactive ethanolic extract. Waste Biomass Valorization 11:1917–1928. https://doi.org/10.1007/s12649-018-0503-7
5. Godvin Sharmila V, Dinesh Kumar M, Pugazhendi A, Bajhaiya AK, Gugulothu P, Rajesh BJ (2021) Biofuel production from Macroalgae: present scenario and future scope. Bioengineered 12(2):9216–9238
6. Fernand F, Israel A, Skjermo J et al (2017) Offshore macroalgae biomass for bioenergy production: environmental aspects, technological achievements and challenges. Renew Sust Energ Rev 75:35–45
7. Peteiro C, Sánchez N, Dueñas-Liaño C et al (2013) Open-sea cultivation by transplanting young fronds of the kelp Saccharina latissima. J Appl Phycol 26(1):519–528
8. Sebök S, Herppich WB, Hanelt D (2017) Development of an innovative ring-shaped cultivation system for a land-based cultivation of marine macroalgae. Aquac Eng 77:33–41
9. Yong YS, Yong WTL, Anton A (2013) Analysis of formulae for determination of seaweed growth rate. J Appl Phycol 25(6):1831–1834
10. Khoo KS, Chew KW, Yew GY et al (2020) Integrated ultrasonication assisted liquid biphasic flotation for efficient extraction of astaxanthin from *Haematococcus pluvialis*. Ultrason Sonochem 105052:19

11. Mahmood T, Hussain N, Shahbaz A, Mulla SI, Iqbal HMN, Bilal M (2023) Sustainable production of biofuels from the algae-derived biomass. Bioprocess Biosyst Eng 46(8):1077–1097
12. Singh SP, Singh P (2015) Effect of temperature and light on the growth of algae species: a review. Renew Sust Energ Rev 50:431–444
13. Chowdury KH, Nahar N, Deb UK (2020) The growth factors involved in microalgae cultivation for biofuel production: a review. Comput Water Energy Environ Eng 09(04):185–215
14. Tang D, Han W, Li P, Miao X, Zhong J (2011) CO_2 fixation and fatty acid composition of *Scenedesmus obliquus* and *Chlorella pyrenoidosa* in response to different CO_2 levels. Bioresour Technol 102(3):3071–3076
15. Juneja A, Ceballos RM, Murthy GS (2013) Effects of environmental factors and nutrient availability on the biochemical composition of algae for biofuels production: a review. Energies 6(9):4607–4638
16. Parthasarathy P, Narayanan SK (2014) Effect of hydrothermal carbonization reaction parameters on. Environ Prog Sustain Energy 33(3):676–680
17. Mehrabadi A, Farid MM, Craggs R (2017) Potential of five different isolated colonial algal species for wastewater treatment and biomass energy production. Algal Res 21:1–8
18. Hansen AC, Zhang Q, Lyne PWL (2005) Ethanol–diesel fuel blends—a review. Bioresour Technol 96:277–285
19. Yanagisawa M, Kawai S, Murata K (2013) Strategies for the production of high concentrations of bioethanol from seaweeds. Bioengineered 4(4):224–235
20. Habig C, DeBusk TA, Ryther JH (1984) The effect of nitrogen content on methane production by the marine algae Gracilaria tikvahiae and Ulva sp. Biomass 4:239–251
21. Jung KA, Lim S-R, Kim Y, Park JM (2013) Potentials of macroalgae as feedstocks for biorefinery. Bioresour Technol 135:182–190
22. Fleurence J (1999) Seaweed proteins: biochemical, nutritional aspects and potential uses. Trends Food Sci Technol 10:25–28
23. Trivedi N, Baghel RS, Bothwell J, Gupta V, Reddy CRK, Lali AM, Jha B (2016) An integrated process for the extraction of fuel and chemicals from marine macroalgal biomass. Sci Rep 6:1–8
24. Ye Lee J, Li P, Lee J, Ryu HJ, Oh KK (2013) Ethanol production from *Saccharina japonica* using an optimized extremely low acid pre-treatment followed by simultaneous saccharification and fermentation. Bioresour Technol 127:119–125
25. Rhein-Knudsen N, Ale MT, Meyer AS (2015) Seaweed hydrocolloid production: an update on enzyme assisted extraction and modification technologies. Mar Drugs 13:3340–3359
26. Adams JMM, Schmidt A, Gallagher JA (2015) The impact of sample preparation of the macroalgae Laminaria digitata on the production of the biofuels bioethanol and biomethane. J Appl Phycol 27:985–991
27. Enquist-Newman M, Faust AME, Bravo DD, Santos CNS, Raisner RM, Hanel A, Sarvabhowman P, Le C, Regitsky DD, Cooper SR et al (2014) Efficient ethanol production from brown macroalgae sugars by a synthetic yeast platform. Nature 505:239
28. Lee KT, Ofori-Boateng C (2013) Biofuels: production technologies, global profile, and market potentials. In: Sustainability of biofuel production from oil palm biomass. Springer, Singapore, pp 31–74
29. Meinita MDN, Marhaeni B, Winanto T, Setyaningsih D, Hong YK (2015) Catalytic efficiency of sulfuric and hydrochloric acids for the hydrolysis of *Gelidium latifolium* (Gelidiales, Rhodophyta) in bioethanol production. J Ind Eng Chem 27:108–114
30. Ra CH, Jeong GT, Shin MK, Kim SK (2013) Biotransformation of 5-hydroxymethylfurfural (HMF) by *Scheffersomyces stipitis* during ethanol fermentation of hydrolysate of the seaweed Gelidium amansii. Bioresour Technol 140:421–425
31. Hargreaves PI, Barcelos CA, da Costa ACA, Pereira N (2013) Production of ethanol 3G from Kappaphycus alvarezii: evaluation of different process strategies. Bioresour Technol 134:257–263
32. Tan IS, Lee KT (2014) Enzymatic hydrolysis and fermentation of seaweed solid wastes for bioethanol production: an optimization study. Energy 78:53–62

33. Karray R, Hamza M, Sayadi S (2015) Evaluation of ultrasonic, acid, thermo-alkaline and enzymatic pre-treatment on anaerobic digestion of Ulva rigida for biogas production. Bioresour Technol 187:205–213
34. Gupta A, Prakash J (2015) Sustainable bio-ethanol production from agro-residues: a review. Renew Sust Energ Rev 41:550–567
35. Kumar S, Gupta R, Kumar G, Sahoo D, Kuhad RC (2013) Bioethanol production from *Gracilaria verrucosa*, a red alga, in a biorefinery approach. Bioresour Technol 135:150–156
36. Borines MG, de Leon RL, Cuello JL (2013) Bioethanol production from the macroalgae *Sargassum* spp. Bioresour Technol 138:22–29
37. Kim HM, Wi SG, Jung S, Song Y, Bae HJ (2015) Efficient approach for bioethanol production from red seaweed *Gelidium amansii*. Bioresour Technol 175:128–134
38. Abbi M, Kuhad RC, Singh A (1996) Bioconversion of pentose sugars to ethanol by free and immobilized cells of Candida shehatae (NCL-3501): fermentation behaviour. Process Biochem 31:555–560
39. Takagi T, Uchida M, Matsushima R, Kodama H, Takeda T, Ishida M, Urano N (2015) Comparison of ethanol productivity among yeast strains using three different seaweeds. Fish Sci 81:763–770
40. Adams JMM, Toop TA, Donnison IS, Gallagher JA (2011) Seasonal variation in *Laminaria digitata* and its impact on biochemical conversion routes to biofuels. Bioresour Technol 102:9976–9984
41. Ota A, Kawai S, Oda H, Iohara K, Murata K (2013) Production of ethanol from mannitol by the yeast strain Saccharomyces paradoxus NBRC 0259. J Biosci Bioeng 116:327–332
42. Yoza BA, Masutani EM (2013) The analysis of macroalgae biomass found around Hawaii for bioethanol production. Environ Technol 34(13–14):1859–1867

Valorization of Marine Waste Towards the Production of High-Value-Added Products, Bioplastics, and Other Industrial Applications

8

Deepshikha Datta, Sayantan Sarkar, Soheli Biswas, Esha Mandal, and Bimal Das

Abstract

The maritime environment is a largely unknown and underappreciated source of functional components that can be employed in a variety of applications such as food preparation, storage, and enrichment. Marine microorganisms could provide new bioactive compounds with potential human use. Marine biomass embodies a biotechnological resource characterized by a wide range of compositions and functional attributes stemming from diverse bioactive elements such as polyphenols, peptides, and polysaccharides. Covering over 70% of the Earth's surface, oceans and seas are vital contributors to the global economy, serving as both critical zones for the fishing industry and essential logistical pathways for mineral extraction and fossil fuel utilization. These activities all result in waste. The creation of high-value-added products from these by-products has accelerated recently due to the discovery of a wide range of biopolymers, nutrients, and functional chemicals that may be used for human consumption or in the pharmaceutical, animal feed, and other sectors. The concept of valorization, a sustainable method of turning marine garbage into useful resources, is explored in this book chapter. Additionally, the book chapter adopts a thorough thematic approach

D. Datta
Department of Chemistry, Brainware University, Barasat, Kolkata, West Bengal, India

S. Sarkar · B. Das (✉)
Department of Chemical Engineering, National Institute of Technology, Durgapur, West Bengal, India
e-mail: bdas.che@nitdgp.ac.in

S. Biswas
Department of Biotechnology, Brainware University, Barasat, Kolkata, West Bengal, India

E. Mandal
Department of Biotechnology, National Institute of Technology, Durgapur, West Bengal, India

M. Rafatullah et al. (eds.), *Multidisciplinary Applications of Marine Resources*,
https://doi.org/10.1007/978-981-97-5057-3_8

to the valorization of marine trash, covering the main contributions made by marine biotechnology to the creation of a circular economy. These contributions range from energy and valorization for biomedical enrolments to bio remedial enrolments for pollution. The chapter goes on to discuss the manufacture of various value-added goods such as fish meal, biodegradable plastics, chitin and chitosan, biofuels, marine collagen, and so on. Besides the different industrial applications of marine waste to fabricate bioactive compounds, construction materials, pharmaceuticals, agricultural fertilizers, etc., are intensively elaborated. The recovery of marine waste represents a promising way to solve environmental problems and, at the same time, creates economic opportunities using innovative technologies and sustainable practices. This work highlights efficient and environmentally conscious use of marine resources, supporting the preservation of our oceans and the well-being of future generations.

Keywords

Waste valorization · Marine biomass · High-value-added products · Industrial applications · Biopolymers

8.1 Introduction

Valorization of marine debris and biopolymers is a promising avenue in the never-ending search for sustainable solutions. It offers a transformative approach to the manufacture of high-value-added products, bioplastics, and a variety of industrial uses. Using the potential of marine trash and biopolymers is an important step in reducing ecological issues, as environmental concerns and knowledge of the harmful effects of typical plastics are expanding [1]. The greatest potential for valorization is seen in marine polysaccharides, which are present in many different marine species. Extended chains of molecules of repetitive monosaccharide units joined by glycosidic linkages define the molecular structure of marine polysaccharides. The seas represent a huge resource that is just waiting to be explored because they are teeming with a variety of biodegradable materials. In addition to addressing the pressing need for waste reduction, this project offers a chance to develop creative, environmentally beneficial substitutes. The process of valorization aims to transform underutilized marine resources into valuable materials by utilizing cutting-edge technologies and sustainable techniques, hence promoting a circular economy. Untapped potential exists in marine garbage to transform the food sector in a number of ways, including enrichment, processing, and storage. First off, marine waste extracts are a source of bioactive components that can be added to food processing to improve nutritional profiles and offer health advantages. Functional foods that support well-being include omega-3 fatty acids, peptides, and antioxidants made from marine debris. Their primary purpose is energy storage, but they also have structural roles. They come from a variety of marine resources, such as marine algae and crustaceans [1, 2]. For the fabrication of innovative biosystems, such as 3D scaffolds, nanofibers, membranes, hydrogels, and bio-inks for tissue engineering,

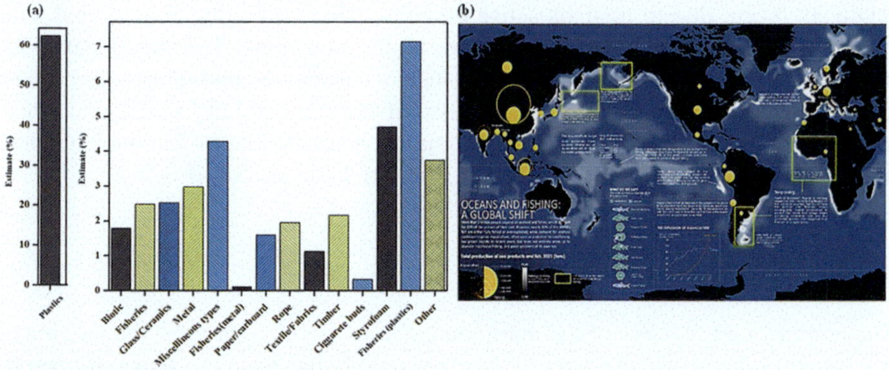

Fig. 8.1 (**a**) Global consumption of marine waste; (**b**) global marine activities [5]

drug delivery, and wound dressing applications, they are great options for inexpensive, renewable, non-toxic, and ubiquitous biomaterials. They manifest an eclectic range of bioactivities, inclusive of anticoagulant, antioxidant, antimicrobial, anticancer, immunomodulatory, and antiviral properties [3]. Because aquatic biomass contains a wide range of bioactive chemicals, such as polyphenols and peptide-polysaccharides, it offers ecological resource with striking variation in both functional attributes and composition [4]. Materials derived from marine garbage can be used in packaging solutions for storage. Packaging that is biodegradable and made from marine resources provides a less harmful option for the environment than traditional plastics. Furthermore, marine biopolymers have barrier qualities that extend the freshness and quality of packaged goods by improving their shelf life. Moreover, marine trash has the ability to improve food goods. Adding components from the sea to recipes adds new textures and nutritional aspects in addition to a wider variety of flavors. Seaweeds, for example, are a great addition to fortified foods due to their high vitamin and mineral content.

This chapter investigates the marine biomass diversity, impact of human activities, marine environment and biomass as a resource, and some sources of unused biomass. Then, it also discusses about the valorization concept, production and industrial applications of value-added products, and their impacts on environment. The global consumption estimation of marine litter waste is illustrated in Fig. 8.1a. Figure 8.1b highlights the extent of global marine activities.

8.2 Marine Biomass Diversity

With its extraordinary biodiversity, the aquatic bionetwork is the ironic origin of life. It stands for unexplored sources of functional ingredients connected to a variety of biological processes. When compared to bioactive substances originating from terrestrial sources, their discovery and incorporation into food products are relatively recent due to their bioactive qualities. There are a lot of neglected and underappreciated marine animals that are highly nutritious. The functional food industry can benefit greatly from a wide range of substances, including polysaccharides, phenolics, pigments, proteins, lipids (polyunsaturated fatty acids), and other secondary metabolites from a variety of marine sources, like prokaryotes, micro- and macroalgae, seaweeds, invertebrates, and vertebrate species (Fig. 8.2).

8.2.1 Marine Microorganisms

The most vibrant features of the maritime environs are thought to be the microbial life found in the ocean. As a result, marine ecology has been the focus of several studies and is well recognized for its extremely diverse character. Furthermore, it has drawn increasing interest as a possible source of bioactive chemicals, mostly because of its distinct qualities linked to deep-sea adaptations to harsh environments [8]. Unique metabolic and physiological capacities have been made possible by the capability to survive in harsh aquatic conditions with an eclectic range of parameters, including temperature, pressure, and salinity; in situations where symbiosis

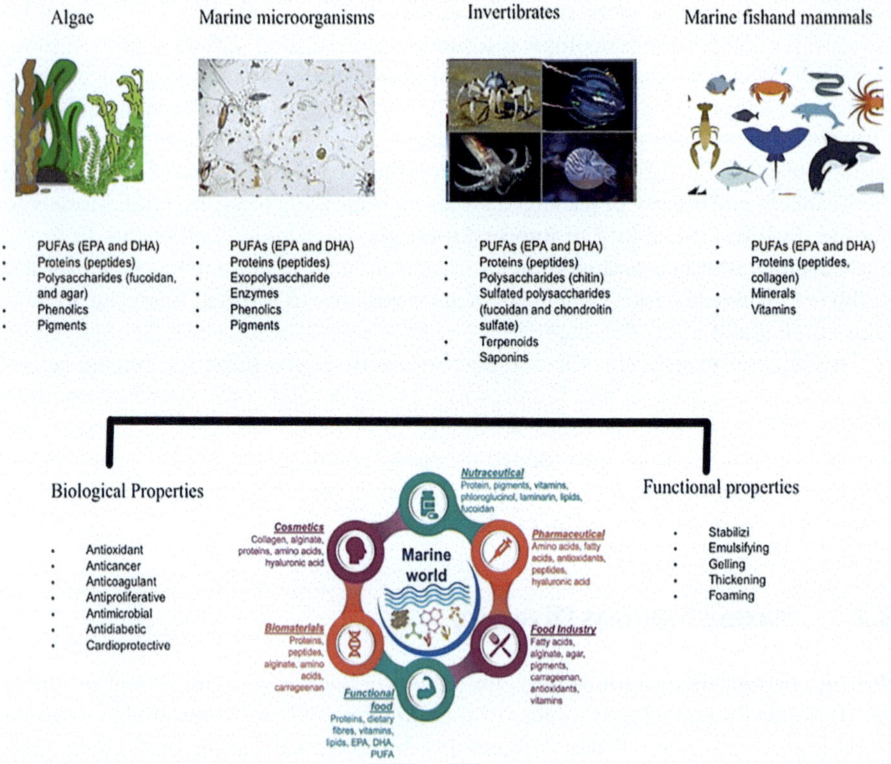

Fig. 8.2 Marine biomass diversity [8, 21]

and competition for resources are present; and in highly oxidative conditions. Many bioactive substances, including vitamins, enzymes, antibiotics, biosurfactants, and bio-emulsifiers, have been produced as a result of these abilities and have been identified as beneficial to the food and pharmaceutical industries [9].

8.2.1.1 Bioactive Polysaccharides from Marine Microbiota

The food industry frequently uses marine microbially produced polysaccharides as emulsifiers, stabilizers, gelling, and water-thickening agents. These microbial polysaccharides can be divided into three main classes based on where they are located within the microbial cell: cell wall, intercellular, and exocellular polysaccharides. Because of their purity and simplicity of isolation, exopolysaccharides, also known as exocellular polysaccharides (EPSs), are particularly attractive forms of polysaccharides for industrial applications. The majority of microbes release these extracellular polymeric substances into their exocellular milieu, where their principal function is to safeguard the cell within its surroundings [6, 9].

8.2.1.2 Proteins and Bioactive Peptides Derived from Microbiota

Microorganisms, such as Chlorella, Dunaliella, Micractinium, Spirulina, Oscillatoria, Scenedesmus, Chlamydomonas, and Euglena provide high-quality protein sources [10]. Furthermore, because of their advantageous amino acid profiles, these microalgal species are thought to provide alternate sources of protein for culinary applications. Specifically, it has been determined that S. platensis is a significant producer of bioactive proteins since it can extract over 75% of the protein nitrogen from biomass that has been defatted with hexane [11].

8.2.1.3 Other Bioactive Compounds from Microbiota

Marine microorganisms that create polyunsaturated long-chain fatty acids (PUFAs) in the marine environment include bacteria, protists, yeast, dinophyllates, and filamentous fungi. Furthermore, it has been discovered that several marine-originating bacterial species, such as Vibrio marinus, Moritella marina, and Shewanella marinintestina, contain long-chain polyunsaturated fatty acids (PUFAs), predominantly those in the omega-3 and omega-6 series, which include eicosapentaenoic acid (EPA) and docosahexaenoic acid (DHA) [6].

8.2.2 Marine Algae

With marine algae making up more than 90% of all marine plants, they are an important source of bioactive substances for the environment. These photosynthetic organisms, which range in size from unicellular microscopic microalgae to multicellular macroalgae of different sizes, are characterized by the absence of genuine stems or roots [12]. The latter are more widely known as seaweed and are divided into three families according to the color they have: green (Chlorophyta), brown (Pheophyta), and red (Rhodophyta). Numerous factors, including as their geographic origin, species, seasons, and environmental circumstances, physiological fluctuations, and processing techniques, might affect their nutritional and chemical contents [6].

8.2.2.1 Algae Polysaccharides

Polysaccharides, primarily SPs, found in seaweed are abundant and have promising biological and physicochemical properties, indicating their potential use across various industries, including food production. Seaweed has a broad variety of polysaccharides, most of which are found in cell wall structures and as storage polysaccharides. On a dry weight basis, the proportion of Polysaccharides in seaweed generas can reach up to 75%; green seaweeds, primarily Ulva, have the highest concentration of polysaccharides [13]. For instance, ulvan, a functional, structural, and storage polysaccharide, makes up the majority of the over 80% polysaccharides found in U. lactuca [14]. The majority of these polysaccharides are found in food, and seaweed's total fiber content can range from 29% to 62% [15]. Dietary fibers support gut health by preserving the intestinal environment.

8.2.2.2 Protein, Peptides, and Amino Acids from Algae

A substantial supply of proteins, amino acids, and peptides with diverse biological activity derived from marine sources include macroalgae. These proteins can frequently be found in cell walls with colors, polysaccharides, and enzymes [12]. The species, geographic location, season, and growing conditions of macroalgae all affect their protein content. In general, the protein content of red and green seaweeds is higher (9–47%) than that of brown seaweeds (4–24%). Studies reveal that macroalgae contain every essential amino acid, even in spite of variations in their amounts. Seaweed may be used as a substitute source of protein because its amino acid composition matches the FAO/WHO's recommendations for dietary protein [6]. Furthermore, while seaweed may be low in graded limiting amino acids like cysteine, tryptophan, lysine, and methionine, it is high in acidic amino acids such as glutamic and aspartic acids [12].

8.2.2.3 Other Bioactive Components from Algae

The two main lipid classes found in seaweed, phospholipids and glycolipids, account for 1–5% of their cell walls. Seaweed is rich in polyunsaturated fatty acids (PUFAs) even though its lipid content is smaller than that of terrestrial plants like sunflower and soy. Different factors, including species, season, location, salinity, temperature, and light intensity, affect the lipid composition of seaweed [12]. Other useful substances derived from algae include vitamins, phenolic compounds, and pigments. Autotrophic organisms need pigments that are photosynthetic to capture sun energy during photosynthesis, which is why algae have a large number of these chemicals [16]. Phycobiliproteins, carotenoids, and chlorophylls are the three main categories of photosynthetic pigments found in marine organisms. Strong antioxidants, carotenoids neutralize reactive oxygen species (ROS).

8.2.3 Marine Invertebrates

As potential sources of new bioactive substances, marine invertebrates, such as sponges, mollusks, echinoderms, and crustaceans, have drawn interest from researchers and businesses all over the world. These multicellular non-chordates come from a variety of species, ranging from the phylum Echinodermata to the phylum Porifera [17]. Their capacity to flourish in harsh halo biotic settings and their metabolic processes have played an important part in the synthesis of a wide range of bioactive substances. Numerous biological actions, such as antioxidant, antibacterial, antiviral, anthelmintic, antifungal, anticancer, antihypertensive, and immune-modulatory qualities, are displayed by these compounds with distinctive structural features [17–19]. Furthermore, a great deal of study on marine bio activities has shown that the most intriguing collection for the discovery of marine natural products is marine invertebrates, or non-chordates.

8.2.4 Marine Fish

The main source of protein in the diet among the many varieties of seafood is marine fish. The global economy depends heavily on marine fish, particularly per capita finfish demand was approximately 13.7 kg in 2010, and it is estimated that by 2030, annual fish consumption will increase by approximately 202 million tons, or 20–22 kg per person [20]. The growing population and consumer concerns over a healthy diet are linked to the rising demand for marine fish for human consumption. Functional materials such as polyunsaturated fatty acids, enzymes, essential fatty acids, minerals, vitamins, and bioactive components may generally be found in abundance in fish. Particularly, marine fish are said to be a great supplier of macromolecules such as premium protein, in addition to fats and a broad range of minerals. A significant portion of the essential amino acid requirements are met by a variety of fish products, including fresh, dried, fermented, and salted fish. Because they are low in calories and saturated fat, they can also be the ideal healthy food [6, 21].

8.3 The Impact of Human Activities

Human activities affect marine life and marine environments through overfishing, habitat loss, invasive species introduction, ocean pollution, ocean acidification, and ocean warming. These have an effect on marine food webs and ecosystems, and they might have unintended repercussions for the persistence and biodiversity of marine life [22].

Marine pollution is the result of things used or dispersed by humans that enter the ocean and have an adverse effect there, such as invading creatures, noise, particles, agricultural, industrial, and residential waste, and excess carbon dioxide. Although marine transportation also makes a considerable contribution, land-based activity accounts for the majority of this waste (80%) [23]. It is a concoction of chemicals and debris, the majority of which is blown or washed into the ocean from terrestrial sources. The ecology, the health of all living things, and global economic systems are all harmed by this pollution [24]. Considering that most of the inputs originate on land, through sewage, continental shelves, rivers, or the atmosphere, are more susceptible to contamination. By releasing dust, pesticides, nitrogen, sulfur, silicon, iron, carbonic acid, and silicon into the water, air pollution also plays a role [25]. Nonpoint causes of pollution include dust, wind-blown debris, and agricultural runoff. The primary cause of these nonpoint sources is runoff that flows into rivers and into the ocean, but dust and debris carried by the wind can also contribute since they can find their way into rivers and the ocean [26]. Pathways of contamination include direct discharge, land runoff, ship pollution, bilge pollution, air pollution, and perhaps deep sea mining.

Marine wreckages pollution, plastic contamination (inclusive of microplastics), acidification of ocean, nutrient pollution, toxin pollution, and undersea noise pollution are the main categories of marine pollution. Ocean plastic pollution is a form of marine pollution caused by plastics, which can range in size from large original materials like bottles and bags to tiny particles created when plastic material breaks into smaller pieces. The majority of marine wreckage is abandoned human trash that drifts or else hangs in the water. Marine life is harmed by plastic pollution. Untreated or poorly treated sewage being dumped into rivers and finally the ocean, as well as the runoff of nutrients (phosphorus and nitrogen) from intensive agriculture, are more causes for concern. These nutrients—nitrogen and phosphorus—which are also found in fertilizers, promote the growth of macroalgae and phytoplankton. This can result in eutrophication, or hazardous algal blooms, which can be dangerous to both humans and marine life. In addition to suffocating delicate coral reefs, excessive algae development can cause a decline in coral health and biodiversity. The decomposition of algal blooms can result in oxygen consumption in coastal waters, which is a serious concern. As a result of reduced vertical mixing of the water column caused by climate change, this scenario could get worse [27].

- Fishing: Ghost nets contribute significantly to marine pollution, entangling marine life and persisting in the oceans [22].
- Shipping: Ballast water discharge introduces invasive species and pollutants, impacting local ecosystems [28, 29].
- Resource Extraction: Drilling and extraction activities generate sediments, chemicals, and occasional oil spills, posing threats to marine environments [30].
- Aquaculture: Excess feed and fecal matter from aquaculture operations can lead to nutrient imbalances and water quality issues [31].
- Coastal Development: Construction debris from coastal development projects includes materials like concrete and plastic, contributing to coastal pollution [32, 33] (Table 8.1).

Table 8.1 Global marine waste production (A table showing statistics on the amount of waste produced by different marine activities, such as fishing, shipping, and resource extraction, with data on various types of waste)

Marine activities	Types of waste	Annual quantity	References
Fishing	Discarded gears (Ghost nets)	Roughly 10% of the debris is made up of fishing gear, with 500,000–1 million tons being lost or thrown into the ocean annually. Currently, almost 46% of the debris in the Great Pacific Garbage Patch is made up of discarded nets, lines, and ropes	[34]
	Plastic packaging	The majority of marine pollution is composed of plastic waste, with 8–10 million metric tons entering the ocean annually	[22]
	Organic waste	The FAO estimates that the annual loss from global fisheries, including byproducts, is about 20 million tons (or 25%)	[35]
Shipping	Ballast water	In order to keep their stability and navigate safely, many ships carry between 30% and 50% of their total cargo in ballast water. This indicates a range of volumes per ship from a few hundred liters to over ten million liters	[28]
	Plastic debris	In addition to the 200 million metric tons of plastic that are thought to be in our marine habitats right now, an additional 11 million metric tons of plastic enter our ocean annually	[29]
Resource extraction	Drilling mud	According to the American Petroleum Institute (API 1995), exploration and production activities produced over 150 million barrels of drilling waste.	[30]
	Metal and chemical waste	One million tons of heavy metals in industrial trash, two to four million tons of acid chemical wastes from pulp mills, 100 million tons of petroleum products, and over 100,000 tons of organic chemical wastes	[36]
	Sediment discharge	The study estimates that there is approximately 3.37 × 108 cubic kilometers of silt in the planet's oceans. That is nearly enough silt to create a layer 2 km deep over all of the continents of Earth	[37]
Aquaculture	Excess feedback and fecal matter	Fecal matter containing microbes is introduced into rivers, lakes, and storm drains that eventually run into the ocean through sewage spills, overflows, and contaminated rainwater. Annually 800,000 tons are released into marine world	[31]

(continued)

Table 8.1 (continued)

Marine activities	Types of waste	Annual quantity	References
Coastal development	Construction debris	Mishandled waste on land eventually makes its way into the ocean through rivers, lakes, and other waterways, and is the main cause of ocean plastic pollution. Our oceans are thought to contain 5.25 trillion bits of marine garbage. Four billion microfibers per km^2 are found below the surface, and 269,000 tons float	[32, 33]

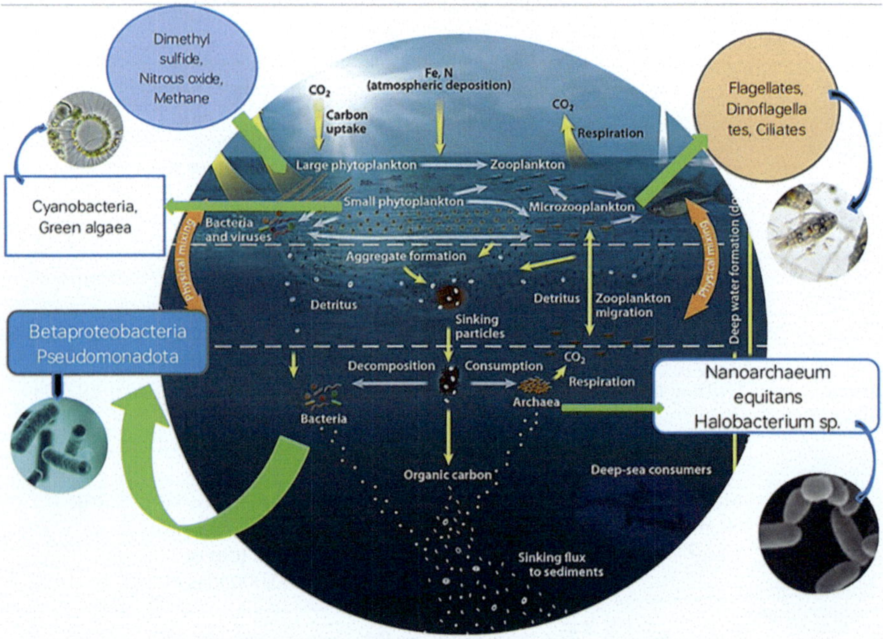

Fig. 8.3 Marine environment components [38]

8.4 Marine Environment and Biomass as a Resource

The maritime environment is a huge, complex ecosystem that is rich in resources and biodiversity, with a great deal of potential for use in a variety of human endeavors. Using marine biomass as a resource is a sustainable strategy that has broad ramifications for many different industries. Figure 8.3 represents an illustrative diagram showing the various components of the marine environment.

8.4.1 The Primary Producers

Phytoplankton, which are microscopic plant organisms suspended in natural waters over 40–60 m of the upper water column, accounts for nearly 0.3% of solar energy incident on the sea surface. This amounts to 75% of the primary productivity of an area of the world oceans approximately 3.5.1014 square meters, with macro algae producing the remaining 25%. Based on primary production by phytoplankton, the total biomass of all consumers ranges from 0.05 to 0.5 gCm^{-2} d^{-1}. However, it can reach as high as 5 gCm^{-2} d^{-1} in some very productive upwelling zones or in specific grass beds [39–41]. Due to this photosynthetic process, the sea produces 1.5 × 1010 mt of carbon annually net, or 15.5 × 1010 mt of carbon gross output, with the majority of that production occurring in shore waters. The amount of fisheries produced can go so far since, after accounting for energetic efficiency, these values equal 8% of the world's aquatic primary production [41, 42]. In a nutshell, the continuance of life as we know it depends on an ongoing energy intake that all living things on Earth constantly disperse after the plants have fixed it and transformed it into chemical energy. All living things degrade or disperse energy, and biological production brought forth by light fixation interacts with this process. Agriculture systems are a prominent illustration of how humans have simplified food webs to channel the output of a few species that are exploited by humans. But this suggests that there is a maximum to the amount of biomass that organisms might potentially produce [41].

8.4.2 The Secondary Producers and Food Webs

Although many other authors used the value of 10%, the carbon transfer efficiency in marine webs may be close to 15%, according to the thermodynamics of biomass flow and secondary production. To survive, all consumers need on chemical energy, which is produced by primary producers, passed to different trophic levels by herbivore feeding, and then passed to various animal levels through predation [42, 43].

The term "zooplankton" refers to suspended, free-living animals in the water column that use primary production. The main component of this food web is the group of copepods. A significant number of first-order predators, such as jellyfish and other crustaceans, which are larval stages of benthic organisms that normally spend a few days to several months suspended in the water column as predators of microzooplankton before being re-engaged as the organisms grow, are also included in zooplankton, despite its primary composition being herbivores [4].

Through the food chain, the caloric value of organisms reveals highly consistent characteristics; it is higher in animals that store fat in their bodies. The caloric value of proteins, carbohydrates, and fats is 4100 cal g^{-1}, 9300 cal g^{-1}, respectively. However, the quantity of calories that are available when these substances are not completely oxidized are almost 90% of their entire caloric values. The herbivores would absorb a large amount of biomass produced by the primary producers and move it up the food chain to higher levels. According to the rule of 10%, a high primary output will therefore indicate a high biomass of consumers; that is, for every ton of top predators, there will be 10 mt of first-order predators, and 100 Mt. of herbivores. The predators' biomass falls between 0.5 and 2 g C m^{-2} and adheres to the 10% rule with regard to the lower limit. Because of their rapid turnover rate, primary producers—primarily phytoplankton—may have a smaller biomass than herbivores. It is important to note that substantial amounts of nutrients are rising from the deep sea into upwelling zones of the sea, such as Peru on the west coast of America and West Africa, enriching the surface waters in the photic zone and promoting primary productivity. Evolution has created short food chains in these zones, which the sardine and anchovies take advantage of, exploiting much of this production. This has led to the growth of large schools that humans exploit, with levels of exploitation reaching over 12 million Mt., as was the case in Peru in the early 1970s [43, 44]. Table 8.2 delivers a detailed breakdown of the composition of numerous marine biomass types, emphasizing the percentages of key bioactive compounds.

8.5 Valorization Concept

Concerns about rising global temperatures, shifting climatic conditions, and massive emissions of greenhouse gases (GHGs) from anthropogenic and industrial sources are growing these days [53]. Efforts to implement several economic and social reforms to meet the climate change objectives are underway [54]. One such possible answer is to embrace the circular economy growth paradigm. The circular economy promises to reduce greenhouse gas emissions by changing the traditional linear production and consumption model to a more sustainable circular model that is based on the reuse of biological and technical resources for a longer length of time in closed-loop systems [55]. Waste management has become a popular part of circular economy in recent years. Sustainable waste management, together with circularity, has emerged as the key to environmental protection and conservation [56]. As a result, it is critical to designate numerous methods for processing any waste in order to achieve long-term economic growth and circularity. The main objective of valorization of marine waste is to convert into valuable products that offers the reduction of carbon footprints, and also provides a viable environmental solution [57].

8.6 Importance of Marine Waste Valorization

The expeditious evolution of worldwide residents has persuaded financial growth implementing competitive market policies coupled with cumulative appetite for food, feed, fuel, and varied end-products boosted the requirements for fossil-based resources [58]. It is imperative to suggest the valorization of marine biomass resources considering the scarcity of land resources [1]. The advancement of high-value-added products from marine waste has been prioritized in current decades owed to the detection of several kinds of biopolymers, many nutrients that could

Table 8.2 Marine biomass composition

Marine Biomass	Composition	Polyphenols (%)	Peptides (%)	Polysaccharides (%)	Other bioactive compounds	References
Macroalgae	Diverse seaweeds	Around 12% (Where two polyphenols - fucoxanthin (9.27 ± 2.28 mg/kg DW) and phloroglucinol (17.96 ± 2.80 mg/kg DW) were found to be in greater numbers)	5–47%	15–76%	Carrageenans, alginates, antioxidants, phenolics	[45, 46]
Phytoplankton	Microscopic algae	Around 15%	Around 5–10%	Up to 90%	Essential fatty acids, chlorophyll	[47]
Fishery by-products	Fish and seafood processing waste	6–10%	15–20%	2–5%	Omega-3 fatty acids, collagen	[48]
Shellfish residues	Shells from mollusks and crustaceans	4–5%	Up to 10%	15–20%	Chitin, minerals, terpenoids, PUFAs (Poly-unsaturated fatty acids)	[49]
Deep-sea microbes	Microbial biomass from ocean depths	15–30%	10–15%	20–40%	Enzymes, exopolysaccharides, phenolics	[50]
Jellyfish	Gelatinous marine organisms	Up to 10%	Up to 50% of their dry weight (The high proportion of protein in jellyfish bodies (82 ± 4% of macromolecules)	15–20%	Collagen, mucin	[51, 52]

find application for human consumption, pharmaceuticals, and others [1]. Valorization of marine waste can be classified in two sections i.e. (i) direct valorization of marine biomass and (ii) valorization of marine biopolymers.

8.6.1 Direct Valorization of Marine Biomass

8.6.1.1 Biosorbents

For the purpose of eliminating radionuclides and other organic and inorganic pollutants from residual waters, for example, the majority of materials derived from biotic waste can effectively and directly utilized as easily accessible biosorbents [59, 60]. With the right physical or chemical treatments, the adsorption efficiency can be improved [61, 62]. It is necessary to properly dispose of the depleted biosorbents, which may involve regeneration and repeated biosorption cycles. Following complete exhaustion, the biosorbents can be pyrolyzed to create biochar or utilized as fertilizers for soil restoration [63]. Seagrasses, algae have been broadly considered as biosorbents for the exclusion of various aquatic contaminants (such as organic compounds) [64–66] as well as industrial dyes [67–69]. Because of their special qualities—such as their high alginate content, greater absorption capabilities, and unrestricted accessibility in the oceans—brown algae are highly effective biosorbents for the eradication of various metal ions from water. Biomass from marine macroalgae and seagrass can be obtained in huge numbers at affordable prices, provided that their harvesting is sustainable and doesn't disrupt the ecosystem balance in coastal areas [1].

8.6.1.2 Biofuel

Fossil fuels supply almost 80% of the energy used worldwide [70]. Global energy consumption is rising as a result of urbanization, industrialization, and population expansion, which is overusing the finite supply of natural fossil fuels [71]. Because greenhouse gases (GHGs) are the primary cause of global warming, these fuels pose a serious threat to the environment. This promotes research on biomass-based bioenergy production [72]. Because biofuel and related technologies have sustainable qualities that help meet the world's energy needs, they are regarded as renewable alternatives to fossil fuels [73]. Energy generation from waste biomass using biorefinery technology is an alternate option because biomass production can be highly expensive to meet energy needs alone [74].

8.6.1.3 Fertilizers

Beach debris can be used as a biofertilizer to promote plant culture and growth. Nowadays, biofertilizers are chosen over chemical fertilizers due to their environmental friendliness and financial effectiveness. Biofertilizers are microorganisms that fix nitrogen, solubilize phosphate, and encourage plant development. The shells of many bivalves, e.g., blue mussels and oysters, are rich in $CaCO_3$, a mineral currently mined from limestone, representing a widely exploited resource for numerous industrial applications in agriculture, as a biofilter media for wastewater treatment, or even cement manufacturing [75–77]. Marine organic waste, such as seagrasses washing ashore, can also be regarded an alternative and sustainable fertilizer source because of its abundance of critical macro- and microelements [63]. However, the low concentration of phosphorous and other hazardous substances in the biomass, particularly if the sampling arena is subjected to the high levels of anthropogenic pressure can be a concern for utilizing the beach debris as fertilizer (Fig. 8.4).

8.7 Production of Value-Added Products

8.7.1 Exploration and Production of Various High-Value Products from Marine Waste

Emerging contaminants originating from human activities and industry have wreaked devastation on public health and biodiversity deterioration on many levels in recent years [78]. Industries and public sewer systems handling such waste streams face a variety of issues in effluent treatment and solid waste disposal due to many reasons ranging from production to the adoption of appropriate technology [78]. Therefore, transformation of marine waste into the useful value-added products has been gaining utmost importance to dodge the bottlenecking of this marine waste.

8.7.1.1 Production of Chitin and Chitosan

After cellulose, chitin is the amino-polysaccharide that is most prevalent in the universe of nature. Acetaglucosamine is used in the chitin synthesis pathway seen in

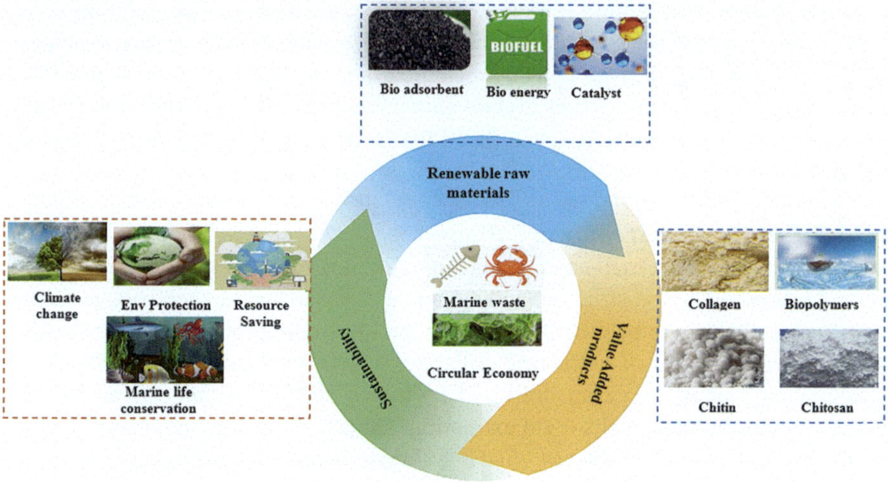

Fig. 8.4 Circular economy in marine biotechnology

many species, where it is generated by the primary key enzyme chitin synthase. Despite not being present in the majority of cellulose-producing organisms, chitin is occasionally regarded as a cellulose derivative because of structural similarities [79]. Furthermore, tunicata, mollusks, arthropods, algae, and protozoa are among the many natural entities that contain chitin. However, the most popular animals utilized to extract low-molecular-weight polymers are arthropods and mollusks [80]. While marine critters like crabs, shrimp, and lobsters are among the most prized seafood delicacies, their shells are sometimes thrown away and dumped into the deep ocean because of their stubborn nature, which pollutes both land and water. Therefore, a variety of biopolymers and molecules exhibiting a range of biological activity can be extracted from the shells. It has been demonstrated that chitin isolated from shell waste has high activity, low toxicity, high biodegradability, and biocompatibility [81]. However, chitosan, the deacetylated derivative of chitin, is among the primary polymers that may be extracted from this waste in addition to chitin. Chitosan is a copolymer of N-acetyl D-glucosamine and D-glucosamine units that exhibits several bioactivities, including antimalarial, anticancer, and antibacterial properties for the production of biofilms, depending on the degree of deacetylation [82]. It is also useful in a variety of scientific domains, including agriculture, food processing, and pharmaceuticals [79]. Several techniques, including (i) chemical treatment and (ii) enzymatic treatment, can be used to remove chitin [83]. The chemical treatment process consists of two steps: demineralization, which uses a strong alkali treatment to remove minerals like calcium and magnesium, and deproteinization, which uses an acid treatment to remove proteins and fatty acids. In addition, chitin can be extracted via lactic acid or non-lactic acid fermentation, which is another method of fermentation. The shells are first enzymatically deproteinized with the help of alkaline and bacterial proteases [84]. By deacetylating chitin with 40–50% strong alkali, such as sodium hydroxide, chitosan is produced. The process of deacetylation involves removing acetyl groups from the structure of chitin and substituting them with amino groups. Furthermore, the degree of acetylation varies with temperature and time, and repeating the alkali treatment is necessary to achieve total deacetylation [85]. Although chemical treatment is the most often used method for this procedure, the chitin deacetylase enzyme, which catalyzes the hydrolysis of N-acetamino links in chitin, is also utilized for the deacetylation of chitin to produce chitosan [86].

8.7.1.2 Production of Marine Collagen

Acid extraction or enzymatic extraction are main methodology of extraction of collagen. Depending on the structure and chemical composition of each part, a distinct extraction technique is needed [87]. A poor yield of collagen was produced using the traditional method of extracting collagen, which involved utilizing EDTA for decalcification of the material, followed by 0.5 M acetic acid and 0.1 acetic acid dialysis to eliminate the salts (acid extraction) [88]. But the enzymatic approach of pepsin has the special ability to cleave collagen at the tropocollagen telopeptide region without jeopardizing the triple helix structure [87].

8.7.1.3 Production of Biopolymers

The valorization of marine waste into a variety of value-added products has grown in importance since it leads to proper fisheries and aquaculture management, is low cost, and ecologically beneficial. A few biopolymers have been extracted from the waste components of certain marine organism species, such as their skeletons, skins, viscera, and heads.

8.7.1.3.1 Salts of 3-Alkylpyridinium Polymer

3-Alkylpyridinium salts are made from the marine sponge species Reneirasarai, which exhibits antifouling properties by preventing the growth of fungi and algae in the marine microenvironment since it is a water-soluble marine toxin. The extraction of this biopolymer involves passing the ethanolic extract supernatant obtained from centrifuging a homogenized marine sponge over an Amicon Y3 membrane [89] (Fig. 8.5) (Table 8.3).

8.7.1.4 Different Applications of the Value-Added Products

Marine biopolymers are ideal for biomedical applications due to their high biocompatibility, non-antigenicity, chelating ability, and bioavailability. Marine polysaccharides are of great interest to the biomedical sector because of their resilience, safeguarding, and high accessibility at a relatively inexpensive rate, in addition to their unique biological and physicochemical properties and natural origin [93].

Polysaccharides found in marine environments have been utilized in pharmaceuticals as matrix materials, emulsifiers, thickeners, binders, stabilizers, and suspending agents. They have been employed in numerous formulations over time, including hydrogels and gels, films and membranes, nanofibers, micro/nanoparticles (MPs/NPs), and three-dimensional porous structures [94].

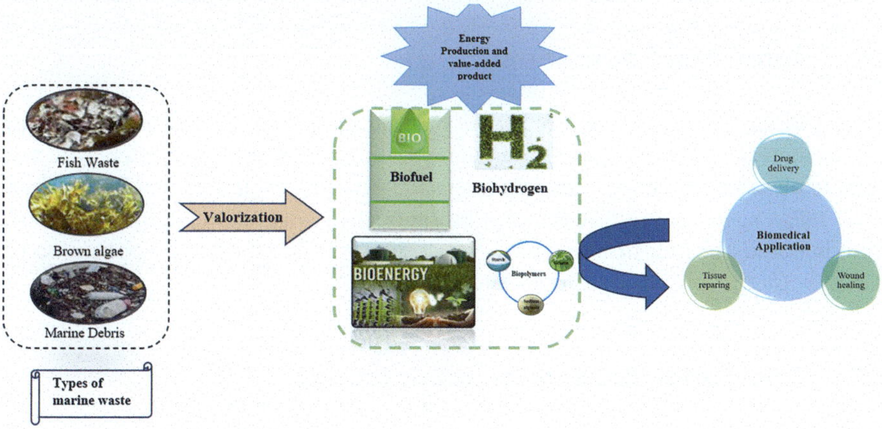

Fig. 8.5 Different value-added products production

Table 8.3 List of collagens from different sources of various species

Name of the species	Applications	References
Lateolabrax japonicus, Trachurus japonicus	Food, cosmetic, and biomedical materials	[88]
Takifugu rubripes skin	Tissue engineering	[90]
Pagrus major and Oreochromis niloticas scales	Biomaterials	[91]
Pogonia cromis, Archosargus probatocephalus bone	Biomedicals and pharmaceuticals	[92]
Sardinops melanostictus, Lateolabrax japonicus scales	Tissue engineering	[90]

8.7.1.4.1 Hydrogels

Because of their very porous structure, biodegradability, and biocompatibility, gels and hydrogels are acknowledged as potential biomaterials for drug delivery, tissue engineering, biosensors, self-healing, and hemostasis systems [95, 96]. Alginate hydrogels are widely employed in wound dressings and other biomedical applications [97]. Catechol, hydrocaffeic acid, and poly(ethylene glycol) have been added to chitosan-based hydrogels to enhance their mechanical, bioadhesive, and antibacterial qualities. Soft tissue engineering can make use of the hydrogel patches and injectable gels that were produced [98].

8.7.1.4.2 Polymeric Nanofibers

Polymeric nanofibers have attracted a lot of attention in recent years because of their distinctive characteristics, including high surface-to-volume area, high and regulated porosity, and mechanical flexibility [99]. Because of their biocompatibility and biodegradability, marine biopolymers are thought to be good candidates for the creation of multifunctional nonwovens.

8.8 Industrial Applications

Marine waste, commonly referred to as marine debris or ocean litter, consists of various materials that enter and accumulate in the oceans, ranging from plastic and metal to abandoned fishing gear. While marine waste is a significant environmental issue, there have been efforts to explore innovative ways to repurpose or utilize these materials in industrial applications. Figure 8.6 illustrates the application of marine waste in various industrial sectors (Table 8.4).

8.9 Environmental and Economic Impact

Recovering and recycling marine plastic waste can help mitigate the global plastic pollution crisis. By turning discarded plastic items into new products or materials, the amount of plastic entering oceans and harming marine ecosystems can be

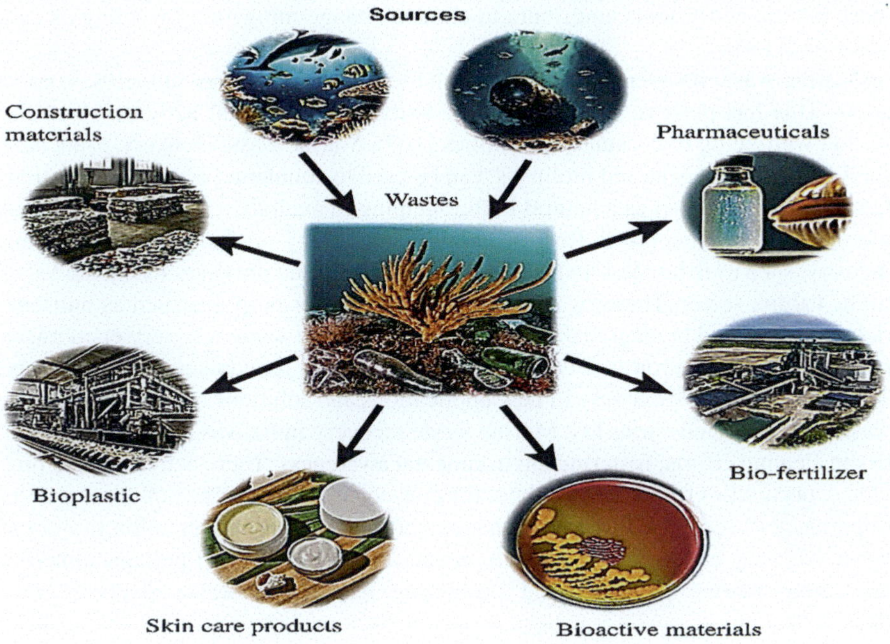

Fig. 8.6 Industrial applications

Table 8.4 Application and the prospects of marine waste in various industrial sectors

Application	Prospects	Purpose	References
Energy production	Waste-to-energy	Certain types of marine waste, particularly non-recyclable plastics, can be used as a source of energy through waste-to-energy technologies	[100]
Construction materials	Eco-friendly building blocks	Some innovative projects aim to use recycled marine plastics in the production of building blocks or construction materials. These materials can be employed in the construction industry, contributing to sustainable and eco-friendly building practices	[101]
Textile industry	Recycled fibers	Fishing nets and other marine debris made of nylon or other fibers can be repurposed into recycled fibers for the textile industry	[102]
Fertilizers and soil conditioners	Organic amendments	Organic materials found in marine waste, such as seaweed and other plant matter, can be processed and used as organic amendments for soil	[103]

reduced [104]. The removal of marine debris helps prevent habitat degradation and protects marine biodiversity. Abandoned fishing gear, for example, can entangle and harm marine animals, and its recovery can mitigate the negative impact on ecosystems [105]. Some marine waste, such as seagrasses and algae, has the potential to sequester carbon dioxide from the atmosphere. The recovery and conservation of these marine ecosystems contribute to climate change mitigation by acting as carbon sinks [106].

Marine waste recovery efforts contribute to cleaner and more attractive coastal areas. This has positive implications for tourism, recreational activities, and the overall well-being of coastal communities [107]. Marine waste recovery, especially shells from crustaceans and mollusks, can be used to counteract ocean acidification. Crushed shells can act as a natural buffer, helping to neutralize acidic conditions in marine environments [108]. Recovering and utilizing by-products from the fishing industry, such as fish scales and bones, reduces waste, and provides additional value to the fishing sector. These by-products can be used for purposes such as nutrient-rich fertilizers and biodegradable plastics [109]. Marine waste recovery encourages research and innovation in developing technologies to repurpose waste materials. This can lead to the discovery of new applications and solutions for utilizing marine waste in various industries [1]. Marine waste recovery initiatives engage communities in cleanup efforts, fostering environmental awareness. These activities also provide opportunities for education about the consequences of marine pollution and the importance of responsible waste management [110]. Increased attention to marine waste recovery can lead to the development and enforcement of policies aimed at preventing marine pollution and promoting responsible waste disposal practices [110].

8.10 Conclusion

The efficient and sustainable utilization of marine resources is crucial for environmental preservation and the well-being of future generations. Exploring innovative technologies and sustainable practices for marine waste recovery not only addresses environmental challenges but also creates significant economic opportunities. Marine waste, including plastic, metal, and abandoned fishing gear, poses a substantial threat to the oceans. However, ongoing efforts focus on repurposing these materials for various industrial applications, ranging from recycling plastic waste and producing eco-friendly building materials to extracting bioactive compounds for pharmaceuticals. Marine waste recovery aligns with circular economy principles, offering a waste-to-resource approach that mitigates plastic pollution and aids biodiversity conservation. Additionally, the removal of marine debris contributes to climate change mitigation, cleaner coastal areas, and the prevention of ocean acidification. The economic opportunities stemming from marine waste recovery are diverse and impactful. They include job creation in waste recovery and recycling industries, the emergence of circular economy businesses, and the development of bioactive compounds for pharmaceuticals. Sustainable farming practices, innovative water filtration technologies, and the promotion of cleaner coastal areas further contribute to economic growth. The journey towards efficient marine resource utilization requires a holistic approach, involving collaboration between governments, industries, communities, and individuals. By valuing and protecting our oceans, we not only address environmental challenges but also pave the way for a more resilient and harmonious coexistence with the natural world, ensuring a positive legacy for future generations.

References

1. Rudovica V, Rotter A, Gaudêncio SP et al (2021) Valorization of marine waste: use of industrial by-products and beach wrack towards the production of high added-value products. Front Mar Sci 8
2. Nitta SK, Numata K (2013) Biopolymer-based nanoparticles for drug/gene delivery and tissue engineering. Int J Mol Sci 14(1):1629–1654
3. Manivasagan P, Bharathiraja S, Moorthy MS, Oh YO, Seo H, Oh J (2017) Marine biopolymer-based nanomaterials as a novel platform for theranostic applications. Polym Rev 57(4):631–667
4. Tennakoon P, Chandika P, Yi M, Jung WK (2023) Marine-derived biopolymers as potential bioplastics, an eco-friendly alternative. iScience 26(4):106404
5. Lincoln S, Chowdhury P, Posen PE et al (2023) Interaction of climate change and marine pollution in Southern India: implications for coastal zone management practices and policies. Sci Total Environ 902:166061
6. Senadheera TRL, Hossain A, Shahidi F (2023) Marine bioactives and their application in the food industry: a review. Appl Sci 13(21):12088
7. Puri M, Gupta A, McKinnon RA, Abraham RE (2022) Marine bioactives: from energy to nutrition. Trends Biotechnol 40(3):271–280
8. Shahidi F, Ambigaipalan P (2015) Novel functional food ingredients from marine sources. Curr Opin Food Sci 2:123–129
9. Jo C, Khan FF, Khan MI, Iqbal J (2017) Marine bioactive peptides: types, structures, and physiological functions. Food Rev Int 33(1):44–61
10. Hossain A, Dave D, Shahidi F (2020) Northern sea cucumber (Cucumaria frondosa): a potential candidate for functional food, nutraceutical, and pharmaceutical sector. Mar Drugs 18(5)
11. Ameen F, AlNadhari S, Al-Homaidan AA (2021) Marine microorganisms as an untapped source of bioactive compounds. Saudi J Biol Sci 28(1):224–231
12. Satheesh S, Ba-Akdah MA, Al-Sofyani AA (2016) Natural antifouling compound production by microbes associated with marine macroorganisms—a review. Electron J Biotechnol 21:26–35
13. Lee SY, S. J. (2017) Hur antihypertensive peptides from animal products, marine organisms, and plants. Food Chem 228:506–517
14. Rioux LE, Beaulieu L, Turgeon SL (2017) Seaweeds: a traditional ingredients for new gastronomic sensation. Food Hydrocoll 68:255–265
15. Ovando CA, de Carvalho JC, de Melo V, Pereira G, Jacques P, Soccol VT, Soccol CR (2018) Functional properties and health benefits of bioactive peptides derived from Spirulina: a review. Food Rev Int 34(1):34–51
16. Madhusudan S, Baskaran R (2023) The sea lettuce Ulva sensu lato: future food with health-promoting bioactives. Algal Res 71:103069
17. Thiviya P, Gamage A, Gama-Arachchige NS, Merah O, Madhujith T (2022) Seaweeds as a source of functional proteins. Phycology 2(2):216–243
18. Zhou L, Li K, Duan X et al (2022) Bioactive compounds in microalgae and their potential health benefits. Food Biosci 49:101932
19. Datta D, Talapatra SN, Swarnakar S (2015) Bioactive compounds from marine invertebrates for potential medicines—an overview. Int Lett Nat Sci 34:42–61

20. Hossain A, Yeo JD, Dave D, Shahidi F (2022) Phenolic compounds and antioxidant capacity of Sea Cucumber (Cucumaria frondosa) processing discards as affected by high-pressure processing (HPP). Antioxidants 11(2)
21. Romano G, Almeida M, Coelho AV et al (2022) Biomaterials and bioactive natural products from marine invertebrates: from basic research to innovative applications. Mar Drugs 20(4)
22. Kurtela A, Antolović N (2019) The problem of plastic waste and microplastics in the seas and oceans: impact on marine organisms. Ribar Croat J Fish 77(1):51–56
23. Sheppard C (2002) World seas an environmental evaluation. Elsevier
24. Dasgupta S, Sarraf M, Wheeler D (2022) Plastic waste cleanup priorities to reduce marine pollution: a spatiotemporal analysis for Accra and Lagos with satellite data. Sci Total Environ 839:156319
25. Duce RA, Galloway JN, Liss PS (2009) The impacts of atmospheric deposition to the ocean on marine\recosystems and climate. WMO Bull 58:61–66
26. Gates A (2023) Polluted earth: the science of the earth's environment. John Wiley & Sons
27. Breitburg D, Levin LA, Oschlies A et al (2018) Declining oxygen in the global ocean and coastal waters. Science 359(6371)
28. Minton MS, Miller AW, Ruiz GM (2015) Implications of ship type on delivery and management of Ballast water. In: Biological invasions in changing ecosystems. De Gruyter Open, pp 343–364
29. Zaman MS, Zaman R, Sizemore R (2019) Plastic pollution of the oceans: a review of marine plastic pollution and its environmental impacts. Adv Sci Technol 13(1):1–8
30. Njuguna J, Siddique S, Bakah Kwroffie L et al (2022) The fate of waste drilling fluids from oil and gas industry activities in the exploration and production operations. Waste Manag 139:362–380
31. Shuval H (2003) Estimating the global burden of thalassogenic diseases: human infectious diseases caused by wastewater pollution of the marine environment. J Water Health 1(2):53–64
32. Solomon B (2021) Plastic waste materials management and reuse in the construction industry: the case of selected first grade contractors in Addis
33. Mihai FC, Gündoğdu S, Khan FR, Olivelli A, Markley LA, Van Emmerik T (2022) Plastic pollution in marine and freshwater environments: abundance, sources, and mitigation. In: Emerging contaminants in the environment challenges and sustainable practices, pp 241–274
34. Kershaw P, Katsuhiko S, Lee S, Samseth J, Woodring D, Smith J (2011) Plastic debris in the ocean. UNEP Year B, pp 20–33
35. Renuka V, Jha AK, Joseph TC (2019) Nature and composition of fish processing industrial waste and handling protocols. In: Fishery waste management: challenges and business opportunities in Gujarat, pp 25–37
36. Greenberg MI, Sexton KJ, Vearrier D (2016) Sea-dumped chemical weapons: environmental risk, occupational hazard. Clin Toxicol 54(2):79–91
37. Straume EO, Gaina C, Medvedev S et al (2019) GlobSed: updated total sediment thickness in the World's Oceans. Geochem Geophys Geosyst 20(4):1756–1772
38. Amthor J, Drell D, Dahlman R, Weatherwax S (2008) Carbon cycling and biosequestration integrating biology and climate through systems science
39. McMahon RF, Hunter RD, Russell-Hunter WD (1974) Variation in aufwuchs at six freshwater habitats in terms of carbon biomass and of carbon: nitrogen ratio. Hydrobiologia 45(4):391–404
40. Pentecost A, Ros J, Prat N (1993) Homage to Ramon Margalef, or why is there such pleasure in studying nature? J Ecol 81(4):833
41. Walsh JJ (1980) Energy-related perturbations of the Northeast Coastal Zone: five years (1974–1979) of oceanographic research
42. Pauly D, Christensen V (1995) Primary production required to sustain global fisheries. Nature 374:255–257
43. Christensen V, Pauly D (1993) Trophic models of aquatic ecosystems, vol 390. WorldFish

44. Aaron Chavez EA, Chvez-Hidalgo A (2013) Biomass from the Sea. In: Biomass now—sustain growth use. IntechOpen
45. Kumar Y, Tarafdar A, Kumar D et al (2022) Polyphenols of edible macroalgae: estimation of in vitro bio-accessibility and cytotoxicity, quantification by LC-MS/MS and potential utilization as an antimicrobial and functional food ingredient. Antioxidants 11(5)
46. Patel AK, Vadrale AP, Singhania RR et al (2023) Algal polysaccharides: current status and future prospects. Phytochem Rev 22(4):1167–1196
47. Mühlenbruch M, Grossart HP, Eigemann F, Voss M (2018) Mini-review: Phytoplankton-derived polysaccharides in the marine environment and their interactions with heterotrophic bacteria. Environ Microbiol 20(8):2671–2685
48. Awuchi CG, Chukwu CN, Iyiola AO et al (2022) Bioactive compounds and therapeutics from fish: revisiting their suitability in functional foods to enhance human wellbeing. Biomed Res Int 2022
49. Lee HJ, Saravana PS, Cho YN, Haq M, Chun BS (2018) Extraction of bioactive compounds from oyster (Crassostrea gigas) by pressurized hot water extraction. J Supercrit Fluids 141:120–127
50. Suleria HAR, Osborne S, Masci P, Gobe G (2015) Marine-based nutraceuticals: an innovative trend in the food and supplement industries. Mar Drugs 13(10):6336–6351
51. Teng L, Wang X, Yu H et al (2023) Jellyfish peptide as an alternative source of antioxidant. Antioxidants 12(3)
52. Kogovšek T, Tinta T, Klun K, Malej A (2014) Jellyfish biochemical composition: importance of standardised sample processing. Mar Ecol Prog Ser 510:275–288
53. Singh P, Dhanorkar M, Sharma S (2024) Valorization of aquatic plant biomass resource to fortified biochar and paper pulp: a strategic approach towards closed-loop technologies, circular economy, and sustainability. Resour Conserv Recycl 202:107385
54. Yang M, Chen L, Wang J et al (2023) Circular economy strategies for combating climate change and other environmental issues. Environ Chem Lett 21(1):55–80
55. Stahel WR (2021) Circular economy. Nature 64
56. Kaza S, Yao LC, Bhada-Tata P, Van Woerden F (2018) What a waste 2.0: a global snapshot of solid waste management to 2050. World Bank, Washington, DC
57. Macreadie PI, Trevathan-Tackett SM, Baldock JA, Kelleway JJ (2017) Converting beach-cast seagrass wrack into biochar: a climate-friendly solution to a coastal problem. Sci Total Environ 574:90–94
58. Ghisellini P, Cialani C, Ulgiati S (2016) A review on circular economy: the expected transition to a balanced interplay of environmental and economic systems. J Clean Prod 114:11–32
59. Magesh N, Annam Renita A, Senthil Kumar P (2020) Practice on treating pharmaceutical compounds (antibiotics) present in wastewater using biosorption techniques with different biowaste compounds. A review. Environ Prog Sustain. Energy 39(6)
60. de Freitas GR, da Silva MGC, Vieira MGA (2019) Biosorption technology for removal of toxic metals: a review of commercial biosorbents and patents. Environ Sci Pollut Res 26(19):19097–19118
61. Bulgariu D, Bulgariu L (2016) Potential use of alkaline treated algae waste biomass as sustainable biosorbent for clean recovery of cadmium(II) from aqueous media: batch and column studies. J Clean Prod 112:4525–4533
62. Safarik I, Baldikova E, Prochazkova J, Safarikova M, Pospiskova K (2018) Magnetically modified agricultural and food waste: preparation and application. J Agric Food Chem 66(11):2538–2552
63. Bădescu IS, Bulgariu D, Ahmad I, Bulgariu L (2018) Valorisation possibilities of exhausted biosorbents loaded with metal ions—a review. J Environ Manag 224:288–297
64. Escapa C, Coimbra RN, Nuevo C et al (2017) Valorization of microalgae biomass by its use for the removal of paracetamol from contaminated water. Water 9(5):1–9
65. Navarro AE, Lim H, Chang E, Lee Y, Manrique AS (2014) Uptake of sulfa drugs from aqueous solutions by marine algae. Sep Sci Technol 49(14):2175–2181

66. Peng FQ, Ying GG, Yang B et al (2014) Biotransformation of progesterone and norgestrel by two freshwater microalgae (Scenedesmus obliquus and Chlorella pyrenoidosa): transformation kinetics and products identification. Chemosphere 95:581–588
67. Mullerova S, Baldikova E, Prochazkova J, Pospiskova K, Safarik I (2019) Magnetically modified macroalgae Cymopolia barbata biomass as an adsorbent for safranin O removal. Mater Chem Phys 225:174–180
68. Ozudogru Y, Merdivan M, Göksan T (2016) Biosorption of Methylene Blue from an aqueous solution by iron oxide-coated Cystoseira Barbata. J Turkish Chem Soc Sect A Chem 3(3):551–564
69. Bouzikri S, Ouasfi N, Benzidia N, Salhi A, Bakkas S, Khamliche L (2020) Marine alga 'Bifurcaria bifurcata': biosorption of Reactive Blue 19 and methylene blue from aqueous solutions. Environ Sci Pollut Res 27(27):33636–33648
70. Balachandar G, Khanna N, Das D (2013) Biohydrogen production from organic wastes by dark fermentation. In: Biohydrogen. Elsevier, pp 103–144
71. Kumar M, Thakur IS (2018) Municipal secondary sludge as carbon source for production and characterization of biodiesel from oleaginous bacteria. Bioresour Technol Rep 4:106–113
72. Karkal SS, Kudre TG (2020) Valorization of fish discards for the sustainable production of renewable fuels. J Clean Prod 275:122985
73. Klavins M, Bisters V, Burlakovs J (2020) Small scale gasification application and perspectives in circular economy. Environ Clim Technol 22(1):42–54
74. Khoo CG, Dasan YK, Lam MK, Lee KT (2019) Algae biorefinery: review on a broad spectrum of downstream processes and products. Bioresour Technol 292:121964
75. Morris JP, Backeljau T, Chapelle G (2019) Shells from aquaculture: a valuable biomaterial, not a nuisance waste product. Rev Aquac 11(1):42–57
76. Oso AO, Idowu AA, Niameh OT (2011) Growth response, nutrient and mineral retention, bone mineralisation and walking ability of broiler chickens fed with dietary inclusion of various unconventional mineral sources. J Anim Physiol Anim Nutr (Berl) 95(4):461–467
77. Scialla S, Carella F, Dapporto M et al (2020) Mussel shell-derived macroporous 3D scaffold: characterization and optimization study of a bioceramic from the circular economy. Mar Drugs 18(6):1–15
78. Jacob BS, Ramani K, Kumar VV (2023) Applied biotechnology for emerging pollutants remediation and energy conversion. Springer
79. Elieh-Ali-Komi D, Hamblin MR (2016) Chitin and chitosan: production and application of versatile biomedical nanomaterials. Int J Adv Res 4(3):411–427
80. Shamshina JL, Berton P, Rogers RD (2019) Advances in functional chitin materials: a review. ACS Sustain Chem Eng 7(7):6444–6457
81. Mohammed MA, Syeda JTM, Wasan KM, Wasan EK (2017) An overview of chitosan nanoparticles and its application in non-parenteral drug delivery. Pharmaceutics 9(4)
82. Rubini D, Farisa Banu S, Veda Hari BN et al (2018) Chitosan extracted from marine biowaste mitigates staphyloxanthin production and biofilms of Methicillin-resistant Staphylococcus aureus. Food Chem Toxicol 118:733–744
83. Younes I, Rinaudo M (2015) Chitin and chitosan preparation from marine sources. Structure, properties and applications. Mar Drugs 13(3):1133–1174
84. Ghorbel-Bellaaj O, Younes I, Maâlej H, Hajji S, Nasri M (2012) Chitin extraction from shrimp shell waste using Bacillus bacteria. Int J Biol Macromol 51(5):1196–1201
85. Kurita K (2006) Chitin and chitosan: functional biopolymers from marine crustaceans. Mar Biotechnol 8(3):203–226
86. Zhao Y, Park RD, Muzzarelli RAA (2010) Chitin deacetylases: properties and applications. Mar Drugs 8(1):24–46
87. Nazeer RA, Joshi I, Balde A, Chatterjee A, Narayanasamy A (2023) Exploitation of marine waste for value-added products synthesis. In: Applied biotechnology for emerging pollutants remediation and energy conversion. Springer Nature, Singapore, pp 333–364
88. Nagai T, Suzuki N (2000) Preparation and characterization of several fish bone collagens. J Food Biochem 24:427–436

89. Elersek T, Kosi G, Turk T, Pohleven F, Sepcic K (2008) Influence of polymeric 3-alkylpyridinium salts from the marine sponge Reniera sarai on the growth of algae and wood decay fungi. Biofouling 24(2):137–143
90. Nagai T, Araki Y, Suzuki N (2002) Collagen of the skin of ocellate puffer fish (Takifugu rubripes). Food Chem 78(2):173–177
91. Ikoma T, Kobayashi H, Tanaka J, Walsh D, Mann S (2003) Physical properties of type I collagen extracted from fish scales of Pagrus major and Oreochromis niloticas. Int J Biol Macromol 32(3–5):199–204
92. Ogawa M, Portier RJ, Moody MW, Bell J, Schexnayder MA, Losso JN (2004) Biochemical properties of bone and scale collagens isolated from the subtropical fish black drum (Pogonia cromis) and sheepshead seabream (Archosargus probatocephalus). Food Chem 88(4):495–501
93. Bilal M, Iqbal HM (2020) Biologically active macromolecules: extraction strategies, therapeutic potential and biomedical perspective. Int J Biol Macromol 151:1–18
94. Ruocco N, Costantini S, Guariniello S, Costantini M (2016) Polysaccharides from the marine environment with pharmacological, cosmeceutical and nutraceutical potential. Molecules 21(5):1–16
95. Chai Q, Jiao Y, Yu X (2017) Hydrogels for biomedical applications: their characteristics and the mechanisms behind them. Gels 3(1)
96. Venkatesan J et al (2015) Seaweed polysaccharides and their potential biomedical applications. Starch/Staerke 67(5–6):381–390
97. Liao J, Jia Y, Wang B, Shi K, Qian Z (2018) Injectable hybrid poly(ϵ-caprolactone)-b-poly(ethylene glycol)- b-poly(ϵ-caprolactone) porous microspheres/alginate hydrogel cross-linked by calcium gluconate crystals deposited in the pores of microspheres improved skin wound healing. ACS Biomater Sci Eng 4(3):1029–1036
98. Du X et al (2020) Anti-infective and pro-coagulant chitosan-based hydrogel tissue adhesive for sutureless wound closure. Biomacromolecules 21(3):1243–1253
99. Cheng H, Yang X, Che X, Yang M, Zhai G (2018) Biomedical application and controlled drug release of electrospun fibrous materials. Mater Sci Eng C 90(2017):750–763
100. Shields MA et al (2011) Marine renewable energy: the ecological implications of altering the hydrodynamics of the marine environment. Ocean Coast Manag 54(1):2–9
101. Li Z (2011) Development and application of eco-friendly concrete. Adv Mater Res 250–253:3827–3836
102. Patti A, Cicala G, Acierno D (2021) Eco-sustainability of the textile production: waste recovery and current recycling in the composites world. Polymers 13(1):1–22
103. Ihejirika CE, Onwudike SU, Nwaogu LA, Emereibeole EI, Ebe TE, Ejiogu CC (2012) Assessment of aquaculture sediment for agricultural fertilizer supplement and soil conditioner in Owerri Urban, Nigeria. J Res Agric 1:034–038
104. Kibria MG, Masuk NI, Safayet R, Nguyen HQ, Mourshed M (2023) Plastic waste: challenges and opportunities to mitigate pollution and effective management. Int J Environ Res 17(1):20
105. Crain CM, Halpern BS, Beck MW, Kappel CV (2009) Understanding and managing human threats to the coastal marine environment. Ann N Y Acad Sci 1162:39–62
106. Chow A (2014) Ocean carbon sequestration by direct injection. In: CO_2 Sequestration valorization. IntechOpen, London, UK, p 2
107. Ghermandi A, Nunes PALD, Portela R, Rao N, Teelucksingh SS (2010) Recreational, cultural and aesthetic services from estuarine and coastal ecosystems
108. Topić Popović N, Lorencin V, Strunjak-Perović I, Čož-Rakovac R (2023) Shell waste management and utilization: mitigating organic pollution and enhancing sustainability. Appl Sci 13(1)
109. Kristbergsson K, Arason S (2007) Utilization of by-products in the fish industry. In: Utilization of by-products and treatment of waste in the food industry, pp 233–258
110. Herdiansyah H, Saiya HG, Afkarina KII, Indra TL (2021) Coastal community perspective, waste density, and spatial area toward sustainable waste management (case study: Ambon bay, Indonesia). Sustain For 13(19)

Potential of Marine Microalgae for the Production of Food and Feed Industry

9

Anwesha Patel, Divyashree Arun, and Shristi Ram

Abstract

Marine microalgae are a versatile solution to global food security and environmental sustainability challenges. They are diverse and known to offer multifarious health benefits, positioning them as a viable alternative to traditional food and feed sources. This chapter underscores the potential of microalgae-derived carbohydrates, proteins, lipids, vitamins, minerals, and bioactive compounds for food and feed applications and as nutraceuticals in the aquaculture sector and livestock production. Emphasis on their cultivation techniques and environmental considerations are discussed providing insights into the recent methodologies used for marine microalgae cultivation and production. Various physicochemical parameters responsible for affecting their growth, such as light, temperature, and nutrient availability, are analysed. Additionally, SWOT (strengths, weaknesses, opportunities, threats) analysis is discussed. Lastly, the integration of marine microalgae into supply chains, including regulatory and market dynamics is evaluated.

A. Patel · D. Arun · S. Ram (✉)
Department of Microbiology, School of Basic and Applied Sciences, Dayananda Sagar University, City campus, Kudlu Gate, Bengaluru, India
e-mail: shristi-sbas@dsu.edu.in

M. Rafatullah et al. (eds.), *Multidisciplinary Applications of Marine Resources*,
https://doi.org/10.1007/978-981-97-5057-3_9

Keywords

Aquaculture · Microalgae · Food and feed industry · Functional food · SWOT analysis

9.1 Introduction

Microalgae are primitive living creatures that form the foundation of aquatic food systems. Encompassing several phyla and types of organisms, including cyanobacteria in certain instances, they represent a phylogenetically varied group. Microalgae thrive on wet soils and rocks as well as freshwater, ocean, and hypersaline settings.

"Blue Biotechnology" also known as Marine Biotechnology, is a growing global field that harnesses marine resources for various biotechnological applications,

Fig. 9.1 Various applications of marine microalgae

serving as both the source and focus of innovative advancements. Microalgae have several industrial applications, such as in food and feed formulations, cosmeceuticals, in agriculture sector as biofertilizers, as well as in wastewater treatment and biofuel generation. Figure 9.1 represents some of the main applications of marine microalgae. Microalgae may be considered a potential food and feed source due to their nutritional properties—a property that is greatly reliant on the microalgae's cellular composition. However, the nutritional value is species-specific which also depends on the physicochemical parameters, such as light, temperature, pH, and available nutrients. Compared to freshwater microalgae cultivation, marine microalgae exhibit significantly lower water footprint in biomass production. Their cultivation is considered highly sustainable for feed and fuel production due to the minimal requirement for potable water.

A diverse range of physiologically active substances have been discovered in microalgal biomass, including protein, polyunsaturated fatty acids (PUFAs), pigments, vitamins, and minerals, as well as extracellular molecules such as oligosaccharides. Interestingly, the (bio) synthesis and accumulation of the desired product in microalgae cells is significantly higher than that of terrestrial plants [1].

9.1.1 Taxonomic Diversity of Marine Microalgae

In aquatic ecosystems, marine algae play a crucial role as primary producers, contributing significantly to the high primary production in coastal regions. They serve as a direct food source for various grazing species such as peracaridas, gastropods, crustaceans, sea urchins, and fishes. The vast biodiversity of marine algae remains largely untapped, representing a valuable and largely unexplored resource. Different taxonomic groups of microalgae and their reported health benefits are represented in Table 9.1.

Table 9.1 Microalgae and their health benefits

Classification	Class	Some of the members	Health benefits	References
Prokaryotic	*Cyanophyceae* (Blue-green algae)	*Nostoc* *Oscillatoria* *Rivularia* *Trichodesmium*	Beneficial effects on blood glucose level and serum lipid concentrations	[2]
Eukaryotic	*Chlorophyceae* (Green algae)	*Chlamydomonas* *Volvox* *Ulothrix* *Spirogyra* *Chara*	Anticancer and antioxidant activities	[3]
	Rhodophyceae (Red algae)	*Polysiphonia* *Porphyra* *Gracilaria* *Gelidium*	Anticoagulant and anti-viral activities	[4]
	Dinoflagellates	*Oodinium* *Pfiesteria*	Antibacterial, antimicrobial, antioxidant nature	[5]
	Euglenoids	*Euglena* *Phacus* *Trachelomonas*	Antioxidant activity	[6]
	Prymnesiophyceae (Golden-brown algae)	*Coccolithales.* *Coccosphaerales.* *Isochrysidales.* *Phaeocystales*	Anti-inflammatory	[7]
	Cryptophyceae (Brown algae)	*Chroomonas* *Rhodomonas*	Decreases risk of cardiovascular diseases, potential benefits in mitigating renal diseases, depression, dementia, rheumatoid arthritis, and asthma	[8]
	Chrysophyceae (Golden algae)	*Synurophyceae* *Pedinellophyceae*	Antioxidant activity, anti-diabetes, anti-obesity, and cancer prevention	[9]
	Dictyophyceae	*Padina*	Cholesterol-lowering effects, antimicrobial, hypoglycaemic, cytotoxic, cardioprotective, hepatoprotective, and immunostimulatory activities	[10]
	Raphidophyceae	*Gonyostomum* *Merotrichia capitata*	Antioxidant, antitumor, and anticancer activities	[11]
	Xanthophyceae (Yellow-green algae)	*Olisthodiscus,* *Ophiocytium* *Botrydium*	Anti-inflammatory effects	[12]
	Bacillariophyceae (Yellowish-brown algae)	*Tabellaria* *Amphipleura*	Antioxidant, antibacterial, antifungal, antiviral, and antiproliferative activities	[11]
	Eustigmatophyceae (Orange-red algae)	*Nannochloropsis*	Antibacterial and anti-oxidant effects	[13]
	Phaeophyceae (Brown algae)	*Ectocarpus* *Fucus*	Antitumor, antioxidant, antiviral, and cytotoxic activities	[14]

9.2 Nutritional Components

The nutritional content of microalgae is species-specific and can even differ within the same species based on physicochemical factors impacting growth conditions. Key nutritional components such as protein, lipids, vitamins, and minerals are crucial considerations, as they have positive effects on the health of living organisms.

9.2.1 Carbohydrates in Marine Microalgae

Marine microalgae primarily store carbohydrates in the form of cellulose and starch which makes up approximately 20% w/w of microalgae biomass [15]. These carbohydrates are composed of various monosaccharides such as glucose, galactose, mannose, xylose, and ribose to name a few, with glucose being the predominant one. The synthesis of carbohydrates begins with carbon dioxide fixation through the Calvin cycle utilizing the Rubisco enzyme [16]. The exploration of carbohydrate-containing natural products derived from microalgae has emerged as a prominent research area. This endeavour aims to uncover novel insights into the structures, functions, and taxonomic distribution of compounds like glycolipids, diverse glycosides, glycosylated toxins. etc. [17].

9.2.2 Proteins as a Valuable Food Source

Microalgae present a promising alternative to animal-based proteins due to their elevated protein levels and rich essential amino acid profiles. Typically, microalgae exhibit remarkably high protein content, reaching up to 70%, surpassing traditional plant protein and animal sources. Certain microalgae species, such as *Spirulina sp.*, *Chlorella* sp., *Arthrospira* sp., *Scenedesmus* sp., and *Synechococcus* sp., show total protein contents surpassing 50% w/w [15]. Marine microalgae harbour a multitude of active proteins with diverse physiological functions exhibiting activities such as anti-tumour, anti-inflammatory, anti-oxidant, and immune stimulation [18]. Their applications range from medicinal, cosmetics to being used as adsorbents for heavy metals in the ocean. Consequently, microalgal proteins offer a potential supplementary source of dietary protein, particularly beneficial in developing countries. Moreover, these proteins can be utilized as ingredients in the production of healthy and functional foods [19].

9.2.3 Lipids and Their Applications

Marine microalgae can be recognized as rich sources of lipids that possess both high nutritional value and significant bioactive properties. Noteworthy biomolecules derived from these organisms include n-6 and n-3 polyunsaturated fatty acids (PUFAs), as well as glycolipids and phospholipids [20].

In microalgae, the average lipid content ranges from 1% to 40%, depending on the species and the growth condition. Microalgae can adjust their membrane fluidity in response to changes in temperature, thereby altering their fatty acid composition [21]. The primary components of Omega-3 (ω3) fatty acids are alpha-linolenic acid, eicosapentaenoic acid (EPA), and docosahexaenoic acid (DHA). Humans are dependent on the diet for these three polyunsaturated fatty acids (PUFAs), they must be obtained from the diet [19]. The fatty acid (FA) composition is a critical factor when choosing microalgal strains for aquaculture. Commonly utilized strains include *Nannochloropsis sp.* (known for EPA production) and *Isochrysis sp.* (*Isochrysis aff. Galbana*: DHA producer; *Isochrysis galbana* (Parke): rich in both EPA and DHA) [22]. *Isochrysis galbana* and *Phaeodactylum tricornutum* are microalgae species that produce EPA at levels exceeding two times the concentration found in cod liver oil. Similarly, *Crypthecodinium cohnii,* another microalga, is capable of producing docosahexaenoic acid (DHA) at levels nearly six times higher than those found in cod liver oil. These microalgae represent promising sources of these ω3 fatty acids, showcasing their potential as alternatives to traditional fish-derived oils. Polyunsaturated fatty acids (PUFAs) derived from microalgae offer several advantages. They provide a more palatable option, addressing taste concerns associated with fish oil. Additionally, microalgae-based PUFAs have the advantage of reduced or no accumulation of toxins commonly found in fish. This makes them a cleaner and safer source of essential fatty acids. Furthermore, microalgae-derived PUFAs align with vegan principles, offering a plant-based and sustainable alternative to traditional animal-based oils. It has also been reported to have numerous beneficial effects on cardiovascular diseases, neurological disorders, and certain cancers [23]. ω3 and ω6 fatty acids are crucial for modulating physiological functions in the organism. ω3 fatty acids exhibit anti-inflammatory, antithrombotic, anti-arrhythmic, blood lipid-lowering, and vasodilatation properties. They are essential for foetal and infant development, particularly contributing to the development of the brain and vision [19]. The vasoconstrictive and aggregative effects of platelets, as well as the vasodilative and anti-aggregative actions on the endothelium, are attributed to the ability of ARA (Arachidonic acid) and EPA. Additionally, they exhibit a chemostatic effect on neutrophils [24].

9.2.4 Vitamins and Minerals

Microalgae offer a valuable source of vitamins, surpassing certain terrestrial plants in their provision of essential nutrients for humans. Notably, microalgae are rich in vitamins D and K, which are often less abundant in many traditional plants or fruits. Additionally, microalgae serve as a source of various B vitamins, including B12, B9, and B6 and other essential vitamins such as A, C, and E. This diversity in the vitamin content of microalgal cells presents an advantageous platform for utilizing them as natural producers of vitamins for human consumption [25]. When compared to freshwater algae, marine algae typically exhibit elevated levels of inorganic ions, attributed to their higher osmotic potential, this distinction is particularly pronounced in the concentrations of sodium and chloride [25].

9.2.5 Bioactive Compounds and Their Health Benefits

Marine microalgae are prolific producers of diverse bioactive compounds with applications spanning from food supplements to bioactive substances used in medical therapy (Fig. 9.2). The wide array of compounds they generate holds promise for various industrial and health-related purposes [26]. They serve as an essential feedstock for producing bioactive compounds such as carotenoids, phenolics, vitamin B12, and peptides with transformative applications in the pharmaceutical, cosmetic, and food industries. Bioactive elements like phytosterols, polysaccharides, fatty acids, and phenolic and volatile compounds from microalgae can be strategically utilized to create functional foods and pharmaceutical products [27]. These marine microalgae have the potential to function as "cell factories" for the synthesis of diverse bioactive substances. However, the challenge lies in improving our ability to efficiently access, identify, and exploit these microorganisms.

The carotenoids' antioxidant capabilities contribute to their potential in preventing ageing, coronary heart disease, cancer, atherosclerosis, and degenerative diseases by mitigating oxidative stress. Antioxidants safeguard against lipid peroxidation and prevent damage to DNA and proteins. Also, they act in signal modulation and transcriptional regulation for catalase, lipid peroxidase, and superoxide dismutase (SOD). Microcolin-A, a peptide originating from *Lyngbya majuscula*, demonstrates immunosuppressive effects by hindering the innate immune system. *Gambierdiscus toxicus* and *Prorocentrum lima*, both marine dinoflagellates, produce ciguatoxin and okadaic acid, respectively. These compounds have also

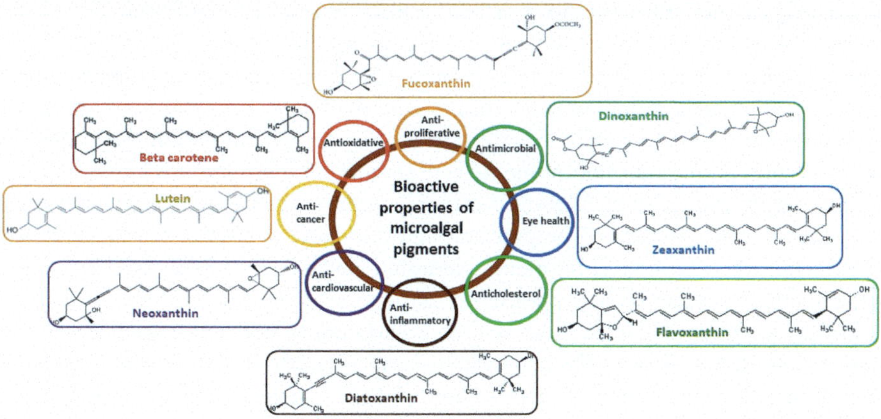

Fig. 9.2 Bioactivity of various microalgal carotenoids

been found to exhibit antifungal activity. Gambieric acids obtained from Dinoflagellate *G. toxicus* have also demonstrated its effectiveness in inhibiting fungal growth. However, microalgal blooms can be hazardous, For instance, when *Alexandrium lusitanicum,* dinoflagellates release neurotoxic compounds that cause paralytic shellfish poisoning (PSPs). Diarrheic shellfish poisoning (DSPs) can occur due to substances like okadaic acid and dynophysotoxin from dinoflagellate Dinophysis. Amnesic shellfish poisoning compounds (ASPs), from the diatom *Nitzschia pungens*, also pose risks. Interestingly, these cytotoxic compounds have garnered interest in the anticancer treatments. Phycocyanin sourced from *Arthrospira,* has not only been known for its strong antioxidant properties but also established roles in modulating the immune system and anti-inflammatory activities. Cyanobacteria like *Arthrospira* and unicellular marine algae like *Porphyridium* and *Rhodella* also synthesize sulphated polysaccharides which show antiviral activities [24].

9.3 Application in the Food and Feed Industry

The attractiveness of growing algae as renewable energy in response to global economic factors such as fossil fuel shortages in the late 1970s and mid-1990s has stimulated interest in microalgae for food and feed products. Microalgae (green or blue-green algae) are among the most significant organisms on the planet, having a flexible and adaptable metabolism. They can synthesize bioactive molecules (primarily secondary metabolites such as unsaturated fatty acids, pigments, and amino acids) with biomedical applications, enhancement of the nutritional value of food, animal feed/aquaculture, and environmental impact (as raw materials for biofuels) [1]. Large-scale microalgae production facilities have been constructed throughout Asia, including India, the United States, Israel, and Australia, since 1980. Around 200 varieties of microalgae are now employed globally. There is a well-established global market for microalgae-based food and feed products, and have been utilized to produce ethanol (fermentation) or biodiesel (conversion), and research into the use of GM (genetically modified) algae for pharmaceutical manufacture is underway. Thus, the application of microalgae in the food and feed industry is becoming more important, to compete with the existing commercially available sources. Since the last decade, advancements in wild microalgae strains cultivation methodology has been improved using photobioreactors. *Dunaliella salina*, *Spirulina platensis*, *Hematococcus pluvialis*, or *Chlorella vulgaris* are now grown on an industrial scale and are important sources of polyunsaturated fatty acid (PUFA), sterols, proteins, polysaccharides, carotenoid pigments, vitamins, and minerals with antiviral, antioxidant, and polysaccharide properties [1].

In the upcoming decades, there will be a particularly noticeable increase in the need for microalgae due to the growing amount of arable area devoted to soybean and corn food crops, which are the traditional feed for animals. Small amounts of microalgal biomass in feed improve animal physiology by improving immune response, disease resistance, and gut function, enhancing the nutritional quality of milk as well as improving antiviral and antibacterial protection and increasing reproductive performance, feed conversion, and weight gain [28]. *Chlorella vulgaris* is used to feed dairy cattle which helps alter the fatty acid profile of milk, *Arthrospira platensis* in feed for pigs and poultry has improved weight and apart from this, there are many other microalgae like *Dunaliella sp., Schizochytrium sp., Tetraselmis chuii, Haematococcus pluvialis* etc. have been employed by feed industry and has shown improved quality of meat, its flavour, and texture [1]. Recently, there has been a trend to incorporate in animal feeds, microalgae or derived products such as biomass extruded of high moisture, desiccated powders, and such other extracts as are applicable [29].

9.3.1 Utilization of Microalgae as Food for Human Consumption

Humans have been using microalgae for thousands of years as food, and adding microalgae as a supplement may relieve pressure on terrestrial food crops that need a lot of resources. Microalgae production requires less area and resources than traditional crops and yields more than terrestrial crops. Its abundant protein content suggests that it has a tremendous ability to address the nutritional needs of an expanding population. Some countries have been using microalgae as food and animal feed for hundreds of years. Due to the busy and modern lifestyles, the population at large in several developed nations are dependent on high-calorie meals leading to the prevalence of obesity, hypertension, diabetes, and other cardiovascular physiological complications. The use of microalgae as food is not a new concept. The Chinese have been reported to consume *Nostoc*, highly rich in Fasciclin motif protein, Vitamin C, and amino acids, for 2000 years. Other Asian countries such as Taiwan, Japan, and Mexico also use *Chlorella* and *Spirulina* species, which are beneficial for human health as they have effective antioxidant, pain-alleviating, anti-inflammation boosting properties, and neuro-enhancing effects.

The increased interest in the utilization of microalgae as an alternative source for human consumption is due to both population growth and awareness of healthier food alternatives. Microalgae are microcell factories that can accumulate high levels of macro and micronutrients. By fine-tuning the optimum physicochemical parameters, the microalgae growth and the desired product accumulation can be obtained. *Arthrospira platensis, Botryococcus braunii, Chlorella vulgaris, Dunaliella salina, Haematococcus pluvialis, Nannochloropsis sp.,* and *Phaeodactylum tricornutum* are prominent examples of the microalgae species commercially available and are considered "safe" for human and animal consumption due to the ability to accumulate high protein (20–40% w/w/), carotenoids (0.3–4%), and lipid content (1–15% w/w) such as ω6 and ω3 [30]. Numerous microalgae species have been studied to have high concentrations of carbohydrates, lipids, proteins, and other compounds with nutritional properties. Dominant in current industrial usage are species of *Chlorella* and *Spirulina* [31]. Microalgae-derived foods are commercialized in the form of capsules, pills, powders, and liquids (Table 9.2).

Table 9.2 Commercially available microalgae-based food products for human consumption [modified from 32]

Microalgae	Food product	Bioactive compound	Health benefits	Commercial form
Dunaliella salina	Candy, Beverages, and Pasta	β-carotene	Protection against gastric ulcers, reduced anaemia, diabetes	Powder
Chlorella vulgaris	Soyabean oil	Carotenoids	Antioxidant activity, efficient on gastric ulcers relief and constipation; preventive measure in atherosclerosis treatment	Oil
Aphanizomenon flos-aquae	Cheese	Protein, Phycocyanin, β-carotene	Alleviation of hyperlipidaemia, protection against renal failure	Tablets and powder
Spirulina arthrospira	Beverages, Spirulina biscuits	C-phycocyanin	Improved immune system	Capsules, powder, or liquid

9.3.1.1 Culinary Potential

Microalgae have contribution in rural cuisines, although this culinary history is being eroded due to cultural changes in eating patterns. Phycigastronomy is gradually being incorporated into Western cutting-edge restaurants. Edible algae form the structure of quite a few dishes due to its unique flavour and nutrient makeup. Microalgae's culinary potential has been far less studied than macroalgae's, most likely because of a deficit in texture or visual presentation vis a vis macroalgae, and also because adverse effects are less documented [33]. Notwithstanding Avant Garde cuisine seems to be utilizing the unicellulars *Tetraselmis chui* and *Nannochloropsis gaditana*, as well as the colonial cyanobacterium *Nostoc sphaericum* [34].

9.3.1.2 Economic and Environmental Benefits of Using Microalgae

Microalgal biomass is shown to be a potential source of alternative to tradition energy sources. The manufacture of biodiesel from microalgae has been widely explored in the literature, and one key use for biomass is in the production of ethanol. Although biodiesel and biohydrogen are seen to be the greatest alternatives to fossil fuels, bioethanol, which can be created from carbohydrate biomass, is the most extensively utilized biofuel in nations throughout the world. Microalgal biomass can be used in bioethanol production via fermentation methods involving fungus or bacteria. These organisms can be cultivated in industries due to their fast pace of development and great photosynthetic efficiency. The reliability of microalgae as an alternative source has been beneficial in aspects of economic as well as environmental perspectives. The requisite is a simple growth medium, with a small influx of nitrogen and phosphate sources, and relatively little water, CO_2, and sunshine, is required. Also, it doesn't require supporting structures like roots, stems, fruits, etc., thus there's no need to spend energy moving storage molecules like starch around or between tissues. Reproduction is continuous as most species reproduce vegetatively through cell division. There is the added convenience in harvesting and subsequent handling as microalgae can grow in non-arable land, lakes, and offshore oceanic waters that are unsuitable for the propagation of other traditional crops. It should be noted that economies of scale are easy to implement despite differences in conditions. It is cost-effective, sustainable, and has a high yield. Overall, the benefits and advantages of microalgae outweigh its demerits, there's no doubt that it can be used as an alternative source and with the scientifically proven facts this field is bound to grow enormously.

9.3.2 Utilization of Microalgae as Feed for Aquaculture and Livestock

Traditional aquaculture is an essential source of food for humans but threatens global sustainability because of the large amount of wastewater it produces. The rampant antibiotics use causes health-safety issues and the water replacement or water treatment costs also need to be factored into the overall process. An increased aquaculture production resulting in the higher requirement for aqua feed instigated the researchers to look for an alternative source that is cost-effective and readily available. Fishes are adversely affected due to organic waste, fish pathogens, and poor water quality leading to the prevalence of diseases in fish [35]. Some marine microalgae strains like diatoms, *Spirulina* have been applied as fish feed and have been reported to not only strengthen their immune system but also enhance their growth promotion, enhance antimicrobial capability, and ameliorate stress-related effects on shrimps, fishes, etc. [36]. Microalgae play a crucial role in providing high-quality nutrition for animals in aquaculture and on farms playing important role in the prevention of many diseases. Extensive assessments of their nutritional content and safety have confirmed that algal biomass is a viable and valuable supplement or alternative to traditional animal feed sources.

Microalgae are invaluable in the development of marine animals, freshwater animals, basically aquaculture. The life cycle of most marine invertebrates, fish hatcheries, and molluscs is dependent on microalgae. Bivalve molluscs like oysters, scallops, clams, and mussels also depend on microalgae as their live feed. Microalgae are essential for larva nutrition for molluscs and penaeida shrimps, as well as being prey in the food pyramid of species of fish with small larvae. *Diatoms* and *haplotypes* are nutrition-rich microalgae that are widely utilized as oyster feed, provide food for zooplankton, and assist in enhancing the quality of culture medium. Phytoplankton can be introduced to rearing ponds resulting in many superior outcomes in terms of survival, growth, and transformation index for both seawater and freshwater animal species than the traditional clear water strategy [37]. Recent studies suggest that microalgae show promising potential as alternatives to fish-based meal and oil in animal diets, leading to improved growth and meat quality in pigs, poultry, and rabbits. Including marine microalgae in ruminant diets has also demonstrated positive effects on the composition of fatty acids in milk and meat. Microalgae, rich in bioactive compounds, contribute to heightened defence mechanisms and improved health in various animal species. *Spirulina* which is a planktonic photosynthetic filamentous cyanobacteria was shown in samples of Labio Rohita to boost growth and phosphatase activity when administered along with copper exposure. Phosphatase activities are involved in active transport, metabolism of glycogen, synthesis of proteins, secretory functions, and certain enzyme synthesis in fish [38].

Moreover, these microalgae-derived bioactive compounds offer protective and antioxidant effects for tissues and influence pigmentation in fish [39]. Thus, apart from the traditional feed, these facts provide us with solid evidence of using microalgae as an alternative source.

9.3.2.1 Microalgae as a Feed Supplement

A consequence of industrial aquaculture's explosive global growth is the scarcity of resources for collecting wild fish, making the availability of essential feedstuffs like fishmeal and fish oil increasingly constrained. This has led to the investigation of alternative feedstuff sources, such as plant-based alternatives for fishmeal and fish oil for aquafeed. *Haematococcus* cultivation for the synthesis of the carotenoid astaxanthin, which gives salmon meat its scarlet colour. Aquaculture feeds based on fish oil may eventually be replaced with long-term microalgae biomass rich in lipids (omega-3 fatty acids). Although it can be challenging to maximize algal productivities, native algal blooms in prawn ponds provide some of the animals' food needs. For many aquaculture species, microalgae provide both their natural food source and the foundation of the natural food chains that support them in the wild. Microalgae are rich in nutrients and include vitamins, vital fatty acids, colours, and other speciality feed ingredients. Microalgae may often be used immediately in the aquaculture process without the requirement for harvesting, processing, or storage. In aquaculture as opposed to terrestrial animal husbandry, the conversion efficiency of feed to animal meat is greater. There are also considerable restrictions to the utilization of microalgae as aquaculture feeds like its mass production in outdoor ponds, and expenses are considerably high. More regulated cultures, such as microalgae production in fermenters and tubular reactors, are only suited for very high-value specialist feeds but if we overcome these hurdles it still stands as a promising field.

9.3.2.2 Benefits for Aquaculture and Livestock Health

The following are the economically beneficial effects and advantages to the technological process wherein microalgae are used in pisciculture: (1) Oxygen generation by microalgae minimizes the dangers of depletion of oxygen while also lowering the energy consumed by typical aeration systems. (2) The ability of microalgae to survive in a pond or other fish-rearing rigs might inhibit the growth of harmful or unfavourable bacteria, improving the environment for aquatic life. As a result, the frequency of water replenishment in a fish tank or other such environments is decreased, as is the associated expense. (3) Since microalgae feed strengthens aquatic animals' immune systems, misuse of antibiotics and other medications for aquaculture might be prevented, improving the safety of aquaculture consumables and enhancing their market potential. (4) Advanced microalgae biotechnology might be used to remediate the aquaculture wastewater at a reasonable cost. This would eliminate the wastewater treatment costs that the aquaculture business has previously paid to sewage treatment plants, therefore lessening the industry's financial burden. (5) Since the gathered biomass is utilized to partially substitute the conventional aquaculture feed, fish-raising expenses may be managed. Taking into account the mentioned benefits, it can be firmly believed that microalgae will help to improve aquaculture as a viable economic sector benefiting factories, consumers, and society as a whole.

9.4 Cultivation Techniques

Microalgae cultivation is typically carried out in two main systems: open raceways (ORW) or photobioreactors (PBR). Open raceways are shallow ponds with depths ranging between 10 and 50 cm. They can be constructed using concrete or carved from the ground, often lined with materials like high-density polyethylene (HDPE) or polyvinyl chloride (PVC). These ponds may be open or covered by a greenhouse. Open raceway systems are commonly employed in the industry for the production of microalgae used in various applications, including food production [40].

Several factors must be considered when choosing the cultivation system:

- The biology of the microalga.
- Costs associated with land, energy, water, and nutrients.
- Local climatic conditions.
- Desired final product characteristics.

9.4.1 Open Pond Systems

Different pond systems are utilized for marine algae production [41]. Microalgae, under phototrophic growth conditions, harness solar energy and assimilate carbon dioxide and nutrients for growth. Commercial production seeks to replicate and optimize these conditions. Reactor choice is crucial, with open tanks, including raceways and circular ponds with rotating agitators, being common. Raceway tanks, widely used for microalgal cultivation, are cost-effective, requiring low power and easy maintenance. However, they may face challenges such as contamination control and environmental variations affecting growth parameters. These systems utilize runoff water and sewage for nutrients, making them economically viable and environmentally friendly. Effective illumination, gas-liquid transfer, and low contamination are crucial for successful algae cultivation in raceway ponds. Deep ponds, aerated for mixing, are employed for small-scale production of *Nannochloropsis salina*, typically with areas less than 10 m^2 and depths exceeding

50 cm. Circular ponds with a centrally pivoted rotating agitator are used in Japan and Taiwan for *Chlorella* [41].

9.4.2 Closed Photobioreactor Systems

Various bioreactor types can be used for microalgae cultivation, with *Spirulina* and *Chlorella* being widely grown in open ponds globally. Closed photobioreactors have garnered significant attention due to their high productivity. Various configurations have been tested at laboratory or pilot scales, including vertical reactors, flat plates, annular systems, plastic bags, green wall panels (GWP), and different forms of tubular reactors, either mechanically stirred or airlifted. These closed systems excel in CO_2 bio fixation owing to better medium homogeneity and mass transfer. However, they face limitations due to excess O_2 production. While closed photobioreactors can maintain sterile conditions to control contamination, this approach increases production costs. Scaling up open photobioreactors typically involves increasing tube diameter, but this can result in inadequate light exposure for cell growth. The high costs associated with closed photobioreactors should also be taken into consideration.

9.4.3 Innovations in Cultivation Methods

A closed cultivation device was created by transforming the tumble movement into a circular pattern within a ring-shaped vessel. This innovative system balances adaptability and manageability while achieving growth rates comparable or superior to standard tanks. Notably, the ring-shaped vessel reduces the variable costs by minimizing the required medium volume for cultivation. The closed system enhances control over the cultivation process and eliminates potential environmental interferences and interfaces. However, it is also crucial to recognize that different vessel designs may vary significantly in terms of structure, environment, and process conditions.

Integrated Multi-Trophic Aquaculture (IMTA) systems present a promising approach to sustainable aquaculture by combining nutrient recycling with enhanced biomass production. In France, an innovative land-based Recirculating Aquaculture System (RAS) underwent a 60-day study. This system integrated a RAS for European sea bass (*Dicentrarchus labrax*) with two additional production components: High-Rate Algal Ponds (HRAP) featuring natural marine polyspecific algal communities and oyster cultivation in distinct open tanks [42].

9.4.4 Challenges and Advancements in Scaling Up Production

Scaling up algal cultures for commercial production is an intricate task that demands expertise. The process involves optimizing the production of sufficient inoculum for large ponds or photobioreactors to minimize time and costs. Effective management of large-scale cultures is crucial to prevent contamination or collapse, thereby reducing the need for re-inoculation. Maintaining stable, high-productivity cultures on a large scale, especially under variable outdoor conditions, presents unique challenges not encountered in small-scale laboratory cultures. Specific methods and protocols tailored to the microalgae, culture system, and production plant location are essential and can only be developed at a large scale. Over time, operators tend to gain experience, leading to improvements in both productivity and reliability in large-scale operations.

9.5 Physicochemical Growth Parameters

Microalgal growth is more influenced by physical parameters such as light intensity and temperature than nutritional factors.

9.5.1 The Role of Light in Growth

Light intensity stands as a crucial factor in microalgal cultivation, significantly impacting their growth and biomass production. The photosynthesis process of microalgae, directly influenced by light intensity, determines their growth rate. Typically, elevated light intensity enhances the growth rate of microalgae, reaching an optimal point specific to each microalgal species. Beyond this point, however, excessive light intensity, up to saturation, may result in photo-inhibition. Conversely, if light intensity falls below the saturation point, the growth of microalgae becomes constrained. The lighting intensity plays a regulatory role in the synthesis of various compounds, including carotenoids such as β-carotene, astaxanthin, lutein, and fatty acids. Also, adjusting the duration of the photoperiod and incorporating pulsed light can prevent photodamage to microalgae cells and establish conditions for optimal light photon absorption [43].

In a study on microalgae *Scenedesmus sp.*, the highest growth rates across different light intensities were observed at 4000 lux, resulting in a biomass production of 2.28×10^6 cells/ml. From this study, it has also been concluded that there is a positive correlation between an increase in lux and elevated growth rates and biomass production [44]. *Chlorococcum sp.* demonstrates optimal growth in a growth medium utilizing saline water at a specified light intensity of 3500 lux. In conclusion, low light intensity prolongs the physical adaptation process of microalgae cells to a new environment, while high light intensity reduces the number of growth phase stages that microalgae undergo [45].

Most microalgae cultivation predominantly occurs in open ponds, harnessing solar energy as the primary light source. When cultivating marine microalgae outdoors, sunlight plays a central role, while for indoor cultivation, artificial light sources like optical fibres or LEDs are more suitable. Research indicates that traditional artificial light sources result in increased biomass production, attributed to their expansive illumination area, stable light source, and cost-effectiveness. On the other hand, LEDs offer advantages such as lower energy consumption, longer lifespan, reduced heat generation, economical construction, and the ability to endure frequent on-off switching.

9.5.2 Temperature and Its Effect on Microalgae

The temperature during cultivation significantly influences the growth and lipid accumulation of microalgae. According to the study by Chaisutyakorn, as the temperature increases, there is a decrease observed in the specific growth rate, biomass, and lipid content across all the microalgae studied [46]. Temperatures that surpass the optimal range for growth are considered more harmful than lower temperatures. This can be understood by an asymmetrical growth curve concerning temperature. Beyond the optimum, the decrease in growth rate follows a linear pattern, leading to lethal temperatures that are reached more or less abruptly, depending on the specific characteristics of the species.

Nannochloropsis oculata is utilized for EPA production, and *Isochrysis galbana* is employed for DHA production. Their optimal productivities are achieved in the temperature range of 14–20 °C, ensuring nutritional indices within the required range for nutritional benefits [47]. At 25 °C, the biomass and growth rate of *P. tricornutum* are seen to peak, reaching 12.23×10^6 cells/L and 2.14×10^6 cells/L/h, respectively. *C. vulgaris* exhibited the highest biomass and growth rate at 20 °C, reaching 12.16×10^6 cells/L and 1.17×10^6 cells/L/h, respectively [48]. *C. malina*, a polar marine microalga, thrives at a favourable growth temperature of 4 °C. Stress is induced at temperatures equal to or greater than 8 °C, prompting a shift in metabolism towards the synthesis of lipids and carbohydrates [49].

9.5.3 Nutrient Availability and Composition

Microalgae exhibit higher growth rates and biomass production when cultivated in media with elevated levels of nitrogen and phosphorus. Conversely, conditions with limited nutrients prove to be more effective for the generation of lipids and fatty acids [50]. Manipulating both biotic and abiotic factors in the conditions of algae culture, such as temperature, nutrient availability, and salinity, has the potential to induce alterations in the growth and biochemical composition of microalgae. The arrangement of essential metabolic pathways varies significantly among different classes of algae, and this diversity, coupled with structural differences in the photosynthetic machinery, can influence crucial processes including nutrient uptake, assimilation, and carbon allocation. Consequently, these factors collectively determine how microalgae respond to environmental changes. In *N. gaditana*, treatments with low N:S result in elevated carbon: sulfur (C:S) and N:S molar ratios, increasing by up to 95% and 81%, respectively. This was manifested in the lipid and protein contents, which showed increases of up to 67% and 55%, respectively [51]. In a study, it was concluded the most suitable medium for the photoautotrophic cultivation of *O. aurita* consisted of 17.65 mmol/L nitrogen, 1.09 mmol/L phosphorus, 0.42 mmol/L silicon, and 24.51 mmol/L sulfur. This formulation resulted in maximum biomass production of 6.1–6.8 g/L under low-light conditions and 6.7–7.8 g/L under high-light conditions [52]. The introduction of nitrate to the medium in *P. tricornutum* cultivation resulted in a notable rise in the specific fucoxanthin concentration, reaching a maximum specific concentration of 59.2 ± 22.8 mg g^{-1} [53].

9.5.4 pH and Salinity

The optimal pH varies among algae species based on their physiological reactions to pH variations [54]. Although the majority of microalgae typically flourish in environments with a neutral pH, certain microalgal species have showcased the capability to thrive in highly acidic ($pH < 5$) or alkaline ($pH > 9$) conditions. Intriguingly, the growth of microalgae has the potential to influence the pH of their surrounding medium. An investigation into proton imbalance during microalgae cultivation highlighted that alkalinity levels were either generated or depleted, depending on the nitrogen source and its metabolic processes. This phenomenon exerted a more pronounced influence on the medium's pH compared to CO_2 assimilation. In a study, the maintenance of a neutral pH was found to be favourable for achieving substantial biomass production. Conversely, controlling the culture pH below 8 resulted in a preference for microalgal growth, marked by limited bacterial coexistence or contamination [55]. Lower pH promotes algal growth through the presence of CO_2 species, but it can negatively impact nutrient uptake or introduce metal toxicity. Controlling microalgae culture pH involves optimizing mineral acid or base content in the medium along with automated CO_2 supply [54]. It has been sustained that exposure to elevated CO_2 concentrations inhibited the growth of *Isochrysis galbana* [56].

Salinity plays a crucial role in microalgal production, particularly in outdoor cultivation systems with unpredictable environments. In a study on *Tetraselmis suecica*, variations in salinity did not significantly impact total protein, carbohydrate, and lipid content. However, an increase in salinity from 30 to 50 and 60 ppt resulted in a notable improvement of almost 22% in total lipid productivity [57].

9.6 Environmental Impact

9.6.1 Sustainability and Ecological Impact

Microalgae processes necessitate a large amount of energy, which must be clean and sustainable. Microalgae are microorganisms that can be found in both marine and freshwater environments. Their presence is essential in aquatic ecosystems as they are the primary building blocks of food chains. The magnitude of their biomass is substantial enough to exert an impact on global climate systems. Approximations speculate that there may be more than 800,000 microalgae species in existence, although only 50,000 have been officially recorded. This clearly illustrates the immense potential for the modification of microalgae, which in turn may provide an alternative reservoir of resources for the present-day world that faces limitations in its available resources. To combat food scarcity as well as to ensure a nutritional food supply for the sustainable living of the increasing population, microalgae are considered innovative sources for adequate nutrition. Since microalgae activities demand a lot of energy, the need for clean, sustainable, economically effective, and efficient energy production systems has become a major concern during the last century. Even if there are still many obstacles in the way, life cycle assessment (LCA) is becoming a more systematic projective analytic method for evaluating a process's overall sustainability. Sustainability is still a difficult worldwide aim today. This technique enables us to gather input inventories, assess the potential effects of a process or product, and translate requests into fair environmental costs linked to sustainable criteria. Recent applications of LCA in microalgae have mostly addressed the environmental concerns of products used for energy [58].

9.6.2 Carbon Footprint and Environmental Benefits

Surprisingly, biological carbon capture and sequestration utilizing microalgae has been recognized as one of the most important and successful carbon sequestration strategies in the world. To address the CO_2 issue, several carbon capture and conversion strategies have been proposed. CO_2 can be extracted from a flue gas mixture using one of the following methods: (i) solvent absorption, (ii) physical adsorption, (iii) membrane separation, or (iv) cryogenic distillation [59]. However, employing a microalgae-based biological carbon-capture strategy promotes carbon footprint reduction and bioenergy generation, making the notion a CO_2-neutral replacement for fossil fuels [60]. This is because microalgae are used as a lignocellulosic biomass feedstock for biofuel production and value-added chemicals. Microalgae is a form of micro-cell factory with a remarkable ability to repair carbon dioxide. Another important characteristic of microalgae is their great photosynthetic efficiency. Microalgae may repair carbon dioxide 10–50 times faster than typical terrestrial plants. It has a significantly faster growth and multiplication rate (a few hours) than higher plants. The capacity of microalgae to withstand high CO_2 concentrations is a concerning factor in choosing appropriate microalgae species for carbon capture, as CO_2-sensitive microalgae do not help capture CO_2. Only CO_2-resistant microalgae strains, which can withstand high CO_2 concentrations and high temperatures, are recommended. *Chlorella, Scenedesmus,* and *Chlorococcum* have been successfully employed to collect CO_2 from effluents and industrial emissions [59].

9.6.3 Biodiversity and Ecosystem Considerations

The algae are a polyphyletic, artificial collection of O_2-evolving, photosynthetic creatures that comprise seaweeds (macroalgae) and a varied group of microorganisms known as microalgae. Microalgae have a large and diverse range of species, covering taxonomic groupings such as *Chlorophyta, Bacillariophyta, and Cyanobacteria.*

Microalgae exhibit morphological, genetic, and functional diversity and may be found in a variety of habitats ranging from freshwater to extreme conditions. Their versatility in ecological niches ranging from surface waters to deep marine settings adds to their ecological functions as primary producers and symbiotic partners. Furthermore, microalgae have economic relevance due to its uses in industries such as food, biofuels, and medicines. However, the prevalence of toxic algal blooms emphasizes the necessity of monitoring and understanding microalgal biodiversity for both ecological health and biotechnological innovation. Ongoing study continues to unearth new species, highlighting the variety and promise of this microscopic world.

9.7 SWOT Analysis

SWOT analysis is crucial for assessing the potential of marine microalgae in the food and feed industry. It identifies strengths such as high nutritional value and rapid growth, highlighting opportunities for innovation and market growth. Simultaneously, weaknesses like cultivation challenges are recognized, guiding strategies to enhance feasibility. The analysis also addresses threats such as regulatory hurdles and market competition, enabling proactive risk mitigation. In summary, SWOT analysis aids in decision-making, optimizing the integration of marine microalgae for sustainable and nutritious food and feed solutions (Table 9.3).

Table 9.3 SWOT analysis

Parameters	Observation
Strengths of large-scale marine microalgae cultivation	• No competition with arable land • Fast growth of microalgae • Sustainable and ecofriendly alternative to traditional crops • Biorefinery approach and recyclability of water, nutrient, and energy • Also contain supplementary vitamins, bioactive compounds, and minerals • Traceability of product • High nutrient content (%w/w protein and %w/w lipid) can help overcome malnutrition • Can minimize the impact of overgrazing and other traditional farming practices • Natural CO_2 sequester
Weaknesses and challenges faced	• Large demand and supply gap • Contamination by other microbes under open pond cultivation • Controlling attack of microalgal predators such as ciliates/ rotifers / cladocerans and copepods • Competition with existing counterparts • Lack of awareness and information among consumer • Lack of consumer acceptability • Lack of authorization support • Need for trained personnel for microalgae cultivation and maintenance • Insufficient funding and investment from the Government • Uncertainties of strain selection and technological immaturity

(continued)

Table 9.3 (continued)

Parameters	Observation
Opportunities in the microalgae industry	• Improvement in up-stream processing can boost production • Improving in downstream processing can ensure its cost effectiveness • Advancement in biorefinery approach may underscores circular economy commitments • Growing market demand for microalgae
Threats to sustainable growth and market competition	• Low R&D investments • Associated risk and challenges • Competition with existing market • High production cost • More strict regulations for algae products in pharmaceutical, food, and feed sector • Highly energy intensive and costly downstream processing

9.8 Conclusion and Future Prospects

Extensive research is underway to explore the potential of microalgae in producing not only nutrients but also biofuels, biochemicals, and various bio-products. The active research and development landscape is focused on leveraging biotechnology, with genetic engineering standing out as a promising avenue for innovation in algae and microalgae. Two main categories of products derived from microalgae exist. The first category includes dried algae, particularly *Chlorella* and *Spirulina*, known for their rich nutrient content, including vitamins B_{12}, C, and D_2 and sold directly as dietary supplements. The second category comprises speciality products isolated from microalgae, such as pigments, antioxidants, proteins, and fatty acids. These high-value compounds can be incorporated into food and feed to enhance their nutritional value. However, it is essential to note that while the technology holds significant potential, it is still in its early infancy stage and has yet to reach commercial applications.

The potential for increased production and commercialization of high-value nutrients from microalgae is influenced by various market and economic factors. However, the lack of publicly available economic and market data poses a challenge in accurately assessing their true industrial potential, making it difficult to support funding and policy decisions. Results indicate that while the market for microalgae-based food and feed products has significant potential for global growth, the current contribution of these products to human and animal nutrition is limited. This limitation arises from the small-scale production and the comparatively high operational and production costs, infrastructure and maintenance expenses especially when compared to traditional commodities as well as the selection of algal strains with high protein contents, and efficient dewatering and harvesting at a commercial scale. Long-term research efforts are crucial for developing sustainable systems for the sustainable production of algal-based products, for applications in food and feed.

Even though the utilization of microalgal and cyanobacterial biomass in the food/feed market is on the rise, however, in the European Union, these organisms are mostly regarded as novel foods and are subject to the regulatory and legislative frameworks. Due to this, it is essential to ensure that their consumption poses no risks of using risk assessments as well as considering factors such as intrinsic characteristics of the biomass (such as secondary metabolites, toxic compounds, allergens), potential contaminants during cultivation (organic and inorganic substances or contaminants), and risks during downstream processing of the harvested biomass. Further research is required to comprehensively address all these aspects.

References

1. Camacho F, Macedo A, Malcata F (2019) Potential industrial applications and commercialization of microalgae in the functional food and feed industries: a short review. Mar Drugs 17(6):312
2. Takenaka H, Yamaguchi Y, Sakaki S, Ishibashi G (2015) Beneficial effect of edible blue-green alga Nostoc flagelliforme (Cyanophyceae) on blood glucose level and serum lipid concentrations in spontaneously non-insulin-dependent diabetic rats. Algal Resour 8(2):129–135
3. Balaji M, Thamilvanan D, Vinayagam SC, Balakumar BS (2017) Anticancer, antioxidant activity and GC-MS analysis of selected micro algal members of chlorophyceae. Int J Pharm Sci Res 8:3302–3314
4. Udayangani RMAC, Somasiri GDP, Wickramasinghe I, Kim SK (2020) Potential health benefits of sulfated polysaccharides from marine algae. Encycl Mar Biotechnol 1:629–635
5. Ferrara L (2020) Dinoflagellates important marine producers of natural bio-compounds with high biotechnological and pharmacological potential. J Food Chem Nanotechnol 6:138–149
6. Kottuparambil S, Thankamony RL, Agusti S (2019) Euglena as a potential natural source of value-added metabolites. A review. Algal Res 37:154–159
7. Bonfanti C, Cardoso C, Afonso C, Matos J, Garcia T, Tanni S, Bandarra NM (2018) Potential of microalga Isochrysis galbana: bioactivity and bioaccessibility. Algal Res 29:242–248
8. Peltomaa E, Johnson MD, Taipale SJ (2017) Marine cryptophytes are great sources of EPA and DHA. Mar Drugs 16(1):3
9. Střížek A, Přibyl P, Lukeš M, Grivalský T, Kopecký J, Galica T, Hrouzek P (2023) Hibberdia magna (Chrysophyceae): a promising freshwater fucoxanthin and polyunsaturated fatty acid producer. Microb Cell Factories 22(1):73
10. Rushdi MI, Abdel-Rahman IA, Saber H, Attia EZ, Madkour HA, Abdelmohsen UR (2021) A review on the pharmacological potential of the genus Padina. South Afr J Bot 141:37–48
11. Morançais M, Mouget JL, Dumay J (2018) Proteins and pigments. In: Microalgae in health and disease prevention. Academic Press, pp 145–175
12. Sansone C, Pistelli L, Del Mondo A, Calabrone L, Fontana A, Noonan DM, Albini A, Brunet C (2022) The microalgal diatoxanthin inflects the cytokine storm in SARS-CoV-2 stimulated ACE2 overexpressing lung cells. Antioxidants (Basel, Switzerland) 11(8):1515
13. Wang F, Cao Y, Guo Y, Zhu Z, Zhang C (2021) Evaluation of antioxidant and antibacterial activities of lipid extracts from Eustigmatos cf. polyphem (Eustigmatophyceae) and preliminary identification of bioactive compound. Algal Res 59:102446
14. Poza AM, Fernández C, Gauna MC, Parodi ER (2018) Biochemical properties and culture optimization of Leathesia marina (Phaeophyceae). Algal Res 33:379–388
15. Lucakova S, Branyikova I, Hayes M (2022) Microalgal proteins and bioactives for food, feed, and other applications. Appl Sci 12(9):4402
16. Ma R, Wang B, Chua ET, Zhao X, Lu K, Ho SH et al (2020) Comprehensive utilization of marine microalgae for enhanced co-production of multiple compounds. Mar Drugs 18(9):467
17. Stonik VA, Stonik IV (2023) Carbohydrate-containing low molecular weight metabolites of microalgae. Mar Drugs 21(8):427
18. Lauritano C, Andersen JH, Hansen E, Albrigtsen M, Escalera L, Esposito F et al (2016) Bioactivity screening of microalgae for antioxidant, anti-inflammatory, anticancer, anti-diabetes, and antibacterial activities. Front Mar Sci 3:68

19. Wu J, Gu X, Yang D, Xu S, Wang S, Chen X, Wang Z (2021) Bioactive substances and potentiality of marine microalgae. Food Sci Nutr 9(9):5279–5292
20. Domingues MR, Calado R (2022) Lipids of marine algae—biomolecules with high nutritional value and important bioactive properties. Biomol Ther 12(1):134
21. Remize M, Brunel Y, Silva JL, Berthon JY, Filaire E (2021) Microalgae n-3 PUFAs production and use in food and feed industries. Mar Drugs 19(2):113
22. Fernandes T, Fernandes I, Andrade CA, Cordeiro N (2016) Changes in fatty acid biosynthesis in marine microalgae as a response to medium nutrient availability. Algal Res 18:314–320
23. Barta DG, Coman V, Vodnar DC (2021) Microalgae as sources of omega-3 polyunsaturated fatty acids: biotechnological aspects. Algal Res 58:102410
24. de Jesus Raposo MF, de Morais RMSC, de Morais AMMB (2013) Health applications of bioactive compounds from marine microalgae. Life Sci 93(15):479–486
25. Fábregas J, Herrero C (1986) Marine microalgae as a potential source of minerals in fish diets. Aquaculture 51(3–4):237–243
26. Mahata C, Das P, Khan S, Thaher MI, Abdul Quadir M, Annamalai SN, Al Jabri H (2022) The potential of marine microalgae for the production of food, feed, and fuel (3F). Fermentation 8(7):316
27. Singh DV, Upadhyay AK, Singh R, Singh DP (2021) Health benefits of bioactive compounds from microalgae. In: Phytomedicine. Academic Press, pp 291–319
28. Saadaoui I, Rasheed R, Aguilar A, Cherif M, Al Jabri H, Sayadi S, Manning SR (2021) Microalgal-based feed: promising alternative feedstocks for livestock and poultry production. J Anim Sci Biotechnol 12(1):76
29. Ahmad A, Ashraf SS (2023) Sustainable food and feed sources from microalgae: food security and the circular bioeconomy. Algal Res:103185
30. Molino A, Iovine A, Casella P, Mehariya S, Chianese S, Cerbone A et al (2018) Microalgae characterization for consolidated and new application in human food, animal feed and nutraceuticals. Int J Environ Res Public Health 15(11):2436
31. Koyande AK, Chew KW, Rambabu K, Tao Y, Chu DT, Show PL (2019) Microalgae: a potential alternative to health supplementation for humans. Food Sci Human Wellness 8(1):16–24
32. Spolaore P, Joannis-Cassan C, Duran E, Isambert A (2006) Commercial applications of microalgae. J Biosci Bioeng 101(2):87–96
33. Pérez-Lloréns JL (2020) Microalgae: from staple foodstuff to avant-garde cuisine. Int J Gastron Food Sci 21:100221
34. Mouritsen OG, Rhatigan P, Pérez-Lloréns JL (2018) World cuisine of seaweeds: science meets gastronomy. Int J Gastron Food Sci 14:55–65
35. Chandan NK, Kumari R, Siddaiah GM (2020) Role of nutraceuticals in fish feed. In: Fish nutrition and its relevance to human health. CRC Press, pp 229–243
36. Turchini GM, Trushenski JT, Glencross BD (2019) Thoughts for the future of aquaculture nutrition: realigning perspectives to reflect contemporary issues related to judicious use of marine resources in aquafeeds. North Am J Aquac 81(1):13–39
37. Guedes AC, Malcata FX (2012) Nutritional value and uses of microalgae in aquaculture. Aquaculture 10(1516):59–78
38. James R (2010) Effect of dietary supplementation of Spirulina on growth and phosphatase activity in copper-exposed carp (Labeo rohita). Israeli J Aquac-Bamidgeh 62
39. Valente LM, Cabrita AR, Maia MR, Valente IM, Engrola S, Fonseca AJ et al (2021) Microalgae as feed ingredients for livestock production and aquaculture. In: Microalgae. Academic Press, pp 239–312
40. Morales M, Collet P, Lardon L, Hélias A, Steyer JP, Bernard O (2019) Life-cycle assessment of microalgal-based biofuel. In: Biofuels from Algae, pp 507–550
41. Sreekumar N, Nandagopal G, Vasudevan A, Antony R, Selvaraju N (2016) Marine microalgal culturing in open pond systems for biodiesel production—critical parameters. J Renew Sustain Energy 8(2)

42. Li M, Callier MD, Blancheton JP, Galès A, Nahon S, Triplet S et al (2019) Bioremediation of fishpond effluent and production of microalgae for an oyster farm in an innovative recirculating integrated multi-trophic aquaculture system. Aquaculture 504:314–325
43. Maltsev Y, Maltseva K, Kulikovskiy M, Maltseva S (2021) Influence of light conditions on microalgae growth and content of lipids, carotenoids, and fatty acid composition. Biology 10(10):1060
44. Ahmad Latiffi NA, Radin Mohamed RMS, Apandi NM, Tajuddin RM (2017) Experimental assessment on effects of growth rates microalgae Scenedesmus sp. in different conditions of pH, temperature, light intensity and photoperiod. Key Eng Mater 744:546–551
45. Putri DS, Sari DA, Astuti SP, Wangiyana IGAS (2021) Effect of medium type, light intensity, and photoperiod on the growth rate of microalgae Chlorococcum sp. local isolate. In: IOP conference series: earth and environmental science, vol 913, No. 1. IOP Publishing, p 012071
46. Chaisutyakorn P, Praiboon J, Kaewsuralikhit C (2018) The effect of temperature on growth and lipid and fatty acid composition on marine microalgae used for biodiesel production. J Appl Phycol 30:37–45
47. Aussant J, Guihéneuf F, Stengel DB (2018) Impact of temperature on fatty acid composition and nutritional value in eight species of microalgae. Appl Microbiol Biotechnol 102:5279–5297
48. Zhao T, Han X, Cao H (2020) Effect of temperature on biological macromolecules of three microalgae and application of FT-IR for evaluating microalgal lipid characterization. ACS Omega 5(51):33262–33268
49. Morales-Sánchez D, Schulze PS, Kiron V, Wijffels RH (2020) Temperature-dependent lipid accumulation in the polar marine microalga Chlamydomonas malina RCC2488. Front Plant Sci 11:619064
50. Yaakob MA, Mohamed RMSR, Al-Gheethi A, Aswathnarayana Gokare R, Ambati RR (2021) Influence of nitrogen and phosphorus on microalgal growth, biomass, lipid, and fatty acid production: an overview. Cells 10(2):393
51. Fernandes T, Fernandes I, Andrade CA, Cordeiro N (2020) Assessing the impact of sulfur concentrations on growth and biochemical composition of three marine microalgae. J Appl Phycol 32:967–975
52. Xia S, Wan L, Li A, Sang M, Zhang C (2013) Effects of nutrients and light intensity on the growth and biochemical composition of a marine microalga Odontella aurita. Chin J Oceanol Limnol 31(6):1163–1173
53. McClure DD, Luiz A, Gerber B, Barton GW, Kavanagh JM (2018) An investigation into the effect of culture conditions on fucoxanthin production using the marine microalgae Phaeodactylum tricornutum. Algal Res 29:41–48
54. Filali R, Tian H, Micheils E, Taidi B (2021) Evaluation of the growth performance of microalgae based on fine pH changes. Austin J Biotechnol Bioeng 8(1)
55. Yu H, Kim J, Rhee C, Shin J, Shin SG, Lee C (2022) Effects of different pH control strategies on microalgae cultivation and nutrient removal from anaerobic digestion effluent. Microorganisms 10(2):357
56. Xia Y, Sekine M, Hirahara M, Kishinami H, Yusoff FM, Toda T (2023) Effects of concentration and frequency of CO2 supply on productivity of marine microalga Isochrysis galbana. Algal Res 70:102985
57. Pugkaew W, Meetam M, Yokthongwattana K, Leeratsuwan N, Pokethitiyook P (2019) Effects of salinity changes on growth, photosynthetic activity, biochemical composition, and lipid productivity of marine microalga Tetraselmis suecica. J Appl Phycol 31:969–979
58. Depra MC, Severo IA, dos Santos AM, Zepka LQ, Jacob-Lopes E (2020) Environmental impacts on commercial microalgae-based products. Sustain Metrics Indic Algal Res 51:102056
59. Onyeaka H, Miri T, Obileke K, Hart A, Anumudu C, Al-Sharify ZT (2021) Minimizing carbon footprint via microalgae as a biological capture. Carbon Capture Sci Technol 1:100007
60. Siddiki SYA, Mofijur M, Kumar PS, Ahmed SF, Inayat A, Kusumo F et al (2022) Microalgae biomass as a sustainable source for biofuel, biochemical and biobased value-added products: an integrated biorefinery concept. Fuel 307:121782

Potential Seaweed-Derived Bioactive Compounds for Pharmaceutical Applications

10

Asmita Saha, Bhoomika Sridhar, and Shristi Ram

Abstract

Seaweeds, also known as marine macroalgae, have been recognized for centuries as a valuable resource with diverse applications. The pharmaceutical industry is increasingly seeking novel, natural compounds for drug development. Seaweeds are rich sources of bioactive compounds, such as polysaccharides, polyphenols, peptides, and lipids, which exhibit a wide range of biological activities. This chapter explores the potential of seaweed-derived bioactive compounds for pharmaceutical applications and their antiviral, antibacterial, anti-inflammatory, antioxidant, and anticancer capabilities, underscoring their potential in combatting a spectrum of ailments. For diabetes, these compounds have been shown to possess hypoglycemic properties, contributing to the regulation of blood sugar levels and aid in weight management. Further, they control cell growth and initiate angiogenesis, cutting off blood supply to tumors showing potent anticancer effects. The immunomodulatory and neuroprotective effects have also been discussed, broadening the scope of their application. Finally, the chapter advocates for further research in the field to fully realize the potential of seaweed-derived bioactive compounds as groundbreaking drugs and transformative therapies, revolutionizing the landscape of modern medicine.

Keywords

Bioactive compounds · Nutrition · Pharmaceutical industry · Seaweed · Therapeutics

A. Saha · B. Sridhar · S. Ram (✉)
Department of Microbiology, School of Basic and Applied Sciences, Dayananda Sagar University, Bengaluru, India
e-mail: shristi-sbas@dsu.edu.in

M. Rafatullah et al. (eds.), *Multidisciplinary Applications of Marine Resources*,
https://doi.org/10.1007/978-981-97-5057-3_10

10.1 Introduction

Seaweeds have recently emerged as a valuable and sustainable source of numerous bioactive compounds. Their historical significance in nutrition, supplements, and medicine spans centuries, paving the way for exploring their therapeutic potential [1]. This chapter navigates through the rich landscape of bioactive compounds present in seaweeds, shedding light on their nutritional abundance and diverse applications in pharmaceuticals, cosmetology, and the evolution of anti-diabetes and anti-cancer drugs [2, 3]. In the pursuit of extracting these valuable compounds, traditional methods like solvent extraction and maceration are giving way to innovative "green" extraction techniques. Approaches such as microwave-assisted extraction, ultrasound-assisted extraction, and enzyme-assisted extraction are ushering in a new era of efficiency and environmental sustainability [4]. Advanced techniques, including pressurized solvent extraction and various chromatography methods like liquid chromatography and size exclusion chromatography, play pivotal roles in isolating and purifying these bioactive compounds. Seaweed bioactive compounds showcase a diverse range of health-enhancing properties, spanning from antioxidant and anti-inflammatory to antimicrobial, anticancer, and antidiabetic [2]. Their versatility allows them to address various health challenges, from managing diabetes to making substantial contributions to anti-cancer strategies. The formulation and delivery systems of seaweed-derived bioactive compounds have also been discussed. Emulsions, encapsulation, hydrogels, and nanoparticles are the major formulation methods and contribute to optimizing the therapeutic potential of seaweed bioactive compounds [5–7]. Various delivery routes, including oral, topical, inhalation, and injection methods, have been scrutinized offering insights into their respective benefits and applications. In essence, this chapter provides a holistic understanding of the vast potential that seaweeds hold in the realm of health and wellness.

10.2 Seaweed Biochemistry and Composition

10.2.1 Overview of Seaweed Biochemistry

Seaweeds are vital members of the aquatic ecosystem, being multicellular, photosynthetic macroalgae. Their significance spans the globe, as they have been utilized for food, supplements, and medicinal purposes since ancient times. Numerous studies underscore the nutritional richness of seaweeds, emphasizing their potential in addressing chronic diseases. These marine organisms are categorized into three major classes: Rhodophyta (red algae), Phaeophyta (brown algae), and Chlorophyta (green algae). Substances like carrageenans and agars which are used as gelling agents to hydrocolloids used for pharmaceuticals are obtained from seaweeds [1].

10.2.1.1 Carbohydrates

Macroalgae contain polysaccharides that provide structure and store energy, with concentrations ranging from 4% to 76% of dry weight. High levels are found in *Ascophyllum, Porphyra, Palmaria,* and green species like *Ulva*. Key polysaccharides, including alginate, carrageenan, fucoidan, and laminarin, find applications in food and feed industries. Alginates, primarily from brown seaweeds, are linear polysaccharides of β-D-mannuronic acid and α-L-guluronic acid. *Durvillaea potatorum* and *Macrocystis pyrifera* are notable sources. Carrageenans, sulfated linear galactans in red seaweeds such as *Chondrus crispus* and *Kappaphycus sp*., and fucoidans, found in brown macroalgae like *Fucus vesiculosus* and *Undaria pinnatifida*, contribute to environmental stress protection. Laminarins, energy storage polysaccharides with 1,3-linked β-D-glucose monosaccharides and variable branching at β-(1,6), are abundant in *Laminaria sp., Saccharina sp., Ascophyllum, Fucus*, and *Undaria sp*., reaching up to 32% depending on the season [8]. Seaweed polysaccharides have the potential to regulate intestinal metabolism, hinder pathogen growth, and may have therapeutic applications for inflammatory bowel disease. Yet, due to their high molecular weight, it is necessary to convert them into oligosaccharides to enhance their fermentability. These polysaccharides are commonly categorized as dietary fiber, resisting digestion in the upper gastrointestinal tract and acting as nourishment for beneficial bacteria in the distal gut [9].

10.2.1.2 Proteins

Brown seaweed species, such as *Chnoospora minima, Dictyota menstrualis, Padina gymnospora*, and *Sargassum vulgare*, can exhibit high protein content (10–15%). Common brown seaweeds like *Laminaria digitata, Ascophyllum nodosum, Fucus vesiculosus*, and *Himanthalia elongata* have varying protein levels, with Laminaria digitata showing an average of 6.8%. They are also rich sources of amino acids like threonine, valine, leucine, lysine, and other essential amino acids. Green seaweeds like Ulva spp., having high concentrations of proline, may have protein content ranging from 8% to 10%. Red seaweeds, such as *Phorphyra tenera* and *Palmaria palmata*, can have varying protein concentrations from 11% to 47% [10]. Mycosporine-like amino acids, like palaythine, asterina, porphyra −334 and usujirene, are aminic metabolites that protect cells from oxidative stress and UV damage. These are extracted from *Alaria esculenta, Ulva intestinalis* and *C. crispus* [10]. Protein content in seaweed can fluctuate based on species, seasonal variations, and geographical factors. Lectins and phycobiliproteins are noteworthy among seaweed-derived proteins. Lectins, glycoproteins with carbohydrate-binding properties, have applications in immunology, histochemistry, agriculture, and medicine owing to their antimicrobial, antitumor, and antiviral activities. Phycobiliproteins are brightly colored and fluorescent proteins utilized in various fluorescence-based techniques, such as fluorescence-activated cell sorting and immunofluorescence [8].

10.2.1.3 Lipids

Seaweed lipids act as reservoirs for omega-3 polyunsaturated fatty acids (PUFAs), including phospholipids (PLs) and glycolipids (GLs) such as eicosapentaenoic acid

(EPA) and docosahexaenoic acid (DHA). These lipids demonstrate notable bioavailability, resistance to oxidation, and elevated omega-3 PUFA levels. With recognized antioxidant, antimicrobial, antitumor, and anti-inflammatory properties, marine lipids are increasingly valued in pharmaceuticals, nutraceuticals, and cosmeceuticals. Specific lipid classes, like monogalactosyldiacylglycerol (MGDG), display inhibitory effects against cancer cells. For example, MGDG derieved from *Fucus evanescence* is reported to be a potent inhibitor against malignant melanoma. Compounds such as 7-methoxy-9-methylhexadeca-4,8-dienoic acid (MMHDA) from *Ishige okamurae* inhibit phospholipase A2 activity, thereby alleviating inflammation in rat models. Glycolipids like capsofulvesin A and B, derived from the green seaweed *Capsosiphon fulvescens*, show potential as anti-diabetic agents. Additionally, red seaweeds like *Osmundaria obtusiloba* and brown seaweeds such as *Sargassum vulgare* contain sulfoquinovosyldiacylglycerol (SQDG), exhibiting antiviral effects against HSV-1 [9].

10.2.1.4 Vitamins

Seaweeds are abundant sources of water-soluble vitamins, including A, B_{12}, C, ß-carotene, pantothenate, folate, riboflavin, and niacin, surpassing fruits and vegetables in vitamin content. Notably, the class *Phaeophyceae* (*Porphyra umbilicalis* and *Crassiphycus changii*) have high quantities of essential and non-essential vitamins, particularly vitamin C and B [11]. Red seaweeds are rich in vitamin A, facilitated by the interaction between prokaryotes synthesizing vitamin B12 and the seaweed surface. While microalgae like *Chlorella* and *Spirulina* have higher vitamin B12 content, seaweeds like Nori and Chlorella supplements still contribute significantly to vitamin B12 coenzymes. Studies also highlight *Grateloupia turuturu* from the French Atlantic Coast as a source of α-tocopherol (vitamin E) and phytonadione (vitamin K1), showcasing seaweeds' notable nutraceutical potential [12].

10.2.1.5 Minerals

Seaweeds, rich in minerals essential for cellular functions and metabolism, offer up to 30% of their dry weight in minerals. Notably, sodium (Na), potassium (K), copper (Cu), calcium (Ca), and magnesium (Mg) levels in seaweeds surpass those in terrestrial foods, with examples like *Ulva lactuca* providing substantial calcium content. Seaweeds, such as *Ascophyllum nodosum* and *Chondrus crispus*, are crucial sources of iodine, meeting recommended daily intake amounts [1]. Seaweeds provide a higher mineral bioavailability of sodium, potassium, iodine, magnesium, fluorine, calcium, iron, arsenic, copper, zinc, and other elements. Brown seaweeds, including *Sargassum* and *Laminaria* species, contain more minerals than red seaweeds [11]. *Palmaria palmata* stands out for its iron content, offering a potential resource to address iron deficiencies. Additionally, zinc, manganese, and other trace minerals in seaweeds contribute to various metabolic processes and are crucial for maintaining good health. These findings emphasize the importance of seaweeds as rich sources of diverse minerals with significant nutritional and nutraceutical potential [12].

10.2.2 Other Bioactive Compounds

10.2.2.1 Pigments

Red seaweeds have high concentrations of phycobiliproteins and carotenoids. Phycobiliproteins play a crucial role in light harvesting during photosynthesis. They enable seaweeds to thrive in deep waters by absorbing light which is poorly captured by chlorophyll. Carotenoids are abundant in brown seaweeds and contribute to light harvesting, filtering harmful light radiations, and exhibit antioxidant properties. These pigments serve as photoprotective agents, defending against reactive oxygen species induced by UV radiation. Phycoerythrin imparts the red color to red seaweeds, with extraction and purification methods evolving for industrial applications. Phycocyanin, a blue pigment, shows immunomodulatory and bioactive properties. Allophycocyanin, located at the core of phycobilisomes, has antioxidative, anti-inflammatory, and hepatoprotective properties. Carotenoids, such as α and β-carotenoids, from red seaweeds are biologically active compounds with antioxidative, anti-inflammatory, and antitumor properties. They are also used in food, cosmetics, and pharmaceutical industries [12].

10.2.2.2 Terpenoids

Terpenoids, a diverse class of natural products found in plants, constitute a significant portion of drugs used for infectious diseases and cancer treatment. They are non-polar hydrocarbons composed of isoprene units which exhibit diverse bioactivities with therapeutic and nutraceutical applications. Over 23,000 terpenoid molecules, including the antimalarial drug Artemisinin and anticancer drug paclitaxel (Taxol®), are crucial in pharmaceuticals. Marine organisms, particularly red seaweeds like those in the Rhodomelaceae family, are rich sources of terpenoids. Seaweeds contain sesquiterpenoids with antimicrobial and cytotoxic potential, exemplified by genera like *Laurencia, Plocamium, and Chondrococcus*. The red seaweeds, in particular, showcase terpenoids with significant antitumor bioactivity. This class of compounds, with variations in structure, plays a vital role in global healthcare for treating various illnesses.

10.2.2.3 Alkaloids

Seaweeds, particularly the genus Caulerpa, are rich sources of alkaloids. *Caulerpa mexicana* extracts containing Caulerpin, demonstrated anti-inflammatory effects. Other indolic alkaloids from Caulerpa, such as racemosin A, B, and C, and caulersin, also showed bioactivities. The anti-inflammatory properties of caulerpin were evident in a murine model of DSS-induced ulcerative colitis, attenuating clinical signs, reducing cytokine levels, and modulating NF-κB activation. Additionally, red algae like *Gracilaria* demonstrated anti-inflammatory effects, inhibiting COX-2 activity, PGE2 synthesis, and proinflammatory cytokine gene expression. An alkaloid from *Gracilaria opuntia* selectively inhibits COX-2 and 5-LOX. Similarly, red algae of the genus *Laurencia*, like *Laurencia obtuse,* reduces TNF-α production in LPS-stimulated cells and decreases inflammatory exudate in a murine paw edema model [13].

10.2.2.4 Phenols

Phenols are compounds containing an aromatic ring with one or more hydroxyl groups. Hydroxybenzoic acids (HBAs) and hydroxycinnamic acids (HBCs), which condense into quinic acid and malic acids, are phenolic acids that are involved in nutrient absorption and enzymatic activities. These compounds are derived from *Ascophyllum nodosum, Fucus vesiculosus,* and *Bifurcaria bifurcate*. Phlorotannins, derivatives of phloroglucinol, are found mainly in brown seaweeds like *Ascophyllum nodosum, Cystoseira sedoises,* and *Ecklonia stolonifera*. They have antioxidant and antibacterial properties. Bromophenols are secondary metabolites, first isolated from red algae *Neorhodomela larix*, and are known to show antimicrobial activity and are responsible for characteristic flavors of seafood. Flavonoids, characterized by heterocyclic oxygen, are present in various species of seaweeds like *Chondrus crispus, Sargassum vulgare,* and *Nitell hookeri*. Certain terpenoids like plastoquinone and chromanol are phenolic in nature and solely found in the *Sargassaceae* family [12]. Various seaweed bioactive compounds and their functions are described in Table 10.1.

10.3 Extraction and Purification Techniques

Extraction and purification are two crucial processes in the isolation of specific compounds or substances from a complex mixture, such as plant extracts, seaweed, or other biological sources. These processes are particularly important in various industries, including pharmaceuticals, food, and natural product research. Extraction is the initial step in obtaining desired compounds from a raw material. It involves the selective removal of target substances from a complex matrix, such as plant tissues or seaweed. The goal is to transfer the desired compounds from the original source to a solvent or extraction medium. In the context of seaweed-derived bioactive compounds, extraction would involve obtaining substances like polysaccharides, polyphenols, or pigments from the seaweed biomass. The extracted solution is then further processed to isolate and purify the specific compounds of interest. Purification is the subsequent step where the crude extract obtained through extraction is refined to isolate and obtain a more concentrated and purified form of the target compound. Since the crude extract may contain a mixture of various compounds, purification aims to separate and eliminate impurities. The selection of purification method depends on the characteristics of the desired compounds. Multiple purification steps may be employed to achieve a higher purity.

Seaweeds are abundant in carbohydrates, proteins, lipids, phenolic compounds, pigments, minerals, and vitamins. Macroalgal protein content ranges from 1.3% to 47% (w/w) of the dry weight; the content is species-specific, also depending on physicochemical factors, growth season, and harvest location. Some red seaweeds surpass common protein sources like eggs, fish, soybean, and cereals. However, protein extraction yields from algae are lower than those from crops like soybean. Efficiently utilizing macroalgae as a sustainable protein source requires significant advancements.

Table 10.1 Bioactive compounds extracted from Seaweeds, their source organism, and functions

Compounds		Source organism	Function
Carbohydrates	Alginate	*Saccharina japonica, Laminaria hyperborea, A. nodosum, Urvillaea potatorum, Macrocystis pyrifera*	– Prebiotics – Thickening and gelling agent – Antimicrobial activity – Scavenging Reactive oxygen species – Anti-inflammatory
	Carrageenan	*Eucheuma denticulatum, Kappaphycus Alvarezii, Mastocarpus stellatus, Chondrus crispus*	– Thickening and gelling – Elevation of plasma – Prebiotic activity – Anti-inflammatory – Anti-cancer – Antihyperlipidemic – Anti-viral
	Fucoidan	*Chorda filum, Laminaria japonica, A. nodosum, F. evanescens*	– Prebiotic agent – Anti-inflammatory – Anti-coagulant – Antioxidant – Antiproliferative – Immunostimulatory
	Laminarin	*Laminaria spp., Saccharina spp.*	– Dietary fibre – Prebiotic – Antioxidant – Immunostimulant – Antimicrobial – Neuroprotective
	Ulvan	*Ulva spp.*	– Antimicrobial – Antihyperlipidemic – Antioxidant – Immunomodulatory

(continued)

Table 10.1 (continued)

Compounds		Source organism	Function
Proteins	Essential amino acids	*Fucus vesiculosus, Alaria esculenta, Laminaria digitata, Saccharina latissima*	– Amino acids supplements – Neurotransmitter – Maintaining immune function – Tissue repair and growth
	Lectins	*Sargassum fusiforme, Chondrus crispus, Eucheuma spinosum, Gracilaria spp.*	– Antimicrobial – Anti-inflammatory – Anti-cancer
	Phycobiliproteins	*Neoporphyra haitanensis, N. yezoensis, Centroceras clavulatum, Kappaphycus alvarezii*	– Fluorescent-based techniques – Nutraceuticals – Antimicrobial – Antioxidant – Anti-inflammatory – Neuroprotective – Hepatoprotective – Immuno-modulatory – Anticancer effects
	Mycosporine-like amino acids	*Alaria esculenta, Ulva intestinalis and C. crispus*	– Protection against UV damage – Protects from oxidative stress
Lipids	Poly unsaturated fatty acids (PUFAs)	*Undaria pinnatifida*	– Antioxidant – Anti-tumor – Neuroprotective effect – Cardioprotective
	Monogalactosyldiacylglycerol (MGDG)	*Fucus evanescence*	– Anti-cancer
	7-methoxy-9-methylhexadeca-4,8-dienoic acid (MMHDA)	*Ishige okamurae*	– Anti-inflammatory
	Sulfoquinovosyldiacylglycerol (SQDG)	*Sargassum vulgare*	– Antiviral effect
	Glycolipids	*Capsosiphon fulvescens*	– Anti-diabetic

(continued)

Table 10.1 (continued)

Compounds		Source organism	Function
Vitamins	A	*Most Red algae*	– Carotenoid supplements
	B	*Chlorella spp., Spiriulina spp., Porphyra spp.*	– Vitamin B supplements
	C	*Crassiphycus changii*	– Antioxidant – Anti-inflammatory
	E	*Grateloupia turuturu*	– Nutraceutical
	K	*Grateloupia turuturu*	– Nutraceutical
Minerals	Iodine	*Ascophyllum nodosum, Chondrus crispus, P. palmata*	– Synthesis of thyroid hormones
	Iron	*Palmaria palmata*	– Iron supplements – Oxygen transportation – DNA synthesis
	Zinc	*C. crispus, Gracilaria spp.*	– DNA and RNA synthesis – Part of metalloenzymes
	Manganese	*C. crispus, P. palmata, Gracilaria spp.*	– Needed for blood clotting – Homeostasis
Pigments	Carotenoids	*Gracilaria spp., Gelidium spp.*	– Antioxidant – Anti-inflammatory – Anti-tumor – Photoprotective effects
	Phycocyanin	*C. crispus, G. amansii*	– Immunomodulatory effect – Antioxidant – Hepatoprotective – Anti-inflammatory
	Phycoerythrin	*Gracilaria gracilis, Grateloupia turuturu*	– Natural colorant – Fluorescent biomarkers
Terpenoids	Terpenoids	*Laurencia, Plocamium, Chondrococcus*	– Antimicrobial – Anti-tumor

(continued)

Table 10.1 (continued)

Compounds		Source organism	Function
Alkaloids	Alkaloids	*Gracilaria opuntia, Laurencia obtuse, Caulerpa mexicana*	– Anti-inflammatory
Phenolic compounds	Phenolic acids (HBAs and HBCs)	*Ascophyllum nodosum, Fucus vesiculosus, Bifurcaria bifurcate*	– Antioxidant – Anti-inflammatory – Antimicrobial – Cardioprotective
	Phlorotannins	*Ascophyllum nodosum, Cystoseira sedoises, Ecklonia stolonifera*	– Antioxidant – Antibacterial
	Bromophenols	*Neorhodomela larix*	– Antimicrobial
	Flavonoids	*Chondrus crispus, Sargassum vulgare, Nitell hookeri*	– Antioxidant – Radical scavenging effect
	Phenolic terpenoids	*Callophycus serratus, Rhodomelaceae family, Sargassaceae family*	– Anti-tumor – Antioxidant – Photoprotection – Wound healing

The extraction of proteins from macroalgae, a crucial step in discovering algal-derived Bioactive Peptides (BAPs), faces challenges due to the rigidity and complexity of the macroalgal cell wall, along with the presence of alginates and polyphenols. Furthermore, the extraction efficiency of proteins from various macroalgal phyla and species can be influenced by their distinct physiological and biochemical characteristics, particularly the types of proteins they contain, biomass concentration, temperature, and the duration of extraction, affecting protein solubilization or precipitation. Moreover, the choice of method and reagents during protein extraction significantly affects the biological activity of the proteins and peptides. Hence, it is imperative to carefully select the extraction method tailored to each species to attain maximum protein recovery. An overview of the extraction and purification steps is presented in a flowchart in Fig. 10.1.

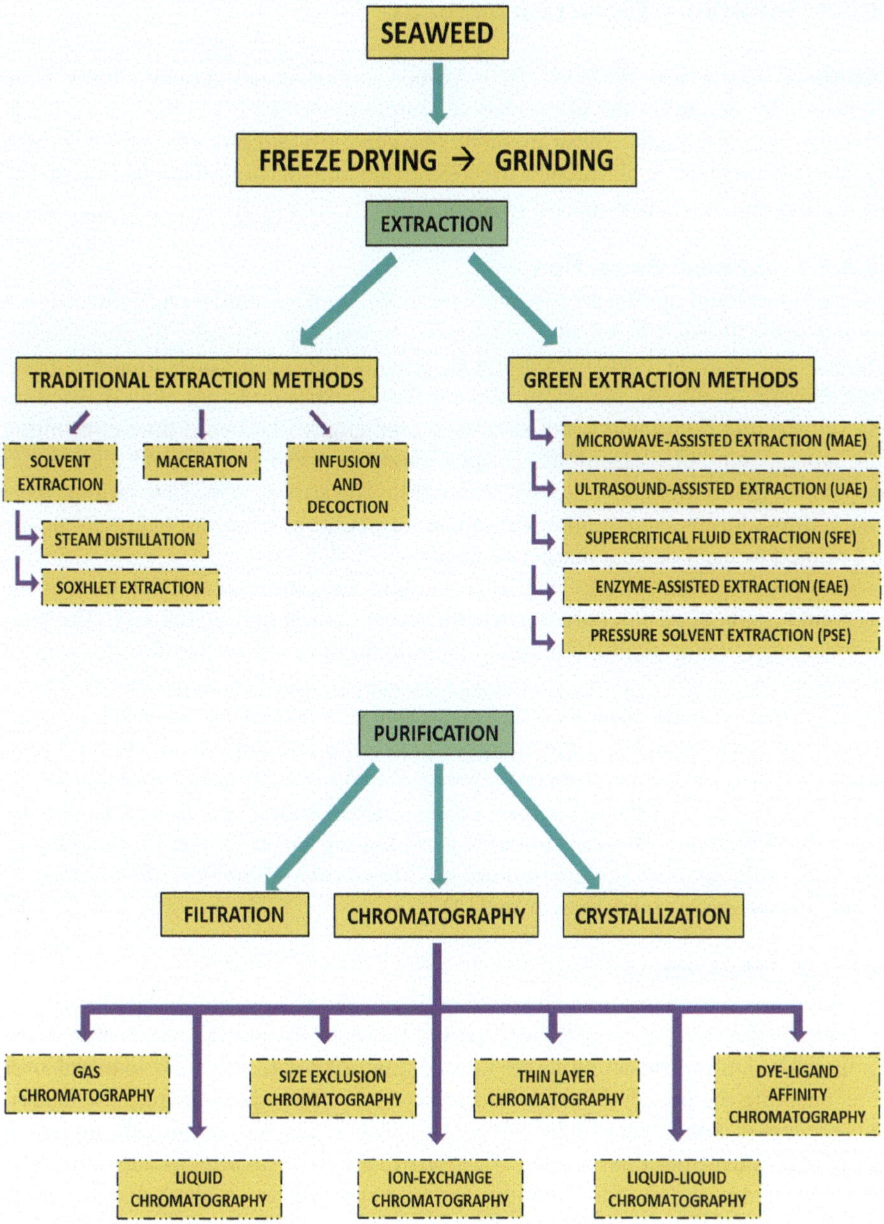

Fig. 10.1 Schematic representation of extraction and purification techniques of Seaweed derived bioactive compounds

10.3.1 Traditional Extraction Methods

Traditional extraction methods for seaweed bioactive compounds have been employed for centuries and often involve simple processes using readily available solvents and equipment. These methods are generally simple, cost-effective, and scalable for large-scale production. However, they may have limitations in terms of selectivity, efficiency, and solvent consumption.

10.3.1.1 Solvent Extraction

The most common approach, solvent extraction, involves immersing seaweed in a suitable solvent, such as ethanol, methanol, or water, to dissolve the target compounds. The solvent is then evaporated, leaving behind a concentrated extract. Conventional methods, including solvent-based extraction, are widely used to recover specific compounds, but they face challenges like being time-consuming and using potentially hazardous. Despite safety concerns, traditional approaches like steam distillation and Soxhlet extraction are still in use. Innovations have emerged to address safety and environmental issues, but conventional methods are still refined to meet current market demands.

Steam distillation, a well-established method for extracting volatile molecules, has found application in conjunction with solvent extraction techniques. The synergy between steam distillation and solvent extraction led to the development of innovative steam distillation extraction (SDE) apparatus. This apparatus integrates the principles of both methods, offering a more comprehensive approach to compound isolation. Moreover, the incorporation of microwave technology further refines and accelerates the extraction process, making SDE a valuable tool in the extraction arsenal. Soxhlet extraction, a classic liquid-solid extraction method introduced in 1879, remains a benchmark for evaluating novel extraction procedures. This technique involves the continuous cycling of solvent between a siphon and a sample, ensuring thorough extraction [14].

10.3.1.2 Maceration

A variant of solvent extraction, maceration, involves soaking seaweed in the solvent at room temperature for an extended period to allow for gradual extraction of the compounds. During maceration, the solvent interacts with the solid material over time, enabling the dissolution and absorption of target compounds from the matrix. This process is commonly employed for extracting a range of compounds, including polysaccharides, from macroalgae. What sets maceration apart is its adaptability to different solvents, agitation conditions, and temperatures, making it a versatile method for extracting various types of compounds from botanical sources. However, several cycles of filtration or centrifugation are needed to separate the desired compounds from the matrix [15]. In contrast to some other extraction methods, maceration is characterized by its simplicity and ease of application. The technique does not require heat or continuous solvent circulation, making it suitable for situations where a gentler extraction process is preferred. Maceration's adaptability to different solvents and conditions makes it a valuable tool in the extraction toolkit, especially when targeting specific compounds from botanical materials [14].

10.3.1.3 Infusion and Decoction

Infusion is a gentle extraction technique involving the steeping of seaweed in hot water to draw out soluble compounds. The infusion duration, typically 5–10 min, depends on seaweed variety and desired potency. While effective for water-soluble components like vitamins and minerals, it may not efficiently extract heat-sensitive compounds such as certain polysaccharides [16].

In contrast, decoction is a more robust method, boiling seaweed to expedite compound release, especially suited for heat-stable components. Decoction duration varies by seaweed type and desired strength. This method excels in extracting heat-stable compounds like polysaccharides and fucoidan but may compromise water-soluble vitamins. Both methods showcase versatile approaches to seaweed extraction, each tailored to specific compound characteristics.

Researchers have extensively employed decoction and infusion techniques for extracting bioactive compounds from diverse sources, including seaweeds. These methods have proven effective in extracting polyphenols and antioxidants from various plant materials [17]. The simplicity and user-friendly nature of traditional extraction methods, such as infusion and decoction, have contributed to their widespread use for extracting bioactive compounds from seaweeds [18]. Decoction, known for its ease of application, has been specifically utilized for obtaining aqueous extracts of bioactive compounds from seaweeds [19]. This exemplifies the versatility of these techniques in extracting valuable compounds, showcasing their relevance in seaweed research and applications.

10.3.2 Green Extraction Methods

"Green" extraction methods for natural products are driven by the pursuit of economical, high-quality, and eco-friendly approaches. Emerging extraction techniques prioritize shorter extraction times, lower operational temperatures, reduced solvent, and process automation, aligning with environmental sustainability. To enhance yields of the target analyte, optimization of various parameters, such as solvent ratio, extraction solvent, extraction time, pH, temperature, and particle size, is crucial [20]. Among the modern methods employed for seaweed bioactive compounds extraction are, microwave-assisted extraction (MAE), enzyme-assisted extraction (EAE), ultrasound-assisted extraction (UAE), pressurized liquid extraction (PLE), and supercritical fluid extraction (SFE) [4].

10.3.2.1 Microwave-Assisted Extraction (MAE)

Microwave-assisted extraction (MAE) has emerged as a cost-effective and widely embraced method, offering both open and closed vessel systems. While the closed system is effective at higher temperature and pressure, it comes with an inherent explosion risk, leading to the development of the open vessel system, operating under atmospheric pressure conditions. [21]. Several critical factors influence MAE efficiency, including microwave power and frequency, solid-to-liquid ratio, solvent concentration, extraction time, extraction pressure, matrix characteristics, and solvent properties. MAE presents compelling benefits over classical methods due to reduced solvent usage, shorter extraction time, higher extraction rates, and lower costs, particularly beneficial for large-scale sample extraction from natural products [22].

For instance, the traditional 2–4 h extraction time for native agar has been significantly shortened with MAE, which selectively and locally heats raw materials. Notable achievements include the successful extraction of seaweed hydrocolloids, yielding high-quality carrageenan from various red seaweed species, such as *Eucheuma cottonii*, *Eucheuma spinosum*, and *Hypnea musiformis* [23]. MAE's efficiency not only reduces solvent and energy consumption but also enhances the quality of end-products, operating on the principle of direct interaction with polar solvent molecules through electromagnetic waves (usually 2.45 GHz) [20]. MAE has demonstrated its effectiveness in extracting various seaweed compounds, including carbohydrates and proteins, further establishing its versatility in the field [24].

10.3.2.2 Ultrasound-Assisted Extraction (UAE)

This technique uses sound waves beyond the audible frequency (>20 kHz) and has gained prominence for its cost-effectiveness and efficiency compared to traditional methods like Soxhlet, maceration, and even advanced techniques such as MAE and supercritical fluid extraction (SFE) [25]. The efficacy of UAE is primarily attributed to acoustic cavitation, facilitating solvent penetration into the sample matrix, thereby enhancing the contact surface area between solid and liquid phases. This mechanical effect expedites solute diffusion, leading to increased extraction yield and reduced extraction time.

Widely employed in food processing, UAE has been applied to extract polysaccharides, carbohydrates, phytochemicals, oils, proteins, aroma, and flavor volatiles [25]. Notably, UAE has been environmentally benign and cost-effective, successfully extracting highly purified nanofibrils from cellulosic materials [26]. While UAE has been utilized for bioactive compound extraction from various marine algae, studies on seaweed hydrocolloids were limited until recently. Recent research showcased enhanced carrageenan yields and purity from the less-known red seaweed *Hypnea musciformis* using UAE-alkaline and aqueous treatments, outperforming conventional methods [27]. Additionally, the successful extraction of laminarin from Irish brown seaweed *Ascophyllum nodosum* and *Laminaria hyperborean* using UAE with HCl modification demonstrated higher yields and antioxidant activity compared to conventional methods [28]. UAE, based on low-frequency and high-power waves, disrupts cells, accelerates diffusion, and increases mass transfer, making it effective for extracting carbohydrates, phenolic compounds, and pigments from various seaweed species [20].

10.3.2.3 Supercritical Fluid Extraction (SFE)

SFE is a highly regarded and eco-friendly method known for its cost-effectiveness and efficiency, utilizing carbon dioxide (CO_2) or water (H_2O) as supercritical fluids. These fluids insure the extraction of highly pure compounds with minimal polar impurities, setting it apart from conventional organic solvent methods [29]. CO_2, especially, is favored for its suitable critical temperature and pressure, making it ideal for extracting thermo-labile compounds at lower temperatures. SFE finds applications in various fields, from sample preparation and analytical purposes to large-scale production for collecting desired products or eliminating unwanted materials. Critical factors influencing SFE efficiency include the choice of fluid, operating temperature (40–90 °C), pressure (100–400 bar), particle size, shape of the sample, moisture content, and solvent flow rate [30]. Seaweed hydrocolloids, known for their sensitivity to high temperatures, benefit from SFE-CO_2, creating a non-oxidizing atmosphere during extraction to prevent degradation. However, the successful extraction of Mexican chia hydrocolloids using SFE with CO_2 at 80 °C and 450 bar successfully demonstrated the versatility and potential applicability of SFE to different types of hydrocolloids [31]. Also, SFE has proven effective in extracting carotenoids, chlorophylls, polyunsaturated fatty acids (PUFAs), and polyphenols from various sources [32].

10.3.2.4 Enzyme-Assisted Extraction (EAE)

EAE stands out as a specific and selective method, leveraging enzymes to degrade algae cell walls. It presents a targeted approach to overcome the complex cell walls of seaweed, which are intricately composed of sulfated and branched polysaccharides associated with proteins, ions, and a chemically diverse matrix. This intricate structure hinders conventional extraction procedures, necessitating a method to break down these barriers. EAE employs digestive enzymes, such as carbohydrases and proteases, to effectively degrade the seaweed cell walls and release active components [33]. EAE offers notable advantages, including eco-friendliness, cost-effectiveness, and the transformation of water-insoluble materials into water-soluble ones, resulting in high yields of bioactive compounds. EAE not only enhances the extraction yield but also maximizes the antioxidant and antiviral activities of the compounds, a feature not often achieved with conventional extraction methods unless optimized [34]. While EAE has been extensively used for extracting pharmaceutical compounds with various health benefits, its application to seaweed hydrocolloids remains relatively limited [35]. Till date, EAE has been successfully employed for extracting proteins, phlorotannin, and hydrocolloids from various seaweeds, showcasing its versatility and suitability for obtaining a range of bioactive compounds [20]. Red seaweed, *Eucheuma cottonii,* has been extracted to produce a high yield of carrageenan with favorable gelling properties [36].

10.3.2.5 Pressurized Solvent Extraction (PSE)

PSE, as a "green" extraction method, has been recognized for extracting nutraceuticals from herbs and foods, significantly enhancing extraction yield, reducing time, and preserving sensitive compounds. Unlike Supercritical Fluid Extraction (SFE),

PSE allows a broader range of solvents to be used. Common solvents such as water, propane, and dimethyl ether are employed, with water being the most widely used. In this method, the extraction solvent, typically water, is heated above its normal temperature and pressure but below the critical temperature and pressure, remaining in a liquid state. PSE works under elevated temperature and pressure, typically ranging from 50 to 200 °C and 35 to 200 bar, respectively, while it may not be suitable for thermolabile compounds, it has demonstrated potential for extracting various bioactive compounds, including polyphenols, antioxidants, and fucoidan, from seaweed and microalgae [37]. Moreover, elevated temperature and pressure enhance the diffusion and solubility of solvents into seaweed, resulting in improved extraction. PSE yields non-hazardous crude extracts that require no further clean-up, reducing both cost and time [38]. Thus, PSE offers high efficiency, shorter extraction time, and low solvent volume requirement (10–50 mL). This stands in stark contrast to traditional methods that demand higher solvent requirement (about 300 mL) and longer extraction times [39].

Successful extraction of fucoidan (sulfated polysaccharides) from brown seaweed *Saccharina japonica* using PSE at an optimal temperature of 140 °C and pressure of 50 bars is reported with sodium hydroxide (NaOH), reinstating the potential of PSE for extracting common seaweed hydrocolloids like alginate, agar, and carrageenan [40]. Despite the high cost of equipment, PSE stands out for its lower environmental impact, reduced solvent requirement, and utilization of non-toxic and biobased solvents, making it a promising green extraction technique for various bioactive compounds, including those from seaweed hydrocolloids [41].

10.3.3 Purification Methods

Purification techniques aim to separate the desired compounds from seaweed extract (containing impurities) to insure their suitability for pharmaceutical applications. The choice of purification method depends on the specific compound being extracted, the properties of the seaweed matrix, the desired purity level, and the scale of production. Some commonly used purification techniques include precipitation, filtration, chromatography, and crystallization.

10.3.3.1 Filtration

In this process, the sample is propelled through a semi-permeable membrane, typically under pressure, resulting in permeate and retentate streams. Ultrafiltration, commonly used for peptides in the 1–10 kDa range, has found application in processing seaweed protein hydrolysates at the pilot scale. The advantages of membrane processing lie in its scalability and established use in large-scale protein processing, e.g., in dairy industry. However, limited information pertaining to the large-scale application of membrane filtration in seaweed protein/peptide processing is available. The selection of permeate or retentate for further processing/analysis depends on the target bioactivity. For instance, studies on *Porphyra spp.* protein hydrolysates revealed that <3 kDa permeates exhibited the lowest α-amylase IC50 value (half-maximal inhibitory concentration). In another case, membrane filtration of *U. lactuca* protein hydrolysates resulted in a 3 kDa permeate fraction having the highest renin inhibitory activity. Antibacterial activity was higher in the >10 kDa retentate of *Saccharina longicruris* hydrolysates, while the <10 kDa permeate fraction demonstrated higher antioxidant and ACE inhibitory activities in hydrolysates from *P. palmata, U. lactuca,* and *S. longicruris* [42].

Although membrane filtration has proven beneficial in hydrolysate fractionation for isolation of peptide identification owing to its simplicity, scalability, and cost-effectiveness, it has its limitations. Firstly, the lack of specificity in membranes allows similar molecular mass compounds to contaminate with the desired peptides. Secondly, interactions among peptides and the membrane, driven by differences in physicochemical properties, pose difficulty in purification. Thirdly, issues like membrane fouling and high dead volumes can impede the transfer of low molecular components to the permeate, particularly in seaweed hydrolysates. Membrane filtration is often utilized as an initial enrichment step followed by chromatographic fractionation [42].

10.3.3.2 Chromatography Techniques

Liquid-liquid partitioning, gas chromatography, liquid chromatography, dye-ligand affinity chromatography, size exclusion chromatography, ion exchange chromatography, and thin-layer chromatography are integral components of the seaweed extraction and purification. In particular, chromatography segregates mixture elements based on their interactions with mobile and stationary phases. The type of chromatographic method employed depends on factors like support or stationary phase and mobile phase characteristics (gas or liquid). The instrumental apparatus, referred to as a chromatograph, generates chromatograms representing the elution profile of compounds. A few key parameters such as retention time, void time, and peak characteristics are vital for identifying and quantifying components [43].

Chromatographic methods play a pivotal role in purifying and extracting a myriad of components from seaweeds. Reverse-phase high-performance liquid chromatography (RP-HPLC), for instance, relies on hydrophobic interactions between mobile and stationary phases for molecule isolation, offering efficiency and versatility in the extraction of diverse molecules. Despite its advantages, RP-HPLC has limitations, such as potential damage to protein samples, necessitating careful consideration of its application. Numerous studies exemplify the application of RP-HPLC in extracting and purifying bioactive molecules from seaweeds. For instance, RP-HPLC coupled with fluorescent and UV detectors was utilized to identify folates in commercial macroalgae, demonstrating its potential in nutritional studies. Additionally, RP-HPLC has been employed to extract antioxidative peptides from green microalgae protein, demonstrating its efficacy against oxidative stress caused by free radicals. Phenolic compounds, specifically oligophlorotannins, extracted from *Ascophylllum nodosum* using RP-HPLC, revealed their distribution and radical scavenging properties. Further experiments have explored the extraction of fucoxanthin, an isomer of all trans-fucoxanthin, from brown seaweeds using RP-HPLC, showcasing its potential applications in health, medicine, and antioxidant functions [43].

10.3.3.3 Crystallization

Crystallization, a widely utilized purification method, involves the controlled formation of crystals from a supersaturated seaweed extract solution. This process enables the isolation of highly pure compounds by leveraging the unique crystalline structures of the target components. The method includes creating a supersaturated solution, initiating controlled nucleation and crystal growth, where the crystals selectively incorporate the desired compounds, leaving impurities in the mother liquor. Through meticulous adjustments of factors like temperature, solvent composition, and concentration, crystallization allows for the separation of individual components based on their distinct crystal lattice arrangements. This technique holds significant value in obtaining purified and well-defined substances from complex mixtures, contributing to advancements in various fields, including pharmaceuticals, food, and biochemical research. In the case of Phycoerythrin (PE) and Phycocyanin (PC) extracted from *P. yezoensis*, initial crystallization, use of the sitting-drop vapor-diffusion method at different temperatures resulted in the growth of suitable crystals under specific conditions, typically appearing within 1 day and reaching full size in 4–5 weeks [44].

10.3.4 Advanced Extraction and Purification Technologies

Advanced extraction and purification methods offer enhanced efficiency and environmental sustainability. The choice of method hinges on factors such as the nature of the desired compounds, seaweed matrix properties, desired purity levels, production scale, and economic considerations. In the alginate extraction process, the alkaline extraction step, a key phase in traditional batch reactors, often demands extensive time and resource consumption. The innovative approach of reactive extrusion addresses these challenges by significantly reducing reaction time, as well as water and reactant (sodium carbonate) usage.

For instance, reactive extrusion reduced extraction time from hours to just 5 min, with water and reactant usage halved compared to traditional methods, achieving comparable yield, purity, and rheological properties for alginate [45]. This method has shown promising results in polymer science and food engineering and holds potential for extracting other seaweed hydrocolloids like agar and carrageenan. While equipment cost remains a consideration, further research is needed to assess the efficiency of this method for other hydrocolloids such as alginate and carrageenan [46].

10.4 Pharmacological Activities of Seaweed-Derived Bioactive Compounds

Seaweeds, abundant in diverse marine ecosystems, are gaining recognition for their unique biochemical composition, housing numerous bioactive compounds with significant pharmacological properties. This topic explores the diverse pharmacological activities of these compounds, focusing on their potential roles in managing diabetes, boosting antioxidant defenses, and contributing to anti-cancer strategies. Understanding the pharmacological potential of seaweed-derived bioactive compounds not only enriches natural pharmaceutical resources but also highlights the importance of marine biodiversity in shaping innovative therapeutic interventions. This exploration holds the promise of uncovering intricate connections between marine ecology and human health, paving the way for advancements in drug discovery and development.

10.4.1 Antioxidant Properties

Antioxidants can be divided into primary and secondary categories based on their mechanisms, with multifunctional antioxidants possessing both primary and secondary properties. Primary antioxidants, known as chain-breaking antioxidants, directly engage with free radicals, converting them into stable, nonradical products. This plays a pivotal role in hindering lipid decomposition, particularly in lipid oxidation processes. Phenolic compounds, featuring multiple hydroxyl groups (dOH), serve as effective primary antioxidants by providing H-atoms to free radicals, resulting in relatively unreactive phenoxyl radicals due to resonance stabilization. Synthetic phenolic compounds like butylated hydroxyanisole (BHA) and butylated hydroxytoluene (BHT) function as efficient chain-breaking antioxidants and are commonly used in food preservation. Naturally occurring phenolic compounds, such as tocopherol, ascorbic acid, or caffeic acid, also act as chain-breaking antioxidants, albeit typically with lower efficiency compared to their synthetic counterparts, depending on the food product type [47].

Fucoidans, a subgroup of sulfated polysaccharides found in brown algae cell walls such as *Laminaria japonica,* possess a polymer of $\alpha 1 \rightarrow 3$-linked 1-fucose backbone with sulfated groups. Demonstrating excellent in vitro antioxidant activity, the sulfate content/fucose ratio and molecular weight serve as indicators of their effectiveness. Galactose-rich fractions from fucoidans exhibit potent scavenging activity against superoxide and hydroxyl radicals, emphasizing their antioxidant potential [48].

Secondary antioxidants, or preventive antioxidants, indirectly control lipid oxidation through mechanisms like transition metal chelation, singlet-oxygen quenching (in photooxidation), and oxygen scavenging. Certain secondary antioxidants work synergistically by replenishing primary antioxidants, ensuring continuous antioxidant activity. Notably, ascorbic acid illustrates this synergy, and the metal-chelating ability of secondary antioxidants is crucial in food systems, especially in the presence of iron, with EDTA being an exemplary metal-chelating antioxidant used in the food industry [49].

While synthetic antioxidants like EDTA and BHT are often more cost-effective and easier to process than natural alternatives, concerns about health risks and toxicity have led to restrictions on their use. Consequently, there is a growing interest in replacing synthetic antioxidants with natural plant-based alternatives, driven not only by safety considerations but also by increased consumer awareness and interest in natural products and the potential health benefits of natural antioxidants [50].

Seaweed emerges as a promising natural source of antioxidants, offering both water- and fat-soluble substances to protect functional food products rich in unsaturated fatty acids from oxidative damage. Key antioxidants derived from algae include phlorotannins, sulfated polysaccharides (fucoidans), and carotenoids, with potential applications in human health and nutrition [51]. The subsequent sections delve into the structure and characteristics of these potent antioxidant compounds derived from seaweed.

10.4.2 Anti-inflammatory Properties

Certain marine organisms, particularly seaweeds, have been observed to possess anti-inflammatory properties, extending beyond brown and green algae to various species of red algae. Derived from seaweeds, these agents have been identified for their ability to impede the production of reactive oxygen species (ROS), H_2O_2-induced lipid peroxidation, and inducible nitric oxide synthase. For example, the red alga *Gracilaria verrucosa* yields enone fatty acids that effectively inhibit the generation of inflammatory markers such as TNF-α, IL-6, and nitric oxide. The extensive array of novel bioactive metabolites from marine organisms, exceeding 1000 compounds, demonstrates potent cytotoxicity against various tumors. Marine macroalgae, or seaweeds, emerge as rich sources of bioactive secondary metabolites endowed with antioxidant, anti-inflammatory, and anticancer activities [52].

A recent study isolated two anti-inflammatory omega-3 polyunsaturated fatty acids (PUFAs), stearidonic acid (SA), and EPA, along with one pro-inflammatory omega-6 PUFA, arachidonic acid (AA), from *Undaria pinnatifida* (Phaeophyceae). SA and EPA showed efficacy in reducing edema, erythema, and blood flow in induced mouse ear inflammation, suggesting the potential of *Undaria pinnatifida* as a remedy for inflammation-related symptoms. Further investigations are necessary to comprehend the environmental and seasonal impacts on fatty acid profiles in seaweeds for optimal biotechnological applications [53].

Marine algae stand out as substantial sources of non-animal polysaccharides, with their structures varying based on algae type, life cycle phase, and environmental conditions during harvesting. Polysaccharides (PSs), carbohydrate biopolymers comprising simple sugars linked by glycoside, exhibit potent biological activities. For instance, isolated PSs (fucoidans) from Mediterranean brown seaweeds demonstrate anti-inflammatory and gastroprotective activities. Porphyran, a sulfated carbohydrate derived from red algae of the genus Porphyra, showcases anti-inflammatory, antioxidant, antihyperlipidemic, and antitumor activities. Fucoxanthins, predominantly found in brown algae (*Phaeophyceae*), are orange-colored xanthophyll pigments with diverse biological activities, including anti-inflammatory, anti-obesity, antiangiogenic, and anticancer properties, rendering them valuable for pharmaceutical purposes [2]. Phlorotannin-rich fractions extracted from *Cystoseira sedoides, Cladostephus spongeosis, and Padina pavonica* exhibit in vivo antioxidant activity, with *Cystoseira sedoides* displaying higher efficacy. Phlorotannins, with robust antioxidant activity against radicals, manifest anti-inflammatory effects by reducing paw edema and ear thickness in mice in a dose-dependent manner. The anti-inflammatory potential of phlorotannins stems from inhibiting oxidative stress by decreasing malondialdehyde (MDA) production and increasing free radicals. Phloroglucinol, a monomer of phlorotannins derived from *Ecklonia cava*, has reported anti-inflammatory effects attributed to its antioxidant mechanism, involving three hydroxyl groups reacting with reactive oxygen species (ROS) [2].

10.4.3 Antimicrobial Activities

Seaweeds are rich sources of essential omega-3 fatty acids, such as α-linolenic acid (ALA), eicosapentaenoic acid (EPA), and docosahexaenoic acid (DHA), known for their nutritional, antimicrobial, and antioxidant properties. EPA and DHA, key polyunsaturated fatty acids (PUFAs), hold potential benefits for conditions ranging from cardiovascular disorders and Alzheimer's disease to hypertension, coronary artery disease, arthritis, and cancer. Distinct fatty acid profiles are observed among brown, red, and green seaweeds, with notable high levels of ALA in green seaweeds [2].

Fucoidans extracted from edible algae, like *Cladosiphon okamuranus*, exhibit antimicrobial activity by inhibiting Helicobacter pylori infection through urease enzyme blockage. Similarly, fucoidans from *Spatoglossum asperum* showcase antimicrobial and antioxidant properties, evident in DPPH radical scavenging activities. The antibacterial effects of crude and purified fucoidans from Fucus vesiculosus extend to various bacteria, with depolymerized fucoidans demonstrating significant antibacterial activity against *Escherichia coli* and *Staphylococcus aureus*, causing membrane disruption and cell death [2].

Polyphenols, exemplified by extracts from *Padina australis*, demonstrate potent antimicrobial effects by damaging bacterial membranes and hindering microbial growth. Dieckol, isolated from *Ecklonia stolonifera*, displays antibacterial activity against methicillin-resistant and methicillin-susceptible *Staphylococcus aureus*. The breakdown of bacterial phospholipid molecules by phenols and flavonoids contributes to growth inhibition and eventual bacterial death. Further research is essential to comprehensively understand the correlation between fucoidan structure and antimicrobial activity [2]. Additionally, the study indicates the potential of seaweeds, particularly *Hormophysa cuneiformis* and *Actinotrichia fragilis*, as sources for developing antimicrobial and anti-diabetic drugs, urging the need for more in-depth investigations to isolate and identify bioactive compounds responsible for these effects [3].

10.4.4 Anticancer Effects

The exploration of bioactive compounds derived from seaweeds has unveiled their remarkable anti-cancer properties, a feature extensively investigated through both in vivo and in vitro assays, revealing promising outcomes. The significance of these anti-cancer properties lies in the potential they hold for revolutionizing cancer therapy and prevention. Fucoxanthin, found in *Laminaria japonica* and *Ishige okamurae*, demonstrated significant inhibition of lung cancer and B16-F10 melanoma cells in both in vivo and in vitro studies. A research on fucoxanthins from *Laminaria japonica* showcased reduced metastasis and inhibited lung cancer cell growth. Scientists have explored fucoxanthin from *Colpomenia sinuosa* and *Sargassum prismaticum*, revealing its in vitro anticancer activity against various cell lines and hepatoprotective potential in preventing cancer metastasis. They have also investigated fucoxanthins from *Undaria pinnatifida*, unveiling substantial growth inhibition in melanoma and cervix squamous carcinoma cells [2].

Beyond fucoxanthins, fucoidans in brown algae, such as *Undaria pinnatifida*, exhibit pharmacological properties, including anticancer activities against lung adenocarcinoma. *Codium decorticatum*, a green alga, demonstrated anti-inflammatory and anticancer activity, inhibiting pro-inflammatory cytokines and displaying cytotoxicity against the HeLa cell line. Seaweeds present a rich source of potential anticancer agents, encompassing a variety of bioactive compounds like fatty acids, sterols, carotenoids, polysaccharides, agar, carrageenan, and phycocolloids. Sterols, such as fucosterol and desmosterol, play crucial roles in cellular function, while agar, a sulfated polysaccharide from red seaweeds, shows promise in suppressing inflammation and inhibiting nitric oxide production. Carrageenan, another seaweed-derived compound, exhibits potent antiviral, antitumor, and immunomodulatory activities [52].

Seaweeds, with their diverse bioactive compounds, present a rich source of potential anti-cancer agents, as evidenced by multiple studies highlighting their inhibitory effects on cancer cell growth, metastasis, and inflammation. The mechanisms of action involve a range of pathways, including cell cycle arrest, inhibition of metastasis, antiangiogenic effects, and antioxidant activity, underscoring the multifaceted nature of seaweed-derived compounds in cancer therapy and prevention.

10.4.5 Antidiabetic Properties

Seaweeds emerge as promising allies in the search for effective diabetes management, given their rich reservoir of anti-diabetic properties. Components such as polyphenols, unsaturated fatty acids, and dietary fibers exhibit diverse anti-diabetic effects by forming protein complexes.

Seaweeds exhibit comprehensive inhibition of crucial enzymes such as α-amylase, α-glucosidase, aldose reductase, and protein tyrosine phosphatase 1B (PTP1B), essential for maintaining glucose balance. In addition to their enzyme-inhibiting properties, seaweeds influence incretin hormones, promote glucose uptake, combat obesity, and protect β-cells. The focus on reducing glucose levels involves inhibiting α-glucosidase and α-amylase activities crucial for mammalian starch digestion, with extracts from various seaweed species, including *Caulerpa racemosa, Spatoglossum schroederi, Padina sulcata, Sargassum binderi, and Turbinaria conoides*, displaying notable inhibitory effects. Compounds like dieckol, phlorofucofuroeckol-A, and bromophenols further contribute to these inhibitory effects, with extracts from *Ascophyllum nodosum, Fucus vesiculosus,* and *Padina arborescens* showing promise in targeting α-glucosidase and α-amylase enzymes, emphasizing the diverse anti-diabetic potential of seaweed compounds [54].

Seaweed-derived compounds target PTP1B, a key player in insulin resistance, with inhibitory effects demonstrated by phlorotannins from *Ecklonia stolonifera and Eisenia bicyclis*, as well as bromophenols from red seaweed species. Notably, dieckol-rich extracts from *Ecklonia cava*, administered in clinical trials, exhibit a significant reduction in insulin resistance and post-prandial hyperglycemia among pre-diabetic individuals. These seaweed compounds also showcase protective effects against hyperglycemia-induced oxidative stress, exemplified by compounds like dieckol and enzymatic digest from *Ecklonia cava*. The variability in potency is attributed to environmental factors, seasonal variations, and experimental differences [55].

Seaweed compounds, with their multifaceted anti-diabetic activities, involve specific protein interactions and broader mechanisms such as anti-inflammatory responses and the upregulation of antioxidant enzymes. In conclusion, the substantial potential of seaweed and its bioactive compounds in managing type 2 diabetes underscores the need for continued and rigorous research efforts, presenting a promising frontier in prevention and management, whether integrated into dietary habits or developed into pharmacological agents.

10.4.6 Neuroprotective and Cardiovascular Effects

Neurodegenerative disorders manifest as structural, chemical, and electrophysiological abnormalities, leading to various symptoms such as muscle weakness, poor coordination, seizures, pain, paralysis, cognitive alterations, and dementia. Alzheimer's disease and Parkinson's disease are prevalent examples.

Marine algae stand out as an ideal source for extracting polysaccharides due to their diverse biological properties, encompassing antioxidant, antiviral, antibacterial, anticancer, anti-inflammatory, immunomodulation, blood coagulation, hypolipidemic, hypocholesterolemic, and osteoprotective effects, making them essential for pharmaceutical and biomedical research. The key seaweed-derived polysaccharides with neuroprotective benefits include alginate, carrageenan, fucoidan, and laminarin. Alginate, a natural linear polysaccharide, is extracted from brown algae, with *Laminaria hyperborea, Laminaria digitata, Macrocystis pyrifera, Ascophyllum nodosum*, and *Laminaria japonica* being common sources. Carrageenan, a family of hydrophilic high molecular weight biopolymers, is derived from red algae, with κ-, ι-, and λ-carrageenan being commercially significant. Seaweed species, geographic location, and extraction season influence the properties of these polysaccharides. Fucoidan, particularly from Ecklonia cava, has demonstrated significant neuroprotective activities, including antioxidant effects and protection against hydrogen peroxide-induced cytotoxicity. Laminarin, a linear polysaccharide extracted from brown seaweeds, has shown promise in preventing neuronal injuries following ischemia-reperfusion. These seaweed-derived polysaccharides offer a unique and valuable resource for pharmaceutical and biomedical applications.

Despite numerous attempts to enhance the industrial utilization of bioactive compounds from seaweed for preventing cardiovascular disease (CVD), several limitations persist. Understanding the comprehensive biological activities of seaweed's diverse bioactive substances in metabolic disorders closely associated with CVD remains a challenge. Phlorotannins, polyphenol molecules derived from seaweed, are generated as secondary metabolites through the acetate-malonate pathway. Some phlorotannins, such as those found in brown seaweeds like Ecklonia species, exhibit exclusive characteristics. Phlorotannins are categorized into four subclasses based on the linkage between phloroglucinol units, each demonstrating various biological activities, including antioxidant, anticancer, antibacterial, anti-allergic, anti-diabetic, anti-inflammatory, anti-proliferative, anti-hypertensive, and anti-obesity effects [56].

Hypertension, a major CVD risk factor, is often evaluated based on the IC50 value representing the angiotensin-converting enzyme (ACE) inhibitor concentration causing 50% ACE activity inhibition. Phlorotannins, including eckol, dieckol, and phlorofucofuroeckol A, derived from Ecklonia cava, exhibit non-competitive ACE inhibition, comparable to or even surpassing captopril. Furthermore, these phlorotannins show effectiveness in inhibiting monocyte-associated vascular inflammation and dysfunction. Studies have also explored the anti-diabetic potential of seaweed-derived polysaccharides, with fucoidan being widely studied for its anti-diabetic activity by inhibiting α-amylase and α-glucosidase. Polysaccharides from various seaweed species, such as Undaria pinnatifida and Laminaria japonica, demonstrate cardioprotective effects through different mechanisms, including PI3K/Akt/Nrf2 signaling [56].

Bioactive peptides from seaweeds, particularly those from Undaria pinnatifida and Palmaria palmata, exhibit anti-hypertensive effects by inhibiting ACE. Carotenoids like fucoxanthin from brown seaweed contribute to anti-atherogenic effects and may disrupt increases in plasma cholesterol and LDL. While in vitro and in vivo studies have actively explored the impact of these dietary components from seaweeds on conditions like diabetes, obesity, hypertension, and hyperlipidemia related to CVD, clinical trials are limited [56]. Table 10.2

Table 10.2 Beneficial effects of seaweed-derived bioactive compounds

Pharmacological activity	Bioactive compounds	Organism examples
Antioxidant properties	Phenolic compounds	*Laminaria japonica*
Anti-inflammatory properties	Enone fatty acids	*Gracilaria verrucosa*
Antimicrobial activities	Omega-3 fatty acids	*Cladosiphon okamuranus, Padina australis, Ecklonia stolonifera*
Anticancer effects	Fucoxanthin	*Laminaria japonica, Ishige okamurae, Colpomenia sinuosa, Sargassum prismaticum, Undaria pinnatifida*
Antidiabetic properties	Polyphenols, unsaturated fatty acids, dietary fibers	*Caulerpa racemosa, Spatoglossum schroederi, Padina sulcata, Sargassum binderi, Turbinaria conoides*
Neuroprotective properties	Polysaccharides	*Laminaria hyperborea, Laminaria digitata, Macrocystis pyrifera, Ascophyllum nodosum, Laminaria japonica, Ecklonia cava, Undaria pinnatifida*
Cardiovascular properties	Polysaccharides, phlorotannins, bioactive peptides, carotenoids	*Ecklonia cava, Undaria pinnatifida, Palmaria palmata*

summarizes different seaweeds and its associated health benefits. Further research is required to bridge the gap between these experimental findings and practical applications, considering the potential influence of these dietary components on epigenetic gene expression in humans. Understanding the mechanisms underlying these interactions will aid in designing effective dietary guidelines for maintaining overall health.

10.5 Formulation and Delivery Systems

Formulation and delivery systems play a crucial role in drug development. Selecting the right formulation and delivery technique is important to harness the maximum therapeutic potential for the seaweed-derived bioactive compounds. Factors like optimal absorption, sustained release, and targeted delivery need to be carefully considered when designing the drug. The correct formulation technique like emulsion or encapsulation determines the stability of the drug in the system. Oral tablets and capsules need to be tested for their biocompatibility and specific site of drug release. Nanocarriers and microencapsulation techniques are recently gaining prominence as they provide protection against degradation, enhance solubility, and offer controlled release profiles. Such systems contribute to overcoming challenges associated with stability and delivery efficiency. Schematic representation of various formulation techniques and modes of delivery are depicted in Fig. 10.2.

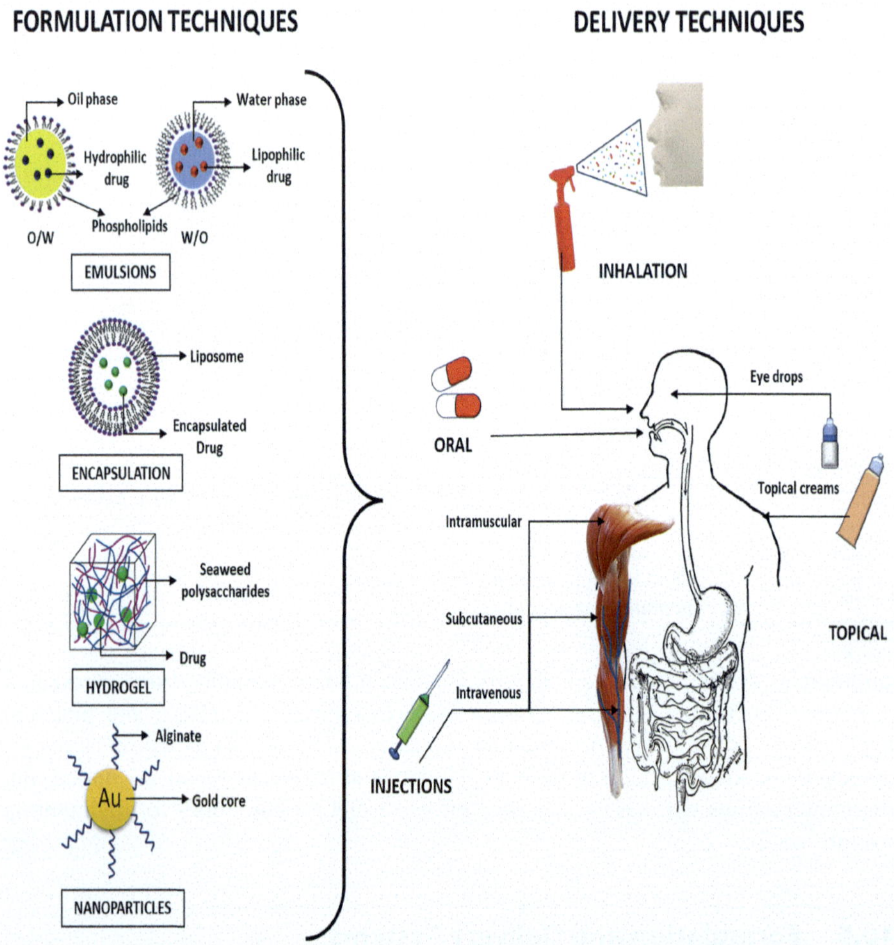

Fig. 10.2 Various formulations and delivery systems of seaweed-derived bioactive compounds in the human body

10.5.1 Formulations

10.5.1.1 Emulsions

This technique involves forming a diphasic colloidal solution of two immiscible liquids, i.e., oil-in-water (O/W) or water-in-oil (W/O), forming uniform droplets. Emulsifiers are added to stabilize the droplets due to their amphiphilic property. In O/W emulsions, the dispersed phase is oil and the dispersal medium is water. The quantity of water governs the stability and lifespan of the formulation. Lipophilic drugs are formulated by this method, whereas W/O emulsions easily release lipophilic compounds [6]. Seaweed polysaccharides namely alginate and carrageenan are formed into an O/W emulsion with dodecenylsuccinic anhydride (DSA) as the stabilizer. Alginate and carrageenan are highly hydrophilic compounds, so formulating them into O/W emulsions helps in stabilizing them and increasing their shelf life.

10.5.1.2 Encapsulation

Seaweed bioactives are generally hydrophilic molecules, hence encapsulating them in colloidal drug carriers like liposomes increases their delivery efficiency and aids in sustained release due to the hydrophobic shells. Limitations of this technique involve leakage and instability. Phlorotannins, bioactive compounds having inherent antioxidant and anti-inflammatory properties, are widely used as anti-cancer drugs. Preparation of phlorotannins drug involves liposomal encapsulation followed by freeze-drying or spray-drying [7].

10.5.1.3 Hydrogels

Seaweed polysaccharides form hydrogel through physical cross-linking with weak interactions such as hydrogen bonding and Van der Waals forces. A hydrogel is characterized by a 3D network of hydrophilic polymer chains exhibiting high water-swelling capacity without water solubility making them appealing product for biomedical use. Seven such hydrogel-forming polysaccharides have been identified, including alginate, carrageenans, ulvan, starch, agarose, porphyran, and (nano) cellulose.

10.5.1.4 Nanoparticles

Due to the small size (1–100 nm) and large surface area to volume ratio, nanoparticles serve as perfect drug carriers and delivery systems. Nanoparticles can be engineered to target specific tissues and controlled drug release. Gold nanoparticles (AuNPs), made using the aqueous extracts of brown algae *Sargassum longifolium,* were seen to have anti-cancer and antioxidant properties. Evaluation of the synthesized AuNPs involved examining their impact on MG-63 human osteosarcoma cells. The results indicated a significant anticancer effect, demonstrating the potential utility of these nanoparticles in cancer treatment. Additionally, the nanoparticles displayed antioxidant activity, as evidenced by their ability to quench free radicals, suggesting a potential role in mitigating oxidative stress [5].

10.5.2 Delivery Systems

10.5.2.1 Oral Delivery

The bioactive compounds can be formulated into oral tablets and capsules to deliver the drug through the digestive system. It is essential to insure the drug's oral and gastrointestinal bioavailability and bioaccessibility. Various in vitro methods, such as solubility, dialysability, and gastrointestinal models, are employed to measure bioaccessibility, providing preliminary insights into the compound's release under simulated digestive conditions. Notably, dynamic gastrointestinal models, like the TNO gastro-Intestinal Model (TIM) system, offer a more realistic representation by simulating the physicochemical conditions and peristaltic forces. Fuicoidans for example have low oral bioavailability due to its high polarity and cannot pass easily through the intestinal epithelial cells. Alginates can be degraded by the human gut microbiome and produce short-chain fatty acids which improve the overall gut health. Regarding polyphenols, human clinical studies reveal challenges in their oral bioaccessibility due to factors like host-related and intestinal factors. Despite this, polyphenols exhibit high biological activity, attributed to biotransformation in the liver and gut microbial metabolism [57].

10.5.2.2 Topical Delivery

Cosmetology is a major domain where seaweed bioactive compounds are widely used and formulated as creams and lotion for topical treatment. Apart from cosmetics, several dermal issues can be treated using marine extracts. Radiation dermatitis (RD) is a common side effect of radiotherapy in cancer patients. Despite various attempts to alleviate RD, it remains largely unresolved. Phlorotannins (PTNs), compounds found in brown seaweeds like Ecklonia cava, can be topically applied to mitigate RD. The PTNs activate the NRF2 pathway, known for its role in oxidative stress response, and suppress the NF-κB pathway, associated with inflammation. This dual action of PTNs suggests their effectiveness in reducing RD through antioxidant and anti-inflammatory mechanisms and highlights the potential of PTNs as a novel and promising cosmeceutical for managing RD, offering a natural and effective alternative for improving skin health during radiotherapy [58].

10.5.2.3 Inhalation

Bioactive compounds can be formulated into aerodynamic microparticles for delivery through the nasal route directly to the respiratory system, targeting various lung and respiratory infections. This approach is particularly beneficial for addressing infections where macrophages, crucial for the immune response, can be evaded by bacterial strains like *Mycobacterium tuberculosis*. Carrageenan, extracted from red seaweed, is suggested as a carrier for anti-tubercular drugs due to its reported macrophage-activating properties. The method involves spray-drying carrageenan with starch, resulting in the creation of uniform aerodynamic microparticles. The biocompatibility of these carriers is assessed through cell viability tests and LDH release assays, revealing low toxicity. Furthermore, the microparticles show an affinity for macrophages, with phagocytosis levels ranging from 36% to 65%. Cytokine secretion assays indicate potential macrophage activation. This highlights the inhalation of seaweed bioactive microparticles as an effective treatment for respiratory issues [59].

10.5.2.4 Injections

Injectable seaweed-based drug delivery systems represent a promising technique, albeit less prevalent currently due to their invasive nature. Administration routes include intravenous, intramuscular, or subcutaneous, based on the target tissue. By combining Fucoidans with thermo-responsive polymers like chitosan, hyaluronic acid (HA), gelatin, or xyloglucan, these formulations acquire hydrogel properties upon injection. The integration of Fucoidan into the hydrogel not only imparts cell-binding capabilities but also demonstrates an immune-modulatory effect, effectively reducing the diffusion of inflammatory mediators. This underscores the potential of injections as an efficacious delivery method for exploiting the therapeutic attributes of seaweed-derived compounds [60].

10.6 Conclusion

In summary, this chapter provides a comprehensive exploration of seaweed biochemistry and its applications in pharmaceuticals and healthcare. Starting with a detailed analysis of seaweed composition and diverse bioactive compounds, the discussion extends to traditional extraction methods, emphasizing the need for refinement. The pharmacological properties of seaweed-derived compounds, including their potential in diabetes treatment, curing neurological diseases, antioxidant capabilities, and anti-cancer contributions, are highlighted, showcasing significant promise in pharmaceutical applications. Additionally, the chapter delves into extraction and purification intricacies, underscoring the importance of refining methods for efficiency and paving the way for novel therapeutic interventions. The final segment addresses formulation and drug delivery, emphasizing the need to understand challenges and opportunities, with extensive potential applications ranging from optimizing formulations to innovating drug delivery systems.

In conclusion, this chapter reveals the dynamic field of seaweed biochemistry and its profound implications for innovative research and advancements in medicine. The preceding chapters provide a foundation for understanding seaweed-derived bioactive compounds and their transformative potential, showcasing the richness of marine biodiversity and inviting further exploration and innovation in healthcare solutions.

References

1. Xie C, Lee ZJ, Ye S, Barrow CJ, Dunshea FR, Suleria HA (2023) A review on seaweeds and seaweed-derived polysaccharides. Nutrition, chemistry, bioactivities, and applications. Food Rev Intl:1–36
2. Lomartire S, Gonçalves AM (2022) An overview of potential seaweed-derived bioactive compounds for pharmaceutical applications. Mar Drugs 20(2):141
3. Osman NA, Siam AA, El-Manawy I, Jeon YJ (2019) Anti-microbial and anti-diabetic activity of six seaweeds collected from the Red Sea, Egypt, Catrina. Int J Environ Sci 19(1):55–60
4. Perez-Vazquez A, Carpena M, Barciela P, Cassani L, Simal-Gandara J, Prieto MA (2023) Pressurized liquid extraction for the recovery of bioactive compounds from seaweeds for food industry application: a review. Antioxidants 12(3):612. https://doi.org/10.3390/antiox12030612
5. Rajeshkumar S, Parameswari RP, Jayapriya J, Tharani M, Ali H, Aljarba NH et al (2022) Apoptotic and antioxidant activity of gold nanoparticles synthesized using marine brown seaweed: an in vitro study. BioMed Res Int 2022(1):5746761

6. Sawant A, Kamath S, Kg H, Kulyadi GP (2021) Solid-in-oil-in-water emulsion: an innovative paradigm to improve drug stability and biological activity. AAPS PharmSciTech 22:1–14
7. Simón L, Arazo-Rusindo M, Quest AF, Mariotti-Celis MS (2023) Phlorotannins: novel orally administrated bioactive compounds that induce mitochondrial dysfunction and oxidative stress in cancer. Antioxidants 12(9):1734
8. Lafarga T, Acién-Fernández FG, Garcia-Vaquero M (2020) Bioactive peptides and carbohydrates from seaweed for food applications: natural occurrence, isolation, purification, and identification. Algal Res 48:101909
9. Lopez-Santamarina A, Miranda JM, Mondragon ADC, Lamas A, Cardelle-Cobas A, Franco CM, Cepeda A (2020) Potential use of marine seaweeds as prebiotics: a review. Molecules 25(4):1004
10. Echave J, Otero P, Garcia-Oliveira P, Munekata PE, Pateiro M, Lorenzo JM et al (2022) Seaweed-derived proteins and peptides: promising marine bioactives. Antioxidants 11(1):176
11. Kumar Y, Tarafdar A, Badgujar PC (2021) Seaweed as a source of natural antioxidants: therapeutic activity and food applications. J Food Qual 2021:1–17
12. Cotas J, Leandro A, Pacheco D, Gonçalves AM, Pereira L (2020) A comprehensive review of the nutraceutical and therapeutic applications of red seaweeds (Rhodophyta). Life 10(3):19
13. Souza CR, Bezerra WP, Souto JT (2020) Marine alkaloids with anti-inflammatory activity: current knowledge and future perspectives. Mar Drugs 18(3):147
14. Garcia-Vaquero M, Rajauria G, Tiwari B (2020) Conventional extraction techniques: solvent extraction. In: Sustainable seaweed technologies. Elsevier, pp 171–189
15. Zhang QW, Lin LG, Ye WC (2018) Techniques for extraction and isolation of natural products: a comprehensive review. Chin Med 13:1–26
16. Fleurence J, Levine IA (eds) (2016) Seaweed in health and disease prevention. Academic Press
17. Timotius KH, Simamora A, Santoso AW (2018) Chemical characteristics and in vitro antidiabetic and antioxidant activities of premna serratifolia l. leaf infusion and decoction. Pharm J 10(6):1114–1118. https://doi.org/10.5530/pj.2018.6.189
18. Monton C, Suksaeree J, Luprasong C (2020) Validated reversed-phase ion-interaction high-performance liquid chromatography for quantitation of nitrate content of clausena anisata (willd.) hook. f. ex benth. leaves. Scientifica 2020:1–9. https://doi.org/10.1155/2020/6424682
19. Coelho MCR, Duarte AP, Pinto S, Botelho HM, Reis CP, Serralheiro ML et al (2023) Edible seaweeds extracts: characterization and functional properties for health conditions. Antioxidants 12(3):684. https://doi.org/10.3390/antiox12030684
20. Mena-García A, Ruiz-Matute AI, Soria AC, Sanz ML (2019) Green techniques for extraction of bioactive carbohydrates. TrAC Trends Anal Chem 119:115612
21. Kadam SU, Tiwari BK, O'Donnell CP (2013) Application of novel extraction technologies for bioactives from marine algae. J Agric Food Chem 61:4667–4675. https://doi.org/10.1021/jf400819p
22. Grosso C, Valentão P, Ferreres F, Andrade PB (2015) Alternative and efficient extraction methods for marinederived compounds. Mar Drugs 13:3182–3230. https://doi.org/10.3390/md13053182
23. Uy SF, Easteal AJ, Farid MM, Keam RB, Conner GT (2005) Seaweed processing using industrial singlemode cavity microwave heating: a preliminary investigation. Carbohydr Res 340:1357–1364. https://doi.org/10.1016/j.carres.2005.02.008
24. Carreira-Casais, A., Lourenço-Lopes, C., Otero, P., Carpena, M., Pereira, A. G., Echave, J., et al. (2021) Application of green extraction techniques for natural additives production
25. Chandrapala J, Oliver CM, Kentish S, Ashokkumar M (2013) Use of power ultrasound to improve extraction and modify phase transitions in food processing. Food Rev Intl 29:67–91. https://doi.org/10.1080/87559129.2012.692140
26. Cheng Q, Wang S, Han Q (2010) Novel process for isolating fibrils from cellulose fibers by high-intensity ultrasonication. II. Fibril characterization. J Appl Polym Sci 115:2756–2762. https://doi.org/10.1002/app.30160

27. Rafiquzzaman SM, Ahmed R, Lee JM, Noh G, Jo G, Kong I-S (2016) Improved methods for isolation of carrageenan from Hypnea musciformis and its antioxidant activity. J Appl Phycol 28:1265–1274. https://doi.org/10.1007/s10811-015-0605-6
28. Kadam SU, O'Donnell CP, Rai DK, Hossain MB, Burgess CM, Walsh D, Tiwari BK (2015) Laminarin from Irish brown seaweeds Ascophyllum nodosum and Laminaria hyperborea: Ultrasound assisted extraction, characterization and bioactivity. Mar Drugs 13:4270–4280. https://doi.org/10.3390/md13074270
29. Sapkale GN, Patil SM, Surwase US, Bhatbhage PK (2010) Supercritical fluid extraction: a review. Int J Chem Sci 8:729–743
30. Heng MY, Tan SN, Yong JWH, Ong ES (2013) Emerging green technologies for the chemical standardization of botanicals and herbal preparations. Trends Anal Chem 50:1–10. https://doi.org/10.1016/j.trac.2013.03.012
31. Campos-Segura M, Acosta-Chi Z, Rosado-Rubio G, Chel-Guerrero L, Betancur-Ancona D (2014) Whole and crushed nutlets of chia (Salvia hispanica) from Mexico as a source of functional gums. Food Sci Technol 34:701–709. https://doi.org/10.1590/1678-457X.6439
32. Crampon C, Boutin O, Badens E (2011) Supercritical carbon dioxide extraction of molecules of interest from microalgae and seaweeds. Ind Eng Chem Res 50:8941–8953. https://doi.org/10.1021/ie102297d
33. Khalil HA, Davoudpour Y, Saurabh CK, Hossain MS, Adnan AS, Dungani R, Paridah MT, Sarker MZ, Fazita MN, Syakir MI, Haafiz MK (2016) A review on nanocellulosic fibres as new material for sustainable packaging: process and applications. Renew Sust Energ Rev 64:823–836. https://doi.org/10.1016/j.rser.2016.06.072
34. Kulshreshtha G, Burlot A-S, Marty C, Critchley A, Hafting J, Bedoux G, Bourgougnon N, Prithiviraj B (2015) Enzyme-assisted extraction of bioactive material from Chondrus crispus and Codium fragile and its effect on herpes simplex virus (HSV-1). Mar Drugs 13:558–580. https://doi.org/10.3390/md13010558
35. Wijesinghe WAJP, Jeon Y-J (2012) Enzyme-assistant extraction (EAE) of bioactive components: a useful approach for recovery of industrially important metabolites from seaweeds: a review. Fitoterapia 83:6–12. https://doi.org/10.1016/j.fitote.2011.10.016
36. Blanco-Pascual N, Alemán A, Gómez-Guillén MC, Montero MP (2014) Enzyme-assisted extraction of κ/ι-hybrid carrageenan from Mastocarpus stellatus for obtaining bioactive ingredients and their application for edible active film development. Food Funct 5:319–329. https://doi.org/10.1039/c3fo60310e
37. Nieto A, Borrull F, Pocurull E, Marcé RM (2010) Pressurized liquid extraction: a useful technique to extract pharmaceuticals and personal-care products from sewage sludge. TrAC Trends Anal Chem 29(7):752–764
38. Plaza M, Turner C (2015) Pressurized hot water extraction of bioactives. Trends Anal Chem 71:39–54. https://doi.org/10.1016/j.trac.2015.02.022
39. Ballesteros-Vivas D, Ortega-Barbosa JP, del Pilar Sánchez-Camargo A, Rodríguez-Varela, L. I., Parada-Alfonso, F. (2021) Pressurized liquid extraction of bioactives
40. Saravana PS, Cho Y-J, Park Y-B, Woo H-C, Chun B-S (2016) Structural, antioxidant, and emulsifying activities of fucoidan from Saccharina japonica using pressurized liquid extraction. Carbohydr Polym 153:518–525. https://doi.org/10.1016/j.carbpol.2016.08.014
41. Calvo-Flores FG (2021) Green processes in foodomics. Green solvents for sustainable processes. In: Reference module in food science
42. Cermeño M, Kleekayai T, Amigo-Benavent M, Harnedy-Rothwell P, FitzGerald RJ (2020) Current knowledge on the extraction, purification, identification, and validation of bioactive peptides from seaweed. Electrophoresis 41(20):1694–1717
43. Batool A, Menaa F (2020) Concentration and purification of seaweed components by chromatography methods. In: Sustainable seaweed technologies, pp 315–370
44. Cai C, Wu L, Li C, He P, Li J, Zhou J (2011) Purification, crystallization and preliminary X-ray analysis of phycocyanin and phycoerythrin from Porphyra yezoensis Ueda. Acta Crystallogr Sect F: Struct Biol Cryst Commun 67(5):579–583

45. Baron R, Vauchel P, Arhaliass A (2010) Modeling of twinscrew reactive extrusion: application to alginate extraction. IFAC Proc Vol 43:341–346. https://doi.org/10.3182/20100707-3-BE-2012.0094
46. Li H, Huang J, Xin Y, Zhang B, Jin Y, Zhang W (2009) Optimization and scale-up of a new photobleaching agar extraction process from Gracilaria lemaneiformis. J Appl Phycol 21:247–254. https://doi.org/10.1007/s10811-008-9358-9
47. Akoh CC (ed) (2017) Food lipids: chemistry, nutrition, and biotechnology. CRC Press
48. Wang J, Zhang Q, Zhang Z, Li Z (2008) Antioxidant activity of sulfated polysaccharide fractions extracted from Laminaria japonica. Int J Biol Macromol 42(2):127–132
49. Haahr AM, Jacobsen C (2008) Emulsifier type, metal chelation and pH affect oxidative stability of n-3-enriched emulsions. Eur J Lipid Sci Technol 110(10):949–961
50. Halliwell B (1996) Commentary oxidative stress, nutrition and health. Experimental strategies for optimization of nutritional antioxidant intake in humans. Free Radic Res 25(1):57–74
51. Cornish ML, Garbary DJ (2010) Antioxidants from macroalgae: potential applications in human health and nutrition. Algae 25(4):155–171
52. Sakthivel R, Devi KP (2019) Antioxidant, anti-inflammatory and anticancer potential of natural bioactive compounds from seaweeds. Stud Nat Prod Chem 63:113–160
53. Khan MNA, Cho JY, Lee MC, Kang JY, Park NG, Fujii H, Hong YK (2007) Isolation of two anti-inflammatory and one pro-inflammatory polyunsaturated fatty acids from the brown seaweed Undaria pinnatifida. J Agric Food Chem 55(17):6984–6988
54. Sharifuddin Y, Chin YX, Lim PE, Phang SM (2015) Potential bioactive compounds from seaweed for diabetes management. Mar Drugs 13(8):5447–5491
55. Admassu H, Gasmalla MAA, Yang R, Zhao W (2018) Bioactive peptides derived from seaweed protein and their health benefits: antihypertensive, antioxidant, and antidiabetic properties. J Food Sci 83(1):6–16
56. Cho CH, Lu YA, Kim MY, Jeon YJ, Lee SH (2022) Therapeutic potential of seaweed-derived bioactive compounds for cardiovascular disease treatment. Appl Sci 12(3):1025
57. Shannon E, Conlon M, Hayes M (2021) Seaweed components as potential modulators of the gut microbiota. Mar Drugs 19(7):358
58. Yang K, Kim SY, Park JH, Ahn WG, Jung SH, Oh D et al (2020) Topical application of phlorotannins from Brown seaweed mitigates radiation dermatitis in a mouse model. Mar Drugs 18(8):377
59. Rodrigues S, Cunha L, Rico J, Rosa da Costa AM, Almeida AJ, Faleiro ML et al (2020) Carrageenan from red algae: an application in the development of inhalable tuberculosis therapy targeting the macrophages. Drug Deliv Transl Res 10:1675–1687
60. Haggag YA, Abd Elrahman AA, Ulber R, Zayed A (2023) Fucoidan in pharmaceutical formulations: a comprehensive review for smart drug delivery systems. Mar Drugs 21(2):112

11 Potential of Seaweeds as Antioxidants and Their Role in Animal Health and Nutrition

Rekha Rani, Payal Karmakar, and Bhopal Singh

Abstract

Rise in awareness among consumers has resulted in a shift of their choices towards "green" alternatives or plant-based options. Seaweeds are photoautotrophic organisms which are available in high abundance due to the fact of oceans covering almost two thirds of our planet. They are of great nutritional value, beside having beneficial ecological roles. Seaweed provides several applications in food, pharmacological, and cosmetics industry and is of utmost importance to animal nutrition. They are a storehouse of different nutrients like vitamins, polysaccharides, proteins, carbohydrate, minerals, and phenolic compounds which exhibit antioxidant properties, thereby maintaining redox balance in animals by scavenging excess free radicals which can prevent and treat infections. Thus, supplementing animal diet with seaweed can help to boost immunity. This can also improve the meat quality as the antioxidants from seaweed can help in retarding lipid oxidation which is an important factor to be considered in meat industries. Other positive roles include imparting probiotic, anti-inflammatory, and immunomodulatory properties, beside improving productivity in animals. The current chapter shows the benefits of antioxidants derived from seaweed on animal health including the mechanism of their actions. However, further research should be performed for understanding the different mechanisms of antioxidant actions on specific animals.

R. Rani (✉)
Department of Dairy Chemistry, College of Dairy and Food Technology, AU, Jodhpur, Rajasthan, India

P. Karmakar
Food Science and Nutrition Department, University of Minnesota, St. Paul, MN, USA

B. Singh
Department of Dairy Technology, Faculty of Sciences, Dayalbagh Educational Institute, Agra, Uttar Pradesh, India

M. Rafatullah et al. (eds.), *Multidisciplinary Applications of Marine Resources*, https://doi.org/10.1007/978-981-97-5057-3_11

Keywords

Marine · Animal · Nutrition · Oxidative stress · Health

11.1 Introduction

The metabolic activity of the animal body is the main cause of production of reactive oxygen species (ROS) which leads to oxidative stress to the body cells and tissues. The secondary metabolites which quench the free radicals by donating hydrogen and electrons are called antioxidants. They also chelate transition metals and dissolve peroxidation compounds, thereby preventing oxidation [1]. The planet earth has 72% water, that is, almost two-thirds of our planet is covered by water. Interestingly, oceans are the storehouses of almost all the elements of the periodic table. Thus, it would not be wrong to say that every life form emerged from the seas as they contain all the elements necessary for life. Algae are the pioneers of photosynthesis. The sea offers a harsh range of environmental stress related to osmosis and desiccation, rapid fluctuations in temperatures and light, which can cause formation of free radicals, but it was observed that the seaweeds do not suffer severe photodynamic damage. This could possibly be due to some compounds present in them and their inherent protective mechanisms. The marine organisms, including seaweeds and macroalgae, are some of the richest sources of nature [2–6]. Few examples of such compounds are mainly polyphenolic compounds like gallic acid, cinnamic acid, benzoic acid, flavonoids, quercetin, and phlorotannins [7]. Their mechanism of action is chain-breaking by donation of hydrogen to free radicals, whereas carotenoids function by trapping free radicals and quenching singlet oxygen [8]. They are also an excellent source of both water-soluble and fat-soluble vitamins such as A, Bl, B12, riboflavin, niacin, pantothanic acid, folic acid, C, D and E, as well as minerals such as K, Na, Ca, P, and I [9]. The polysaccharides derived from seaweeds have proved to be of high medicinal value [10]. The Plantae kingdom includes green algae, termed as Chlorophytas; the red algae, the Rhodophytas; and the Ochrophytas, also called brown seaweed; these are classified in the Phaeophyceae class which belongs to the microalgae kingdom, as well as the Protista kingdom, as shown in Fig. 11.1. The antioxidant seaweeds have several applications in food, pharmaceutical, biomedical, cosmetic, and agricultural industry [11], but their use in animal feed and health is still to gain more recognition as they have innumerable benefits [12, 13], which include antioxidant, antimicrobial, anti-inflammatory, prebiotic, and immunomodulatory effects [13, 14]. This chapter focuses on the importance of antioxidants to animals and the components of seaweed that can provide that antioxidative benefits. Special attention has been paid to the mode of action of the marine-derived antioxidants and their effects on animal health and development.

Fig. 11.1 Classification of seaweed

11.2 Importance of Antioxidants to Animals

Due to exposure of animals to unfavorable environment, there is generation of reactive oxygen species (ROS) which leads to oxidative stress. Several stress factors, both biotic and abiotic (e.g., bacteria, temperature, environmental pollutants, chemicals, radiation, food, diseases, and medications), cause production of unnecessary ROS which leads to oxidative damage [15]. The weakening of antioxidant defense mechanism is one of the major reasons for the initiation and spread of diseases in animals like interdigital dermatitis diseases and foot and mouth viral infection [16, 17].

Antioxidant machinery can be classified into two varieties: enzymatic and nonenzymatic. The enzymatic category includes catalase (CAT), superoxide dismutase (SOD), glutathione peroxidase (GPX), and several other enzymes [15], whereas the nonenzymatic one includes carotenoids, natural flavonoids, melatonin, vitamins A, C, E, thiol antioxidants like thioredoxin, glutathione (GSH), and lipoic acid [18–22].

The concentration and activity of antioxidant enzymes are inversely proportional to the rate of free radical generation which increases the virulence of pathogenic microorganisms released in huge amount during bacterial infections. GPX, GSH, and CAT are some of the valuable components of the cell which help them to provide protection from oxidative stress, thereby making the cells immune towards several diseases [18, 19, 23]. It was observed in a study that the functioning of the antioxidant enzymatic machinery got reduced as a result of viral infection in cattle suffering from foot and mouth disease, which establishes the effect of oxidative stress in animal health [16]. Another study showed that oxidative stress in sick heifers was responsible for reduced levels of antioxidants (enzymatic and nonenzymatic) [17]. Antioxidant enzymes such as SOD, GSH, CAT, and MDA activity might play a crucial role in dairy heifers in the pathogenesis of acute laminitis [17]. It was observed that there is a strong correlation between antioxidant machinery in sheep and foot infectious diseases. Increased ROS levels and oxidative stress lead to peroxidation of lipid and damage to protein and DNA, which, in turn, deteriorates the infection [24, 25]. Body cells also contain antioxidants in very small amounts, which can prevent oxidation of several crucial components of cell like DNA, RNA, or proteins. Ascorbic acid is a good antioxidant agent, which, on account of being a cofactor for several enzymes, can neutralize reactive oxygen species [26, 27]. The brown seaweed contains the highest amount of this vitamin, followed by green seaweeds, and finally red seaweeds with values of 0.815 mg/g, 0.781 mg/g, and 0.720 mg/g d.m., respectively [26]. The stronger the antioxidant, the more readily it is absorbed and more effectively it can remove free radicals and redox metals at appropriate amounts [27–29]. Continuous exposure to oxidative stress can significantly lead to chronic inflammation, which is the main cause of various chronic diseases in animals [30]. Certain genetic mutations or DNA injury could be responsible for oxidative stress generated by tumor initiation and progression [31], which can even lead to severe cases. Free radicals are also found to be related to degradation of neuropathological conditions in animals, which can be prevented by antioxidants to work against oxidative damage, thereby preventing disease progression in animals. Marine algae compounds, by virtue of their antioxidant properties in several animal diseases, can be used for treating hyperinflammatory responses due to their multifaceted and complex properties [32]. Seaweed chemical components, such as carotenoids, fatty acids, sterols, polysaccharides, phycocolloids, and phycobilins, have the ability to reduce proinflammatory cytokines. Studies have shown that the signalling mechanisms relating to inflammatory enzymes activation were reduced by antioxidant properties of many seaweed extracts [32].

The algal biomass also contains a diverse array of vitamins, including A, B1, B2, B3, B5, B6, and B8 [14, 33], vitamin C (ascorbic acid) [12, 26, 33, 34], and vitamin E (α-tocopherol) [12, 33, 35]. Beside these, carotenoids present in seaweeds in the form of β-carotene also exhibit provitamin A activity [36]. However, the vitamin content of every algae species differs from one depending on several internal and external factors which include genetic variant, life cycles, growth rate, concentration of nutrients in their habitat, the exposure to light, and the processes by which they are collected. In short, several factors are present which can be amended to obtain the preferred variety of algae [37]. It is generally recommended to harvest seaweed when the functional compounds are at their highest concentration. This can help in reduction of algal biomass supplemented to the feed [38]. Figure 11.2 shows utilization of seaweed biomass.

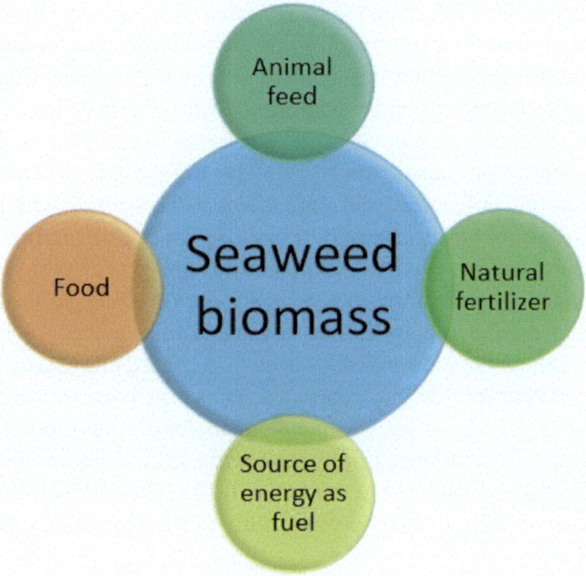

Fig. 11.2 Utilization of seaweed biomass

11.3 Major Components of Seaweeds Acting as Antioxidants

The composition of seaweeds is not as well-known as that of the land plants, but there are several varieties of metabolic compounds found in the aquatic flora that are unique to them and are not produced by terrestrial plants [39]. Beside carbohydrates, protein, and minerals, seaweeds are home for a diverse range of bioactive compounds that comprise a variety of antioxidants, the most significant of them includes polysaccharides, pigments, terpenoids, certain water-soluble vitamins and phenolic compounds, such as polyphenols, and apolar compounds like carotenoids, and tocopherols [40]. The bioactive compounds identified and isolated from seaweeds include polyphenols such as phlorotannins [41], carotenoid pigments such as fucoxanthin [42] and astaxanthin, sterols, sulphated polysaccharides (laminarins and fucoidans), and mycosporine-like amino acids (MAAs).

The first in the list are the natural pigments called carotenoids. Till date, about 600 types of carotenoids have been found. They have structural similarity with hydrophobic tetraterpenes that include branched hydrocarbons with a C-40-bound methyl group [43, 44]. The conjugated double bonds in the carbonate skeleton are mainly responsible for the pigmentation of carotenoids as they have the capability to capture photons at different wavelengths of the visible spectrum. Carotenoids are necessarily found in all organisms that perform photosynthesis [45]. Carotenoids can further be divided into two large groups based on molecular structure: Carotenes and xanthophylls. Carotenes have linear or cyclic hydrocarbons at one or both ends of the molecule (like β-carotene), whereas xanthophylls are oxygenated derivatives of carotenes (such as violaxanthin) and can be found in green algae, as well as in more developed plants, although there are compounds (i.e., canthaxanthin and loroxanthin) that are exclusive from green algae [46, 47]. Another type of xanthophylls, such as diatoxanthin or fucoxanthin, are produced by brown algae [48]. As mentioned earlier, carotenoids function by preventing the photosynthetic system from damage, thereby playing the role of a photoprotector. In order to get adapted to changes, carotenes are produced by several microalgae in response to external stimuli that cause stress [49]. The microalgae species which produce the highest amounts of carotenoids are *Dunaliella salina*, which produces β-carotene, and *Haematococcus pluvialis*, which produces astaxanthin [50]. Carotenoids also show provitamin A activity. Astaxanthin is also a protector of skin and eye, immune modulator, muscle strengthener, and coloring agent. It has been claimed by researchers that regular dietary intake of astaxanthin scavenges free radicals by virtue of its regenerative capacity, which, in turn, retards the aging process [51].

Other microalgae-generated secondary pigments are phycobiliproteins. They act as antioxidants by the mechanism involving protection from harmful radiation alongside exploiting the light energy [52]. These compounds are mainly generated by red and blue algae (cyanobacteria), namely Spirulina, (which produces phycocyanin, blue) and Porphyridium (which produces phycoerythrin, red), respectively [53].

The components of seaweeds acting as antioxidants, next to pigments, are phenolic compounds. These are mainly secondary metabolites which can be of both terrestrial and aquatic origin [54]. One of the common characteristics in the structures of phenolic compounds is the presence of a benzene ring which has a hydroxyl group that had substituted one hydrogen. They act as antioxidative agents by preventing the formation of free radicals by virtue of their metal ion chelating capacity [55, 56]. They are further subdivided into five categories: Flavonoids, lignans, tannins, tocopherols, and phenolic acids [57]. Among them, largest subgroup is that of flavonoids that exhibit several biological activities including antioxidant activity [58, 59]. These are mainly found in brown algae because of the presence of polymers of phloroglucinols (1,3,5-trihydroxybenzene), namely, phlorotannins [60]. They have a unique property of having eight rings in their structures which are interconnected with each other, which make them different from polyphenols obtained from terrestrial plants having only three or four rings. This renders much higher antioxidative properties in the seaweed [60].

Another major component of seaweed acting as antioxidant is polysaccharide. There are few polysaccharides which are specifically found in marine life. Brown seaweeds and red seaweeds have higher content of polysaccharides, as compared to that of green seaweeds containing 380–610 g/kg of dry mass (d.m.), 360–660 g/kg d.m., and 150–650 g/kg d.m., respectively [61]. Polysaccharides unique to brown seaweeds are alginate, fucoidan, and laminarin; carrageenan and ulvan are mainly found in red seaweeds and green seaweeds, respectively [13, 14, 61, 62], out of which fucoindan, carragenan, and ulvan exhibit antioxidative activity in animals [13, 63]. Polysaccharides, namely fucoidan and laminarin, play a beneficial role on animal health by modulating the gut environment, strengthening the immune system, and enhancing growth rate and productivity [62, 64, 65].

The algal biomass also contains a diverse array of vitamins, including A, B1, B2, B3, B5, B6, and B8 [14, 33], vitamin C (ascorbic acid) [12, 26, 33, 34], and vitamin E (α-tocopherol) [12, 33, 35]. Beside these, carotenoids present in seaweeds in the form of β-carotene also exhibit provitamin A activity [37]. However, the vitamin content of each species of algae differs from one another depending on several internal and external factors which include genetic variant, life cycles, growth rate, amount of nutrients in their habitat, the exposure to light, and the processes by which they are collected. In short, several factors are present which can be amended to obtain the preferred variety of algae [36].

Seaweeds also act as storehouse of several micro- and macroelements (sodium, potassium, calcium, magnesium, manganese, selenium, zinc, iron, and iodine) which provide benefits to animal health, and they are present in a highly bioavailable form. The mineral content of seaweed is 10–20 times more than terrestrial plants by virtue of their unique property of bioaccumulation of minerals from seawater [34, 66]. These elements exhibit antioxidant properties as well [13, 14, 33, 61]. It was observed that the feed containing seaweed had higher ash content than the feed obtained from other sources [67]. However, the accumulation of toxic metal ions and high iodine content are few of the negative attributes which limit the consumption of seaweeds [14, 61]. Seaweeds can also act as source of proteins in animal nutrition [68]. Out of the three categories of seaweeds, the red algae have the maximum level of protein, followed by the green and then the brown ones containing crude protein level of 64–376 g/kg d.m., 32–352 g/kg d.m., and 24–168 g/kg d.m., respectively [13, 33, 34]. The seaweeds are also major sources of both essential and nonessential amino acids, with the former in *Sargassum latifolium* representing about 42% of the total amino acids (12 mg/g d.m.).

However, there are few things to be considered before choosing the seaweeds for being used as antioxidants such as the type of species, the time and location of their harvesting and their composition, and their growing conditions which involve nutrient density in water, light intensity, temperature, etc. [65, 66]. Thus, seaweeds are preferred to be harvested when abundant active compounds are available so that a comparatively lesser amount of feed supplementation can suffice for the animals [38]. The components of seaweeds and their effect on animal health are shown in Table 11.1.

11.4 Mode of Action of Antioxidants on Animals

Oxidation is the main reason for the development of reactive oxygen species. This is a free radical chain mechanism which involves three steps: initiation which leads to the formation of free peroxyl radicals, propagation wherein peroxyl radicals react with unsaturated fatty acids to form lipid hydroperoxides, and finally termination wherein two peroxyl radicals react to produce a non-radical species. The mode of action of antioxidants is the inhibition of the first two steps, thereby completely inhibiting the oxidation. The antioxidants found in seaweeds include polyphenols such as phlorotannins, flavonoids, chlorophyll-related compounds, pigments such

Table 11.1 Seaweed components and their effect on animal health

Seaweed components	Effect on animal health	References
Seaweeds' biomass	Dried biomass contains higher amount of antioxidants content and positive effect on animal health	[11, 35]
Polysaccharides (laminarin and fucoidan)	Regulate the gut environment, enhance growth performance, lower the risk of diarrhea, boost the innate immune system, promote productivity and good effect on animal health	[62, 64, 65]
Microelements in seaweed—e.g., zinc, selenium, and manganese	Exhibit powerful antioxidant properties	[33, 61]
Animal feed that contains seaweed	Ash content is higher compared to regular feed/ concentrates	[67]
Sulfated polysaccharides from marine algae (carragenan, Fucoindan, and ulvan)	Have antioxidant activity and can improve antioxidant status in animal's body	[13, 63]
Polysaccharides used in animal feed as prebiotics	Suppress the growth of harmful bacteria like *Escherichia coli* and *Clostridium difficile* while having a positive impact on the cellulolytic and lactic acid bacteria already present in the gastrointestinal tract (GIT). Reducing the risk of inflammation, higher nutrient absorption from feed, maintaining a healthy gut flora	[62, 64]
Phenolic compounds from seaweed	Possess a wide range of biological activities, including antioxidant properties (prevention and/or reduction in oxidative damage), antibacterial, antiviral, anti-inflammatory, antidiabetic, antimethanogenic, etc.	[11, 14, 38, 69]
Seaweed phenolic compounds	Phenolic substances can be used as an alternative to antimicrobials in feed, lowering the antibiotic resistance in cattle	[38, 69]
Prebiotics based on seaweed biomass to animal feed	Beneficial impact on immune system and gut health	[70]
Seaweeds as a supplement	Strengthen the animals' antioxidant defense system, should be strengthened, regulate inflammatory and thermo-respiratory reactions	[68, 71]
Fucoidan and laminarin extracted from *Laminaria digitata*	Longissimus thoracis at lumborum muscle has been strengthened in pork meat because of the deposition of marine-derived bioactive antioxidant components	[72]
Polysaccharides—laminarin, fucoidan, and alginate	Positive effects of polysaccharides on animal development, performance, immunoglobulin production, intestinal environment	[65]

as astaxanthin and fucoxanthin, phospholipids, bromophenols, and polysaccharides. The polyphenolic compounds use their phenol rings to trap electrons and scavenge peroxy radicals, hydroxyl radicals, or superoxides. Their unique phenolic structure and the location and number of the hydroxyl groups control the metal chelating potency of phenolic compounds [73]. The property which makes algal phlorotannins better antioxidants than plant polyphenols is the presence of eight interconnected rings [74]. *Grateloupia filicina* extracts use their phenolic compounds to donate electrons and scavenge hydrogen peroxide, thereby resulting in neutralization [75]. The bioactive compounds from *Enteromorpha* sp. (green seaweeds) have similarity to reductones in their actions of donating electrons and reacting with free-radicals to produce more stable products [76].

The antioxidant activity and the total phenolic content are positively correlated [75, 77–80]. This is because the phlorotannins, which are polyphenols obtained from seaweeds, are capable of scavenging free radicals, peroxyl radical [80], superoxide radicals [81], chelating ferrous ions [77], and nitric oxide [82]. *Eisenia bicyclis*, *Ecklonia cava*, and *Ecklonia kurome* (Japanese brown algae) are also the sources of eckol, dieckol, phlorofucofuroeckol A and 8,8′-bieckol, which showed 2–10 times more antioxidant activities than that of catechin, α-tocopherol, and ascorbic acid [83]. They also prevented phospholipid peroxidation in a liposome system [83]. Ethanol extracts of *Sargassum siliquastrum* were reported to show a 95% scavenging effect of DPPH radicals at 0.5 mg/ml or greater [84]. RSA of Taondiol (extracted from the brown alga *Taonia atomaria* of Aegean Sea) was found to be 1.7 and 3 times stronger than trolox and ascorbic acid, respectively [85]. A strong antioxidant activity was shown in water extract of *Scytosiphon lomentaria* as per linoleic acid peroxidation assay (about 22 mg catechin equivalents/g dry sample) [81]. Higher DPPH RSA (98 ± 0.005%) was exhibited by the benzene fraction of *Gelidiella acerosa* seaweed extract (500 μg/ml) as compared to BHT (96.9 ± 0.047%) [86]. Even, methanol (69.69%) and benzene (64%) fractions showed strong activities on $NO^{\bullet}$ scavenging which were like those of BHT [86].

The extracts of *H. fusiformis* showed higher antioxidant activities, especially against $NO^{\bullet}$ and DPPH radical as compared to commercial antioxidants by virtue of their free-radical, reactive oxygen species scavenging (O_2^-, H_2O_2, HO, NO), metal chelation, reducing power, and lipid peroxidation capacity [87]. The aqueous chloroform fraction demonstrated O_2-scavenging and metal chelating activities like those of BHT. The antioxidant activities of phloroglucinol, eckol, and dieckol were investigated [88]. DPPH, alkyl, hydroxyl, and superoxide radical scavenging activities were observed in all the phlorotannins that were purified from *E. cava* collected from Jeju Island. The eckol samples scavenged about 93% of DPPH at 0.25, 0.5, 1 mg/ml of concentrations, which was higher than phloroglucinol and dieckol. It was found that the phenolic compounds as well as the antioxidant capacities of the three tested species of edible brown seaweeds from Ireland increased due to heat treatment [78]. Similar results were found in case of low-molecular weight fraction of water-soluble extract from *Porphyra yezoensis* which exhibited an increase in their DPPH-RSA at temperatures over 100 °C [89]. This is because a dehydrated compound of porhyra-334 (which is a typical mycosporine-like amino acid) was present in the extract which is the reason behind showing high-IC50 value of 10.1 μg/ml in DPPH-RSA, in comparison to that of porphyra-334 (N1000 μg/ml). A strong RSA (82%) from methanolic extract of *G. filicina* was reported which was higher than that of α-tocopherol, BHA and BHT. It showed a higher hydroxyl RSA

(69%) than that obtained for BHA (37%) and BHT (38%) [75]. Sulphated polysaccharides of *Sargassum fulvellum* were found to scavenge NO (nitric oxide) more efficiently than BHA and α-tocophorol [90]. The role of seaweed as antioxidants is shown in Table 11.2.

Table 11.2 Seaweeds as a source of antioxidants for animal feeding

Seaweeds and species	Effect on animal	Effect on health	References
Brown seaweeds (*Ascophyllum nodosum*, *Sargassum* sp. and *Laminaria* sp.)	Animals	Helps animal growth performance, immunoglobulin production and improve the intestinal environment	[65, 91]
Brown seaweeds (*Sargassum latifolium*)	Animal	Source of protein for animal feed. The biomass is rich with phenolic components like flavonoids, kaempherol, and carotenes	[68]
Seaweeds (*Sargassum* spp.)	Goat diets	Strengthening the antioxidant system, controlling infections, stimulating immunity, dietary supplements and their derivatives improved the immunological responses and antioxidant capabilities	[92, 93]
Dietary seaweed extract (from *Laminaria digitata* and *Laminaria hyperborea*)	Male-rabbit semen	Containing polysaccharides, enhancing the seminal plasma's antioxidant capacity and the antioxidant state of animals effect improved the animals' growth, health, and development	[94]
Seaweeds (*Ascophyllum nodosum*)	Feed additive for pigs	Helps in pathogen reduction and decrease in fecal shedding of the pathogen *E. coli* O157:H7	[95–97]
Seaweeds (*Ascophyllum nodosum extract*)	Pigs	Improves the immunity and fight against the enteric disease by *Salmonella Typhimurium*	[98]
Seaweeds (*Ascophyllum nodosum*)	Minks	Improves the kidney function, fight against Aleutian mink disease virus (AMDV)	[99]
Seaweed powder (*Sargassum latifolium*)	Sheep	Defend against heat stress and stress brought on by bacterial endotoxins, and lessen their negative consequences	[68, 71]
Brown seaweeds (alginate oligo-saccharide)	Pigs	Possibility of treating intestinal digestion problems, capacity to alter weaning-induced oxidative stress	[100, 101]
Brown seaweed (*Ascophyllum nodosum*)	Animals	Phlorotannins possess antibacterial properties	[102]
Brown seaweeds (*Sargassum latifolium*)	Animals	*Sargassum latifolium* contains 12 mg/g d.m. of essential amino acid that is 42% of the total amino acids content and used as alternative source of proteins for animal feeding	[68]

11.5 Effect of Seaweed-Derived Antioxidants on Animal Health

The content of malondialdehyde and other primary antioxidant enzymes (glutathione, superoxide dismutase, and catalase) can be used as a measure of the antioxidant capacity of seaweeds [12, 33, 68, 101]. The incorporation of animal diet with seaweeds can aid in maintaining the redox balance in animals, their good health, and productivity. Seaweed supplementation in the diet of animals can also boost antioxidant activity and their immunity. However, the effectiveness of the antioxidants depends on the animal species, the species of seaweed, and the doses in their diet. Few compounds extracted from seaweeds can also reduce the effect of stress in animals. Higher the antioxidant activity in animals, lower is the oxidative stress in animals [103]. The immunocompetence of heat-stressed lambs exposed to suboptimal production conditions was optimized by using extracts from brown seaweeds [104]. The minerals obtained from seaweeds, for example, selenium and zinc, have been known to provide protection to animals against oxidant stress by acting as antioxidants [105]. Antioxidants contained in the biomass of *A. nodosum* (phlorotannins, vitamins—ascorbic acid, tocopherol, and pigments—carotenoids) can scavenge peroxyl radicals and may prevent or diminish oxidative stress in weaned piglets [106]. Alginate oligosaccharide, which is extracted from brown seaweeds, has successfully reduced weaning-induced oxidative stress and intestinal digestive disorders in pigs [107]. Sheep can be protected from stress induced by bacterial endotoxins [68] as well as heat stress using seaweed powder derived from *Sargassum latifolium* [71], and their harmful effects can be reduced effectively. Supplementing animal diet with seaweeds may improve their antioxidant defense system and improve their thermo-respiratory responses [68, 71]. Table 11.3 shows the health benefits of animals fed with seaweed supplementation in diet.

11.6 Applications of Seaweeds

11.6.1 Anti-biofilm Application

Biofilms have been responsible for the occurrence of various diseases. Anti-biofilm activity was tested based on their action on Gram-positive and Gram-negative bacteria. *Salmonella enterica* is one of the most prominent causal factors of bacterial food-borne diseases. Biofilms formed by these bacteria were adversely affected by brominated furanones obtained by extraction from the red algae *Delisea pulchra*. *Sesbania grandiflora* extract has anti-biofilm and antibacterial activity against *Staphylococcus aureus*. Sulfated polysaccharides from marine macroalgae showed antimicrobial, anti-biofilm, and antifouling properties on dental plaque bacteria.

Table 11.3 Seaweed as an antioxidants and health benefits of animals

Health benefits	References
Reduces the problems caused by oxidative stress and helps in treating various diseases	[12, 13]
Anti-inflammatory, antioxidant, immunomodulatory, antimicrobial, and prebiotic	[13, 14]
Ability to accumulate hazardous metal ions like arsenic, cadmium, and lead as well as micro- and macroelements, which may restrict their usage as an ingredient in animal feed	[13, 61]
Provide animals to all essential and a many nonessential amino acids	[61]
Malondialdehyde (MDA), a by-product of lipid peroxidation in animal blood, SOD, GPX, CAT, and GSH are essential elements of the cell's mechanism for protection against oxidative stress, which support the immune cells' functions to fight infectious disease	[23]
Antioxidant enzymes decreased viral infection in cattle and oxidative stress has a role in clinical symptoms	[16]
SOD, GSH, CAT, and MDA act as antioxidant enzymes and help in the pathogenesis of severe laminitis in dairy heifers	[17]
In various diseases such as interdigital dermatitis, footroot infections, and interdigital pouch infections, CAT levels were lower in the serum compared to the control group. In case of interdigital dermatitis, infections were found to have reduced serum concentrations of CAT, GPX, and GSH	[24]
In sheep with foot disease, MDA and NO values were considerably greater	[24]
CAT concentrations were considerably lower in serum. Antioxidants have positive role in reducing foot infectious diseases in animals	[24]
Seaweeds can serve as a source of vitamin C as an antioxidant and in red seaweeds (0.720 mg/g d.m), brown seaweeds (0.815 mg/g), green seaweeds (0.781 mg/g)	[26]
Strong antioxidants have the ability to eliminate free radicals and redox metals easily absorbed by the body. The endogenic antioxidant plays a crucial role in improving the efficiency of the cells	[27–29]
Proinflammatory cytokines and certain inflammatory cell mediators, such as cytokines, chemokines, and fatty acid derivatives, can also react by increasing the number of immune cells present at the injury or disease site, thereby increasing the production of ROS	[108]
During neuropathological problems in animals, free radicals are commonly found. Therefore, protecting against oxidative damage may help the evolution of diseases like neurodegenerative in animals	[32]
ROS generation leads to inflammation. Due to their intricate and diverse qualities, marine algae possess antioxidant as well as anti-inflammatory properties and may be ideal options for treating hyper-inflammatory problem in animals	[32]
The biomass collected from the coastal zone is washed, sodium and other impurities were removed, dried, grounded and used as a seaweed powder in animal diet which acts as an antioxidant in feed	[68, 100]
Polysaccharide obtained from ulvan from green seaweed (*Ulva* sp.) used as a feed additive for laying hens and acts as an antioxidant	[109]
Pelleting of seaweeds is suggested for animals feeding so that animals cannot reject and antioxidant source can be incorporated	[67]

11.6.2 Biofuel Application

Seaweeds also find application in the production of biofuels. One of the biggest advantages of using seaweed as a fuel is that, depending on the treatment applied, a single algal biomass can produce several different kinds of renewable biofuels.

11.6.3 Fish Feed Application

The algal biomass acts as a good source of micronutrients along with other nutritional properties or as a bulk feedstuff for fish as well as for other animals and used in ruminants nutrition. However, the presence of toxic metals, such as arsenic, limits their use to a certain concentration. The consumption of microalgae is carried out directly, in the case of mollusks and crustaceans, and indirectly in urine from previous ingestion by zooplankton species.

11.6.4 Food Applications

Algae and seaweeds also serve as source of compounds acting as food ingredients. One example is phycocolloids, which are a group of natural polymers constituted by polysaccharides with the ability to form gels (hydrocolloid). They find application in food, drugs, paintings, and cosmetics industries due to their physical–chemical characteristics. Mainly red and brown algae are used in the production of three hydrocolloids: Agar, alginate, and carrageenan. Other functional ingredients present in algae are carotenoids (used as food colorants, feed supplements, and nutraceuticals), lipids, proteins, polysaccharides, and phenolic compounds.

11.6.5 Source of Iodine and Dietary Fibers

Iodine deficiency is one of the most prevalent diseases found among the population of developing countries. Seaweeds are one of the best natural sources of iodine. Edible seaweeds contain iodine levels between 4300 and 2,660,000 μg/kg. Regular consumption of seaweeds is related to extremely low incidence of breast cancer due to high availability of iodine in the diet. The Food and Nutrition Research Institute of the Department of Science and Technology (FNRI-DOST), Philippines, has developed various types of noodles supplemented with squash, seaweeds, and saluyot. The addition of seaweeds to noodles was used as a means to provide added flavor, color, and functionality, as well as increased nutritional value due to their iodine content. Seaweeds contributed 170 μg iodine/100 g of the product.

Seaweeds have a high content of fibers and minerals. The rising awareness in consumers has made them aware of the relationship between health and diet which led to an increase in the demand for functional foods. Seaweeds help in overcoming the technological problems associated with low-salt meat products, including fat- and water-binding properties. In addition, efforts are also being carried out to take advantage of the nutraceutical benefits of seaweeds by adding them to various other food products as well.

11.6.6 Pharmacology and/or Medical Applications

The bioactive compounds produced by algae are rich in pharmacological benefits. Environmental conditions are responsible for the production of these bioactive compounds. However, geographical parameters like environment and season affect the bioactivity of the species and may result in their variation.

11.6.7 General Cosmeceuticals Applications

In this aspect, brown and red marine seaweeds play a major role in cosmeceutical bioactivity, that is, being used as ingredients for skin care products. Phlorotannins, polysaccharides, carotenoids, fatty acids, as well as bioactive proteins, vitamins, and minerals are the main seaweed extracts rich in potential cosmeceutical ingredients exhibiting bioactive activities like antioxidant, photo-protection, anti-wrinkling, anti-cellulite, moisturizing, whitening, and even, UV protection.

11.6.8 Other Applications

Seaweeds have always been used as fertilizers by coastal communities by virtue of their property of enriching the soil with N, P, and K, favoring agriculture, and they have the ability to promote seed germination, increase frost resistance, and improve resistance to fungal and insect pests. Algae possess an interesting ability to take many minerals from the sea, with no substantial differences in the quantity among different types of algae. The most important macroelements they contain are sodium, calcium, magnesium, potassium, chloride, sulfur, and phosphorus. Regarding oligo-elements, iodine, iron, zinc, selenium, copper, molybdenum, fluorine, manganese, boron, cobalt, and nickel are the most relevant ones. Figure 11.3 gives a pictorial representation of various seaweeds having beneficial role on animal health.

11.7 Current Status

As per the report given by Food and Agriculture Organization (FAO) in 2018, 30.4 million tons of seaweed were obtained till 2015. Although China, Chile, and Norway are the highest producers of wild species of seaweed, the countries that are leading in their cultivation are China, Philippines, Indonesia, and the Republic of Korea [110]. The world market of seaweed definitely comes under the section which witnesses a high rate of growth each year, which is approximately 8.9%. It is expected

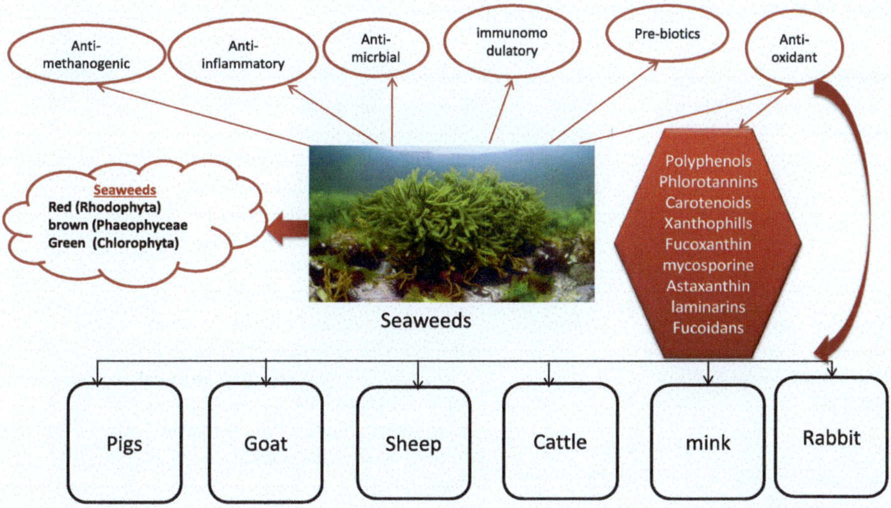

Fig. 11.3 Pictorial representation of various seaweeds having beneficial role on animal health

that, by 2024, the value of the market of seaweed will be around 22.13 billion USD [11, 111], out of which 85% of the market will be provided by China and Indonesia. However, they are cultivated and extracted and are mainly put to use for the food and hydrocolloid industry [11, 112]. Till date, 221 different varieties of seaweeds have been discovered that have successfully been put to commercial use [113]. The different seaweeds available in the market for animals are shown in Table 11.4.

11.8 Conclusion

Thus, it can be summarized that seaweeds are natural sources of antioxidants that help to improve animal's health and productivity, beside enhancing their ability to resist several stresses induced by diseases, unfavorable temperatures, or other unpleasant environmental conditions. Seaweed can be used as a feed/feed additive either in intact form or just the bioactive extract. The biologically active molecules of seaweed that play a vital role in improving the antioxidant system in the animal body are several phenolic compounds like flavonoids, pigments, vitamins, minerals, and even certain polysaccharides. Dietary intervention of these compounds has successfully proved to enhance the antioxidant potential that helped to improve animal health, development, and productivity. Incorporating animal diet with seaweeds or compounds derived from seaweeds can boost their immunity, antioxidant activity, and the overall gut environment. There is a need to identify the harvests and the products should be regularized as the seaweeds from different locations produce different levels of active compounds. The implementation of the practice to induce seaweed in animal feed also calls for initial economic support from the countries that lack sufficient marine spaces. However, more elaborate research is required on

Table 11.4 Seaweeds available in markets for animals

Seaweeds	Designed for	Benefits	References
OceanFeedVR (Milltown, Ireland)	Wine, equine, and bovine	In postweaning pigs, growing-finishing pigs, gestating and lactating sows—improved feed intake, digestive function, feed conversion ratio, nutrient use, weight gain, enhanced performance, gut health, and lowered the occurrence of diarrhea in pigs	[65, 70]
TascoVR application (Acadian AgriTech™, Dartmouth, Nova Scotia Canada)	Animals	Polysaccharide and oligosaccharide alginate 52% (mannitol 5%, alginic acid 25%, fucose containing sulfated polysaccharide 15%, laminarin 3%, other carbohydrates 4%). Improve immune system and carcass characteristics	[95]
Seaweed biomass is OceanFeedVR Equine, OceanFeedVR Swine, OceanFeedVR Bovine, Ocean Harvest Technology Company Ltd. (Tuam, Co Galway, Milltown, Ireland)	Swine, equine, bovine	Contain a different variety of green, brown, and red seaweed species and have entirely natural macroalgal blend	[113]
Sun-dried *Ascophyllum nodosum* (Tasco-14™)	Pigs	Decrease in the pathogen *E. coli* O157:H7's fecal shedding	[96]

the different compounds obtained from seaweed exhibiting antioxidant activity and their unique biological actions. This could help in targeting the specific problems faced by animals with precise solutions. Resources can be utilized at a large scale is the pollution of seaweed with heavy metals or other industrial wastes.

Author Contributions All authors (PK and RR) contributed equally to the conceptualization, investigation, visualization, writing—original draft preparation, writing—review and editing.

Conflicts of Interest The authors declare no conflicts of interest.

References

1. Enrique C, Lester P (2002) Handbook of antioxidants. Marcel Dekker, New York
2. Barrow C, Shahidi F (2008) Marine nutraceuticals and functional foods. CRC Press, New York
3. Chen JH, Lim JD, Sohn EH, Choi YS, Han ET (2009) Growth-inhibitory effect of a fucoidan from brown seaweed Undaria pinnatifida on Plasmodium parasites. Parasitol Res 104:245–250
4. Cox S, Abu-Ghannam N, Gupta S (2010) An assessment of the antioxidant and antimicrobial activity of six species of edible Irish seaweeds. Int Food Res J 17:205–220

5. Rupérez P, Ahrazem O, Leal JA (2002) Potential antioxidant capacity of sulphated polysaccharides from edible brown seaweed Fucus vesiculosus. J Agric Food Chem 50:840–845
6. Taskin E, Ozturk M, Taskin E, Kurt O (2007) Antibacterial activities of some marine algae from the Aegean Sea (Turkey). Afr J Biotechnol 6:2746–2751
7. Krishnaiah D, Sarbatly R, Bono A (2007) Phytochemical antioxidants for health and medicine—a move towards nature. Biotechnol Mol Biol Rev 1:97–104
8. Sachindra NM, Sato E, Maeda H, Hosokawa M, Niwano Y, Kohno M et al (2007) Radical scavenging and singlet oxygen quenching activity of marine carotenoid fucoxanthin and its metabolites. J Agric Food Chem 55:8516–8522
9. Dhargalkar VK, Pereira N (2005) Seaweed promising plant of the millennium. Sci Cult 71:60–66
10. Smit AJ (2004) Medicinal and pharmaceutical uses of seaweed natural products: a review. J Appl Phycol 16:245–262
11. Cotas J, Leandro A, Monteiro P, Pacheco D, Figueirinha A, Gonçalves AMM, da Silva GJ, Pereira L (2020) Seaweed phenolics: from extraction to applications. Mar Drugs 18(8):384
12. de Quiros ARB, Frecha-Ferreiro S, Vidal-Perez AM, Lopez Hernandez J (2010) Antioxidant compounds in edible brown seaweeds. Eur Food Res Technol 231(3):495–498
13. Liu ZW, Sun X (2020) A critical review of the abilities, determinants, and possible molecular mechanisms of seaweed polysaccharides antioxidants. Int J Mol Sci 21(20):7774
14. Corino C, Modina SC, Di Giancamillo A, Chiapparini S, Rossi R (2019) Seaweeds in pig nutrition. Animals 9(12):1126
15. Kaur G, Sharma A, Guruprasad K, Pati PK (2014) Versatile roles of plant NADPH oxidases and emerging concepts. Biotechnol Adv 32(3):551–563
16. Khoshvaghti A, Askari A, Nazifi S, Ghane M (2014) Evaluation of some antioxidant enzymes in cattle infected with Foot and Mouth Virus. Istanbul Univ Vet Fak Derg 40(1):70–75
17. Hayat MA, Ding J, Li Y, Zhang X, Zhang J, Li S, Wang HB (2020) Determination of the activity of selected antioxidant enzymes during bovine laminitis, induced by oligofructose overload. Med Weter 76(5):289–295
18. Rahman K (2007) Studies on free radicals, antioxidants, and co-factors. Clin Interv Aging 2(2):219–236
19. Valko M, Leibfritz D, Moncol J, Cronin MT, Mazur M, Telser J (2007) Free radicals and antioxidants in normal physiological functions and human disease. Int J Biochem Cell Biol 39(1):44–84
20. Kurutas EB (2016) The importance of antioxidants which play the role in cellular response against oxidative/nitrosative stress: current state. Nutr J 15(1):71
21. Changxing L, Saeed M, Kamboh AA, Alagawany MA, El-Hack ME, Arain MA, Babazadeh D, Dhama K, Mo C (2018) Reconsidering a citrus flavonoid naringin as a promising nutritional supplement and its beneficial health applications in humans, animals and poultry. Int J Pharm 12(8):874–883
22. Changxing L, Chenling M, Alagawany M, Jianhua L, Dongfang D, Gaichao W, Wenyin Z, Syed SF, Arain MA, Saeed M et al (2018) Health benefits and potential applications of anthocyanins in poultry feed industry. World Poult Sci J 74(2):251–264
23. Rahal A, Kumar A, Singh V, Yadav B, Tiwari R, Chakraborty S, Dhama K (2014) Oxidative stress, prooxidants and antioxidants: the interplay. Biomed Res Int 2014:761264
24. Yurdakul I, Yildirim BA (2018) Assessment of oxidative status in foot diseases of sheep. Acta Sci Vet 46(1):6
25. Eryilmaz F (2017) Vitamin C: an antioxidant agent. IntechOpen, Rijeka, Croatia. https://www.intechopen.com/chapters/56013. Accessed 21 Jul 2021
26. Nielsen CW, Rustad T, Holdt SL (2021) Vitamin C from seaweed: a review assessing seaweed as contributor to daily intake. Food Secur 10(1):198
27. Arain MA, Mei Z, Hassan FU, Saeed M, Alagawany M, Shar AH, Rajput IR (2018) Lycopene: a natural antioxidant for prevention of heat-induced oxidative stress in poultry. World Poult Sci J 74(1):89–100

28. Waheed Janabi AH, Kamboh AA, Saeed M, Xiaoyu L, BiBi J, Majeed F, Naveed M, Mughal MJ, Korejo NA, Kamboh R et al (2020) Flavonoid-rich foods (FRF): a promising nutraceutical approach against lifespan-shortening diseases. Iran J Basic Med Sci 23:1–14
29. Naiel MAE, Alagawany M, Patra AK, El-Kholy AI, Amer MS, Abd El-Hack ME (2021) Beneficial impacts and health benefits of macroalgae phenolic molecules on fish production. Aquaculture 534:736186
30. Juarez-Portilla C, Olivares-Banuelos T, Molina-Jiménez T, Sánchez-Salcedo JA, Moral DID, Meza-Menchaca T, Flores-Muñoz M, López-Franco Ó, Roldán-Roldán G, Ortega A, Zepeda RC et al (2019) Seaweeds-derived compounds modulating effects on signal transduction pathways: a systematic review. Phytomedicine 63:153016
31. Visconti R, Grieco D (2009) New insights on oxidative stress in cancer. Curr Opin Drug Discov Devel 12(2):240–245
32. Barbalace MC, Malaguti M, Giusti L, Lucacchini A, Hrelia S, Angeloni C (2019) Anti-inflammatory activities of marine algae in neurodegenerative diseases. Int J Mol Sci 20(12):3061
33. Morais T, Inacio A, Coutinho T, Ministro M, Cotas J, Pereira L, Bahcevandziev K (2020) Seaweed potential in the animal feed: a review. J Mar Sci Eng 8(8):559
34. Matos GS, Pereira SG, Genisheva ZA, Gomes AM, Teixeira JA, Rocha CMR (2021) Advances in extraction methods to recover added-value compounds from seaweeds: sustainability and functionality. Foods 10(3):516
35. Farvin KHS, Jacobsen C (2013) Phenolic compounds and antioxidant activities of selected species of seaweeds from Danish coast. Food Chem 138(2–3):1670–1681
36. Holdt SL, Kraan S (2011) Bioactive compounds in seaweed: functional food applications and legislation. J Appl Phycol 23(3):543–597
37. García-Casal MN, Ramírez J, Leets I, Pereira AC, Quiroga MF (2009) Antioxidant capacity, polyphenol content and iron bioavailability from algae (Ulva sp., Sargassum sp. and Porphyra sp.) in human subjects. Br J Nutr 101:79–85
38. Ford L, Curry C, Campbell M, Theodoridou K, Sheldrake G, Dick J, Stella L, Walsh PJ (2020) Effect of phlorotannins from brown seaweeds on the in vitro digestibility of pig feed. Animals 10(11):2193
39. Plaza M, Cifuentes A, Ibañez E (2008) In the search of new functional food ingredients from algae. Trends Food Sci Technol 19:31–39
40. Holck J, Hjernø K, Lorentzen A, Vigsnæs LK, Hemmingsen L, Licht TR, Mikkelsen JD, Meyer AS (2011) Tailored enzymatic production of oligosaccharides from sugar beet pectin and evidence of differential effects of a single DP chain length difference on human faecal microbiota composition after in vitro fermentation. Process Biochem 46:1039–1049
41. Zou Y, Qian ZJ, Li Y, Kim MM, Lee SH, Kim SK (2008) Antioxidant effects of phlorotannins isolated from Ishige okamurae in free radical mediated oxidative systems. J Agric Food Chem 56:7001–7009
42. Airanthi MKWA, Hosokawa M, Miyashita K (2011) Comparative antioxidant activity of edible Japanese brown seaweeds. J Food Sci 76:C104–C111
43. Roriz CL, Barros L, Prieto MA, Morales P, Ferreira ICFR (2017) Floral parts of Gomphrena globosa L. as a novel alternative source of betacyanins: optimization of the extraction using response surface methodology. Food Chem 229:223–234
44. Albuquerque BR, Prieto MA, Barreiro MF, Rodrigues A, Curran TP, Barros L, Ferreira ICFR (2016) Catechin-based extract optimization obtained from Arbutus unedo L. fruits using maceration/microwave/ultrasound extraction techniques. Ind Crop Prod 95:404–415
45. Del Campo JA, García-González M, Guerrero MG (2007) Outdoor cultivation of microalgae for carotenoid production: current state and perspectives. Appl Microbiol Biotechnol 74:1163–1174
46. Baroli I, Niyogi KK, Barber J, Heifetz P (2000) Molecular genetics of xanthophyll-dependent photoprotection in green algae and plants. Philos Trans R Soc Lond Ser B Biol Sci 355:1385–1394

47. Grünewald K, Hirschberg J, Hagen C (2001) Ketocarotenoid biosynthesis outside of plastids in the unicellular green alga Haematococcus pluvialis. J Biol Chem 276:6023–6029
48. Lohr M, Wilhelm C (2001) Xanthophyll synthesis in diatoms: quantification of putative intermediates and comparison of pigment conversion kinetics with rate constants derived from a model. Planta 212:382–391
49. Bhosale P (2004) Environmental and cultural stimulants in the production of carotenoids from microorganisms. Appl Microbiol Biotechnol 63:351–361
50. Raja R, Hemaiswarya S, Rengasamy R (2007) Exploitation of Dunaliella for β-carotene production. Appl Microbiol Biotechnol 74:517–523
51. Dufossé L, Galaup P, Yaron A, Arad SM, Blanc P, Murthy KNC, Ravishankar GA (2005) Microorganisms and microalgae as sources of pigments for food use: a scientific oddity or an industrial reality? Trends Food Sci Technol 16:389–406
52. Van Den Berg H, Faulks R, Granado HF, Hirschberg J, Olmedilla B, Sandmann G, Southon S, Stahl W (2000) The potential for the improvement of carotenoid levels in foods and the likely systemic effects. J Sci Food Agric 80:880–912
53. Jespersen L, Strømdahl LD, Olsen K, Skibsted LH (2005) Heat and light stability of three natural blue colorants for use in confectionery and beverages. Eur Food Res Technol 220:261–266
54. Duan XJ, Zhang WW, Li XM, Wang BG (2006) Evaluation of antioxidant property of extract and fractions obtained from a red alga, Polysiphonia urceolata. Food Chem 95:37–43
55. Manach C, Scalbert A, Morand C, Rémésy C, Jiménez L (2004) Polyphenols: food sources and bioavailability. Am J Clin Nutr 79:727–747
56. Pinela J, Prieto MA, Barreiro MF, Carvalho AM, Oliveira MBPP, Vázquez JA, Ferreira ICFR (2016) Optimization of microwave-assisted extraction of hydrophilic and lipophilic antioxidants from a surplus tomato crop by response surface methodology. Food Bioprod Process 98:283–298
57. Thomas NV, Kim SK (2011) Potential pharmacological applications of polyphenolic derivatives from marine brown algae. Environ Toxicol Pharmacol 32:325–335
58. Kähkönen MP, Hopia AI, Vuorela HJ, Rauha JP, Pihlaja K, Kujala TS, Heinonen M (1999) Antioxidant activity of plant extracts containing phenolic compounds. J Agric Food Chem 47:3954–3962
59. Pinela J, Prieto MA, Barreiro MF, Curran TP, Ferreira ICFR (2017) Valorisation of tomato wastes for development of nutrient-rich antioxidant ingredients: a sustainable approach towards the needs of the today's society. Innov Food Sci Emerg Technol 41:160–171
60. Wang T, Jónsdóttir R, Liu H, Gu L, Kristinsson HG, Raghavan S, Ólafsdóttir G (2012) Antioxidant capacities of phlorotannins extracted from the brown algae Fucus vesiculosus. J Agric Food Chem 60:5874–5883
61. Øverland M, Mydland LT, Skrede A (2019) Marine macroalgae as sources of protein and bioactive compounds in feed for monogastric animals. J Sci Food Agric 99(1):13–24
62. Kraan S (2012) Algal polysaccharides, novel applications and outlook. In: Chang C-F (ed) Carbohydrates—comprehensive studies on glycobiology and glycotechnology. IntechOpen, Rijeka, Croatia
63. Salehi B, Sharifi-Rad J, Seca AML, Pinto DCGA, Michalak I, Trincone A, Mishra AP, Nigam M, Zam W, Martins N (2019) Current trends on seaweeds: looking at chemical composition, phytopharmacology, and cosmetic applications. Molecules 24:4182
64. Lynch MB, Sweeny T, Callan JJ, O'Sullivan JT, O'Doherty JV (2010) The effect of dietary laminaria derived laminarin and fucoidan on nutrient digestibility, nitrogen utilisation, intestinal microflora and volatile fatty acid concentration in pigs. J Sci Food Agric 90(3):430–437
65. Del Tuffo L, Laskoski F, Vier CM, Tokach MD, Dritz SS, Woodworth JC, DeRouchey JM, Goodband RD, Constance LA, Niederwerder M et al (2019) Effects of Oceanfeed Swine feed additive on performance of sows and their offspring. Kansas Agricult Exp Stat Res Rep 5(8):1–22
66. Makkar HPS, Tran G, Heuze V, Giger-Reverdin S, Lessire M, Lebas F, Ankers P (2016) Seaweeds for livestock diets: a review. Anim Feed Sci Technol 212:1–17

67. Rjiba SK, Chermiti A, Mahouachi M (2010) The use of seaweeds (Ruppia maritima and Chaetomorpha linum) for lamb fattening during drought periods. Small Rum Res 91(1):116–119
68. Ramadan G, Fouda WA, Ellamie AM, Ibrahim WM (2020) Dietary supplementation of Sargassum latifolium modulates thermo-respiratory response, inflammation, and oxidative stress in bacterial endotoxin-challenged male Barki sheep. Environ Sci Pollut Res 27(27):33863–33871
69. Campbell M, Ortuno J, Ford L, Davies DR, Koidis A, Walsh PJ, Theodoridou K (2020) The effect of ensiling on the nutritional composition and fermentation characteristics of brown seaweeds as a ruminant feed ingredient. Animals 10(6):1019
70. Ruiz AR, Gadicke P, Andrades SM, Cubillos R (2018) Supplementing nursery pig feed with seaweed extracts increases final body weight of pigs. Austral J Vet Sci 50(2):83–87
71. Ellamie AM, Fouda WA, Ibrahim WM, Ramadan G (2020) Dietary supplementation of brown seaweed (Sargassum latifolium) alleviates the environmental heat stress-induced toxicity in male Barki sheep (Ovis aries). J Thermal Biol 89:102561
72. Moroney NC, O'Grady MN, Robertson RC, Stanton C, O'Doherty JV, Kerry JP (2015) Influence of level and duration of feeding polysaccharide (laminarin and fucoidan) extracts from brown seaweed (Laminaria digitata) on quality indices of fresh pork. Meat Sci 99:132–141
73. Santoso J, Yoshie-Stark Y, Suzuki T (2004) Anti-oxidant activity of methanol extracts from Indonesian seaweeds in an oil emulsion model. Fish Sci 70:183–188
74. Wang Y, Xu Z, Bach SJ, McAllister TA (2009) Sensitivity of Escherichia coli to seaweed (Ascophyllum nodosum) phlorotannins and terrestrial tannins. Asian Austral J Anim Sci 22:238–245
75. Athukorala Y, Lee KW, Song CB, Ahn CB, Shin TS, Cha YJ (2003) Potential antioxidant activity of marine red alga Grateloupia filicina extracts. J Food Lipids 10:251–265
76. Ganesan K, Suresh Kumar K, Subba Rao PV (2011) Comparative assessment of antioxidant activity in three edible species of green seaweed, Enteromorpha from Okha, northwest coast of India. Innov Food Sci Emerg Technol 12:73–78
77. Chew YL, Lim YY, Omar M, Khoo KS (2008) Antioxidant activity of three edible seaweeds from two areas in South East Asia. LWT Food Sci Technol 41:1067–1072
78. Rajauria G, Kumar A, Abu-Ghannam N, Gupta S (2010) Effect of hydrothermal processing on colour, antioxidant and free radical scavenging capacities of edible Irish brown seaweeds. Int J Food Sci Technol 45:2485–2493
79. Siriwardhana N, Lee KW, Kim SH, Ha JW, Jeon YJ (2003) Antioxidant activity of Hizikia fusiformis on reactive oxygen species scavenging and lipid peroxidation inhibition. Food Sci Technol Int 9:339–347
80. Wang T, Jónsdóttir R, Ólafsdóttir G (2009) Total phenolic compounds, radical scavenging and metal chelation of extracts from Icelandic seaweeds. Food Chem 116:240–248
81. Kuda T, Tsunekawa M, Hishi T, Araki Y (2005) Antioxidant properties of dried 'kayamonori', a brown alga Scytosiphon lomentaria (Scytosiphonales, Phaeophyceae). Food Chem 89:617–622
82. Valentão P, Trindade P, Gomes D, de Pinho PG, Mouga T, Andrade PB (2010) Codium tomentosum and Plocamium cartilagineum: chemistry and antioxidant potential. Food Chem 119:1359–1368
83. Shibata T, Ishimaru K, Kawaguchi S, Yoshikawa H, Hama Y (2008) Antioxidant activities of phlorotannins isolated from Japanese Laminariaceae. J Appl Phycol 20:705–711
84. Cho SH, Kang SE, Cho JY, Kim AR, Park SM, Hong YK et al (2007) The antioxidant properties of brown seaweed (Sargassum siliquastrum) extracts. J Med Food 10(2007):479–485
85. Nahas R, Abatis D, Anagnostopoulou MA, Kefalas P, Vagias C, Roussis V (2007) Radical-scavenging activity of Aegean Sea marine algae. Food Chem 102:577–581
86. Suganthy N, Nisha SA, Pandian SK, Devi KP (2010) Antioxidant and metal chelating potential of the solvent fractions of Gelidiella acerosa, the red algae inhabiting South Indian coastal area. Biomed Pharmacother. https://doi.org/10.1016/j.biopha.2010.09.022

87. Karawita R, Siriwardhana N, Lee KW, Heo MS, Yeo IK, Lee YD et al (2005) Reactive oxygen species scavenging, metal chelation, reducing power and lipid peroxidation inhibition properties of different solvent fractions from Hizikia fusiformis. Eur Food Res Technol 220:363–371
88. Ahn GN, Kim KN, Cha SH, Song CB, Lee JH, Heo MS et al (2007) Antioxidant activities of phlorotannins purified from Ecklonia cava on free radical scavenging using ESR and H_2O_2-mediated DNA damage. J Eur Food Res Technol 226:71–79
89. Yoshiki M, Tsuge K, Tsuruta Y, Yoshimura T, Koganemaru K, Sumi T et al (2009) Production of new antioxidant compound from mycosporine-like amino acid, porphyra-334 by heat treatment. Food Chem 113:1127–1132
90. Kim SH, Choi DS, Athukorala Y, Jeon YJ, Senevirathne M, Rha CK (2007) Antioxidant activity of sulfated polysaccharides isolated from Sargassum fulvellum. J Food Sci Nutr 12:65–73
91. Dierick N, Ovyn A, De Smet S (2009) Effect of feeding intact brown seaweed Ascophyllum nodosum on some digestive parameters and on iodine content in edible tissues in pigs. J Sci Food Agric 89(4):584–594
92. Telles CBS, Mendes-Aguiar C, Fidelis GP, Frasson AP, Pereira WO, Scortecci KC, Camara RBG, Nobre LTDB, Costa LS, Tasca T et al (2018) Immunomodulatory effects and antimicrobial activity of heterofucans from Sargassum filipendula. J Appl Phycol 30(1):569–578
93. Angulo C, Chavez-Infante L, Reyes-Becerril M, Angulo M, Romero-Geraldo R, Llinas-Cervantes X, Cepeda-Palacios R (2020) Immunostimulatory and antioxidant effects of supplemental feeding with macroalga Sargassum spp. on goat kids. Trop Anim Health Prod 52(4):2023–2033
94. Vizzari F, Massanyi M, Knızatova N, Corino C, Rossi R, Ondruska Ľ, Tirpak F, Halo M, Massanyi P (2021) Effects of dietary plant polyphenols and seaweed extract mixture on male-rabbit semen: quality traits and antioxidant markers. Saudi J Biol Sci 28(1):1017–1025
95. Allen VG, Pond KR, Saker KE, Fontenot JP, Bagley CP, Ivy RL, Evans RR, Schmidt RE, Fike JH, Zhang X et al (2001) Tasco: influence of a brown seaweed on antioxidants in forages and livestock—a review. J Anim Sci 79(E-Suppl):E21–E31
96. Bach SJ, Wang Y, McAllister TA (2008) Effect of feeding sun-dried seaweed (Ascophyllum nodosum) on fecal shedding of Escherichia coli O157:H7 by feedlot cattle and on growth performance of lambs. Anim Feed Sci Technol 142(1–2):17–32
97. Gardiner GE, Campbell AJ, O'Doherty JV, Pierce E, Lynch PB, Leonard FC, Stanton C, Ross RP, Lawlor PG (2008) Effects of Ascophyllum nodosum extract on growth performance, digestibility, carcass characteristics and selected intestinal microflora populations of grower-finisher pigs. Anim Feed Sci Technol 141(3–4):259–227
98. Turner JL, Dritz SS, Higgins JJ, Minton JE (2002) Effects of Ascophyllum nodosum extract on growth performance and immune function of young pigs challenged with Salmonella typhimurium. J Anim Sci 80(7):1947–1953
99. Farid AH, Smith NJ (2020) Dietary supplementation of Ascophylum nodosum improved kidney function of mink challenged with Aleutian mink disease virus. BMC Vet Res 16(1):465
100. Abdoun KA, Okab AB, El-Waziry AM, Samara EM, Aa A-H (2014) Dietary supplementation of seaweed (Ulva lactuca) to alleviate the impact of heat stress in growing lambs. Pak Vet J 34(1):108–111
101. Wan J, Jiang F, Xu QS, Chen DW, He J (2016) Alginic acid oligosaccharide accelerates weaned pig growth through regulating antioxidant capacity, immunity and intestinal development. RSC Adv 6(90):87026–87035
102. Nagayama K, Iwamura Y, Shibata T, Hirayama I, Nakamura T (2002) Bactericidal activity of phlorotannins from the brown alga Ecklonia kurome. J Antimicrob Chemother 50(6):889–893
103. Fike JH, Allen VG, Schmidt RE, Zhang X, Fontenot JP, Bagley CP, Ivy RL, Evans RR, Coelho RW, Wester DB (2001) Tasco-forage: I. Influence of a seaweed extract on antioxidant activity in tall fescue and in ruminants. J Anim Sci 79(4):1011–1021
104. Saker KE, Fike JH, Veit H, Ward DL (2004) Brown seaweed- (Tasco™) treated conserved forage enhances antioxidant status and immune function in heat-stressed wether lambs. J Anim Physiol Anim Nutr 88(3–4):122–130

105. Rey-Crespo F, Lopez-Alonso M, Miranda M (2014) The use of seaweed from the Galician coast as a mineral supplement in organic dairy cattle. Animal 8(4):580–586
106. Michiels J, Skrivanova E, Missotten J, Ovyn A, Mrazek J, De Smet S, Dierick N (2012) Intact brown seaweed (Ascophyllum nodosum) in diets of weaned piglets: effects on performance, gut bacteria and morphology and plasma oxidative status. J Anim Physiol Anim Nutr (Berl) 96(6):1101–1111
107. Wan J, Zhang J, Chen DW, Yu B, He J (2017) Effects of alginate oligosaccharide on the growth performance, antioxidant capacity and intestinal digestion–absorption function in weaned pigs. Anim Feed Sci Technol 234:118–127
108. Hussain SP, Hofseth LJ, Harris CC (2003) Radical causes of cancer. Nat Rev Cancer 3(4):276–285
109. Li Q, Luo J, Wang C, Ta W, Wang H, Zhang X, Liu K, Jia Y, Lyv X, Wang L et al (2018) Ulvan extracted from green seaweeds as new natural additives in diets for laying hens. J Appl Phycol 30(3):2017–2027
110. FAO (2018) The state of world fisheries and aquaculture 2018—meeting the sustainable development goals. FAO, Rome, Italy. ISBN 978-92-5-130562-1
111. Custódio M, Villasante S, Cremades J, Calado R, Lillebø AI (2017) Unravelling the potential of halophytes for marine integrated multi-trophic aquaculture (IMTA)—a perspective on performance, opportunities and challenges. Aquac Environ Interact 9:445–460
112. Ferdouse F, Holdt SL, Smith R, Murúa P, Yang Z (2018) The global status of seaweed production, trade and utilization. FAO Globefish Res Program 124:120
113. Michalak I, Tiwari R, Dhawan M, Alagawany M, Farag MR, Sharun K, Emran TB, Dhama K (2022) Antioxidant effects of seaweeds and their active compounds on animal health and production—a review. Vet Q 42(1):48–67

Marine-Based Food as Drug: A Novel Approach to Fight Against Infectious Diseases

12

Rekha Rani, Payal Karmakar, and Bhopal Singh

Abstract

In the past few decades, the world has witnessed a rigorous search for different novel compounds of marine origin due to the massive applications of marine resources in food, pharmaceutical, and cosmetic industries. Their minimal growth needs and their immense biodiversity make them suitable for their wide applications. Marine environment is a taxonomically diverse ecosystem which is a rich source of biologically active natural products. Shift in consumer's choice towards "green" alternatives has motivated the use of this sustainable marine culture which is environmentally friendly as well. Moreover, human beings and animals have grown resistance towards the conventional antibiotic drugs, which calls for development of new unconventional alternative drugs. This chapter deals with various marine sources and marine-based foods acting as drugs that have proved themselves as effective solutions towards several diseases, beside showing beneficial effects on human and even animal health. However, this area is still at its infancy and needs further and deeper digging to bring more marine sources into light that can pave a path towards betterment of human and animal life. The rate of progressive research in this field can assure a promising future of marine sources in the science of innovative food and drugs.

R. Rani
Department of Dairy Chemistry, College of Dairy and Food Technology, AU, Jodhpur, Rajasthan, India

P. Karmakar (✉)
Food Science and Nutrition Department, University of Minnesota, St. Paul, MN, USA

B. Singh
Department of Dairy Technology, Faculty of Science, Dayalbagh Educational Institute, Agra, Uttar Pradesh, India

M. Rafatullah et al. (eds.), *Multidisciplinary Applications of Marine Resources*,
https://doi.org/10.1007/978-981-97-5057-3_12

Keywords

Marine · Nutraceuticals · Aquatic · Food · Drugs

12.1 Introduction

The current lifestyle and food habits among the majority of population in developed countries have led to numerous diseases and ailments. Since time immemorial, natural and nature-derived products have been catering to mankind by acting as sources of a wide variety of life-saving drugs which comprise more than 60% of all the drugs available in the market [1]. Antimicrobial compounds are formed and accumulated in micro and macroorganisms as a response to unfavorable environmental conditions, thereby leading to their metabolome evolution. This is the key reason behind the success of natural products as sources of drugs. Increase in the number of pathogen-caused infections calls for the thorough search of antibiotics. However, currently, the world is witnessing an alarming increase in the rate of development of resistance of pathogenic bacteria towards antibiotics, whereas a decrease in the rate of discovery of new antibiotics.

More than 80% of different plant and animal species are found in the oceans, which cover two-third of the earth surface. Oceans are the home to a tremendously large amount of various living organisms having unique property of getting adapted to changing environment by virtue of their different physiologies. However, till now, terrestrial environment has been the prime focus for most of the drug-related affairs, whereas, in comparison to it, very little importance is given to marine sources [2]. For about 50 years, marine habitats have been overlooked by the pharmaceutical industry [3]. The reason underlying this may be due to the assumptions that sea water contains lesser variety of microbes and their recovery and culture are more difficult as compared to the ones derived from soil and earth [4]. However, elaborate health-based research has instigated the discovery and exploration of these new marine molecules and their supposedly enormous benefits.

In the year 1966, the first ever marine antibiotic was discovered and named as *Pentabromopseudilin* [5]. Marine environment is highly rich in microflora. Genetic diversity leads to chemical diversity which is a driving factor for discovery and development of new drugs. The diverse range of organisms it offers can prove to be a highly promising source of antibiotics and the marine secondary metabolites have tremendous potential of being recognized as new drug sources [6]. This chapter thoroughly reviews how the drugs based on marine foods can be effective against several infectious diseases and where we will be heading to in the future by following the path of these marine drugs.

12.2 Marine Sources of Bioactive Molecules

Marine organisms represent a valuable source of bioactive compounds; marine organisms such as sponges, tunicates, bryozoans, molluscs, bacteria, microalgae, macro algae, and cyanobacteria have recently been utilized for biotechnology. Compounds produced from these organisms are effective as therapeutics for infectious and noninfectious disease, with a high specificity for target molecules. Table 12.1 and Fig. 12.1 describes the various marine organisms and general bioactive compounds found in them.

12.3 Current Strategies of Marine Food Developments as Nutraceuticals

Recent trends in functional foods and supplements have demonstrated that bioactive molecules play a major therapeutic role in human disease. Nutritionists and biomedical and food scientists are working together to discover new bioactive molecules that have increased potency and therapeutic benefits. Marine life constitutes almost 80% of the world biota with thousands of bioactive compounds and secondary metabolites derived from marine invertebrates such as tunicates, sponges, molluscs, bryozoans, sea slugs, and many other marine organisms. These bioactive molecules and secondary metabolites possess antibiotic, antiparasitic, antiviral, anti-inflammatory, antifibrotic, and anticancer activities. They are also inhibitors or activators of critical enzymes and transcription factors, competitors of transporters, and sequestrants that modulate various physiological pathways [13].

The concept of "nutraceutical" has been introduced by Stephen DeFelice in 1995 [14] combining the terms *nutrition* and *pharmaceutical*. The nutraceuticals are substances that can be considered as a food or part of a food which provides medical or human health benefits including prevention and treatment of diseases [15]. These nutraceutical active ingredients include certain polysaccharides, phytochemicals, vitamins, peptides, and fatty acids which can be obtained either by natural extraction or chemical and biotechnological synthesis. Certain marine fish species and algae have been identified as sources of polyunsaturated fatty acids (PUFA) which are rich in ω-3 or ω-6 fatty acids. The presence of these unsaturated fatty acids in marine-derived foods increases their applicability as nutraceuticals in the food industry [16].

Marine nutraceuticals might have a positive effect on human health as they can protect human body against damage by reactive oxygen species (ROS), which attack macromolecules such as membrane lipids, proteins, and DNA and lead to many health disorders such as cancer, diabetes mellitus, neurodegenerative, and inflammatory diseases with severe tissue injuries.

Table 12.1 Marine organism and their bioactive compounds

Marine algae (microalgae and macroalgae)	*Microalgae*: the most primary and simply organized members of marine plant life, rich sources β-carotene, Vitamins C, A, E, H, B_1, B_2, B_6, and B_{12}, astaxanthin, polysaccharides, and polyunsaturated fatty acids *Macroalgae*: Rich sources of insoluble and soluble dietary fibers as they are not digested by enzymes in the gut and are mainly composed of indigestible sulfated polysaccharides [7]. Bioactive molecules from microalgae are commercially produced, used as food additives and also incorporated into infant milk formulations and dietary supplements
Marine invertebrates (sponges, mollusks, echinoderms, and crustaceans)	Sources of bioactive peptides, steroids, terpenoids, strigolactones, alkaloids, ether, and phenols [8]. Extracts or compounds isolated from marine invertebrates are known to include antibacterial, antiviral, anthelmintic, antifungal, antihypertensive, anticancer, and immune modulatory properties
Sponges	Sponges are sessile metazoans that consist of a gelatinous material, mesophyl, and are the simplest form of multicellular animals. Their hollow pitcher-like body is reinforced by spicules (a needle-like silica or calcium structure) and spongia (collagen fibers) [9]. There are many chemically diverse compounds like alkaloids, terpenoids, polyketides, macrolides, polyphenolic compounds, peptides, and sterols that are isolated from sponges
Marine fish	Marine fish occupy the highest position in marine animal consumption and are important to the world economy. In 2012, fish provided approximately 16% of the world's protein requirements with herring, salmon, cod, flounder, tuna, mullet, and anchovy being the most common species of fish used for food [10]. One of the largest commercially canned fishery products in the world is tuna
Mollusks, echinoderms, and crustaceans	Consumed as marine foods and considered as natural functional foods Crustaceans are the biggest and most important class of marine arthropods in the global fisheries markets; they also have significant roles in nutraceutical industries [9]. In particular, crabs, prawns, and shrimps have gained great attention due to their effective utilization and health benefits [11]
Marine functional ingredients	Materials extracted from fish and shellfish; and extracts from marine plants and other marine organisms and marine bacteria. Consumer awareness regarding the health and benefit of fish oil in the diet is increased; therefore, the demand for fish oil and a proliferation of fish oil-fortified products increase globally
Fish waste	As a source of bioactive materials, proteins, minerals, amino acids, vitamins, antioxidants, and peptides fish bones (good source of calcium). Fish waste major outlet is fishmeal and oil extraction
Fish protein hydrolysates	Enzymes are used to break down fish proteins into amino acids to produce a highly functional food ingredient. As well as having functional properties, this powdered form of protein is eye-catching to the food processing sector. Current uses include milk replacement in young animal feeding; fermented fish sauces; the production of fish peptone broth as a microbiological growth media; and as food binding and gelling agents

(continued)

Table 12.1 (continued)

Bioactive proteins and peptides	Fish and other foods are recognized to contain antihypertensive peptides known as angiotensin I-converting enzyme (ACE) inhibitors. These bioactive components are found in the muscle flesh of various species. Proteins play a structural role in the marine organisms. Proteins from marine sources including fish (cod, tuna, herring, rout, hake, pollock, and haddock), crustaceans, mollusks, extremophiles, and seaweeds have unique properties such as gel-formation, film and foaming capacity, antioxidant, anticoagulant, and antimicrobial activity. Moreover, bioactive peptides can be generated from proteins using hydrolysis (acid or alkaline), cooking, or fermentation. These peptides have a range of bioactivities including immunomodulatory, antithrombotic, and antihypertensive activity [12]

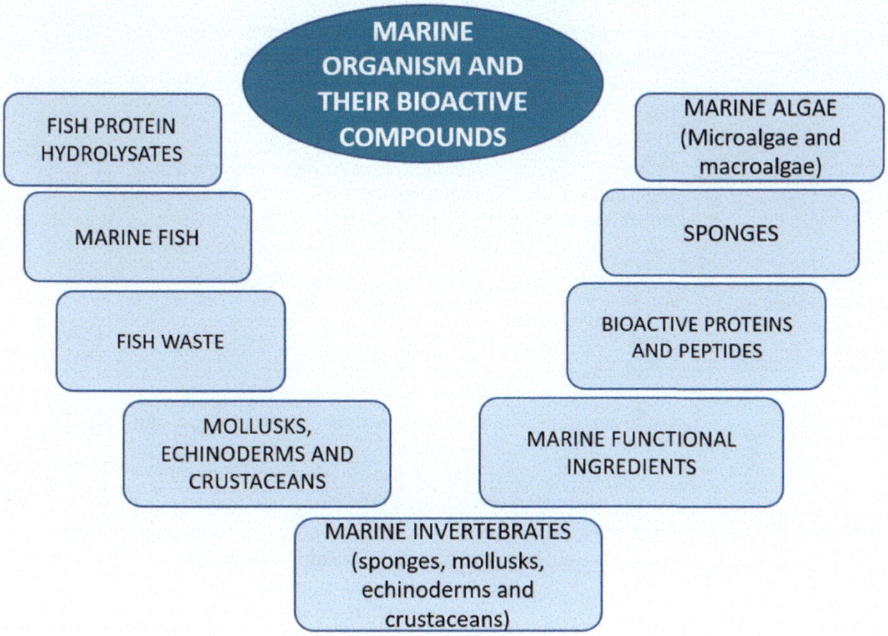

Fig. 12.1 Marine organism and their bioactive compounds

12.3.1 Fish Waste/By-Products Utilization

Following section summarizes the widely available marine-based nutraceuticals and recent research carried out for the purposes of isolation, identification, and characterization of marine-derived bioactive compounds with various therapeutic potentials (Table 12.2).

Table 12.2 Marine-based nutraceuticals and their therapeutic applications

Fish waste	Product	Uses	Reference
Scraps from 5 marine species (white croaker, horse mackerel, flying fish, chub mackerel, sardine)	Fish protein hydrolysate by enzymatic treatment	Used as a cryoprotectant to suppress the denaturation of proteins lizard fish surimi during frozen storage	[17]
Collagens containing yellow tail fish bone and swine skin wastes	Raw materials for production of protein hydrolysates and peptides	Potential use as food ingredients	[18]
Fish waste or by-catch	Enzymes and bioactive peptides	Used for fish silage, fish feed, or fish sauce production	[19]
Fish waste	Auto-hydrolysis of waste fish viscera to produce peptone hydrolysates	Used in microbiological media to support growth and bacteriocin production by lactic acid bacteria	[20]
Shrimp by-products	Preparation of high-quality salt-fermented sauces	Potential source of flavoring such as shrimp sauce	[21]
By-products of matjes (salted)	Production of fish oil	Stability to evaluate the quality characteristics	[20]
Fish collagen	Gelatin	Stabilizer	[22]
Shrimp and crab shell	Chitosan	Applications in cosmetic and pharmaceutical industries	[23]
Fish waste	Chitosan	Chitosan, with its positive charge, can be used for coagulation and recovery of proteinaceous materials present in such food processing operations	[24]
Fish waste	Chitosan	Nontoxic flocculent (in the treatment of organic polluted wastewater) chelating agent (removal of heavy and reactive metals from industrial waste water)	[25]
Fish waste	Chitosan	Food preservative: direct addition into food, in food packaging materials, film-forming ability with antimicrobial properties, wide spectrum antimicrobial activity against bacteria, yeast, and fungi, addition of chitosan sachets into packages	[26]
Fish waste	Fish oil	Source of unsaturated fatty acids (UFA), ozone-treated fish waste oil as a transportation diesel fuel	[27]

(continued)

Table 12.2 (continued)

Fish waste	Product	Uses	Reference
Fish scales, skin, and bone	Marine collagen	Used as a scaffold, carrier due to excellent bioactive properties such as biocompatibility, low antigenicity, high biodegradability, and cell growth potential	[28]
	Marine collagen	Effective scaffolds for biomedical applications and can be served as substitute for mammalian collagen in food, cosmetics, and biomedical materials	[29]

12.4 Recent Development on Marine Food

The consumption of aquatic food has globally increased in recent decades, as a result of better understanding of their health benefits and the good image of seafood among consumers. Recent advances in food technology have opened up a number of novel avenues to develop natural substances as a food or food ingredients [9]. Due to the wide range of their living environments, marine organisms have developed unique properties and bioactive compounds compared to terrestrial sources. The importance of functional food ingredients has been well-recognized in connection with health promotion, disease risk reduction, and reduction in health care costs. These include omega-3 fatty acids from fish, marine mammals and algal/fungal sources, chitosan, chitosan oligosaccharides and glucosamine, carotenoids, enzymes, and protein hydrolysates as well as algal phenolics and carbohydrates [30] (Fig. 12.2).

12.4.1 Bioactive Compounds

Bioactive compounds are molecules that can present therapeutic potential with influence on energy intake, while reducing pro-inflammatory state, oxidative stress, and metabolic disorders. They are present in small quantities in foods, mainly in fruits, vegetables, and whole grains, and provide health benefits beyond the basic nutritional value. Fish bioactive compounds are substances present in fish by-products with biological activity. These constituents are beneficial to human health [31]. Collagen, chitin, enzymes, gelatin, glycosaminoglycans, polyunsaturated fatty acids (PUFA), minerals, protein and peptides, and vitamins are among the most valuable compounds. The marine environment is a vast and relatively untapped source for new bioactive ingredients including polyunsaturated fatty acids, polyphenol, sterols, proteins, sulphated polysaccharides, antioxidants, and pigments [32]. Table 12.3 represents the various food ingredients from marine source and their various therapeutic benefits.

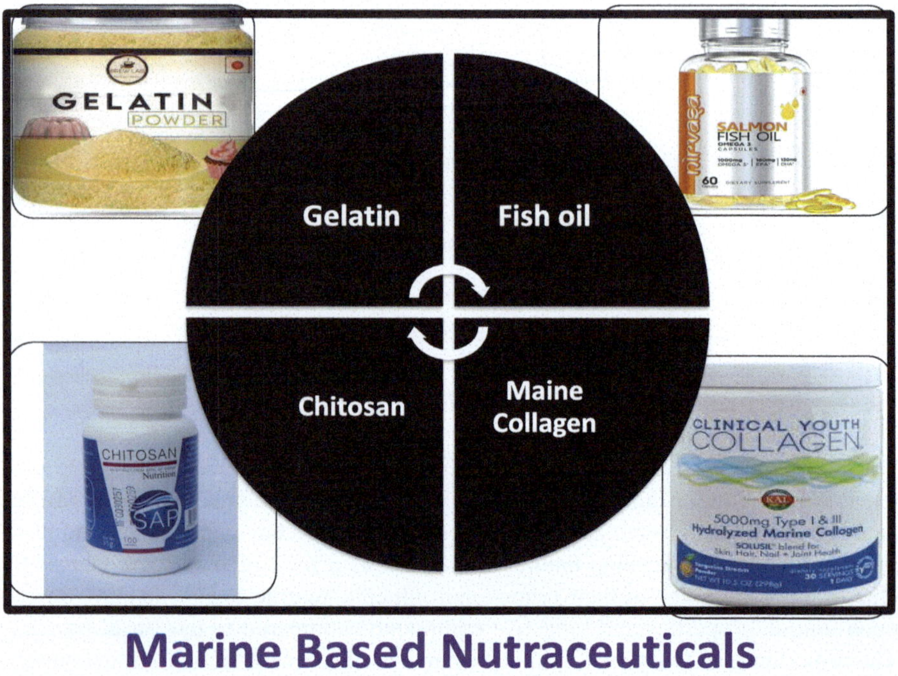

Fig. 12.2 Marine-based nutraceuticals

Marine-based bioactive food ingredients can be derived from a vast array of sources, including marine plants, microorganisms, and sponges, all of which contain their own unique set of biomolecules [33]. Marine bioactive peptides are reported to be involved in various biological functions such as antihypertension, immunomodulatory, antithrombotic, antioxidant, anticancer, and antimicrobial activities, in addition to nutrient utilization [10].

Several investigations have been carried out in laboratories around the world to extract compounds from marine processing industry for the final high-added-value products. Some of these compounds are collagen, omega 3 fatty acids, protein concentrates or chitin/chitosan, among others [34].

12.4.2 Biopolymer

Edible marine invertebrates from crustaceans, molluscs, and echinodermata can be considered as novel and organic-based biopolymers for various functional applications. These biopolymers are made of polysaccharide and protein, which are utilized by humans as food and nutraceutical ingredients for specific diseases in decades. In recent times, the potential application of biopolymers such as of glycosaminoglycans, fucosylated chondroitin sulfate, chitin, and chitosan is increasing in the biomedical industry for treating various ailments. These compounds showed

Table 12.3 Functional food ingredients from marine sources, health benefits and applications

Functional food ingredient	Source	Bioactive compound	Health benefits
Omega-3 fatty acids	Algae, fungi, fish (capelin sardine, Pacific anchovy, salmon mackerel, herring, freshwater fish, tuna, krill, menhaden, halibut, cod), seal blubber	DHA, EPA, DPA	Prevention and treatment of coronary artery disease, hypertension, diabetes, arthritis and other inflammatory, autoimmune disorder, cancer, essential for growth and development of brain and retina [35]
Carotenoids	Octopus, sea cucumber, sea bream, Atlantic herring, Pacific herring, obster, crab, crayfish, trout, salmon, redfish, red snapper, tuna, mussel, squid, shrimp	Fucoidan, agar, furcelleran, ascophyllan, laminarin, polyuronides, phlorotannins, glutathione, alginates, carrageenan	Antitumor, antidiabetic, antibacterial, anticoagulant, cardioprotective, antioxidant, anti-inflammatory, antiallergic [36]
Chitin/chitosan	Shrimp, crawfish, krill, clam/oyster, crab	Glucosamine, Heterochitosan/HeteroCOS, sulphated hetero-COS, LMW chitosan/LMW chitin, COS, HMWCOS	Antimicrobial, anti-inflammatory, antioxidant, anticarcinogenic, antinuclear, renal disease prevention and recovery, dietary fiber, weight reduction, reduce total and LDL cholesterol level [37]
Protein and peptides	Krill, herring, mussel, prawn, oyster, algae, Pacific hake, capelin, scad, mackerel, Alaska pollock, conger eel, yellow fin sole, sardine, salmon, nori, wakame	Bioactive peptides, enzymes, protein hydrolysates, collagen, gelatin, albumin	Anticoagulant, opioid, immunomodulatory, antibacterial, antithrombotic, and antihypertensive activity [38]
Phenolics/carbohydrates	Microalgae, macroalgae (seaweed)	Laminarin, fucoidan, agar, furcelleran, ascophyllan, glutathione, alginates, carrageenan, fucoidan	Anticoagulant, cardioprotective, antioxidant, anti-inflammatory, antiallergic, antitumor, antidiabetic, antibacterial [39]
Marine enzymes	Various fish body viscera like Atlantic cod, carp, harp seals, and tuna, crustaceans, mollusks and short-finned squid, Atlantic cod, seal, salmon, Red Sea bream, rainbow trout	Gastric proteases; pepsins, gastricsins and chymosins, serine and cysteine proteases, lipases, transglutaminase	Use in the fats and oils industry, cold renneting milk and fish feed digestion aid, meat tenderizing, curing of Herring [40]
Vitamins and minerals	All marine sources	Fat and water-soluble vitamins, iron, iodine, manganese, and zinc	Perform many essential functions in body [41]

EPA eicosapentaenoic acid, *DPA* docosapentaenoic acid, *DHA* docosahexaenoic acid, *LMW* low molecular weight, *HMW* high molecular weight, *COS* chitosan oligosaccharides specific functions as antimicrobial, anti-inflammatory, anticancer, wound healing, hypoglycemic actions, and also in animal feed [42].

12.4.3 Marine Enzymes

In the past decade, researchers have explored the marine life, namely, marine bacteria, actinomycetes, fungi, and several other marine microorganisms, to isolate a variety of enzymes having distinct activities. Some of these enzymes have already been used in industrial application. Due to the competition among organisms for space and nutrients in the marine and terrestrial environments, as a selective force, a complex network of reactions has led to evolving organisms and microorganisms to produce diverse enzyme systems. Marine organisms that have been studied include: shrimps, oysters, abalone, sea cucumber, fish, algae or seaweeds, archaea, actinomycetes, and fungi. These have demonstrated several applications in aquaculture, pharmaceuticals, nutraceuticals, cosmeceuticals, and biomaterials [43].

The marine enzymes including Alpha amylase, Protease, Xylanase, Ligninase, Cellulose, Lipase, Keratinase, Amylase, Beta-galactosidase, Esterase, Marino pyrroles have been used for many industrial, biotechnological and pharmaceutical applications. Carrageenase enzymes have been used in the biomedical field; marine enzymes have found use in bioethanol production, prevention of red algal bloom, and for obtaining algal protoplasts [44].

12.4.4 Antioxidants

Many synthetic antioxidants, such as butylated hydroxyanisole (BHA) and butylated hydroxytoluene (BHT), are used as food additives to prevent deterioration. But current trend shows the inclination of human population towards natural ingredients, due to which it has become necessary to find the substitute of synthetic antioxidant. Antioxidant peptides from marine sources found its potential application in food and other industries to retard the oxidation of food and pharmaceutical products [45].

The antioxidant activity of phycocyanin pigment extracted from marine filamentous cyanobacteria *Geitlerinema* sp. TRV57 was investigated. It was found that the phycocyanin pigment possesses antioxidant property and it can be used as a promising pharmaceutical and nutraceutical compound.

A new antioxidant from the marine sponge-derived bacterium *Streptomyces* sp. SBT345 was isolated. The results demonstrate that sponge-associated actinomycetes are rich sources for natural products with new pharmacological activities and relevance to drug discovery.

12.4.5 Marine Life as Source of New Drugs

Available treatments for many infectious diseases are limited. In particular, diseases caused by viral pathogens have demonstrated the need for new medicines, due to the increasing appearance of resistance to these available treatments. Thousands of novel compounds have been isolated from various marine organisms and tested for pharmacological properties, many of which are commercially available. The screening of natural products derived from marine species for antiviral activity has yielded a considerable number of active crude aqueous and organic solvent extracts. Today, over 40 compounds are commercially available in pharmacological markets, including alternative antiviral medicines or those being tested as potential antiviral drugs [46].

12.4.6 Antiviral Drugs

Emerging viruses have become threat to human population. Infectious diseases kill millions of people every year, and for many of these diseases, there is no treatment, cure, or vaccine. Meanwhile, antibiotics and other life-saving drugs used against many diseases are rapidly losing their effectiveness as bacteria and other microbes develop resistance to them [47]. In that sense, it is imperative to find new alternatives to combat microbes that have developed resistance to known drugs. Marine natural products could be an avenue to trigger the exploitation of new antiviral drugs to fight against several infectious diseases.

Marine polysaccharides are attracting increasing attention in medical and pharmaceutical development because of their important biological properties. The seaweed polysaccharides have now become a rich resource of potential antiviral drugs due to their antiviral activities against various viruses. The structural diversity and complexity of marine polysaccharides and their derivatives contribute to their antiviral activities in different phases of many different viral infection processes.

Squalene is applied as an adjuvant in vaccines against influenza and tuberculosis, as a moisturizer and emollient in cosmetic lotions [48], and as an antioxidant, immune system booster, and cholesterol level reducer in food industry. A marine-derived fungus-Eurotium rubrum, F33, was found to exhibit an inhibitory effect against H1N1 virus, on the basis of its effects on cell proliferation, cytotoxicity, and damage to host cells during viral invasion using CellTitre-Glo assay.

The study on the effect of antimalarial peptide and polyketide natural products from the Fijian marine cyanobacterium inferred that kakeromamide B bound several *Plasmodium* actin-like proteins and a sortilin protein, suggesting possible interference with parasite invasion of host cells.

12.4.7 Antimicrobial Properties

The presence of NH_3^+ groups, that is, the polycationic nature of the biopolymer, is mainly responsible for the antimicrobial action of chitosan and chitin. Majority of the microorganisms have negative charge on their surface, which react with the cationic groups, leading to leakage of cellular substances followed by cell death. The teichoic acid in Gram-positive bacteria and lipopolysaccharides in Gram-negative bacteria both play a major role in the interaction with chitosan [49]. The greater the number of positively charged amino groups in the structure of a biopolymer, the greater its antimicrobial activity. Chitosan has more positively charged amino groups as compared to that of chitin, thus, having higher antimicrobial activity. However, one major limitation of chitosan is its poor solubility in organic solvents, which restricts its applications.

12.4.8 Antituberculosis Activity

Marine invertebrates and microbes, particularly marine sponges and actinomycetes, appear to be among the most abundant reserves of marine natural products that are active against drug-resistant pathogens. A total of 1.5 million people died from TB in 2018 (including 251,000 people with HIV). Worldwide, TB is one of the top ten causes of death and the leading cause from a single infectious agent (above HIV/AIDS). Multidrug-resistant TB (MDR-TB) remains a public health crisis and a health security threat. WHO estimates that there were 484,000 new cases with resistance to rifampicin—the most effective first-line drug, of which 78% had MDR-TB [47]. The search for new antituberculosis agents has become increasingly important with the emergence of multidrug-resistant strains of *Mycobacterium tuberculosis*. Despite the small number of investigators looking at marine products as potential leads against *M. tuberculosis*, the research in this area has been productive [50]. Several antimycobacterial chemical compounds had been derived from marine natural products like sponge (*Pachypellina* sp., *Acanthella klethra*) and Okinawan soft coral *Litophyton viridis*, like Axisonitrile-3, Puupehenone, Pseudopteroxazole, Ergorgiaene, Ircinol A, Manzamine A H, and Litosterol.

12.4.9 Antihypertensive Activity

Hypertension is one of the most common causal factors of cardiovascular disease (CVD). Chitosan derivatives—COS, have proved to show antihypertensive effects. Renin plays a major role in the renin-angiotensin system (RAS) apart from ACE. Renin cleaves plasma angiotensinogen to angiotensin-I, then it is converted by ACE to angiotensin-II. Renin inhibition is a potential antihypertensive strategy. Deacetylated COS (90%) shows a higher renin inhibitory activity than 50% deacetylated COS [51].

12.4.10 Anti-allergy and Anti-inflammatory Activity

An antigen and the antigen-specific IgE interact with each other to induce allergies. It was found that COS had anti-inflammatory effects by virtue of regulating Th2 and pro-inflammatory cytokines, and thus, proved to be a promising candidate for the development of a potent therapeutic agent for the treatment of allergic asthma [52].

12.4.11 Anti-obesity and Antidiabetic Activity

In today's world, obesity is one of the prime reasons of concern. However, it is observed that slower incidence of such cases has been observed in populations where seafood is consumed. This is because seafood is the source of certain nutrients which can effectively fight against obesity. Chitosan and its derivatives are used in the treatment of obesity-related diseases. Chitosan has the ability to bind dietary lipids and bile acids by hydrophobic interactions, thereby inhibiting the activity of pancreatic lipase, thus reducing the absorption of intestinal fat in the gastrointestinal tract. The electrostatic attraction between positively charged amino groups of chitosan and negatively charged carboxyl groups of FA and bile salts is the cause of chitosan's hypolipidemic effects [53]. This is the mechanism which helps in the lowering of the level of lipids in the plasma. Oral intake of chitosan in the form of supplements can reduce the level of cholesterol and LDL in the plasma, and the level of triacylglycerol (TG) in the liver as well as plasma. Daily administration of chitosan in the drinking water (0.8%) reduced the levels of serum glucose, urine glucose, and serum triglycerides.

12.4.12 Anticancer and Antitumor Activity

Cancer is prevented by limiting the growth of the malignant cells. Chitosan shows anticancer effect. The stimulating effect on the immune system aids in the antitumor activity, which results in inhibition of angiogenesis and apoptosis from DNA fragmentation [54]. The biopolymer showed antiproliferative activity against the cancer cells. This also showed limiting action against breast cancer cells and reduces chances of recurrence of breast cancer (Fig. 12.3).

12.5 Role of Action of Marine Drugs Against Various Diseases

12.5.1 Arthritis

One of the mostly found diseased conditions mainly prevalent in old-age groups is arthritis. It is simply defined as an inflammation of the joints. Its treatment primarily involves prevention of the inflammation, thereby somewhat reducing pain and easing mobility. Diet is one of the most important factors regulating this disease.

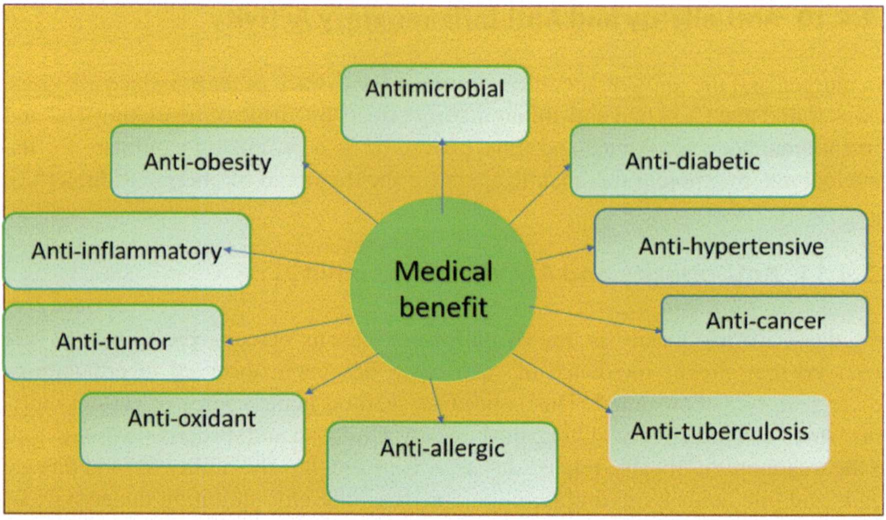

Fig. 12.3 Marine life as a source of drug

Marine sources play a major part in this dietary activity. PUFA obtained from marine sources have shown to act effectively in treating joint pain [55]. PUFA also helped in reducing pain in tender joints or morning stiffness. Researchers found evidences that the ingestion of collagen hydrolysates, obtained from fish waste, has also helped in reducing pain induced by osteoarthritis.

12.5.2 Asthma

It was recorded that children consuming more of fish products (mainly fish oil) have lesser chances of getting asthma as compared to those who rarely eat fish. The primary components involved in this role are omega-3 fatty acids. Supplementation of diet with omega-3 fatty acids helped in reduction of breathing difficulties and other symptoms, along with reduced drug doses required by asthma patients [56].

12.5.3 Diabetes

Diabetes is the causal factor of several other diseases, e.g., CVD, diabetic neuropathy, retinopathy, and nephropathy. The control and regulation of diabetes involves dietary management including both blood glucose and blood lipid concentrations [57]. Also offering protection against heart disease and cancer, some marine extracts have been associated with improvements in glycemic control. This calls for more extensive research of these food ingredients which can help in prevention of insulin resistance and diabetes.

12.5.4 Neuroinflammation

It has been observed that behind the pathology of neurodegeneration, there could possibly be an inflammatory component responsible for most of the neurodegenerative diseases like Alzheimer's, Parkinson's, and motor neuron disease. As a preventative measure, a diet rich in EPA and DHA is highly recommended. Population consuming fish or seafood in their diet which is highly enriched in omega-3 PUFAs, at least once a week, has reduced risks of developing dementia [58]. Evidence is also emerging which suggests that marine algae could possess therapeutic activities for combating neurodegenerative diseases associated with neuroinflammation. An increased intake of omega-3 fatty acids may increase the efficacy of anti-inflammatory medications and perhaps reduce the need for conventional drugs. This dietary choice could be achieved through the use of marine-sourced PUFAs as fortificants in margarine, dairy products (milk, yoghurt, cheese), breads and baked goods, meat-based and fish-based meals, for example.

12.5.5 Cognitive Decline and Depression

Dementia can be described as a restriction of an individual's ability to function independently. Currently, the drugs used in its treatment have a very limited therapeutic value, which calls for new strategies to slow down the progression of predementia syndromes. Consumption of fish and marine omega-3 PUFA is associated with a reduced risk and possibly delaying the onset of dementia. Researchers have observed that Chlorella aids in decreasing oxidative stress, thereby preventing the onset of dementia. DHA also helps in preventing neurodegenerative diseases as it is of particular importance for brain function as it maintains an optimal state of neural membranes, enabling membrane fluidity and thickness, which in turn affects cell signalling [59]. Several studies also suggest that the consumption of omega-3 fatty acids is inversely correlated to the prevalence and severity of depression, while clinical studies have presented evidence relating to the benefits of omega-3 compounds in the treatment of depressive disorders. Even, supplementation with long chain omega-3 PUFAs has been shown to improve mood in patients with major depression and bipolar disorder.

12.6 Future Prospects

Marine natural products (MNPs) represent a handsome percentage of the source of numerous anti-infective agents discovered in the past decade [60]. However, drug development from marine organisms faces few challenges. Different metabolites can be obtained from same organism at different times due to the varying environmental conditions. The main hurdle is the fact that the secondary metabolite to be used is present in the microorganisms residing in the marine animals and not in the host invertebrate. Lack of abundance of raw material supply hinders their smooth designing and purification. However, this hurdle can be met by the controlled use of natural resources and favoring the growth of marine organisms, besides employing more of rapid random screening. Another challenge is the biological and analytical evaluation of minor metabolites. Thus, in future, scientists will utilize technological advancements for better accessibility of minor metabolites and biological assay methods for the extraction and purification of fractions. Molecular methods like fluorescence in situ hybridization using group-specific rRNA can be used for identification of the types of bacteria residing in marine animals. They function by transferring biosynthetic gene clusters to a vector suitable for large-scale application.

Establishing aquaculture of the source organisms has seen noteworthy progress in cancer applications. We can see increasing synthetic interventions which depicts the modification of their integral structure for ultimately improving their biochemical efficacy. Another breakthrough is genome mining, that is, genome-guided method rather than traditional bioactivity-guided method for the search of 'talented microbes' associated with antimicrobial component production. More and more research is being carried out for the identification of novel compounds from marine organisms and their successful extraction and isolation and also assessment of safety and efficiency when exposed to large-scale commercial applications.

12.7 Conclusion

Marine ecosystem has diverse living organisms compared to terrestrial ecosystem. It has tremendous scope in the field of food and drug preparation to uncap many health applications. Marine life is a rich source of bioactive compounds which can help to cure several diseases including parasitic, viral, infectious, and noninfectious diseases. The advancements in biotechnology, extraction methods, and further research are the key factors to exploit the marine life as a source of food which can cure many ailments. Fortification of foods with nutraceuticals is the current trend in functional food industry. Marine-based food ingredients and nutraceuticals can be derived from marine sources. Infectious diseases are threat to human as well as animal life and it is evident from the recent Covid-19 pandemic. As infectious diseases evolve and develop resistance to existing pharmaceuticals, the marine environment provides novel leads against fungal, parasitic, bacterial, and viral diseases. Difficulty in isolation and purification procedures, toxicity associated with the active compound, ecological impacts on natural environment, and insufficient investments by pharmaceutical companies are some of the clear reasons behind market failures and a poor pipeline of marine natural products available to date. Development of novel technologies and collaborations between research institute and industry will be essential to ensure the future success of marine natural health food products. The pharmaceutical industry needs to come forward and engage their resources more in the field of marine food which will open new avenues in both food and pharmaceutical industry.

References

1. Newman DJ, Cragg GM (2016) Natural products as sources of new drugs from 1981 to 2014. J Nat Prod 79(3):629–661
2. Rahman H, Austin B, Mitchell WJ, Morris PC, Jamieson DJ, Adams DR et al (2010) Novel anti-infective compounds from marine bacteria. Mar Drugs 8(3):498–518
3. Ritchie H, Ellis G (2010) 'A system that works for the sea'? Exploring stakeholder engagement in marine spatial planning. J Environ Plan Manag 53(6):701–723
4. Laatsch H (2006) Marine bacterial metabolites. In: Proksch P, Muller WEG (eds) Frontiers in marine biotechnology. Horizon Bioscience, Norfolk, pp 225–288
5. Månsson M (2011) Discovery of bioactive natural products from marine bacteria. Technical University of Denmark, Kongens, Denmark
6. Donia M, Hamann MT (2003) Marine natural products and their potential applications as anti-infective agents. Lancet Infect Dis 3(6):338–348
7. Jiménez-Escrig A, Gómez-Ordóñez E, Rupérez P (2011) Seaweed as a source of novel nutraceuticals: sulfated polysaccharides and peptides. Adv Food Nutr Res 64:325–337
8. Hu GP, Yuan J, Sun L, She ZG, Wu JH, Lan XJ et al (2011) Statistical research on marine natural products based on data obtained between 1985 and 2008. Mar Drugs 9(4):514–525
9. Kim SK, Dewapriya P (2012) Bioactive compounds from marine sponges and their symbiotic microbes: a potential source of nutraceuticals. Adv Food Nutr Res 65:137–151
10. Kim SK, Pallela R (2012) Medicinal foods from marine animals: current status and prospects. Adv Food Nutr Res 65:1–9
11. Smith VJ, Brown JH, Hauton C (2003) Immunostimulation in crustaceans: does it really protect against infection? Fish Shellfish Immunol 15(1):71–90
12. Murray BA, FitzGerald RJ (2007) Angiotensin converting enzyme inhibitory peptides derived from food proteins: biochemistry, bioactivity and production. Curr Pharm Des 13(8):773–791
13. Suleria HAR, Osborne S, Masci P, Gobe G (2015) Marine-based nutraceuticals: an innovative trend in the food and supplement industries. Mar Drugs 13(10):6336–6351
14. DeFelice SL (2005) The nutraceutical health sector: a point of view. In: Regulation of functional foods and nutraceuticals: a global perspective. Blackwell Publishing, pp 201–212
15. Rajasekaran A, Sivagnanam G, Xavier R (2008) Nutraceuticals as therapeutic agents: a review. Res J Pharm Technol 1(4):328–340
16. Harris A (2010) Leading system transformation. School Leadership Manag 30(3):197–207
17. Nikoo M, Benjakul S, Rahmanifarah K (2016) Hydrolysates from marine sources as cryoprotective substances in seafoods and seafood products. Trends Food Sci Technol 57:40–51
18. Ahmed M, Verma AK, Patel R (2020) Collagen extraction and recent biological activities of collagen peptides derived from sea-food waste: a review. Sustain Chem Pharm 18:100315
19. Gildberg AR (2004) Enzymes and bioactive peptides from fish waste related to fish silage, fish feed and fish sauce production. J Aquat Food Product Technol 13(2):3–11
20. Jayathilakan K, Sultana K, Radhakrishna K, Bawa AS (2012) Utilization of byproducts and waste materials from meat, poultry and fish processing industries: a review. J Food Sci Technol 49:278–293
21. Pongsetkul J (2018) Properties and changes during fermentation of salted shrimp paste as affected by quality of row material and selected halophilic bacteria. Doctoral dissertation, Prince of Songkla University
22. Siburian WZ, Rochima E, Andriani Y, Praseptiangga D (2020) Fish gelatin (definition, manufacture, analysis of quality characteristics, and application): a review. Int J Fish Aquat Stud 8(4):90–95
23. Kumari S, Kishor R (2020) Chitin and chitosan: origin, properties, and applications. In: Handbook of chitin and chitosan. Elsevier, pp 1–33
24. Nouj N, Hafid N, El Alem N, Cretescu I (2021) Novel liquid chitosan-based biocoagulant for treatment optimization of fish processing wastewater from a Moroccan plant. Materials 14(23):7133

25. El-Gaayda J, Titchou FE, Oukhrib R, Yap PS, Liu T, Hamdani M, Akbour RA (2021) Natural flocculants for the treatment of wastewaters containing dyes or heavy metals: a state-of-the-art review. J Environ Chem Eng 9(5):106060
26. Kumar S, Mukherjee A, Dutta J (2020) Chitosan based nanocomposite films and coatings: emerging antimicrobial food packaging alternatives. Trends Food Sci Technol 97:196–209
27. Jeon JY, Han Y, Kim YW, Lee YW, Hong S, Hwang IT (2019) Feasibility of unsaturated fatty acid feedstocks as green alternatives in bio-oil refinery. Biofuels Bioprod Biorefin 13(3):690–722
28. Lim YS, Ok YJ, Hwang SY, Kwak JY, Yoon S (2019) Marine collagen as a promising biomaterial for biomedical applications. Mar Drugs 17(8):467
29. Subhan F, Ikram M, Shehzad A, Ghafoor A (2015) Marine collagen: an emerging player in biomedical applications. J Food Sci Technol 52:4703–4707
30. Shahidi F, Ambigaipalan P (2015) Phenolics and polyphenolics in foods, beverages and spices: antioxidant activity and health effects—a review. J Funct Foods 18:820–897
31. Awuchi CG, Chukwu CN, Iyiola AO, Noreen S, Morya S, Adeleye AO et al (2022) Bioactive compounds and therapeutics from fish: revisiting their suitability in functional foods to enhance human wellbeing. BioMed Res Int 2022:3661866
32. Senadheera TR, Hossain A, Shahidi F (2023) Marine bioactives and their application in the food industry: a review. Appl Sci 13(21):12088
33. Rasmussen RS, Morrissey MT (2007) Marine biotechnology for production of food ingredients. Adv Food Nutr Res 52:237–292
34. Al Khawli F, Martí-Quijal FJ, Ferrer E, Ruiz MJ, Berrada H, Gavahian M et al (2020) Aquaculture and its by-products as a source of nutrients and bioactive compounds. In: Advances in food and nutrition research, vol 92. Academic, pp 1–33
35. Shahidi F, Ambigaipalan P (2015) Novel functional food ingredients from marine sources. Curr Opin Food Sci 2:123–129
36. Kumar MS, Buttar HS, Dalal Y, Singh RB (2022) Cardioprotective and neuroprotective effects of nutraceuticals derived from marine origin. In: Functional foods and nutraceuticals in metabolic and non-communicable diseases. Academic, pp 707–723
37. Bastiaens L, Soetemans L, D'Hondt E, Elst K (2019) Sources of chitin and chitosan and their isolation. In: Chitin and chitosan: properties and applications. Wiley, pp 1–34
38. Maestri E, Pavlicevic M, Montorsi M, Marmiroli N (2019) Meta-analysis for correlating structure of bioactive peptides in foods of animal origin with regard to effect and stability. Compr Rev Food Sci Food Saf 18(1):3–30
39. Vidanarachchi JK, Kurukulasuriya MS, Wijesundara WMNM (2013) Biological and biomedical applications of marine nutraceuticals. In: Marine nutraceuticals: prospects and perspectives. CRC Press, Boca Raton, FL, pp 345–392
40. Jeyashakila R, Sivaraman B, Jeyasekaran G (2022) Enzymes from aquatic resources and their application in food and cosmetic industry. In: Advances in fisheries biotechnology. Springer Nature, Singapore, pp 471–494
41. Godswill AG, Somtochukwu IV, Ikechukwu AO, Kate EC (2020) Health benefits of micronutrients (vitamins and minerals) and their associated deficiency diseases: a systematic review. Int J Food Sci 3(1):1–32
42. Ganesan R, Manigandan S, Samuel MS, Shanmuganathan R, Brindhadevi K, Chi NTL et al (2020) A review on prospective production of biofuel from microalgae. Biotechnol Rep 27:e00509
43. Abdul BAA, Alijani S, Thanabal SD, Kim SK (2019) Marine enzymes production tools to the pharmaceutical industry. Indian J Geo Mar Sci 48(10):1656–1666
44. Chauhan PS, Saxena A (2016) Bacterial carrageenases: an overview of production and biotechnological applications. 3 Biotech 6(2):146
45. Olatunde OO, Benjakul S (2022) Antioxidants from crustaceans: a panacea for lipid oxidation in marine-based foods. Food Rev Intl 38(1):1–31
46. Yasuhara-Bell J, Lu Y (2010) Marine compounds and their antiviral activities. Antivir Res 86(3):231–240

47. World Health Organization (1998) WHO/HICARE symposium on radiological accidents and environmental epidemiology: a decade after the chernobyl accident: Proceedings, 24–25 August 1996, Hiroshima, Japan (No. WHO/EHG/98.06). World Health Organization
48. de la Fuente B, Tornos A, Príncep A, Lorenzo JM, Pateiro M, Berrada H et al (2020) Scaling-up processes: patents and commercial applications. In: Advances in food and nutrition research, vol 92. Academic, pp 187–223
49. Wang M, Li Z, Zhang Y, Li Y, Li N, Huang D, Xu B (2021) Interaction with teichoic acids contributes to highly effective antibacterial activity of graphene oxide on Gram-positive bacteria. J Hazard Mater 412:125333
50. Davies-Bolorunduro OF, Osuolale O, Saibu S, Adeleye IA, Aminah NS (2021) Bioprospecting marine actinomycetes for antileishmanial drugs: current perspectives and future prospects. Heliyon 7(8):e07710
51. Tao W, Wang G, Wei J (2021) The role of chitosan oligosaccharide in metabolic syndrome: a review of possible mechanisms. Mar Drugs 19(9):501
52. Hetland G, Tangen JM, Mahmood F, Mirlashari MR, Nissen-Meyer LSH, Nentwich I et al (2020) Antitumor, anti-inflammatory and antiallergic effects of Agaricus blazei mushroom extract and the related medicinal basidiomycetes mushrooms, Hericium erinaceus and Grifola frondosa: a review of preclinical and clinical studies. Nutrients 12(5):1339
53. Zhou DY, Wu ZX, Yin FW, Song S, Li A, Zhu BW, Yu LL (2021) Chitosan and derivatives: bioactivities and application in foods. Annu Rev Food Sci Technol 12:407–432
54. Pathak MP, Pathak K, Saikia R, Gogoi U, Ahmad MZ, Patowary P, Das A (2022) Immunomodulatory effect of mushrooms and their bioactive compounds in cancer: a comprehensive review. Biomed Pharmacother 149:112901
55. Tański W, Świątoniowska-Lonc N, Tabin M, Jankowska-Polańska B (2022) The relationship between fatty acids and the development, course and treatment of rheumatoid arthritis. Nutrients 14(5):1030
56. Sedighiyan M, Abdollahi H, Karimi E, Badeli M, Erfanian R, Raeesi S et al (2021) Omega-3 polyunsaturated fatty acids supplementation improve clinical symptoms in patients with Covid-19: a randomised clinical trial. Int J Clin Pract 75(12):e14854
57. Awuchi CG, Echeta CK, Igwe VS (2020) Diabetes and the nutrition and diets for its prevention and treatment: a systematic review and dietetic perspective. Health Sci Res 6(1):5–19
58. Nozaki S, Sawada N, Matsuoka YJ, Shikimoto R, Mimura M, Tsugane S (2021) Association between dietary fish and PUFA intake in midlife and dementia in later life: the JPHC Saku Mental Health Study. J Alzheimers Dis 79(3):1091–1104
59. Balakrishnan J, Kannan S, Govindasamy A (2021) Structured form of DHA prevents neurodegenerative disorders: a better insight into the pathophysiology and the mechanism of DHA transport to the brain. Nutr Res 85:119–134
60. Blunt JW, Copp BR, Hu WP, Munro MH, Northcote PT, Prinsep MR (2008) Marine natural products. Nat Prod Rep 25(1):35–94

Fueling the Future: Extraction and Production Technological Development of Marine Macroalgae-Seaweed Biodiesel

13

Mardiana Idayu Ahmad, Sami Alsaadi,
Mohammad Aliff Shakir, Fozy Binhweel, Wardah Senusi,
Ehsan Shalfoh, and H. P. S. Abdul Khalil

Abstract

In the contemporary era, the quest for sustainable, eco-friendly energy alternatives has become paramount against the backdrop of escalating environmental concerns and depleting fossil fuel reserves. Biodiesel emerges as a pivotal contender in this context, offering a renewable and cleaner energy alternative. This chapter delves into a potential solution by investigating the viability of third-generation seaweed biomass as a reliable and cost-effective feedstock for biodiesel production, amidst the prevalent challenges posed by first- and

M. I. Ahmad (✉)
Environmental Technology Division, School of Industrial Technology, Universiti Sains Malaysia, Penang, Malaysia

Renewable Biomass Transformation Cluster, School of Industrial Technology, Universiti Sains Malaysia, Penang, Malaysia
e-mail: mardianaidayu@usm.my

S. Alsaadi
Environmental Technology Division, School of Industrial Technology, Universiti Sains Malaysia, Penang, Malaysia

Faculty of Science, Biology Department, Hail University, Hail, Saudi Arabia

M. A. Shakir · F. Binhweel · W. Senusi · E. Shalfoh
Environmental Technology Division, School of Industrial Technology, Universiti Sains Malaysia, Penang, Malaysia

H. P. S. Abdul Khalil
Renewable Biomass Transformation Cluster, School of Industrial Technology, Universiti Sains Malaysia, Penang, Malaysia

Bioresource Technology Division, School of Industrial Technology, Universiti Sains Malaysia, Penang, Malaysia

M. Rafatullah et al. (eds.), *Multidisciplinary Applications of Marine Resources*,
https://doi.org/10.1007/978-981-97-5057-3_13

second-generation feedstocks, such as the food versus fuel debate and land-use dilemmas. A meticulous review of both conventional and advanced techniques for extracting lipids from seaweed and converting them into biodiesel is provided, offering a panoramic view of the current technological landscape in seaweed biodiesel production. While seaweed-derived biodiesel proffers economic benefits and favorable characteristics, it is not without its challenges, notably its lower lipid content compared to first- and second-generation feedstock sources, which may impose constraints on the biodiesel industry. To navigate this limitation, this chapter underscores the importance of expanded research to develop innovative methods that can enhance lipid extraction and biodiesel conversion efficiencies. Harnessing seaweed as a feedstock for biodiesel production unfurls a promising avenue toward sustainable energy resources, harboring the potential to make a significant positive impact on economic and social sustainability.

Keywords

Extraction · Production technology · Seaweed · Marine macroalgae · Biodiesel

13.1 Introduction

The inexorable surge in global population has precipitated a concomitant escalation in energy demand, casting a spotlight on the importance of identifying sustainable energy alternatives [1]. Biodiesel has emerged as a pivotal alternative to conventional diesel fuel, attributed to its renewable, biodegradable, and nontoxic properties, thereby presenting a viable pathway toward mitigating the environmental quandaries associated with fossil fuel consumption [2]. The utilization of fossil fuels, while indispensable in the current energy landscape, harbors a plethora of deleterious consequences. Notably, the prevailing rate of consumption portends the depletion of oil reserves by mid-century, while the combustion thereof significantly augments atmospheric CO_2 concentrations, thereby exacerbating global warming and climate change through the emission of greenhouse gases [3, 4]. Furthermore, with the global demand for primary energy sources projected to burgeon by 1.8% per annum from 2005 to 2035, the extant trajectory of energy utilization foreshadows an impending global energy crisis. Thus, the exploration and optimization of alternative energy sources, such as biodiesel, become paramount to navigate the intricate nexus of energy demand, environmental sustainability, and economic viability.

Biodiesel, renowned for its renewable and carbon-neutral attributes, emerges as a linchpin for both environmental protection and sustainable economic development [5]. This alternative fuel not only presents itself as a viable substitute for traditional fossil fuels due to its commendable characteristics such as reduced emissions and inherent renewability, but also champions the cause of sustainability. Biodiesel can be synthesized from a plethora of readily accessible renewable sources, including vegetable oil, animal fat, and microorganisms, thereby solidifying its stature as a promising and sustainable fuel alternative [6]. Moreover, biodiesel parallels petroleum diesel in terms of physicochemical properties, which facilitates its direct utilization in diesel engines without necessitating modifications to the engine structure [7]. While biodiesel has been substantiated to exert a positive environmental impact, owing to its diminished emissions during combustion and superior biodegradability compared to fossil fuels, the production costs associated with certain biofuels remain a point of contention. Despite offering a reduced environmental footprint, the elevated expenses incurred in biodiesel production, attributed to the diverse methodologies employed, pose a significant hurdle to its widespread adoption [8].

Biodiesel is a renewable fuel that can be sourced from a spectrum of feedstocks, each categorized into one of four generations, each with distinct characteristics and challenges. The first-generation feedstocks encompass food or food-derived crops, such as maize, barley, potato, sugar beet, palm oil, and soybeans oil, which are favored for their high carbohydrate or sugar content, thereby facilitating efficient biofuel production [9–11]. However, the utilization of these feedstocks has been associated with a myriad of socioeconomic and environmental dilemmas, including escalating demand and prices, depletion of natural resources, and soil degradation due to overfertilization [12]. In contrast, second-generation feedstocks, derived from nonedible plants, emerge as a promising alternative, offering several advantages such as adaptability to suboptimal agricultural land, low fertility and humidity requirements, and noncompetition with food supply chains [13]. Despite yielding biodiesel of comparable quality to edible feedstocks and presenting environmental and land rehabilitation benefits, nonedible oils are critiqued for their lower oil yield and higher alcohol consumption during biodiesel production [14].

Algae biomass is designated as the third-generation feedstock, and its genetically modified counterpart, identified as the fourth-generation feedstock, have garnered attention for their potential to enhance hydrogen-to-carbon yields and establish an artificial carbon sink to mitigate carbon emissions. In particular, algae are lauded for their myriad benefits, including widespread availability, stability against population growth-induced biomass fluctuation, significant oxygen production, and superior photosynthetic efficiency (6–8%) compared to terrestrial biomass (1.8–2.2%). Furthermore, algae exhibit low water consumption, noninterference with human food supply, adaptability to diverse environments, including saline and wastewater, and potential for high-value by-product production, all while achieving higher biomass production rates per unit of land area utilized [15].

Seaweed feedstock has emerged as a potential cornerstone for biodiesel production, attributed to a confluence of factors that underscore algal biomasses as promising sources of renewable energy. Notably, algal biomasses boast a high growth rate and exhibit robust photosynthetic activity, which, when coupled with the scalability of sea farming for mass cultivation, mitigate land demand and facilitate high biomass conversion rates. Additionally, the ease of handling and the potential for achieving net-zero CO_2 emissions further elevate the viability of seaweed feedstock in biodiesel production. The contribution of algae to sustainability development is multifaceted and can be amplified by various influences, with its photosynthetic impact on atmospheric CO_2 depletion being paramount. Algal systems necessitate carbon resources ranging from 10 to 50 times higher than other plants for photosynthesis and growth, exerting a pivotal impact on reducing greenhouse gas emissions and mitigating global warming. Furthermore, when juxtaposed with petroleum fuels and other biofuel sources, algae manifest a significant influence on urban land development, thereby contributing to a holistic approach toward environmental and urban sustainability [16].

The present chapter undertakes a meticulous review of recent works concerning the utilization of seaweed feedstock for biodiesel production, exploring its multifaceted environmental and economic advantages. A detailed exploration elucidates both solvent and advanced techniques for lipid extraction, accompanied by a nuanced discussion delineating their respective merits and limitations. Moreover, this chapter delves into an in-depth discourse on synthetic methods of biodiesel production technologies, encompassing transesterification, pyrolysis, and emulsification, providing a granular insight into their operational mechanisms and applicability. A critical analysis is also furnished, scrutinizing the economic feasibility of harnessing seaweed for biodiesel production, thereby providing a balanced perspective on its practicality and potential challenges. The review culminates with a future outlook, projecting the expectations and potential trajectories of seaweed-derived biodiesel. The findings distilled from this review converge on the conclusion that seaweed feedstock heralds promise as a sustainable source for biodiesel fuel production, offering a viable pathway towards cleaner and renewable energy solutions.

13.2 Seaweed

13.2.1 Overview

Seaweed is a macroalgae predominantly located at the seabed of shallow coastal waters, proliferates at depths up to 180 m, often anchoring itself to substrates such as rocks, black watershells, and other marine vegetation [17]. Beyond its ecological role, seaweed serves a pivotal function in marine aquaria, not only providing a primary food source for herbivorous inhabitants but also acting as a natural filtration system. It achieves this by reducing levels of total nitrogen and phosphate while simultaneously oxygenating the environment [18]. Compositionally, seaweed is a rich amalgam of proteins, lipids, and a substantial proportion of carbohydrates, predominantly in the form of polysaccharides. It is also a repository of minerals including potassium, chlorine, sodium, calcium, magnesium, sulfur, phosphorus, iodine, iron, copper, and manganese and is enriched with vitamins, phytohormones, and pigments [19]. Seaweeds are taxonomically categorized based on pigmentation into green algae (Chlorophyceae), red algae (Rhodophyceae), and brown algae (Phaeophyceae). Notably, in India, approximately 624 algal species have been identified, with green seaweed constituting around 72%, red algae 27%, and brown algae approximately 1% of the total [17].

13.2.2 Importance and Benefits

Seaweeds are pivotal in fixing CO_2 and releasing substantial amounts of oxygen through the photosynthetic process, form an integral component of the aquatic photosynthetic group, which also encompasses saltwater plants and phytoplankton. This group is attributed with 50–80% of the oxygen production on earth. When comparing the oxygen production of seaweeds to terrestrial plants, based on the quantity of organic carbon produced per square meter annually, seaweeds can generate between 2 and 14 kg of biomass. In contrast, terrestrial plants, such as trees and grasses in temperate regions, and microalgae typically produce approximately 1 kg. Moreover, seaweeds are estimated to absorb around 175 million tons of carbon, accounting for 10% of global vehicle emissions, and their photosynthesis contributes to an efficient nutrient cycle. The oxygen production through seaweed is vital for aquatic life, especially for more vulnerable fish species.

Beyond their environmental significance through photosynthesis and primary production, seaweeds play a crucial role in marine ecosystem food webs, enhancing the production of consumable marine life such as fish, abalone, and sea urchins. They support the development of herbivores and sustain the food chain [20]. Furthermore, humans rely on seaweed as a food source, with approximately 80% of harvested seaweed being utilized for human consumption. Seaweeds are not only consumed as sea vegetables and incorporated into various food preparations to augment nutritional profiles and flavor, but are also an essential component in the diet of numerous communities [21]. While in Eastern Asia, seaweeds are a staple, in other parts of the world their use may be confined to coastal populations or utilized minimally for purposes such as exotic dietary foods, nutritional supplements, low environmental footprint foods, and animal feed [22]. Despite its limited recognition in various global regions, seaweeds have recently found application as feedstock in the biofuel industry.

13.2.3 Seaweed as a Biomass for Biofuel

Biofuels derived from various biomass-based energy sources, hold a pivotal role in offering alternatives to fossil fuels, with seaweed emerging as a particularly promising candidate. Through the process of photosynthesis, biomass utilizes CO_2 for growth, establishing itself as a renewable resource. Notably, algae exhibit a more pronounced photosynthetic effect compared to other biomasses and can be cultivated in substantial quantities under controlled environmental conditions. Certain algae species boast a high lipid content, and it is estimated that one acre of land can produce approximately 15,000 gallons of oil per year from such sources on a global scale [23].

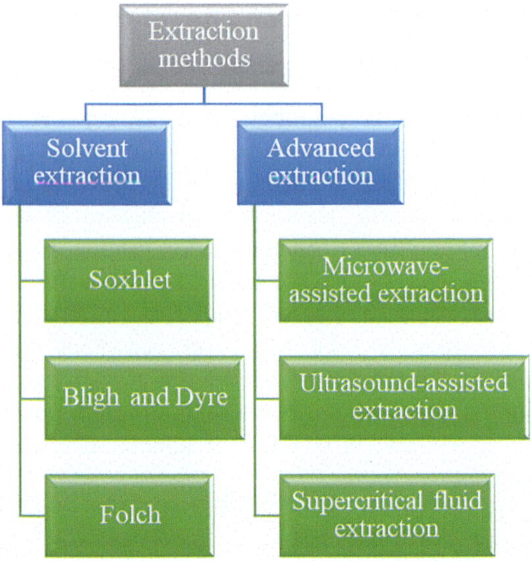

Fig. 13.1 Solvent and advanced extraction methods for seaweed lipids

13.3 Extraction of Seaweed Lipids

Lipid extraction methods hold paramount significance in the realm of biodiesel production, with oil extraction from organic materials being achievable through various conventional and advanced approaches. Traditional methods, such as mechanical extraction and solvent extraction exemplified by the Soxhlet, Folch, and Bligh and Dyer methods, have been predominantly utilized. Moreover, enzymatic extraction has also been a prevalent approach. In recent times, advanced techniques have emerged to extract oils from plant and animal-based feedstocks, including supercritical CO_2 extraction and the application of microwave and ultrasound waves [24]. The selection of an oil extraction technique is significantly influenced by the source of feedstock and the cost of biodiesel production, thereby forming a crucial component of the overall efficiency of the process [25]. Figure 13.1 illustrates the solvent and advanced methods employed to extract lipids from seaweed feedstock.

13.3.1 Solvent Extraction

Employed extensively in the derivation of oil for biodiesel production, solvent extraction stands out for its efficacy, retrieving almost the entirety of the oil and leaving a scant 0.5–0.7% of residual oil in the raw material. This technique involves the migration of a soluble fraction from a solid substrate into a liquid solvent. While it offers the advantage of relatively lower operational costs compared to supercritical fluid extraction, solvent extraction is not without its limitations. These include the suboptimal extraction of polar lipids, extended extraction durations, and the hazards related to the boiling of solvents [26].

13.3.1.1 Soxhlet Extraction Technique

The Soxhlet extraction technique was developed by Franz von Soxhlet in 1879, is a widely utilized laboratory method designed specifically for extracting lipids from solids [27]. It has garnered widespread use in extracting lipids from algae feedstocks due to its efficacy, particularly with dry algae biomass and conventional solvent [28]. The technique proves advantageous for obtaining desirable products with limited solubility in a solvent, especially when insoluble impurities are present. If the compound has high solubility in the solvent, separation from insoluble matter can be achieved through simple filtration. Soxhlet method involves the continuous passage of the solvent through the sample matrix via boiling and condensation, facilitating the transfer of compounds of interest from the solid to the solvent, given adequate contact time [29]. The extraction occurs at temperatures below the boiling point of solvent for prolonged periods, often not requiring harsh conditions. Selecting an appropriate solvent is pivotal to achieving optimal analyte recovery [30]. The polarity of solvent is a critical factor influencing extraction efficiency. The chosen solvent should possess the ability to dissolve desired analytes and have a polarity similar to these compounds, excluding other components of the solid matrix. Commonly used solvents include organic liquids such as acetone, hexane, ethanol, methanol, or ethyl acetate, either mixed in water or used in pure form [31]. The choice of solvent is a crucial parameter that affects the extraction procedure, alongside other factors like extraction time, temperature, sample ratio, and the characteristics of the sample submitted for extraction.

13.3.1.2 Bligh and Dyer Extraction

The Bligh and Dyer extraction technique, proposed by E. G. Bligh and W. J. Dyer in 1959 is primarily utilized to extract lipids from organic materials, offering rapid results. Initially applied to fish tissues, it has since been employed for various feedstocks, including algae- and plant-based tissues [32]. The Bligh and Dyer (B&D) method necessitates the homogenization of living tissues and other feedstocks, such as algae and plant tissues, with a solvent mixture. This mixture comprises chloroform, methanol, and water (H_2O) in a ratio of 2:2:1.8, respectively. Over time, the homogenized mixture separates into two layers: the first layer consists of lipid-chloroform components, while the second layer comprises nonlipid-methanolic components. To isolate the lipid, the first (chloroform) layer must be separated. Although the B&D method was originally developed for lipid extraction from fish tissue, it has evolved to be utilized for wet organic matter, such as fats, meat, and other food products, with modifications like the inclusion of organic solvents, ethanol, and ethyl acetate, in addition to chloroform and methanol [7]. The B&D method presents both advantages and disadvantages. On one hand, it is simple to use and separation can be concurrently achieved with lipid extraction. On the other hand, the method employs solvents with high toxicity, has limited substitutes, and can be costly [33].

13.3.1.3 Folch Extraction

The Folch extraction method established by Jordi Folch in 1957 is a solvent extraction technique utilized to extract lipids from various sources, including seaweeds [34]. This method necessitates the amalgamation of organic solvents, specifically chloroform and methanol, in a volumetric ratio of 2:1 and adheres to a 20:1 (v/w) ratio of solvent volume to sample weight. The dry, ground fat sample is homogenized with these mixed organic solvents. Subsequent to this, the addition of a saline solution or pure water facilitates the separation of lipid and non-lipid contents within the homogenized mixture. Over time, two distinct layers emerge: a non-lipid component layer and a lipid component layer, the latter containing trace, negligible amounts of lipids.

The Folch extraction method has been adapted to extract lipids from various feedstocks, such as food products, algal lipids, and animal tissues, occasionally employing alternative organic solvents while maintaining the original volumetric ratio. For instance, light petroleum and n-hexane can be utilized in a 1:1 (v/v) ratio. The method has been effectively applied to extract lipids from a diverse array of products, including meats, burgers, poultry, eggs, avocados, potatoes, pizzas, French fries, biscuits, cakes, mayonnaise, and ice cream. The findings underscore the Folch extraction technique as an efficient and straightforward tool for the quantitative extraction of lipids [7]. However, like other solvent methods, the Folch method presents both advantages and disadvantages. Notable merits include the capacity to handle multiple samples concurrently, speed, and gentle process completion. Conversely, a significant drawback is the use of toxic reagents, which pose potential adverse impacts on both the environment and human health [33].

13.3.2 Advanced Extraction

13.3.2.1 Supercritical Fluid Extraction

Supercritical fluid extraction (SFE) emerges as a pivotal technique for lipid extraction, applicable in both small-scale analytical research and large-scale industrial contexts. This method which predominantly employs CO_2 as the solvent, is particularly esteemed for its efficacy in lipid extraction processes. The fundamental principle of SFE hinges on the meticulous manipulation of pressure and temperature. Initially, a sample of dried, ground food is placed within the extraction chamber, whereupon the temperature and pressure are systematically elevated. Subsequently, supercritical CO_2 (SC-CO_2) is introduced into the heated, pressurized sample. Under these high-pressure conditions, CO_2 transitions into a liquid state, functioning as a solvent to extract lipids from the sample. Upon completion of the extraction, a reduction in temperature and pressure facilitates the reversion of CO_2 to its gaseous state, yielding pure lipids devoid of solvent fractions [35]. Despite the advanced classification and widespread endorsement of SFE, it is not without limitations. A notable constraint is its proficiency exclusively with nonpolar lipids, such as triglycerides, while polar lipids, namely, phospholipids and glycolipids, cannot be similarly extracted. This limitation is attributed to the inherent capacity of the supercritical technique to dissolve nonpolar lipids. To circumvent this, enhancements to the supercritical method have been proposed. Given that polar lipids are hydrophilic compounds and are not readily dissolved in SC-CO_2, the incorporation of polar cosolvents, such as methanol and ethanol has been suggested. However, it is imperative to consider the toxicity of these cosolvents. Alternatively, since polar lipids are hydrophilic, introducing minimal quantities of water can augment the extraction of polar lipids when SC-CO_2 is modified [7].

13.3.2.2 Microwave-Assisted Extraction

Microwave-assisted extraction (MAE) emerges as a noncontact method of extraction, leveraging microwaves to uniformly decompose target matter. This technique selectively interacts with polar molecules, such as water, inducing intracellular heating. The specific heat and pressure generated by the microwaves facilitate the rupture of the cell membrane, enabling lipid extraction. Notably, the incorporation of microwaves during solvent extraction has demonstrated a remarkable enhancement in lipid production from both wet and dry seaweed biomass, compared to conventional solvent extractions like the Folch and Bligh & Dyer methods. A study by [36] underscored the advantages of employing MAE for lipid extraction in comparison to other solvent extraction methods. The research, which utilized wet *Nannochloropsis* sp., revealed that lipid production with the Folch method was 8.8, the highest among solvent extraction methods. However, when MAE was employed, lipid production escalated to 12.8. MAE presents both merits and demerits. On the one hand, the technique is cost-effective, exhibits low solvent consumption, ensures rapid processing, achieves high oil recovery, and reduces CO_2 emissions [14]. Conversely, a primary limitation of MAE is that the polarity of the solvent can adversely impact the extraction process. Moreover, it is not suitable for heat-sensitive substances in terms of treatment [33].

13.3.2.3 Ultrasonic-Assisted Extraction

Ultrasound-assisted extraction (UAE) has been recognized as a pivotal technology in achieving the objectives of "green" chemistry [14]. This technique operates on the principle of generating acoustic waves within a liquid medium through ultrasonic exposure, thereby creating cycles of high and low pressure. During the high-pressure cycle, the vacuum bubbles formed in the low-pressure cycle are vigorously collapsed, a phenomenon known as cavitation. This process of cavitation generates high-pressure and intense liquid jets, which mechanically disrupt the cellular structure of the substrate, enhancing mass transfer and facilitating lipid extraction [36]. UAE has been validated as an effective technique for lipid recovery from various sources, including seaweeds, yeast, and fungi. Research has demonstrated that UAE achieves comparable levels of lipid extraction to conventional chloroform–methanol (2:1) extraction, but significantly reduces the extraction time from 8–12 h to a mere 15–20 min. Moreover, ultrasonication is frequently employed to concurrently extract oil and produce biodiesel from diverse feedstocks [14]. UAE presents both advantages and disadvantages in the context of lipid extraction. The merits of this method include low-temperature operation, minimal energy requirements, and the ability to perform high-rate extraction in a short timeframe. Conversely, the drawbacks of UAE involve the necessity to meticulously manage both the duration and intensity of ultrasound to avert potential adverse effects [33]. Table 13.1 reviews the literature regarding advanced extraction methods and provides a comparison with solvent methods.

13.4 Synthesis Technologies

Various synthesis technologies, each with distinct advantages and disadvantages are employed in the production of biodiesel, notably including pyrolysis, transesterification, and microemulsion methods. Microemulsion is recognized for its environmentally friendly attributes, notably its capacity to mitigate pollution and its ability to enhance the cetane number (CN) of the resultant biodiesel, with alcohol being instrumental in improving the CN. In contrast, pyrolysis, or thermal cracking, typically yields biodiesel characterized by a low CN, high viscosity, and volatility, which may pose challenges in specific applications. Among the three, transesterification stands out as particularly effective and reliable, attributed to its operational benefits of requiring lower temperatures and pressure, alongside a shorter processing time, and is further commended for its high efficiency and straightforward procedural steps. Consequently, it is often the preferred method in various biodiesel production contexts [39]. Figure 13.2 displays the syntheses methods for biodiesel fuel from seaweed feedstock.

13.4.1 Pyrolysis

Pyrolysis is a process that involves the transformation of a substance into another through heat application or catalyst ignition, utilizes thermal decomposition to convert organic materials into fuel in an oxygen-deprived environment, typically operating at temperatures between 1000 and 1300 °C [39, 40]. This chemical process, which breaks down a material into smaller molecules by cleaving its bonds, yields products similar to those produced during petroleum-diesel synthesis, exhibiting analogous combustion characteristics [19]. Various feedstocks, such as animal lipids, vegetable oils, natural fatty acids, or methyl esters of fatty acids are utilized in pyrolysis which can sometimes produce more products than the widely used transesterification reaction [39]. However, the economic viability of pyrolyzing organic feedstock for synthetic diesel production remains a challenge. Operator parameters categorize thermal cracking into three main types: fast pyrolysis (850–1250 K), flash pyrolysis (1050–1300 K), and conventional pyrolysis (550–900 K) [27]. While conventional pyrolysis involves a vapor residence time between 5 and 30 min, contributing to the total reaction time, flash pyrolysis necessitates a high heat transfer rate, finely ground materials, and brief vapor residence times (more than 2 s) [39]. Although the product of pyrolysis boasts favorable biodiesel properties, such as low viscosity, reduced sulfur and water, and a high cetane number, it tends to contain less charcoal and residual carbon than desired.

Table 13.1 A comparison between solvent and advanced extraction methods

Extraction technique		Overview	Limitation	Advantages	Improvements	References
Solvent extraction	Soxhlet	• Utilizes hot solvent across numerous cycles to extract lipids, followed by lipid separation	• Utilization of toxic solvents • Consuming time	• Simple process • Globally recognized and utilized	• Introduction of pressure variations • Incorporation of ultrasound and microwaves	[35]
	Folch	• Involves homogenizing the fat sample with a solvent mixture (20:1 v/v) and either pure or saline water to separate lipid and non-lipid layers	• Employing toxic solvents • Primarily formulated for fish tissues	• High result accuracy • An optimal tool for quantitative extraction	• Expansion to utilize various solvents for diverse food products	[34]
	Bligh& Dyer	• Engages in homogenizing the fat sample with a mixture solution (chloroform:methanol:water in ratios 2:2:1.8) until two distinct layers form	• Utilization of toxic solvents	• Simplistic technique • Enables solvent recovery	• Inclusion of additional organic solvents to accommodate a wide array of organic materials	[37]
Advanced extraction	Supercritical fluid extraction	• Engages liquid CO_2, transformed under high pressure, which acts as a solvent to extract lipids at varying temperatures	• Limited to nonpolar lipids • Diminished extraction for high-moisture samples	• Eco-friendly technique • Efficient and contemporary • Safeguarded technology • Time efficient	• Employing cosolvents, e.g., methanol, for polar lipid extraction	[35]
	Ultrasound-assisted extraction	• Subjects a mixture of lipids and solvents to ultrasound vibrations at a specified frequency and duration	• Utilization of toxic solvents	• Preserves fatty acids • Minimizes extraction time	• Adjustments in ultrasound duration and frequencies	[7, 35]
	Microwave-assisted extraction	• Involves the intermittent and brief agitation of homogenizing solvents and fat sample via microwave waves	• Potential oxidation of lipids due to excessive waves	• Rapid extraction • Applicable in a humid environment	• Modifications in wave intensity and duration	[7, 38]

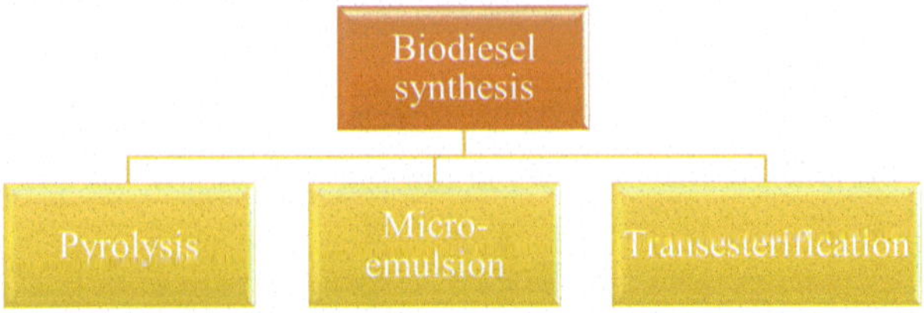

Fig. 13.2 Methods of biodiesel synthesis

13.4.2 Micro-emulsion

Micro-emulsion emerges as a viable method to address the issue of high viscosity inherent in crude natural oil. Defined as a colloidal equilibrium dispersion, a micro-emulsion forms spontaneously, creating optically isotropic fluid microstructures with dimensions typically ranging from 1 to 150 nm. This is achieved by combining two normally immiscible liquids and one or more ionic or nonionic amphiphiles, which usually act as an oil phase, an aqueous phase, and a surfactant, respectively [41]. Biodiesel fuels often undergo treatment with micro-emulsions to mitigate their high viscosity and low fluidity, among other drawbacks. In the creation of micro-emulsion biodiesel fuel, oils are amalgamated with emulsifying compounds, such as alcohols predominantly methanol, ethanol, propanol, or butanol. However, the utilization of microemulsions presents notable challenges, such as the formation of carbon deposits in the engine and insufficient fuel consumption. While the microemulsion method facilitates a straightforward and direct reduction in biodiesel viscosity, its prolonged use for producing combustion engine fuel can introduce issues, including elevated carbon content, incomplete fuel burning, and increased lubricant viscosity [42].

13.4.3 Transesterification

Transesterification is a widely recognized and prevalently utilized on a commercial scale for biodiesel production, pertains to a pivotal category of reactions where an alkoxyl group is transposed, converting one ester type into another. Specifically, in the context of biodiesel production, the transesterification of triacylglycerols (TAGs) is underlined. This reaction also dubbed 'esterification of free fatty acids (FFA)' and 'alcoholysis' due to the cleavage of alcohol during the process, involves reacting vegetable and seaweed oil feedstock with a short-chain alcohol (commonly methanol, though ethanol is also used) to yield biodiesel and glycerin (or glycerol). Methanol, owing to its cost-effectiveness, polarity, and short molecular chain is the predominant alcohol used for transesterification. When methanol is utilized, the resultant biodiesel is termed Fatty Acid Methyl Esters (FAMEs). While the reaction can naturally take a full day to complete, the incorporation of catalysts significantly accelerates the reaction rate and enhances product yield. Given the reversible nature of the reaction, elevating the product yield can also be achieved by shifting the reaction equilibrium through the addition of excess methanol [43]. The transesterification process necessitates 1 mol of triacylglycerol (from vegetable oil or algal lipid) and 3 mol of an alcohol (such as methanol or ethanol), yielding 3 mol of fatty acid alkyl esters (biodiesel) and 1 mol of glycerol through a series of consecutive reversible reactions. Post-reaction, the products segregate into three layers: glycerol at the bottom, biodiesel at the top, and a soapy mixture in between [44]. Catalysts, which can be categorized as heterogeneous, homogeneous, or biocatalysts, play a pivotal role in transesterification reactions by enhancing the reaction rate and thereby increasing product production [45].

13.4.3.1 Homogeneous Catalyst

Homogeneous catalysts, typically employed in transesterification processes are favored for their straightforward reaction mechanics and can be broadly categorized into acidic and basic types, with H_2SO_4, NaOH, and KOH being exemplary catalysts. These catalysts can yield a high biodiesel output in a relatively short time frame. However, the utilization of homogeneous catalysts in transesterification presents a significant challenge concerning the quality of the produced biodiesel. Their application on a large scale for biodiesel production is constrained due to the elevated production costs, attributed to the multiple washing and purification steps necessary to achieve the requisite quality [45].

13.4.3.1.1 Homogeneous Base-Catalyst

Homogeneous base-catalysts, particularly KOH and NaOH are prevalent in biodiesel synthesis due to their ability to yield substantial methyl ester amounts under relatively mild conditions 60 °C and 1 atm for approximately 1 h. Notably, base-catalyst reactions exhibit a reaction rate that is about 4000 times faster than their acid-catalyst counterparts, attributed to a favorable reaction mechanism where the carbonyl group in triglyceride, diglyceride, or monoglyceride is readily susceptible to nucleophilic attack from a catalytically active species, RO [45]. Other pivotal catalysts encompass sodium isopropoxide, potassium methoxide, and sodium ethoxide, with CH_3OK and CH_3ONa being considered particularly apt catalysts for transesterification reactions due to their dissolution into CH_3O^-, K^+, and Na^+. Previous study reported on exploring the basic catalytic activity of these catalysts using sunflower oil as a feedstock [46]. Despite the impressive catalytic activity of base-catalyzed homogeneous catalysts, they present notable challenges, such as the difficulty in catalyst recovery and the generation of excessive wastewater during separation and neutralization processes, complicating disposal and elevating biodiesel production costs, thereby impacting profitability. Additionally, minimal levels of water and free fatty acids are imperative to avoid potential contamination and saponification, which could diminish product yield [47].

13.4.3.1.2 Homogenous Acid Catalyst

Acid-catalyzed transesterification is less commonly employed than its base-catalyzed counterpart, primarily due to its inherently slow reaction rate and the requisite high methanol-to-oil ratio. Moreover, acid catalysts exhibit reduced activity when exposed to elevated temperatures during the transesterification process.

Table 13.2 The advantages and disadvantages of conventional biodiesel production

Technologies	Advantages	Disadvantages	References
Pyrolysis	• Exhibits satisfactory characteristics • The by-product, syngas, possesses utility	• Expensive • Entails complex processes and equipment	[49]
Micro-emulsion	• Does not produce by-products • The emission is low	• Generates substantial carbon residue • Combustion is not fully efficient	[49]
Transesterification	• Widely recognized and utilized synthesis method • By-products have practical applications	• Pretreatments are necessitated in instances of high FFA • Residual glycerol in biodiesel can produce highly hazardous chemicals	[7]

Notably, acid catalysts can directly produce biodiesel from economical lipid feedstocks that contain more than 6% free fatty acids (FFA). Liquid acid catalysts, such as sulfuric acid, demonstrate a robust tolerance to oils with high FFA content, facilitating concurrent esterification and transesterification and thus producing a relatively high yield of esters. The mechanism of acid-catalyzed transesterification involves the protonation of the carbonyl group in esters, which, following the nucleophilic attack of alcohol, generates carbon cations and forms a tetrahedral intermediate. This intermediate then regenerates the catalyst. Importantly, acid-catalyzed transesterification can proceed in the absence of water [48]. Table 13.2 delineates the advantages and disadvantages of various biodiesel synthesis methods.

13.5 Economic Feasibility

Seaweed-derived biodiesel, with its potential to mitigate energy and environmental challenges, presents a promising economic prospect. Seaweed is a rapidly proliferating renewable resource, minimizes competition with food crops by eliminating the need for freshwater and arable land for cultivation [50, 51]. Its high biomass yield and carbon dioxide storage capability during growth render it an attractive feedstock for biodiesel production. Moreover, seaweed cultivation can invigorate local economies by generating employment opportunities in coastal regions [52]. The economic feasibility of seaweed biodiesel is also contingent upon factors such as governmental support, oil prices, and consumer demand. Fluctuations in fossil fuel costs can influence the competitiveness of biodiesels as oil prices escalate, biodiesels become economically appealing. Implementing supportive strategies, such as subsidies, tax incentives, and carbon pricing systems, can significantly enhance the economic viability of seaweed biodiesel industries, thereby fostering adoption and investment. Furthermore, with the global emphasis on reducing carbon emissions intensifying, a market for seaweed biofuel may materialize, driven by the demand for eco-friendly alternatives to fossil fuels [53]. Consequently, persistent research and development are crucial to refine manufacturing processes, augment energy conversion efficiency, and curtail production expenses, enabling the industry to realize its maximal economic potential.

13.6 Conclusion

Navigating through the multifaceted domain of biodiesel production, this review underscores the pivotal role of exploring alternative, sustainable feedstocks, particularly spotlighting the potential harbored by third-generation seaweed biomass. While biodiesel derived from edible oils presents palpable sustainability and ethical dilemmas, seaweed biomass emerges as a promising contender, offering a blend of economic viability and environmental prudence. Through a meticulous exploration of various oil extraction and biodiesel conversion technologies, both conventional and advanced, this review illuminates the myriad of methodologies available, each with its unique advantages and limitations. Notably, while seaweed-derived biodiesel presents a panoply of advantages, such as cost-effectiveness and favorable physicochemical properties, the lower lipid content relative to other sources necessitates innovative technological advancements to optimize production. The imperative for further research is evident, particularly focusing on enhancing extraction efficiencies and developing technologies that can adeptly navigate the challenges presented by the lipid content of seaweed. In a world grappling with the exigencies of climate change and sustainable development, seaweed biodiesel not only presents a viable alternative fuel, but also opens avenues for bolstering economic and social sustainability, particularly in coastal regions. Thus, the insights gleaned from this review not only pave the way for future research endeavors, but also beckon towards a future where energy production harmoniously coexists with ecological and social well-being.

Acknowledgments The authors would like to thank the International Centre for Genetic Engineering and Biotechnology Biotechnology, ICGEB Grant Number of 304.PTEKIND.6501382.I151 (R504-LR-GAL008-0006501382-I151) for funding this study.

References

1. Shakir MA, Ahmad MI, Yusup Y, Rafatullah M (2023) From waste to wealth: converting rubber wood sawdust into green mycelium-based composite. Biomass Convers Biorefinery. https://doi.org/10.1007/s13399-023-05113-9
2. Babadi AA et al (2022) Emerging technologies for biodiesel production: processes, challenges, and opportunities. Biomass Bioenergy 163:106521. https://doi.org/10.1016/j.biombioe.2022.106521

3. Jeliani ZZ, Fazelian N, Yousefzadi M (2021) Introduction of macroalgae as a source of biodiesel in Iran: analysis of total lipid content, fatty acid and biodiesel indices. J Mar Biol Assoc UK 101(3):527–534. https://doi.org/10.1017/S0025315421000382
4. Shakir MA et al (2023) Review on the influencing factors towards improving properties of composite insulation panel made of natural waste fibers for building application. J Ind Text 53:1–33. https://doi.org/10.1177/15280837231151440
5. Singh D, Sharma D, Soni SL, Sharma S, Kumari D (2019) Chemical compositions, properties, and standards for different generation biodiesels: a review. Fuel 253:60–71. https://doi.org/10.1016/j.fuel.2019.04.174
6. Vasi K et al (2020) Biodiesel production using solid acid catalysts. Catalysts 10(2):237
7. Binhweel F, Hossain MS, Ahmad MI (2023) Recent trends, potentials, and challenges of biodiesel production from discarded animal fats: a comprehensive review. Bioenergy Res 16(2):778–800. https://doi.org/10.1007/s12155-022-10527-w
8. Suhud Shote A (2019) Biofuel: an environmental friendly fuel. In: Anaerobic digestion. IntechOpen, pp 1–13. https://doi.org/10.5772/intechopen.82856
9. Choudhary M, Singh A, Gupta M, Rakshit S (2020) Enabling technologies for utilization of maize as a bioenergy feedstock. Biofuels Bioprod Biorefining 14(2):402–416. https://doi.org/10.1002/bbb.2060
10. Dey S, Reang NM, Das PK, Deb M (2021) A comprehensive study on prospects of economy, environment, and efficiency of palm oil biodiesel as a renewable fuel. J Clean Prod 286:124981. https://doi.org/10.1016/j.jclepro.2020.124981
11. Xie W, Wan F (2019) Guanidine post-functionalized crystalline ZIF-90 frameworks as a promising recyclable catalyst for the production of biodiesel via soybean oil transesterification. Energy Convers Manag 198:111922. https://doi.org/10.1016/j.enconman.2019.111922
12. Shakir MA et al (2023) Sandwich composite panel from spent mushroom substrate fiber and empty fruit bunch fiber for potential green thermal insulation. Buildings 13(1):224. https://doi.org/10.3390/buildings13010224
13. Shakir MA, Azahari B, Yusup Y, Yhaya MF, Salehabadi A, Ahmad MI (2020) Preparation and characterization of mycelium as a bio-matrix in fabrication of bio-composite. J Adv Res Fluid Mech Therm Sci 65(2):253–263
14. Pikula K et al (2020) The advances and limitations in biodiesel production: feedstocks, oil extraction methods, production, and environmental life cycle assessment. Green Chem Lett Rev 13(4):11–30. https://doi.org/10.1080/17518253.2020.1829099
15. del Río PG, Gomes-Dias JS, Rocha CMR, Romaní A, Garrote G, Domingues L (2020) Recent trends on seaweed fractionation for liquid biofuels production. Bioresour Technol 299:122613. https://doi.org/10.1016/j.biortech.2019.122613
16. Ghedini E, Taghavi S, Menegazzo F, Signoretto M (2021) A review on the efficient catalysts for algae transesterification to biodiesel. Sustainability 13(18):10479. https://doi.org/10.3390/su131810479
17. Nakhate P, van der Meer Y (2021) A systematic review on seaweed functionality: a sustainable bio-based material. Sustainability 13(11):6174. https://doi.org/10.3390/su13116174
18. Wang S et al (2020) A state-of-the-art review on dual purpose seaweeds utilization for wastewater treatment and crude bio-oil production. Energy Convers Manag 222:113253. https://doi.org/10.1016/j.enconman.2020.113253
19. Singh D, Sharma D, Soni SL, Sharma S, Kumar Sharma P, Jhalani A (2020) A review on feedstocks, production processes, and yield for different generations of biodiesel. Fuel 262:116553. https://doi.org/10.1016/j.fuel.2019.116553
20. Farghali M, Mohamed IMA, Osman AI, Rooney DW (2023) Seaweed for climate mitigation, wastewater treatment, bioenergy, bioplastic, biochar, food, pharmaceuticals, and cosmetics: a review, vol 21(1). Springer International Publishing. https://doi.org/10.1007/s10311-022-01520-y
21. Zhao Y, Bourgougnon N, Lanoisellé JL, Lendormi T (2022) Biofuel production from seaweeds: a comprehensive review. Energies 15(24):9395. https://doi.org/10.3390/en15249395

22. Murakami S et al (2021) The edible brown seaweed Sargassum horneri (Turner) C. Agardh ameliorates high-fat diet-induced obesity, diabetes, and hepatic steatosis in mice. Nutrients 13(2):551. https://doi.org/10.3390/nu13020551
23. Aravind S, Barik D, Ashok N (2022) Optimization of oil yield from the macro algae spirogyra by solvent extraction process using RSM and ANN. Int J Photoenergy. https://doi.org/10.1155/2022/3690635
24. Shakir MA, Yhaya MF, Ahmad MI (2017) Physical effects of pre-hydrolysis and alkali treatment of empty fruit bunch (EFB) fibers. Int J Chem Eng Appl 8(6):387–391. https://doi.org/10.18178/ijcea.2017.8.6.689
25. Bhuiya MMK, Rasul MG, Khan MMK, Ashwath N, Azad AK (2016) Prospects of 2nd generation biodiesel as a sustainable fuel—Part 1: Selection of feedstocks, oil extraction techniques and conversion technologies. Renew Sust Energ Rev 55:1109–1128
26. Abdullah N, Amran NA, Yasin NHM (2017) Algae oil extraction from freshwater microalgae Chlorella vulgaris. Malaysian J Anal Sci 21(3):735–744
27. Karmakar B, Halder G (2019) Progress and future of biodiesel synthesis: advancements in oil extraction and conversion technologies. Energy Convers Manag 182:307–339. https://doi.org/10.1016/j.enconman.2018.12.066
28. Binhweel F, Pyar H, Senusi W, Shaah MA, Hossain MS, Ahmad MI (2023) Utilization of marine ulva lactuca seaweed and freshwater azolla filiculoides macroalgae feedstocks toward biodiesel production: kinetics, thermodynamics, and optimization studies. Renew Energy 205:717–730. https://doi.org/10.1016/j.renene.2023.01.114
29. Shakir MA, Azahari B, Salehabadi A, Yusup Y, Yhaya MF, Ahmad MI (2020) Structural, thermal, and mechanical properties of spent mushroom substrate (SMS) and rubberwood sawdust (RWS) binderless particleboards. J Adv Res Fluid Mech Therm Sci 75(2):113–124. https://doi.org/10.37934/arfmts.75.2.113124
30. Quitério E, Grosso C, Ferraz R, Delerue-Matos C, Soares C (2022) A critical comparison of the advanced extraction techniques applied to obtain health-promoting compounds from seaweeds. Mar Drugs 20(11):677
31. Mussatto SI (2015) Generating biomedical polyphenolic compounds from spent coffee or silverskin. In: Coffee in health and disease prevention. Elsevier, pp 93–106
32. Eg B (1959) A rapid method of total lipid extraction and purification. Can J Biochem Physiol 39:911–917
33. Nadu T (2023) Different methods of lipid. J Sci Res Teaching 2(7):33–43
34. Folch J, Lees M, Sloane Stanley GH (1957) A simple method for the isolation and purification of total lipids from animal tissues. J Biol Chem 226(1):497–509
35. Hewavitharana GG, Perera DN, Navaratne SB, Wickramasinghe I (2020) Extraction methods of fat from food samples and preparation of fatty acid methyl esters for gas chromatography: a review. Arab J Chem 13(8):6865–6875
36. Sati H, Mitra M, Mishra S, and Baredar P (2019) Microalgal lipid extraction strategies for biodiesel production: A review. Algal Res 38:101413
37. Tiwari BK (2015) Ultrasound: a clean, green extraction technology. TrAC Trends Anal Chem 71:100–109
38. Bligh EG, Dyer WJ (1959) A rapid method of total lipid extraction and purification. Can J Biochem Physiol 37(8):911–917
39. Costa DSV, Bragagnolo N (2017) Development and validation of a novel microwave assisted extraction method for fish lipids. Eur J Lipid Sci Technol 119(3):1600108
40. Neupane D (2023) Biofuels from renewable sources, a potential option for biodiesel production. Bioengineering 10(1):29. https://doi.org/10.3390/bioengineering10010029
41. Shakir MA, Yhaya MF, Ahmad MI (2017) The effect of crosslinking fibers with polyvinyl alcohol using citric acid. Imp J Interdiscip Res 3(4):387–391
42. Supriyanto E, Sentanuhady J, Dwiputra A, Permana A, Muflikhun MA (2021) The recent progress of natural sources and manufacturing process of biodiesel: a review. Sustainability 13(10):5599

43. Vilas Bôas RN, Mendes MF (2022) A review of biodiesel production from non-edible raw materials using the transesterification process with a focus on influence of feedstock composition and free fatty acids. J Chil Chem Soc 67(1):5433–5444
44. Dhivya Priya N, Thirumarimurugan M (2020) Biodiesel—a review on recent advancements in production. In: Bioresource utilization and bioprocess. Springer, Singapore, pp 117–129
45. Akram F et al (2022) Current trends in biodiesel production technologies and future progressions: a possible displacement of the petro-diesel. J Clean Prod 370:133479. https://doi.org/10.1016/j.jclepro.2022.133479
46. Nayab R et al (2022) Sustainable biodiesel production via catalytic and non-catalytic transesterification of feedstock materials—a review. Fuel 328:125254. https://doi.org/10.1016/j.fuel.2022.125254
47. Atadashi IM, Aroua MK, Aziz ARA, Sulaiman NMN (2013) The effects of catalysts in biodiesel production: a review. J Ind Eng Chem 19(1):14–26
48. Xie W, Han Y, Wang H (2018) Magnetic Fe_3O_4/MCM-41 composite-supported sodium silicate as heterogeneous catalysts for biodiesel production. Renew Energy 125:675–681
49. Adipah S (2019) Biodiesel preparation from Allanblackia floribunda seed oil. J Environ Sci Public Health 3(1):25–33
50. Tabatabaei M et al (2019) Reactor technologies for biodiesel production and processing: a review. Prog Energy Combust Sci 74:239–303. https://doi.org/10.1016/j.pecs.2019.06.001
51. Shakir MA et al (2023) Effect of fibrillation degree of empty fruit bunch fiber on sandwich composite incorporate with spent mushroom substrate. J Ind Text 53:152808372311536. https://doi.org/10.1177/15280837231153661
52. Milledge JJ, Smith B, Dyer PW, Harvey P (2014) Macroalgae-derived biofuel: a review of methods of energy extraction from seaweed biomass. Energies 7(11):7194–7222
53. Hossain MS et al (2021) Seaweeds farming for sustainable development goals and blue economy in Bangladesh. Mar Policy 128:104469
54. Yong WTL, Thien VY, Rupert R, Rodrigues KF (2022) Seaweed: a potential climate change solution. Renew Sust Energ Rev 159:112222

Utilization of Marine Green Algae for the Development of Bioplastics

14

Mohammad Aliff Shakir, Mardiana Idayu Ahmad, and H. P. S. Abdul Khalil

Abstract

Embarking on a journey through the confluence of sustainability and material science, this chapter unveils the innovative extraction and formulation of biopolymers from marine green algae, a promising and abundant feedstock for bioplastic production. Addressing the limitations of other biomass sources, marine green algae emerge as a biocompatible alternative to conventional plastics, offering solutions to mitigate plastic pollution. This exploration delves into various extraction methods including isolation of biopolymers and both direct and indirect production techniques, the physicochemical properties, applications, and environmental impacts of algae-derived bioplastics. Furthermore, it acknowledges the current production and usage challenges, providing insights into their prospective role in shaping a sustainable future. Through a comprehensive examination of these aspects, this chapter underscores the transformative potential of

M. A. Shakir
Environmental Technology Division, School of Industrial Technology, Universiti Sains Malaysia, Penang, Malaysia

M. I. Ahmad (✉)
Environmental Technology Division, School of Industrial Technology, Universiti Sains Malaysia, Penang, Malaysia

Renewable Biomass Transformation Cluster, School of Industrial Technology, Universiti Sains Malaysia, Penang, Malaysia
e-mail: mardianaidayu@usm.my

H. P. S. Abdul Khalil
Renewable Biomass Transformation Cluster, School of Industrial Technology, Universiti Sains Malaysia, Penang, Malaysia

Bioresource Technology Division, School of Industrial Technology, Universiti Sains Malaysia, Penang, Malaysia

M. Rafatullah et al. (eds.), *Multidisciplinary Applications of Marine Resources*,
https://doi.org/10.1007/978-981-97-5057-3_14

marine green algae-derived bioplastics in advancing sustainable technologies and contributing to global environmental conservation efforts.

Keywords

Marine · Bioplastic · Algae · Film · Green packaging

14.1 Introduction

In contemporary society, plastic has emerged as an indispensable synthetic material, lauded for its durability, flexibility, and resistance to a variety of environmental factors such as water, heat, and electricity, with its production process being notably efficient [1]. Primarily sourced from petroleum, these plastics including familiar types such as polyethylene (PE), polypropylene (PP), and polyethylene terephthalate (PET) are categorized as fossil based. Despite their chemical stability and slow biodegradability, they largely contribute to the escalating problem of nonbiodegradable waste. The environmental implications of plastic waste, impacting land, air, freshwater bodies, and oceans, have become a major focus in environmental studies [2]. Notably, it was revealed that European aquatic environments receive between 307 and 925 million liters of waste annually, predominantly plastic, accounting for about 82% of this pollution [3]. This waste is largely composed of single-use items such as bottles, packaging materials, and bags. Projections suggest that by 2025, the environment could be laden with up to 11 billion tons of plastic, exacerbating ecological concerns. The breakdown of these plastics into smaller particles, ranging from macro- to nanosizes, poses a significant threat to human health, as these particles can infiltrate the food chain, underscoring the urgent need for effective waste management and pollution control strategies [4].

Recent research in sustainable materials has increasingly focused on bioplastics, a promising direction for reducing environmental impact. These bioplastics, derived from renewable sources such as agricultural, algal, and bacterial polymers, currently represent a small yet significant portion of the plastic industry, accounting for about 1% of the over 368 million tons produced annually [5]. As illustrated in Fig. 14.1, the number of publications on bioplastic production has accelerated significantly in the past two decades, reflecting growing scientific and industrial interest in this area. Despite the potential of bioplastics, their production is still in a nascent stage with a global output of 2.11 million tons, of which a little over half is biodegradable. The development of biodegradable plastics faces a major challenge due to their hygroscopic nature, which compromises structural integrity by absorbing moisture. Therefore, enhancing various properties of bioplastics, such as durability, rigidity, heat resistance, and cost-effectiveness, remains a critical focus for researchers [6]. Currently, the primary application of bioplastics is in the packaging industry, which constitutes 52% of the market share. With optimistic projections suggesting an increase in production capacity from 2.11 million tons in 2020 to 2.87 million tons by 2025, the biodegradable segment of bioplastics is expected to see

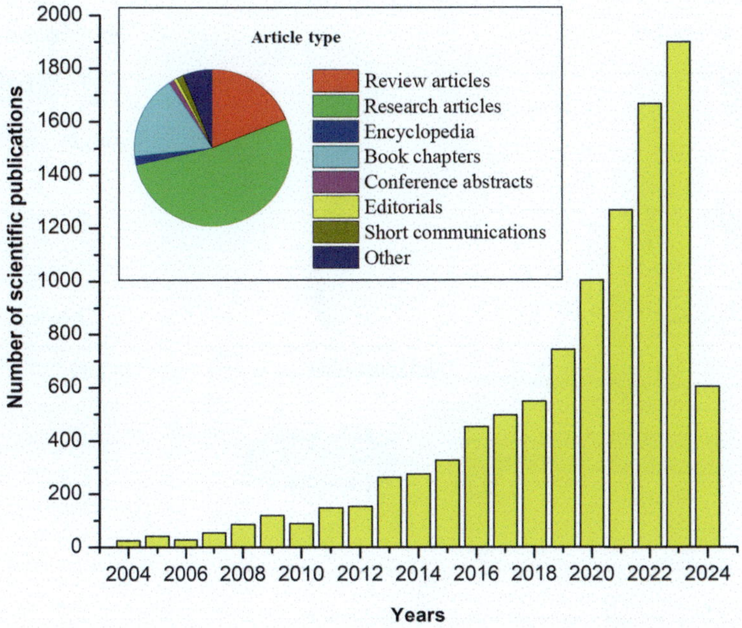

Fig. 14.1 Accelerated number of publications about production bioplastic in the past two decades

significant growth, potentially positioning it as a strong competitor to traditional fossil-fuel-based plastics [7].

14.2 Bioplastic

Bioplastics representing an environmentally friendly shift in material science are synthesized from covalently bonded monomeric units which are predominantly sourced from biological entities like bacteria and microalgae. Integral to these sustainable polymers are cellular components such as proteins, lipids, and polyhydroxyalkanoates (PHA), which are essential in the formation of bio-based degradable plastics [8]. Beyond microbial sources, Fig. 14.2 illustrates the classification of sources for bioplastic production into different generations: the first generation from agricultural products like corn and sugarcane, the second generation from byproducts like sawdust, and the third generation from microorganisms such as algae and bacteria. This categorization helps in understanding the evolution and diversification of raw materials used in bioplastic production. However, the process of converting plant-based materials, particularly those with intricate structures like bamboo into bioplastics, often necessitates additional energy for efficient biomass conversion. The cultivation of biomass for bioplastic production is not without challenges. Factors such as the prolonged growth period of certain plants and susceptibility to environmental and societal changes highlight the need for strategic resource

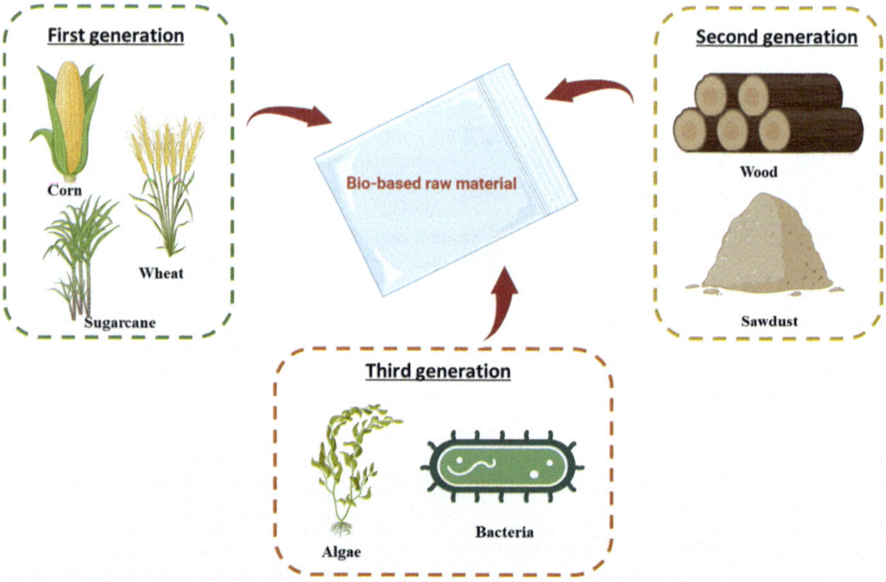

Fig. 14.2 Generational sources of bioplastic production

management. The sustainability of biomass as a raw material is dependent on a balance between its consumption rate and natural replenishment. Therefore, selecting rapidly renewable and agriculturally nondisruptive biomass sources is pivotal in the burgeoning bioplastics industry. As demand for bioplastics continues to escalate, establishing a sustainable and strategic approach to biomass sourcing is increasingly vital to the growth of industry and ecological impact.

The burgeoning field of bioplastic production has recently turned its focus towards the algae-based bioeconomy, identifying algal biomass as a crucial third-generation resource that skillfully avoids the use of arable land. Microalgae, in particular, have garnered attention for their rapid growth and potential as a sustainable alternative to traditional food-based bioplastics. Their versatility in bioplastic production is remarkable as they can be utilized both directly in their raw biomass form and indirectly through extraction processes to obtain essential bioplastic components. Various innovative techniques, such as hot molding, compression with polymers, melt mixing, and solvent casting, are employed to convert microalgae into practical bioplastic materials. The unique composition of algae is low in lignin and rich in long-chain hydrocarbons lead to facilitate the economical extraction of high-quality cellulose that are essential for bioplastic manufacturing [9]. Furthermore, the rapid duplication rate and ability of microalgae to thrive in diverse, often polluted environments such as CO_2 rich gases or nutrient-rich wastewater has underscore their potential as an underutilized yet promising renewable resource. This adaptability of microalgae not only makes them a valuable asset in bioplastic production but also positions them as a key player in addressing the intertwined challenges of water, food, and energy conservation in a sustainable manner.

14.3 Marine Green Algae: An Overview

Marine algae are broadly categorized into three distinct types based on their pigmentation: green algae (*Chlorophyta*), brown algae (*Phaeophyta*), and red algae (*Porphyra*), each group characterized by its unique coloration [10]. Among these, marine green algae such as seaweeds stand out as promising feedstocks for bioplastic production due to their rapid and sustainable growth and lack of competition for land. These algae are advantageous for their high biomass yield, cost-effectiveness, and adaptability to diverse natural environments, enabling year-round cultivation. Historically used in food technology, biotechnology, microbiology, and medicine, the role of marine green algae in the plastic industry is a recent development [11]. They offer an eco-friendly and nontoxic alternative for bioplastics, with properties akin to traditional plastics in terms of strength and chemical resistance. The unique characteristics of marine green algae make them an innovative and viable source for bioplastics, marking their significance in current environmental conservation efforts.

Green algae are typically classified as "green" due to the presence of chlorophyll a and b pigments, which predominantly absorb red and blue wavelengths of light, reflecting green light and thus giving them their characteristic color. These algae are characterized by five main pigments: lutein, zeaxanthin, violaxanthin, neoxanthin, and β-carotene [12]. Additionally, green algae are noted for their high carbohydrate content, making them a valuable source of various polysaccharides such as cellulose, sulfated galactans, sulfuric acid polysaccharides, and xylans. The *Ulva* genus is one of the marine green algae, widely known for causing the disruptive "green tides" phenomena, adversely impacting marine ecosystems and tourism sectors. Figure 14.3 shows the appearance of *Ulva rigida* and light photographs of cells from the middle surface of the thallus of the alga.

Despite this, Ulva biomass is finding utility in the production of compost, fertilizers, and biofuels, capitalizing on its rich nutritional profile that includes proteins, dietary fibers, and complex polysaccharides. Past research has identified Ulva seaweeds as being rich in cellulose and hemicellulose, crucial for forming networks suitable for sustainable packaging [13]. Although these cellulosic components are typically insoluble in water, recent advancements have demonstrated the feasibility of film formation from the water-soluble fraction of *Ulva rigida*. This development signifies a notable progression in utilizing *Ulva rigida* for creating biodegradable films and packaging materials [14]. The conversion of Ulva materials into environmentally friendly products not only aligns with sustainable development goals but also elevates the value of this abundant yet problematic biomaterial, turning an ecological challenge into a resource for sustainable innovation.

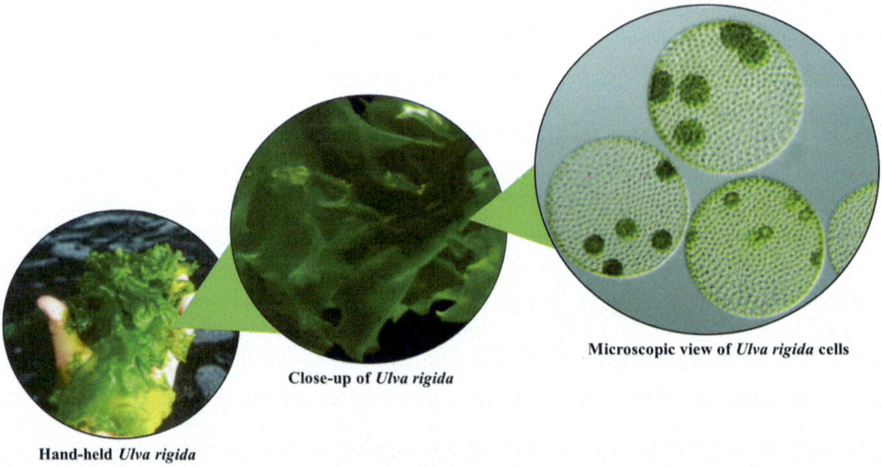

Fig. 14.3 *Ulva rigida* structure and cellular detail

14.3.1 Challenges Using Other Biomass for Bioplastic Production and Advantages of Marine Green Algae

A diverse array of renewable feedstocks, including starch, cellulose, fruit waste, seeds, and oils, are being utilized in the production of bioplastics, often combined with various plasticizers for different applications. However, the use of these terrestrial-based biomasses for bioplastics poses significant challenges due to their intensive requirements for land, freshwater, and energy, along with substantial transportation and processing expenses. In this context, marine green algae stand out as an efficient alternative, offering a more sustainable solution by requiring less land and freshwater, thereby mitigating some of the major constraints associated with land-based biomass [15]. Despite this, the processing of marine green algae involves costs and environmental impacts, particularly in terms of CO_2 emissions during transportation and processing stages. The natural propensity of marine green algae to thrive in varied marine environments without the need for added nutrients or fertile land marks it as a sustainable option [16]. However, land-based algae cultivation is typically more profitable for high-value products. This is less for bioplastic production, which underscoring the economic benefits of marine-based cultivation. The versatility of algae is capable of growing in both marine and freshwater environments has enhanced its scalability for bioplastic production. Establishing bioplastic industries near coastal regions could further reduce transportation costs and bolster the blue bioeconomy. Moreover, the high carbohydrate content in algae accentuates its potential as a valuable resource for bioplastic manufacturing, making it a promising candidate for sustainable and cost-effective bioplastic production [17].

14.4 Extraction of Biopolymers from Marine Green Algae

Algae widely present in our ecosystems serve as a valuable resource for the production of biopolymers, benefiting from easy extraction methods and abundant availability. Recent innovations in biopolymer extraction from algae have led to the development of advanced methods such as microwave-assisted, ultrasound-assisted, and subcritical water-assisted extraction. These methods have proven to be more efficient than traditional solvent extraction, with microwave and ultrasound techniques notably increasing biopolymer yields by up to 33% and 5% more than conventional methods [18–20]. The inherent richness of algae in natural starch and cellulose polymers has spurred significant advancements in biopolymer technology, utilizing various extraction methods. The extensive photosynthetic capacity of algae makes them an effective feedstock for producing diverse biopolymers. Through fermentation, algal biomass can be transformed into different polymeric classes, including polyesters, polyamides, and polyolefins [21]. Notably, the natural polymers extracted from algae are characterized by low toxicity and exhibit beneficial properties like antimicrobial, antitumor, antioxidant, and immunomodulatory activities. Additionally, algal biomass is being utilized to create biodegradable composites that exhibit robust mechanical properties, including strong adhesion and tensile strength. This versatility and efficiency make algae an increasingly important resource in the field of sustainable biopolymer production.

14.4.1 Isolation of Biopolymers

Solvent extraction is an effective method for producing biopolymers from algal biomass, offering a simpler alternative with less need for downstream processing compared to fermentation methods. Although this process is efficient, it heavily depends on mechanical operations such as sifting, filtration, and centrifugation, and its optimization relies on the careful adjustment of both chemical and physical parameters to maximize biopolymer yield [22]. To further streamline this method, selecting algal biomass based on productivity can significantly reduce the necessity for extensive experimentation, leading to time savings and more efficient production.

In addition, microwave extraction has emerged as a novel and eco-friendly approach to recover valuable products from biomass. Particularly useful for extracting biopolymers from microalgal cells, microwave-assisted extraction benefits from the effects of microwave radiation on ions and dipoles. This method is characterized by its rapidity, uniformity, compactness, and reduced experimental times, alongside low solvent consumption and no energy requirements. Enhancing the process through microwave-assisted water extraction, especially with increased temperatures, can lead to higher yields and stronger biopolymers, marking a significant advancement in the field of biopolymer extraction from algae [23].

Ultrasound-assisted extraction utilizes the cavitation phenomenon generated by ultrasound waves, which creates turbulence that agitates microparticles within algal biomass, leading to the disruption of algae cell walls. This process effectively converts ultrasound energy into vibrational energy, enhancing the transfer rate and facilitating the efficient removal of biopolymers from the algae [24]. A significant advantage of this method over conventional techniques is the substantial reduction in extraction time, shortening the process from hours to mere minutes. Additionally, it allows for extraction at room temperatures without affecting the yield, setting it apart from traditional methods [25]. This technique also eliminates the need for membrane separation processes, thereby minimizing material losses and supporting environmentally friendly practices. Remarkably, the extraction yield achieved through ultrasound-assisted methods is 5% higher than that of conventional methods with a notable fourfold decrease in extraction time. Key parameters influencing the extraction process include sonication time, temperature, and ultrasound frequency, where increased sonication and frequency lead to improved biopolymer isolation from algae biomass [26]. This highlights the effectiveness and efficiency of the ultrasound approach in biopolymer extraction from algal sources.

Subcritical water extraction is an innovative and emerging technique for extracting bioactive compounds from biomass, utilizing water that is pressurized below its critical pressure (less than 22.12 MPa) and heated above boiling point, up to 647.14 K [27]. This method is increasingly being recognized for its efficiency in isolating biopolymers from algae biomass as it offers a more environmentally friendly alternative to traditional extraction methods. A significant advantage of subcritical water extraction lies in its use of water as the solvent, which eliminates the need for harmful chemicals, thereby promoting a greener extraction process. This technique is not only advantageous for its high product quality and yield but also for its shorter reaction times and lower energy consumption. Enhancing this process, the addition of ionic liquid catalysts has been shown to improve the dissolution of biopolymers, increasing the efficiency of their removal. The benefits of using ionic liquids include their extensive dissolving range, high thermal stability, negligible vapor pressure, and robust chemical stability [28]. Further advancements have been made by combining deep eutectic solvents with subcritical water extraction, a method that has shown great effectiveness and yield in the removal of biopolymers from algae [29]. These innovations in extraction technology highlight the potential for more sustainable and efficient processes in biopolymer production from algae.

14.4.2 Biopolymers from Algae

Polylactic acid (PLA) stands out as a highly adaptable biopolymer, primarily derived from ecoconscious and renewable resources from algae underscoring its role in sustainable manufacturing. The chemical transformation of PLA from a low-molecular weight polymer into a more functional material is achieved through ring-opening polymerization, specifically involving cyclic lactide, which significantly broadens its application scope [30]. Enhancing versatility of PLA can be blended with various biodegradable polyesters. This approach will improve its physical properties and lead to the development of more efficient and environmentally friendly bioplastic composites [31]. An important distinction in PLA composites lies between biodegradable aliphatic polyesters and nonbiodegradable aromatic polyesters, which is key in understanding their environmental impact and degradation profiles. Additionally, the incorporation of stereo-specific forms of lactic acid, such as L-lactic acid and D-lactic acid, in PLA production highlights the significance of chemical configuration in tailoring materials for diverse industrial applications, ranging from packaging to biomedical fields [32]. This multifaceted approach to PLA production and application reflects a growing trend towards sustainability and efficiency in material science.

The production of PHA is currently confronted with economic challenges due to the reliance on costly traditional substrates, underscoring the urgent need to explore more affordable and renewable resources [33]. The use of agri/food waste as a raw material for PHA production presents a unique challenge, given its complex chemical structure, which makes direct conversion processes difficult and necessitates innovative approaches. Key to overcoming these hurdles is the advancement in genetic engineering, aimed at enhancing the efficiency of microorganisms to convert complex waste into valuable PHAs, offering a sustainable and economically feasible route [34]. Additionally, research involving *Haloferax mediterranei* has shed light on the potential of using macroalgae-derived polysaccharides as an alternative carbon source which significantly improves the efficiency of PHA production [35]. Inherent characteristics such as high biodegradability and recyclability make PHAs an environmentally favorable alternative to conventional polymers. However, the current commercial production of PHAs is impeded by the high costs of raw materials, emphasizing the need for industrial-scale solutions that are both cost-effective and efficient. The ongoing research and industry optimism regarding the potential of algae as a future resource for PHA production reflect a broader commitment to developing sustainable biopolymers that can meet market demands while adhering to environmental sustainability.

The synthesis of polyhydroxybutyrates (PHBs) by bacteria as reserve materials under stress conditions showcases a natural mechanism for producing these significant macromolecules [36]. Recent advancements in metabolic engineering applied to cyanobacteria mark a notable leap in biotechnology, enhancing PHB production efficiency through the incorporation of essential biosynthesis genes [37]. This development positions PHBs as an emerging form of biodegradable plastic, highlighting a shift towards sustainable materials that align with ecological needs due to their rapid degradation in natural environments. Looking to the future, the role of PHBs in various industries as a substitute for traditional petroleum-based plastics indicates a potential paradigm shift in material science and emphasizes the importance of biopolymers in sustainable development. Genomic research in algae, particularly in species such as *Chlamydomonas reinhardtii*, *Cyanidio schyzonmerolae*, *Nannochloropsis* sp., *Ostreococcus tauri*, *Phaeodactylum tricornutum*, and *Thalassiosira pseudonana* are paving new paths for the synthesis of PHBs, reflecting a growing interest in exploiting algal biology for biopolymer production [38]. The investigation of diverse algal species for the integration of the PHB pathway demonstrates an expanding field of biotechnological applications, likely leading to more efficient and varied sources of biodegradable plastics.

14.5 Bioplastic Production and Fabrication Methods

Bioplastic productions from algae is still limited due to lack of knowledge, process optimization, cost of production, and raw material availability, but developed countries are promoting bioplastic production from various renewable resources. Presently, injection molding, compression molding, film casting, blow molding, blown film extrusion, thermoforming, and fiber spinning are the different bioplastic molding methods used. Bioplastic can be produced from direct method, in which micro- and macroalgal inclusions (melted whole biomass) are used as base materials [39]. Alternatively, fermentation of pretreated algal biomass using bacteria/archaea to produce bioplastic could be considered as indirect method for bioplastic [40]. However, both required detailed process flow and techno-economic analysis to attain the sustainable as well as low-cost bioplastic production.

14.5.1 Direct Method of Bioplastic Production Using Algae

Utilizing algal biomass as a base material in bioplastic production represents a holistic approach, incorporating carbohydrates, proteins, fats, and polymers to maximize resource efficiency. This method involves blending the algal biomass with both petroleum-based plastics and other bioplastics, creating a unique hybrid bioplastic [41]. In the literature, various reports on algal biomass blended with petroleum plastics including PE, PP, PVC, poly(butylene succinate) (PBS), and poly(butylene adipate-co-terephthalate) (PBAT) are available [42–44]. This innovative approach not only showcases the potential for reducing petroleum reliance but also allows for customization of the properties of material by adjusting the algae-to-plastic ratio. Advanced microscopic techniques like scanning electron microscopy and Fourier Transform Infrared Spectroscopy have been pivotal in confirming successful encapsulation and chemical bonding of algae within these bioplastic matrices [15, 45]. The direct conversion of algal biomass into bioplastics offers considerable advantages such as notably reduced processing costs and a lower environmental impact compared to conventional plant biomass methods.

Emphasizing recyclability, this production method aligns with contemporary environmental goals and promotes a sustainable cycle of use and reuse. Key properties including tensile strength, water vapor transmission rate, and oxygen transmission rate are carefully considered to tailor these bioplastics for specific applications, particularly in packaging [46]. Algae-based bioplastics are distinguished by their biodegradability and compostability, presenting an eco-friendly alternative to traditional plastics. This direct conversion technique streamlines production by bypassing the need for fermentation by setting it apart from other bioplastic production methods like PLA, PHA, and PHB. Furthermore, combining algae with other biodegradable materials like PLA to create versatile composites opens up a broad spectrum of applications. This enhances the utility and appeal of these sustainable materials.

14.5.2 Indirect Method of Bioplastic Production Using Algae

The indirect approach leverages residual algal biomass from the extraction of key components as a sustainable source of organic feedstock for generating bioplastic precursors. This method exemplifies an innovative integration of biorefinery concepts in efficiently utilizing every part of the algal biomass [47]. Central to this process is the production of precursors like PLA and PHB through fermentation, a burgeoning area of research that explores algae as a viable alternative to synthetic polymers. The crucial role of specific microorganisms in fermenting algal substrates to produce these precursors is gaining attention and opening new pathways for bioplastic manufacturing. Studies highlight the potential for converting seaweed polysaccharides into various bioplastics, indicating significant progress in sustainable material development [48]. The fermentation of algal extracellular polymeric substances by lactic acid bacteria under controlled conditions is a key step towards producing bio-based bioplastics. Additionally, the established industrial presence of lactic acid produced through anaerobic fermentation offers a synergistic opportunity for integrating algal bioplastic production with existing processes [49]. This indirect approach not only aligns with environmental sustainability but also enhances the economic viability of the algal industry, maximizing the utility and value of algal biomass.

The indirect method of bioplastic production from algal biomass presents a blend of economic and operational advantages and challenges. Economically, it capitalizes on leftover algal biomass often rich in sugar compounds, reducing material costs. This approach streamlines production by eliminating the need for biomass pretreatment that is typically required for compound extraction. Additionally, the possibility of integrating this method with existing fermentation systems and algal industries can further reduce material costs by showcasing its potential synergy with current industrial practices. However, this method faces several challenges. One significant limitation is the dependency on algae-based industries that restricts biomass availability, posing a challenge to scalability [50]. Variability in the presence of desired sugar compounds in spent biomass can affect the consistency and effectiveness of the production process. Furthermore, the potential for lower yields due to certain algae biomass types not being suitable for bioplastic precursor development highlights the need for selective biomass sourcing. The time-intensive nature of pretreatment and fermentation processes can lead to extended production cycles, impacting overall efficiency. At last, the seasonal availability of macroalgae introduces an element of unpredictability that affects the consistent supply of raw materials necessary for the process.

14.6 Current Limitations and Challenges in the Production and Use of Algae-Based Bioplastics

The successful scaling up of algae-based bioplastics from laboratory experiments to commercial production is intricately linked to the careful selection of algal species, each with its unique biomass composition and suitability for polymer production. A study involving *Ulva lactuca*, *Cladophora* sp., and *Calothrix scytonemicola* exemplifies this, highlighting the species-specific potential for producing effective bioplastic materials [51]. Notably, *Calothrix scytonemicola*, known for its high PHA content, stands out as a promising candidate, emphasizing the critical role of targeted species selection in this field [52]. The cellulose from *Cladophora* sp. algae features a web-like configuration, in which the fibrils form a highly intertwined network. This unique structure results in cellulose that possesses superior mechanical properties [53]. Meanwhile, *Ulva* spp. boasts a remarkable composition of ulvan and monosaccharides including hexose and pentose sugars, making it ideal for bioplastic production [54]. This illustrates the complexities and nuances of using various algae for bioplastic synthesis. This approach potentially signals a significant shift in the plastics industry, moving away from traditional petroleum-based plastics to innovative, biodegradable alternatives derived from a diverse range of algal sources.

Creating sustainable and resilient bioplastics from algae involves a complex selection of polymers, where factors such as biodegradability, strength, renewability of feedstock, degradation rate, and specific polymer characteristics like size, molecular weight, and moisture content play a crucial role. This process is further complicated by environmental concerns, particularly the release of greenhouse gases such as carbon dioxide and methane during bioplastic degradation, emphasizing the need for a careful assessment of environmental impacts during the manufacturing phase [55]. Additional challenges include the emission of unpleasant odors from bioplastics made from algae lipids and the slow production process attributed to carbohydrate agglomeration in the algal biomass [56]. To address the odor issue in bioplastics derived from high fatty acid microalgae, using absorbents like activated carbon or zeolite is essential for maintaining commercial appeal. Another significant barrier to the broader adoption of algal-based bioplastics is the general lack of consumer awareness and understanding of bioplastics. This issue is often coupled with misconceptions regarding their cost due to the use of organic materials. This gap in knowledge and understanding presents a challenge for market acceptance and usage of bioplastic products.

The advancement to commercial-scale production of algal-based plastics is heavily reliant on the development of effective algae cultivation systems that are capable of mass-producing algae for polymer and compound extraction. Open pond systems offer a cost-effective and scalable solution but they face several challenges, including reduced productivity, high risks of contamination, and difficulties in cultivating specific algal strains at a large scale [57]. In contrast, closed photobioreactor systems provide a controlled environment that reduces contamination risks and supports the selective growth of algae species resulting in higher productivity [58].

One key advantage of closed systems is their enhanced photosynthetic rate that significantly contributes to efficiency despite higher scale-up costs compared to open pond systems. Therefore, selecting between open pond and closed photobioreactor systems involves balancing cost efficiency with the quality and specificity of algae cultivation, a critical factor affecting the feasibility of large-scale algal-based plastic production.

Effective disposal and treatment of bioplastic waste is an essential aspect of the bioplastics industry, with composting standing out as a highly efficient method due to its ability to rapidly decompose biodegradable bioplastics. The array of waste treatment options for bioplastics, which includes recycling, incineration, anaerobic digestion, and landfilling, highlights the necessity for a variety of strategies to manage these materials postuse. Notably, PLA bioplastics demonstrate environmental benefits over traditional plastics, as evidenced by their accelerated biodegradation in industrial composting facilities, typically within 4–6 weeks. Tackling the significant challenges in bioplastic production such as effective algae cultivation and the meticulous selection of appropriate polymers or strains is imperative to enhance the commercial feasibility of bioplastics. Addressing these challenges is not only just crucial for the growth of the bioplastics industry but also pivotal in the broader context of combating plastic pollution and reducing the reliance on fossil fuel resources.

14.6.1 Future Research Directions

The involvement of marine green algae in the biorefinery concept highlights its potential as a sustainable resource for synthesizing polymers essential for bioplastics production and a range of valuable coproducts. This approach offers an innovative solution to waste disposal challenges by creating consumer-grade bioplastics from algal waste residues, reinforcing the principles of a circular economy. Beyond bioplastics, the versatility of marine green algae extends to producing diverse byproducts such as fuel, nutraceuticals, pharmaceuticals, and cosmetics, demonstrating its wide-ranging industrial applications. A notable advancement is the successful use of residual microalgae biomass combined with PBAT to create biocomposites, as evidenced in previous research which signifies a significant improvement in the mechanical properties of bioplastics [59]. The biodegradability of these algae-based biocomposites adds an important environmental dimension to their use, especially as agricultural films that decompose in the soil. This progress encourages collaboration between industry and academia to research and develop algae-based bioplastics, potentially enhancing quality with new production processes and improved properties. As the bioplastic industry grows and demands new biodegradable materials, the need for ongoing innovation becomes apparent especially in the development of bio-based and biodegradable bioplastics. The role of marine green algae in reducing production costs and its promising pathway toward bio-based and biodegradable bioplastics underscores its potential as a key player in the field of sustainable materials science.

14.7 Summary

In conclusion, the exploration of marine green algae as a raw material for bioplastic production presents a promising avenue toward sustainable material development. This chapter has underscored the versatility of algae, particularly in terms of their biochemical composition, which makes them ideal candidates for bioplastic precursors. The advantages of using algae include their biodegradability and the potential to reduce reliance on fossil fuels, which aligns with global sustainability goals. However, challenges such as scalability, seasonal variability, and the need for selective biomass sourcing highlight the complexity of this endeavor. Future research should focus on optimizing production processes, enhancing the material properties of algae-based bioplastics, and integrating these bioplastics into the circular economy. Collaborative efforts between industry and academia are crucial for overcoming these challenges and harnessing the full potential of marine green algae in bioplastic production. As we move forward, it is clear that marine green algae not only offer a path to more sustainable materials but also represent a step towards a more ecoconscious and resource-efficient future.

Acknowledgement The authors would like to thank the International Centre for Genetic Engineering and Biotechnology, ICGEB Grant Number of 304.PTEKIND.6501382.I151 (R504-LR-GAL008-0006501382-I151) for funding this study.

References

1. Ibrahim NI, Shahar FS, Sultan MTH, Shah AUM, Safri SNA, Mat Yazik MH (2021) Overview of bioplastic introduction and its applications in product packaging. Coatings 11(11):1423
2. Weber CJ, Bigalke M (2022) Opening space for plastics—why spatial, soil and land use data are important to understand global soil (micro)plastic pollution. Microplastics 1(4):610–625
3. González-Fernández D, Cózar A, Hanke G, Viejo J, Morales-Caselles C, Bakiu R, Barceló D, Bessa F, Bruge A, Cabrera M (2021) Floating macrolitter leaked from Europe into the ocean. Nat Sustain 4(6):474–483
4. Sipe JM, Bossa N, Berger W, von Windheim N, Gall K, Wiesner MR (2022) From bottle to microplastics: can we estimate how our plastic products are breaking down? Sci Total Environ 814:152460
5. Talan A, Pokhrel S, Tyagi R, Drogui P (2022) Biorefinery strategies for microbial bioplastics production: sustainable pathway towards circular bioeconomy. Bioresour Technol Rep 17:100875
6. Boey JY, Lee CK, Tay GS (2022) Factors affecting mechanical properties of reinforced bioplastics: a review. Polymers 14(18):3737
7. Mal N, Satpati G, Raghunathan S, Davoodbasha M (2022) Current strategies on algae-based biopolymer production and scale-up. Chemosphere 289:133178
8. De Castro TR, De Macedo DC, De Genaro Chiroli DM, Da Silva RC, Tebcherani SM (2022) The potential of cleaner fermentation processes for bioplastic production: a narrative review of polyhydroxyalkanoates (PHA) and polylactic acid (PLA). J Polym Environ 30(3):810–832
9. Zanchetta E, Damergi E, Patel B, Borgmeyer T, Pick H, Pulgarin A, Ludwig C (2021) Algal cellulose, production and potential use in plastics: challenges and opportunities. Algal Res 56:102288

10. Thiruchelvi R, Das A, Sikdar E (2021) Bioplastics as better alternative to petro plastic. Mater Today Proc 37:1634–1639
11. Aroyehun AQB, Razak SA, Palaniveloo K, Nagappan T, Rahmah NSN, Jin GW, Chellappan DK, Chellian J, Kunnath AP (2020) Bioprospecting cultivated tropical green algae, Caulerpa racemosa (Forsskal) J. Agardh: a perspective on nutritional properties, antioxidative capacity and anti-diabetic potential. Food Secur 9(9):1313
12. Sulaiman WSHW, Zain HHM, Othman R, Latif NHM, Hatta FAM, Ramya R (2023) Algae carotenoid pigments as new sources of halal bioactive ingredients. J Halal Sci Technol 2(1):68–82
13. Sonchaeng U, Wongphan P, Pan-Utai W, Paopun Y, Kansandee W, Satmalee P, Tamtin M, Kosawatpat P, Harnkarnsujarit N (2023) Preparation and characterization of novel green seaweed films from Ulva rigida. Polymers 15(16):3342
14. Ludmila M, Veronika L, Alena M, Tatyana M, Svetlana Z, Victor E (2023) Morphological changes and biochemical reaction of Ulva rigida in response to the toxic effect of bisphenol A under experimental conditions. Aquat Bot 184:103579
15. Abdo SM, Youssef M, El Nagar I, Mohamed HE, El-Kholy SA, Youssef A (2024) Processing and characterization of antimicrobial bioplastic films based on green microalgae Scenedesmus obliquus extract-loaded polyurethane. Int J Biol Macromol 257:128711
16. Greene C, Scott-Buechler C, Hausner A, Johnson Z, Lei XG, Huntley M (2022) Transforming the future of marine aquaculture: a circular economy approach. Oceanography 35(2):26
17. Sudhakar M, Venkatnarayanan S, Dharani G (2023) Bioplastic production from renewable biomass (seaweeds). In: Green sustainable process for chemical and environmental engineering and science. Elsevier, pp 439–454
18. Kartik A, Akhil D, Lakshmi D, Panchamoorthy Gopinath K, Arun J, Sivaramakrishnan R, Pugazhendhi A (2021) A critical review on production of biopolymers from algae biomass and their applications. Bioresour Technol 329:124868
19. Shakir MA, Yhaya MF, Ahmad MI (2017) Physical effects of pre-hydrolysis and alkali treatment of Empty Fruit Bunch (EFB) fibers. Int J Chem Eng Appl 8(6):387–391
20. Shakir MA, Azahari B, Salehabadi A, Yusup Y, Yhaya MF, Ahmad MI (2020) Structural, thermal, and mechanical properties of spent mushroom substrate (sms) and rubberwood sawdust (rws) binderless particleboards. J Adv Res Fluid Mech Therm Sci 75(2):113–124
21. Pellis A, Malinconico M, Guarneri A, Gardossi L (2021) Renewable polymers and plastics: performance beyond the green. New Biotechnol 60:146–158
22. Faidi A, Lassoued MA, Becheikh MEH, Touati M, Stumbé J-F, Farhat F (2019) Application of sodium alginate extracted from a Tunisian brown algae Padina pavonica for essential oil encapsulation: microspheres preparation, characterization and in vitro release study. Int J Biol Macromol 136:386–394
23. Georgiopoulou I, Tzima S, Louli V, Magoulas K (2023) Process optimization of microwave-assisted extraction of chlorophyll, carotenoid and phenolic compounds from Chlorella vulgaris and comparison with conventional and supercritical fluid extraction. Appl Sci 13(4):2740
24. Putra VG, Mutiarahma S, Chaniago W, Rahmadi P, Kurnianto D, Hidayat C, Carrera C, Palma M, Setyaningsih W (2022) An ultrasound-based technique for the analytical extraction of phenolic compounds in red algae. Arab J Chem 15(2):103597
25. Shen L, Pang S, Zhong M, Sun Y, Qayum A, Liu Y, Rashid A, Xu B, Liang Q, Ma H (2023) A comprehensive review of ultrasonic assisted extraction (UAE) for bioactive components: principles, advantages, equipment, and combined technologies. Ultrason Sonochem 101:106646
26. Flórez-Fernández N, Domínguez H, Torres M (2019) A green approach for alginate extraction from Sargassum muticum brown seaweed using ultrasound-assisted technique. Int J Biol Macromol 124:451–459
27. Álvarez-Viñas M, Rodríguez-Seoane P, Flórez-Fernández N, Torres MD, Díaz-Reinoso B, Moure A, Domínguez H (2021) Subcritical water for the extraction and hydrolysis of protein and other fractions in biorefineries from agro-food wastes and algae: a review. Food Bioprocess Technol 14:373–387

28. Essien SO, Young B, Baroutian S (2020) Recent advances in subcritical water and supercritical carbon dioxide extraction of bioactive compounds from plant materials. Trends Food Sci Technol 97:156–169
29. Dardavila MM, Pappou S, Savvidou MG, Louli V, Katapodis P, Stamatis H, Magoulas K, Voutsas E (2023) Extraction of bioactive compounds from C. vulgaris biomass using deep eutectic solvents. Molecules 28(1):415
30. Mehariya S, Iovine A, Casella P, Musmarra D, Chianese S, Marino T, Figoli A, Sharma N, Molino A (2020) Bio-based and agriculture resources for production of bioproducts. In: Current trends and future developments on (bio-)membranes. Elsevier, pp 263–282
31. Veziroglu S, Ayna M, Kohlhaas T, Sayin S, Fiutowski J, Mishra YK, Karayürek F, Naujokat H, Saygili EI, Açil Y (2021) Marine algae incorporated polylactide acid patch: novel candidate for targeting osteosarcoma cells without impairing the osteoblastic proliferation. Polymers 13(6):847
32. Rahaman H, Hosen S, Gafur A, Habib R (2020) Small amounts of poly(d-lactic acid) on the properties of poly(l-lactic acid)/microcrystalline cellulose/poly(d-lactic acid) blends. Results Mater 8:100125
33. Obulisamy PK, Mehariya S (2021) Polyhydroxyalkanoates from extremophiles: a review. Bioresour Technol 325:124653
34. Wang J, Liu S, Huang J, Cui R, Xu Y, Song Z (2023) Genetic engineering strategies for sustainable polyhydroxyalkanoate (PHA) production from carbon-rich wastes. Environ Technol Innov 30:103069
35. Ghosh S, Gnaim R, Greiserman S, Fadeev L, Gozin M, Golberg A (2019) Macroalgal biomass subcritical hydrolysates for the production of polyhydroxyalkanoate (PHA) by Haloferax mediterranei. Bioresour Technol 271:166–173
36. Chia WY, Tang DYY, Khoo KS, Lup ANK, Chew KW (2020) Nature's fight against plastic pollution: algae for plastic biodegradation and bioplastics production. Environ Sci Ecotechnol 4:100065
37. Balaji S, Gopi K, Muthuvelan B (2013) A review on production of poly β hydroxybutyrates from cyanobacteria for the production of bio plastics. Algal Res 2(3):278–285
38. Rahman A, Miller C (2017) Microalgae as a source of bioplastics. In: Algal green chemistry. Elsevier, pp 121–138
39. Sudhakar MP, Maurya R, Mehariya S, Karthikeyan OP, Dharani G, Arunkumar K, Pereda SV, Hernández-González MC, Buschmann AH, Pugazhendhi A (2023) Feasibility of bioplastic production using micro and macro algae—a review. Environ Res 240:117465
40. Akash S, Sivaprakash B, Rajamohan N, Vo D-VN (2023) Biotechnology to convert carbon dioxide into biogas, bioethanol, bioplastic and succinic acid using algae, bacteria and yeast: a review. Environ Chem Lett 21:1477
41. Cheah WY, Er AC, Aiyub K, Yasin NHM, Ngan SL, Chew KW, Khoo KS, Ling TC, Juan JC, Ma Z (2023) Current status and perspectives of algae-based bioplastics: a reviewed potential for sustainability. Algal Res 71:103078
42. Park Y-K, Lee J (2022) Achievements in the production of bioplastics from microalgae. Phytochem Rev 22:1–19
43. Onen Cinar S, Chong ZK, Kucuker MA, Wieczorek N, Cengiz U, Kuchta K (2020) Bioplastic production from microalgae: a review. Int J Environ Res Public Health 17(11):3842
44. De Luca S, Milanese D, Gallichi-Nottiani D, Cavazza A, Sciancalepore C (2023) Poly(lactic acid) and its blends for packaging application: a review. Clean Technol 5(4):1304–1343
45. El-Sheekh MM, Alwaleed EA, Ibrahim A, Saber H (2024) Preparation and characterization of bioplastic film from the green seaweed Halimeda opuntia. Int J Biol Macromol 259:129307
46. Ghobashy MM, Mousa SA, Siddiq A, Nasr HM, Nady N, Atalla AA (2023) Optimal the mechanical properties of bioplastic blend based algae-(lactic acid-starch) using gamma irradiation and their possibility to use as compostable and soil conditioner. Mater Today Commun 34:105472
47. Madadi R, Maljaee H, Serafim LS, Ventura SP (2021) Microalgae as contributors to produce biopolymers. Mar Drugs 19(8):466

48. Rajpoot AS, Choudhary T, Chelladurai H, Verma TN, Shende V (2022) A comprehensive review on bioplastic production from microalgae. Mater Today Proc 56:171–178
49. Nagarajan D, Chen C-Y, Ariyadasa TU, Lee D-J, Chang J-S (2022) Macroalgal biomass as a potential resource for lactic acid fermentation. Chemosphere 309:136694
50. Harb TB, Chow F (2022) An overview of beach-cast seaweeds: potential and opportunities for the valorization of underused waste biomass. Algal Res 62:102643
51. Mat Yasin N, Othman N, Adam F (2024) Evaluation of the properties on carrageenan bio-films with Chlorella vulgaris blending. Chem Eng Commun 211(3):400–414
52. Leong YK, Chang J-S (2022) Bioprocessing for production and applications of bioplastics from algae. In: Biomass, biofuels, and biochemicals. Elsevier, pp 105–132
53. Steven S, Fauza AN, Mardiyati Y, Santosa SP, Shoimah SMA (2022) Facile preparation of cellulose bioplastic from Cladophora sp. algae via hydrogel method. Polymers 14(21):4699
54. Manikandan NA, Lens PN (2023) Sustainable biorefining and bioprocessing of green seaweed (Ulva spp.) for the production of edible (ulvan) and non-edible (polyhydroxyalkanoate) biopolymeric films. Microb Cell Factories 22(1):140
55. Lomonaco T, Manco E, Corti A, La Nasa J, Ghimenti S, Biagini D, Di Francesco F, Modugno F, Ceccarini A, Fuoco R (2020) Release of harmful volatile organic compounds (VOCs) from photo-degraded plastic debris: a neglected source of environmental pollution. J Hazard Mater 394:122596
56. Li R, Zhu X, Peng F, Lu F (2023) Biodegradable, colorless, and odorless PLA/PBAT bioplastics incorporated with corn Stover. ACS Sustain Chem Eng 11:8870
57. Katayama T, Nagao N, Kasan NA, Khatoon H, Rahman NA, Takahashi K, Furuya K, Yamada Y, Abd Wahid ME, Jusoh M (2020) Bioprospecting of indigenous marine microalgae with ammonium tolerance from aquaculture ponds for microalgae cultivation with ammonium-rich wastewaters. J Biotechnol 323:113–120
58. Nwoba EG, Parlevliet DA, Laird DW, Alameh K, Moheimani NR (2020) Pilot-scale self-cooling microalgal closed photobioreactor for biomass production and electricity generation. Algal Res 45:101731
59. Ibrahim NN, Ahmad I, Abdullah N, Koji I, Mohamad SE, Akhir FNBM (2023) Biocomposites from microalgae. In: Composites from the aquatic environment. Springer, pp 99–138

Production of Biochar from Marine Algae and Its Use in the Treatment of Industrial Effluent

15

Hamza Mohamed Flafel, Mohd Rafatullah, Japareng Lalung, Riti Thapar Kapoor, Masoom Raza Siddiqui, Moonis Ali Khan, and Mohammad Qutob

Abstract

In the present era, the widespread issue of environmental pollution poses a substantial global challenge. Industrial effluent, containing a range of organic and inorganic pollutants, poses significant hazards to both aquatic ecosystems and human health if released into water bodies. Hence, it is essential to subject these effluents to treatment in order to eradicate or, at the very least, reduce the levels of pollutants present in industrial wastewater (WW) before releasing it into the raw water. Various approaches were explored for WW treatment, with the adsorption technique utilizing biochar derived from marine algae emerging as the most promising method. This method has emerged as a highly promising and sustainable solution owing to several key factors. The availability of marine algae as a renewable resource, coupled with the cost-effectiveness and operational simplic-

H. M. Flafel
Environmental Technology Division, School of Industrial Technology, Universiti Sains Malaysia, Penang, Malaysia

Libyan Center for Studies and Research in Environmental Science and Technology, Brack Al-Shatti, Libya

Department of Environmental Science, Faculty of Environment & Natural Resources, Wadi Al-Shatti University, Brack Al-Shatti, Libya

M. Rafatullah (✉) · J. Lalung · M. Qutob
Environmental Technology Division, School of Industrial Technology, Universiti Sains Malaysia, Penang, Malaysia
e-mail: mrafatullah@usm.my

R. T. Kapoor
Centre for Plant and Environmental Biotechnology, Amity University, Noida, Uttar Pradesh, India

M. R. Siddiqui · M. A. Khan
Chemistry Department, King Saud University, Riyadh, Saudi Arabia

M. Rafatullah et al. (eds.), *Multidisciplinary Applications of Marine Resources*,
https://doi.org/10.1007/978-981-97-5057-3_15

ity of producing biochar, enhances the attractiveness of this approach. Furthermore, the implementation of biochar in industrial effluent treatment yields nontoxic byproducts, contributing to the overall sustainability of the process. Through an in-depth examination of the production processes, operational considerations, and case studies illustrating the successful implementation of marine algae-derived biochar in industrial effluent treatment, this chapter provides valuable insights into the promising realm of eco-friendly and economically viable solutions for mitigating the adverse environmental impacts of industrial activities.

Keywords

Biochar · Bioremediation · Industrial effluent · Marine algae · Pollutants

15.1 Introduction

The growing global population and increased industrial production have led to a continuous increase in the consumption of resources. Both renewable and nonrenewable raw materials undergo a variety of chemical, physical, and biological processes to create a wide range of valuable products, including paper, essential chemicals, fuels, food items, pharmaceuticals, textiles, plastics, and various materials [1]. The manufacturing of products often results in the creation of multiple unwanted waste materials in various forms, including solids, liquids, and gases [2]. Industrial activities produce WW that frequently contains biodegradable organic substances, nutrients, heavy metals (HMs), toxic compounds, and persistent organic compounds. These WW streams, when untreated or only partially treated, might damage aquatic ecosystems and have detrimental effects on human health [3]. Therefore, it is crucial to discover appropriate methods for treating this WW before it is discharged into a water source. To address this challenge, researchers and environmentalists are continuously exploring innovative and eco-friendly technologies for WW treatment. Conventional techniques for treating industrial WW, such as membrane technology, coagulation-flocculation, and filtering, have a number of disadvantages [4]. However, biochar, a carbonaceous substance, has attracted interest because of its potential for use in environmental management. It is a solid biomass product with a surface that contains oxygen functional groups, a porous structure, and a high carbon content (65–90%) [5]. High levels of biomass ash have the ability to adsorb industrial effluents like HMs and organic compounds, etc. [6]. Algae marine consists of four primary categories of algae. The first group, known as Chlorophyceae, consists of green algae containing chlorophyll a and b pigments, including species like *Chlamydomonas*, *Spirogyra*, and *Chara*. The second category is Phaeophyceae, which includes brown algae predominantly found in marine environments. These algae feature chlorophyll a, c carotenoids, and xanthophyll pigments and include species such as *Dictyota*, *Laminaria*, and *Sargassum*. Rhodophyceae, the third category, includes species such as *Porphyra*, *Gracilaria*, and *Gelidium* that are red algae and are distinguished by the presence of the red pigment r-phycoerythrin. Occasionally, the fourth type, also referred to as blue-green algae or cyanobacteria, is classified as seaweed. This type of algae is commonly found in home aquariums, where it rapidly covers surfaces and is often referred to as slime algae or smear algae, produced through photosynthesis by either eukaryotic or prokaryotic microorganisms, holding significant promise as a renewable biomass source which can be used in biochar production. They have garnered substantial research interest due to their versatility, with applications ranging from energy production to WW treatment. The primary chemical constituents found in microalgae biomass include proteins, lipids, and carbohydrates, although their proportions may vary between different microalgae species. The development rate of algae is remarkable; it is 5–10 times faster than that of terrestrial crops. Microalgae are particularly attractive for biotreatment because of their photosynthetic properties, and they are unique in that they can ingest these chemicals as source of carbon [7]. When comparing the literature alongside contemporary research, we discover that algal biochar is used for both adsorption and energy storage through the enhancement of its porous structure via innovative synthesis techniques or the creation of porous carbon derived from algae [8]. In a world facing unprecedented environmental challenges, industrial effluent treatment stands out as a pressing concern. Conventional methods often fall short of effectively addressing the ever-increasing industrial pollution. As we strive for innovative solutions, "Eco-biochar" presents a groundbreaking approach that is not only sustainable but also environmentally conscious. The goal of this chapter is to give a thorough review of the processes involved in producing and activating biochar made from marine algae. Then, it addresses industrial effluent challenges, and industrial effluent treatment methods.

15.2 Production of Biochar from Marine Algae

Producing biochar from marine algae is a fascinating and innovative approach to harnessing the power of nature to address environmental and agricultural challenges. When produced from marine algae, biochar becomes an even more environmentally friendly and valuable resource due to the unique properties of these aquatic plants. Among the methods of producing biochar, marine algae is formed..

15.2.1 Traditional and Pyrolysis Enhanced by Microwave

One particular promising method for turning biomass into valuable biochar is pyrolysis. The various types of pyrolysis, including conventional quick pyrolysis, slow pyrolysis, and innovative microwave-assisted pyrolysis, operate under different conditions. The resulting gaseous, liquid, and solid products vary in composition based on parameters like residence time, temperature, and heating rate. Notably, the most bio-oil yield is produced by fast pyrolysis, while the highest biochar production is produced by slow pyrolysis. Pyrolysis involves subjecting biomass of marine algae to heat in the absence of air or oxygen, usually in the 350–700 °C temperature range [9]. The resulting products are influenced by various factors, with temperature and heating rate being crucial. The production of more biochar is influenced by lowering the heating rate, lengthening the residence period, and lowering the pyrolysis temperature. Additionally, Properties of the feedstock, like as moisture content and particle size, have a big impact on how much biochar is produced by pyrolysis. The traditional method of producing charcoal by slow pyrolysis produces the most quantity of biochar when compared to gasification and fast pyrolysis. Pyrolysis can store around half of the carbon from biomass as stable biochar. The method of slow pyrolysis involves subjecting biomass to a vapor residence period that can vary from several minutes to hours, resulting in the generation of biochar. Vapors are extensively reacted and constrained with the solid phase to enhance biochar yield. Low heating rates (0.1–1 K s^{-1}) and residence times of roughly 450–550 s are typical for slow pyrolysis. The process begins with prepyrolysis, then solid decomposition resembling higher pyrolysis to produce products of pyrolysis. Carbon-rich biochar is produced as a result of the extremely slow breakdown of biochar. In fixed-bed reactors, where heating is delivered by a heated surface, traditional slow pyrolysis is frequently used. However, research has looked into other heating techniques, like microwave heating, for this process.

Additionally, the thermochemical process of microwave-assisted pyrolysis is a very effective way to produce biochar. This process has a number of benefits, including as increased product output, a reduction in the amount of hazardous chemicals present, and significant energy and cost savings. Using electromagnetic radiation to result in molecular heat and oscillation, microwave technology is superior to traditional pyrolysis in a number of technical aspects. These include the capacity to heat larger biomass particles uniformly, the creation of syngas with a higher heating value (HHV) appropriate for using microwaves to generate electricity on-site, the production of cleaner products devoid of fluidization and agitation, and the established, scalable nature of microwave heating technology [10]. Notably, there are differences in the production of biochar products between microwave-assisted pyrolysis and traditional slow pyrolysis. As though microwave pyrolysis requires pretreatment and catalysts prior for heating, traditional pyrolysis depends on regular heating procedures. The pyrolysis of lignocellulosic biomass has been extensively studied, while the creation of algal biochar using microwave heating has received little attention.

15.2.2 Hydrothermal Carbonization

In recent years, hydrothermal carbonization (HTC) has emerged as a feasible thermochemical technique for generating biochar. In this step, the carbohydrates present in macroalgae undergo conversion into solid rich in carbon substance known as hydrochar within an aqueous environment, maintaining temperatures between 175 and 250 °C for 30–120 min under a pressure of 20–60 bar. HTC proves particularly suitable for wet biomass, as there is no requirement for predrying before carbonization. This advantage arises from the utilization of hot compressed water in processing wet biomass, thereby saving energy and costs associated with the drying phase prior to thermochemical conversion [11]. HTC's low carbonization temperature, suitability for wet biomass, affordability, and capacity to produce compounds with added value are its main advantages. Key process parameters for HTC encompass biomass loading, time, catalyst amount, pressure, and temperature. Granulometry, representing the biomass particles' size, also holds significance in macroalgae biomass HTC. Reducing particle size facilitates enhanced water penetration into biomass pores, thereby expediting heat, increasing available surface area, and mass transfer during the HTC process [12]. HTC results in gaseous, solid, and liquid products. One solid byproduct of the HTC process is hydrochar, which demonstrates high stability and possesses a higher carbon content compared with raw biomass. HTC produces treated water that contains minerals including potassium, phosphorus, and nitrogen, along with organic compounds, while the predominant component in the gaseous product is CO_2. Separating the liquid (processed water) and solid (hydrochar) products from the wet slurry is achieved through gravity filtering. Temperature, followed by time and pressure, plays a crucial role in this process. Lowering the carbonization temperature increases hydrochar production. High-temperature HTC of macroalgal biomass yields hydrochar with elevated carbon content, concurrently reducing oxygen and hydrogen levels. These conditions lead to the synthesis of a solid product rich in carbon, exhibiting improved HHV and porosity. It is advisable to keep the residence time brief when converting algal biomass into hydrochar. Utilizing the HTC process offers various advantages, including cost-effectiveness, the ability to use wet biomass, and time efficiency. By effectively harnessing algal biomass and converting it into useful products, this approach eliminates the necessity for a pretreatment step like drying, enabling the transformation of algal biomass into valuable products [13].

15.2.3 Torrefaction

Torrefaction serves as a preliminary step to pyrolysis, aimed at eliminating volatiles. This pretreatment process is applicable in both dry and wet settings. The temperature range for wet torrefaction is 180–260 °C, with elevated pressures 200–700 psi and 5-min duration of time. On the other hand, there are temperature requirements for dry torrefaction of 200–300 °C, atmospheric pressure and a response time of 80 min. The particular torrefaction technique used has a significant impact on the final biochar's quality. The adsorption capacity of the biochar produced is significantly increased by wet torrefaction, which produces large amounts of low ash, very hydrophobic biochar. Additionally, a rise in moisture content enhances the rate of heat transfer, resulting in a reduction in heating costs. Moreover, a wet environment easily triggers hydrolysis attack, leading to the decomposition of biomass [14]. Additionally, changing from N_2 to another carrier gas to other oxidative gases or air can reduce the amount of energy used, the time it takes for torrefaction, and the cost of the carrier gas. With the exception of a somewhat elevated ash content, this enhanced application viability, decreased costs, and boosted energy efficiency [15]. The advantages of torrefaction are as follows: (1) increased heating value or energy density; (2) decreased atomic O/C ratios and moisture content; (3) increased hydrophobicity or water-resistivity; (4) enhanced grindability and reactivity; and (5) more homogeneous biomass features.

15.3 Strategies for Activating Algal Biochar

The effective utilization of biochar for the treatment of industrial effluents hinges significantly on the porosity and surface area of the biochar. Thorough investigation of these critical factors is imperative, or alternatively, appropriate activation strategies must be employed to induce desirable features in the biochar. It is noteworthy that biochar, immediately postproduction, exhibits minimal propensity for absorption or adsorption compared to its activated counterpart. This discrepancy can be seen in the material's surface area, pore size, pore volume, and total number of pores. When biochar is not properly activated, it poses a number of issues, such as (1) numerous intermolecular spaces resulting from the breakage of bonds within organic components; (2) blocked pores that cause the production of tar; (3) negligible pore size, resulting in a smaller surface area distribution; and (4) the presence of contaminants like as ashes and condensates, which diminish volume and pore size. Therefore, in order to increase its active centres, anchor ions, or molecules, algal biochar is usually activated, as well as to enhance its internal stability, structure, or characteristics like porosity and pore size [16]. Nonetheless, greater porosity and specific surface areas do not always equate to superior outcomes. Biochar with limited porosity frequently performs better when used as a capacitor, especially when combined with algal biochar. Physical activation and chemical activation are the two main types of activation techniques. These strategies work well for increasing porosity and specific surface area because they involve microscopic processes like pore expansion and creation, as shown in Fig. 15.1.

15.3.1 Physical Activation

The physical activation procedure, which is often referred to as thermal activation, entails heating biochar in the presence of CO_2, steam, or air between 700 and 900 °C [17]. This high-temperature, oxidative environment initiates pore formation through auto-gasification, reducing volatiles and transforming the biochar into a porous carbon material. The elevated temperature causes pores filled with carbon volatiles to open, leading to diffusion with adjacent pores. Consequently, this sequence of events significantly enhances vigorous micropore formation and the specific surface area, with a lower concentration of mesopores. However, there are restrictions associated with physical activation, and the process is heavily influenced by variables like temperature, the kind of activating agent (steam, CO_2, etc.), the degree of activation, and the precursor material. Despite the economic feasibility of using air for activation, careful monitoring is essential, as prolonged processing times may lead to combustion, reducing the biochar yield. It is essential to sustain the pore volume

Fig. 15.1 Generation and activation of marine algae-derived biochar

and specific surface area created by steam and CO_2 activation at very high temperatures in order to obtain the required pore properties, volume, and surface area. Remarkably, CO_2 activation often results in an increase in pores, which raises the surface area of the resultant biochar [18]. Algal biochar's oxygen-rich functional groups can also be increased by physical stimulation. The activation of ozone holds special significance, given its status as one of the nature's most powerful oxidants. Its robust oxidation capabilities persist even under atmospheric pressure conditions. Following the reaction, the residual atoms have the ability to spontaneously create stable O_2 without causing additional pollution. This process is primarily employed for activating microporous biochar (with a range of 0.37–0.52 $cm^3\ g^{-1}$). Furthermore, the activation process causes the surface of algal biochar to get coated in acidic functional groups that contain oxygen, such as ethers, quinones, anhydrides, and lactones. These improvements enhance its CEC (cation exchange capacity) [19].

15.3.2 Chemical Activation

The biochar requiring activation undergoes a chemical activation process prior to thermal treatment. Chemical activation agents and biochar interact during this process, which typically involves oxidation and dehydration at temperatures between 400 and 950 °C. Microporous carbon surfaces are formed in part by these oxidation and dehydration processes. Choosing chemical activation has a number of benefits over physical activation, such as increased efficiency, a surface area more than 4000 $m^2\ g^{-1}$, controlled porosity, a smaller temperature range, and a larger carbon output. However, it is crucial to acknowledge the elevated risks associated with chemical activation, particularly in terms of managing the type and quantity of chemicals used. Overabundance can cause pollution, have corrosive effects on machinery, and raise questions about general chemical safety, particularly at higher temperatures. Commonly employed activation agents include acids such as H_2SO_4,

HCl, HNO_3, and H_3PO_4, strong bases such as NaOH and KOH, and agents such as H_2O_2. Temperature, the kind of chemical agents utilized, the kind of biomass, and the concentration of the chemical agent all affect the degree of activation and the physical properties of the resulting biochar in chemical activation. In the study conducted by the authors in [20], that using, three types of algae, *Ulva ohnoi* (GA), *Agardhiella subulata* (RA), and *Sargassum hemiphyllum* (BA), collected from Pingtung, Taiwan, were prepared for further analysis. The algae were cleaned with deionized water and then dried in an oven at 60 °C for a day. The dried algae were then pulverized into particles that were no larger than 500 μm. The biomass was activated using a $ZnCl_2$ impregnation process, involving immersion in a $ZnCl_2$ solution (1:3, w:w), followed by drying and calcination at 700 °C. Impurities were eliminated with a further HCl treatment and deionized water washing. The resulting biochars were termed as ZGAB, ZRAB, and ZBAB, referring to their original algal origins. The results demonstrated that the combination of brown algae with $ZnCl_2$ chemical activator (ZBAB) efficiently formed mesoporous biochar with good physicochemical features and delivered the best Ciprofloxacin adsorption capacity. Ciprofloxacin adsorption capacity of the ZBAB was found to be high (190–300 mg^{-1}). Chemical activation has the potential to enhance the porosity of macroalgae-activated carbon. Nevertheless, even with higher porosity, the depletion of activators in algal biochar might simultaneously reduce the production of algal biochar by more than 25%. The type of macroalgae, the activation temperature, the heating rate, the activation period, and the kind and quantity of activator employed all influence the specific microstructure and physicochemical properties [21].

The researchers created magnetic biochar from waste marine macroalgae by incorporating iron oxide particles, making it magnetized for easy separation. While this magnetic biochar had a lower surface area compared to traditional pinewood sawdust biochar, it exhibited superior HM removal capabilities. However, the iron particles used for magnetization partially reduced its adsorption efficiency by clogging surface pores. The study discovered that a concentration of 0.025–0.05 mol L^{-1} for the iron loading solution optimized separation without compromising the efficacy of HM adsorption. This magnetic biochar demonstrated exceptional HM removal, especially for copper, and was attributed to the presence of oxygen-containing functional groups on its surface. The research suggests that this magnetic algae biochar has promising applications in various pollutant removal processes due to its high performance and ease of separation.

15.4 Industrial Effluent Challenges

The swift rise in the global human population has spurred industrial growth, playing a vital part in the advancement of countries' economies. The upsurge in WW in developing regions is directly linked to population expansion. WWs originating from diverse industrial outlets carry substantial organic pollutants, and certain sources additionally harbor oil, grease, HMs, and chemicals [22]. Numerous manufacturers or small industries in rural regions contribute to local economic development, such as vineyards, tanneries, cheese manufacturing, textile dyeing, coffee processing, and other industries that release a variety of pollutants, as shown in Fig. 15.2. The WW produced at these manufacturing locations frequently contains elevated levels of organic materials (challenging to biologically degrade), nutrients, and metals. Operators often face financial constraints in implementing conventional, costly physical, chemical, and biological technologies for treating these effluents. Furthermore, conventional treatment processes frequently yield effluents containing contaminants at concentrations that fail to comply with stringent regulations governing their direct release into natural water bodies [23, 24].

Because of the wide variety of contaminants it carries, the uncontrolled release of contaminated water from these applications poses serious environmental dangers and is becoming an increasingly pressing concern. Many different chemicals are found in WW, some of which are extremely harmful to living things at high amounts. In addition, significant volumes of organic and inorganic fertilizers are released into the environment, which raises the levels of biological oxygen demand (BOD) and chemical oxygen demand (COD). Excessive nitrogen (N) and phosphorus (P) loading causes aquatic habitats to become eutrophic, which leads to environmental problems such as solid waste production and the release of undesirable pollutants into the atmosphere [25]. It encourages the growth of harmful microbes that endanger a range of aquatic species and lower drinking water quality, aggravating chronic health problems in the vicinity of the discharge site. HMs are the most common class of contaminants found in WW. Even at low quantities, breathing in, consuming, or coming into touch with these harmful compounds can be extremely dangerous to one's health and increase the risk of developing cancer [26]. Concern over the existence of new toxins in water bodies, the risks they pose, and the effects they have on aquatic life and human health has grown in recent years. Synthetic organic compounds are referred to as emerging contaminants because their hazards have

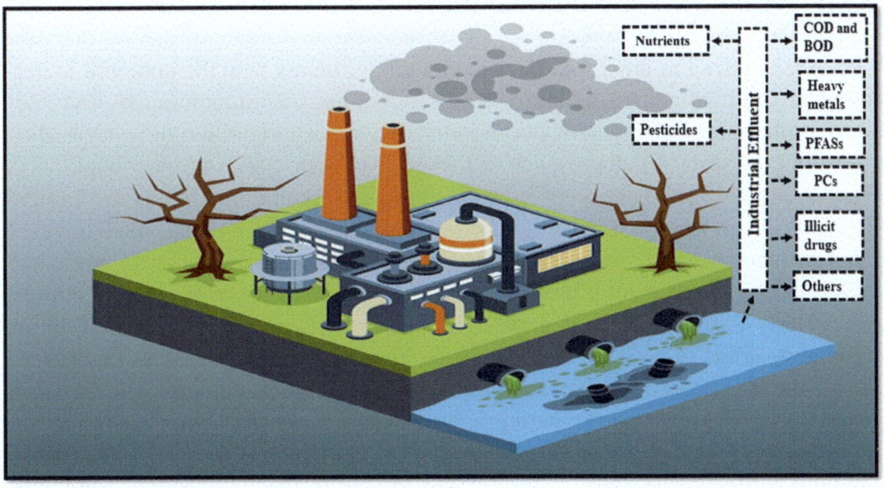

Fig. 15.2 Types of pollutants that are released from industrial activities

lately come to light, even though they have long been present in the environment. These pollutants fall into a number of different categories, including flame retardants, insecticides, artificial sweeteners, pharmaceutical compounds (PCs), and per- and polyfluoroalkyl substances (PFASs). At this point, it is still unclear how much of a harm they have to the environment and human health [27–29]. When water becomes polluted and purification becomes essential, it is crucial to choose the most effective treatment strategy to accomplish the desired purification goals.

15.5 Application of Marine Algae for Industrial Effluent Treatment and Contamination Removal

The release of many types of industrial WW into water bodies causes serious environmental concerns. These WWs add to the environmental problems by including a variety of contaminants, including hydrocarbons, HMs, fertilizers, and others. The environment and human health are seriously at risk from water contamination on a global scale. Marine algae are essential for the elimination of these effluents. The great diversity aids in the screening and identification of novel, promising species and genera by researchers working on the development of marine algae-based waste and WW treatment systems [30]. Over the past few years, microalgae have been employed in the treatment of diverse industrial wastes and WW generated by various industries. One of the most effective adsorbents for removing contaminants from aquatic environments is biochar, as shown in Table 15.1. This includes organic pollutants, HMs, and nitrogen/phosphorus. The unique properties of biochar produced from marine algae allow it to effectively remove pollutants such as HMs, nitrogen, phosphorus, and both organic and inorganic substances [31].

15.5.1 Pharmaceutical Remediation

PCs are recognized as emerging environmental pollutants with the potential to negatively impact both aquatic and terrestrial ecosystems, including humans. Over 200 different pharmaceutically active compounds have been identified in water bodies, illustrating the widespread presence of these substances. For example, quantities of up to 6.5 mg L^{-1} of the antibiotic ciprofloxacin have been found in aquatic habitats [39]. It is clear that pharmaceutical chemicals are being biomagnified to reach higher trophic levels in food chains due to their widespread use for therapeutic purposes. Once released from pharmaceutical industries, a variety of pharmaceuticals are regularly found in water streams. These include common nonsteroidal antiinflammatory drugs (e.g., ibuprofen and diclofenac), antibiotics (such as erythromycin, roxithromycin, ketoconazole, quinolones, and fluoroquinolones), β-blockers (propranolol), antidepressants, antiepileptics (e.g., carbamazepine), cytostatics, and hormones, particularly oral contraceptives. The discharge of these pharmaceutical pollutants into water systems raises toxicity concerns, even at minimal concentrations. The challenges stem from their persistent (nonbiodegradable) nature in the

Table 15.1 Capability of biochar derived from marine algae to remove different pollutants

Type of algae for produced biochar	Type of pollutant	Contact time (min)	pH	Removal efficiency	References
Green marine algal	Arsenic	30	4	99.5	[32]
Laminaria japonica	Phosphate	30	6	97.02	[33]
Ulva fasciata	Methylene blue	–	9	89	[34]
Sargassum duplicatum	Persistent organic pollutants	720	3	93	[35]
Chlorella sp.	Heavy metals	(10, 8, 15 40, 60, 80)	(2, 6, 7)	72–99	[36]
Bloomed *Taihu* blue algae	Nickel	(20, 60)	6	93.26	[37]
Spirulina platensis	Congo red dye	120	2	82.6	[38]

aquatic environment [40]. Hence, the increasing concern is about finding an economical technology to effectively eliminate pharmaceuticals. Limited efforts have been made to address the removal of these compounds using various methods such as adsorption, advanced oxidation, membrane filtration, microextraction, and electrochemical oxidation. The primary limitations of AC stem from its elevated cost and the lack of effective regeneration and reuse. Economical substitutes for these problems have been used to remove several contaminants, including biomass, nonliving algae, clay materials, agricultural wastes, and wastes from processing seafood [41]. Green algae are inexpensive adsorbents because they contain cellulosic polysaccharides that react with a significant amount of proteins to form glycoproteins. These elements are essential to the biosorption process because they contain several functionalized groups, such as amino, carboxyl, sulphate, and hydroxyl. The algal biomass consists of varying proportions of constituents, comprising 53% protein, 21% polysaccharides, and 17% lipids [42]. Recently, the removal of PCs over the algal biomass surface has been investigated, for the adsorptive removal of ciprofloxacin (CIP) and diclofenac (DIC) from water [43]. Kuo et al. [44], explored the removal of the antibiotic sulfamethoxazole with N-doped graphitic biochar made from *Spirulina*. The results showed that the biochar produced demonstrated the ability to capture sulfamethoxazole through electron transfer.

15.5.2 Dyes Remediation

Various industries generate a substantial amount of WW annually. Among the diverse WW types, dye effluents not only contribute to aesthetic issues but also pose significant environmental threats. The discharge of dye WW elevates the toxicity of aquatic environments as well as the levels of COD and biochemical oxygen demand (BOD). Moreover, it has detrimental effects on both human and animal health [45]. Approximately, 10% of synthetic dyes are released into the environment by industries such as textiles, food processing, dye, leather, cosmetics, and paper manufacturing, not receiving the appropriate treatment. Evidently, there is demand for an affordable, environmentally friendly, and effective approach to address WW containing these dyes. Treating water contaminated with dyes is challenging due to their synthetic nature and the complexity of their primarily aromatic compounds based on benzene, naphthalene, and xylene. Because of their synthetic nature and the complexity of their mostly aromatic (benzene, naphthalene, and xylene-based) molecules, these compounds are designed to withstand fading when exposed to light, water, and oxidizing agents. They also show stability in water and resistance to biological degradation. These compounds are designed to withstand fading in the presence of light, water, and oxidizing agents. They also demonstrate stability in water and are resistant to biological deterioration. Many physical, chemical, and biological methods have been developed over the years to remove different hazardous dyes from aqueous solutions, either separately or in combination. (Bio)adsorption is one method of WW decolorization that has proven to be especially effective. This method offers several advantages for controlling water pollution, including lower initial investment costs, significant effectiveness under favourable conditions, versatility in simultaneously removing inorganic, and organic pollutants, minimal generation of disposable sludge, and straightforward regeneration processes. Numerous studies have investigated the elimination of dyes from water using diverse adsorbents. These include both the dead and living biosorbents, such as fungi, algae, and microbial cultures, as well as natural materials such as clays, zeolites, and other siliceous compounds [46]. Additionally, adsorption potential has been investigated for a variety of solid wastes, including agricultural (leaf, fibre, fruit peels, seeds, and waste materials from the forest industries, like sawdust and bark), industrial (sludge, fly ash, and red mud), and miscellaneous adsorbents, like hair, crushed brick, tea waste, cow dung ash, and eggshells.

The literature suggests various minerals, organic substances, and biological materials as potential adsorbents for treating diverse pollutants [47]. Despite the great sorption capacity of commercial activated carbon, its practical usage is often constrained by its relatively elevated cost. Conversely, low-cost adsorbents, though economically favorable, typically have limited sorption capacities. Given the substantial quantity of adsorbents required for treating dye effluents, current research is focused on enhancing the sorption capacity of these cost-effective alternatives. Recently, there has been a lot of interest in marine macroalgae as a possible treatment for dye-contaminated water. Algae are abundant, renewable biological resources that are thought to adsorb dye molecules through interactions with functional groups like phosphate, amino, carboxylate, and hydroxyl found in their cell walls. A range of modification techniques, including physical and chemical pretreatments, have been utilized in an effort to optimize algae's sorption potential. Notably, these alteration techniques have the potential to increase or decrease the algal biomasses' biosorption efficiency. For instance, a study [48] revealed that the Acid Blue 25 removal efficiency for untreated *Jania adhaerens* (red alga), untreated *Padina australis* (brown alga), HCl-treated *P. australis*, and HCl-treated *J. adhaerens* was 77.34%, 71.28%, 50.56%, and 85.19%, respectively.

15.5.3 Heavy Metals Remediation

The increasing demands of the industrial sector have led to a notable increase in worldwide HM pollution in recent years. These metals, originating from both municipal and industrial activities, can infiltrate watersheds, posing a substantial challenge for removal. Their high solubility in water makes them persistent threats to worldwide water resources. Traditional removal approaches may generate substantial quantities of toxic sludge and secondary pollutants, proving ineffective, especially when dealing with lower but still hazardous concentrations of metals present in WW [49]. However, biochar produced from marine algae exhibits the capability to effectively remove and absorb different HMs from aqueous solutions, such as Zn(II), Co(II), Ni(II), Cr(VI), and Hg(II). Several studies in the body of literature have demonstrated how well biochar produced from microalgae removes HMs. In specific investigation, Daneshvar et al. [50], utilized the 500 °C temperature during the pyrolysis process to produce biochar from *Scenedesmus quadricauda*. The study focused on evaluating the biochar's performance in removing Cr(VI), revealing complete elimination of Cr(VI) with initial concentrations between 1 and 10 mg L^{-1}. The authors further identified ionic exchange, electrostatic interaction, and Cr(VI) reduction as the key mechanisms involved in the removal process. Prior to this, Bordoloi et al. conducted research on a related topic [51], conducted several equilibrium investigations, including Freundlich and Langmuir adsorption isotherms, for assessing the effectiveness of Scenedesmus microalgae-derived biochar in removing Co(II) ions. The results showed that the produced biochar has an adsorption capability of 0.672 mg g^{-1}. Ge et al. [52] found that high salinity *Spirulina* microalgae biochar effectively removed Hg(II) ions through immobilization, obtaining a sorption capacity ranging from 6 to 12.7 mg g^{-1} for sustained absorption, especially when *Microcystis* sp. was involved. Chelated nickel from WW produced during alloy electroplating is successfully removed by biochar. In a different study, Zn, Cd, and Cu were extracted from liquid solutions using biochar derived from marine algae, namely, *Sargassum fusiforme* and *Saccharina japonica*. The findings indicated in which the predominant mechanism for the elimination of HMs was attributed to a substantial presence of oxygen functional groups. Furthermore, cation exchange emerged as an additional removal mechanism facilitated by biochar, given the abundant mineral content in marine algae, including Na, K, Mg, Ca, and others [53].

15.6 Conclusion

Through a comprehensive examination of production processes, operational considerations, and real-world case studies, this chapter sheds light on the practical aspects of industrial effluent challenges and utilizing marine algae-derived biochar in industrial effluent treatment. The successful implementation of this approach underscores its potential as an eco-friendly and economically viable solution for mitigating the adverse environmental impacts of industrial activities. As we strive to address the global challenges posed by environmental pollution, the insights provided in this chapter offer valuable contributions to the ongoing efforts in developing sustainable and effective WW treatment methods.

Recommendations

- Explore advanced pyrolysis techniques to enhance biochar yield and quality, investigate the influence of different marine algae species on biochar properties. Research and identify novel marine algae sources for biochar.
- Future studies should focus on optimizing production processes, exploring different marine algae varieties, and assessing the potential for scaling up the application.
- Increase awareness among the public and stakeholders about the benefits of using marine algae-derived biochar in industrial effluent treatment.

References

1. Selvasembian R et al (2022) Recent progress in microbial fuel cells for industrial effluent treatment and energy generation: fundamentals to scale-up application and challenges. Bioresour Technol 346:126462
2. Panda AK, Singh RK, Mishra DK (2010) Thermolysis of waste plastics to liquid fuel: a suitable method for plastic waste management and manufacture of value added products—a world prospective. Renew Sust Energ Rev 14(1):233–248
3. Bhat SU, Qayoom U (2021) Implications of sewage discharge on freshwater ecosystems. In: Sewage—recent advances, new perspectives and applications. IntechOpen
4. Teh CY, Budiman PM, Shak KPY, Wu TY (2016) Recent advancement of coagulation–flocculation and its application in wastewater treatment. Ind Eng Chem Res 55(16):4363–4389
5. Zhang C et al (2019) Biochar for environmental management: mitigating greenhouse gas emissions, contaminant treatment, and potential negative impacts. Chem Eng J 373:902–922
6. Khan AA et al (2022) Recent progress in microalgae-derived biochar for the treatment of textile industry wastewater. Chemosphere 306:135565
7. Veluchamy C, Palaniswamy R (2020) A review on marine algae and its applications. Asian J Pharm Clin Res 13(3):21
8. Hou Z, Luo M, Yang Y, Zhou J, Liu L, Cai J (2021) Algae-based carbons: design, preparation and recent advances in their use in energy storage, catalysis and adsorption. New Carbon Mater 36(2):278–303
9. Chen WH, Lin BJ, Huang MY, Chang JS (2015) Thermochemical conversion of microalgal biomass into biofuels: a review. Bioresour Technol 184:314–327
10. Du Z (2013) Thermochemical conversion of microalgae for biofuel production. University of Minnesota
11. Yu KL et al (2017) Recent developments on algal biochar production and characterization. Bioresour Technol 246:2–11
12. Rasam S, Talebkeikhah F, Talebkeikhah M, Salim AS, Moraveji MK (2021) Physico-chemical properties prediction of hydrochar in macroalgae Sargassum horneri hydrothermal carbonisation. Int J Environ Anal Chem 101(14):2297–2318
13. Law XN et al (2022) Microalgal-based biochar in wastewater remediation: its synthesis, characterization and applications. Environ Res 204:111966
14. Gan YY et al (2018) Torrefaction of microalgal biochar as potential coal fuel and application as bio-adsorbent. Energy Convers Manag 165:152–162

15. Chen WH, Peng J, Bi XT (2015) A state-of-the-art review of biomass torrefaction, densification and applications. Renew Sust Energ Rev 44:847–866
16. Titirici MM et al (2015) Sustainable carbon materials. Chem Soc Rev 44(1):250–290
17. Sajjadi B, Chen WY, Egiebor NO (2019) A comprehensive review on physical activation of biochar for energy and environmental applications. Rev Chem Eng 35(6):735–776
18. Leng L et al (2021) An overview on engineering the surface area and porosity of biochar. Sci Total Environ 763:144204
19. Alvarez PM, García-Araya JF, Beltrán FJ, Masa FJ, Medina F (2005) Ozonation of activated carbons: effect on the adsorption of selected phenolic compounds from aqueous solutions. J Colloid Interface Sci 283(2):503–512
20. Nguyen T-B, Truong QM, Chen CW, Chen WH, Dong C-D (2022) Pyrolysis of marine algae for biochar production for adsorption of Ciprofloxacin from aqueous solutions. Bioresour Technol 351:127043
21. Guo Z, Li D, Liu Y, Wang H, Wang L (2020) Recent advances and challenges in biomass-derived porous carbon nanomaterials for supercapacitors. Chem Eng J 397:125418
22. Sarayu K, Sandhya S (2012) Current technologies for biological treatment of textile wastewater—a review. Appl Biochem Biotechnol 167:645–661
23. Zhang D, Gersberg RM, Ng WJ, Tan SK (2014) Removal of pharmaceuticals and personal care products in aquatic plant-based systems: a review. Environ Pollut 184:620–639
24. Sukumaran D (2013) Phytoremediation of heavy metals from industrial effluent using constructed wetland technology. Appl Ecol Environ Sci 1(5):92–97
25. Wang C et al (2022) Biochar-based slow-release of fertilizers for sustainable agriculture: a mini review. Environ Sci Ecotechnol 10:100167
26. Abidli A, Huang Y, Ben Rejeb Z, Zaoui A, Park CB (2022) Sustainable and efficient technologies for removal and recovery of toxic and valuable metals from wastewater: recent progress, challenges, and future perspectives. Chemosphere 292:133102
27. Hena S, Gutierrez L, Croué JP (2021) Removal of pharmaceutical and personal care products (PPCPs) from wastewater using microalgae: a review. J Hazard Mater 403:124041
28. Sudarshan S et al (2023) Impact of textile dyes on human health and bioremediation of textile industry effluent using microorganisms: current status and future prospects. J Appl Microbiol 134(2):lxac064
29. Koutra E, Mastropetros SG, Ali SS, Tsigkou K, Kornaros M (2021) Assessing the potential of Chlorella vulgaris for valorization of liquid digestates from agro-industrial and municipal organic wastes in a biorefinery approach. J Clean Prod 280:124352
30. Bilal M, Rasheed T, Sosa-Hernández JE, Raza A, Nabeel F, Iqbal HMN (2018) Biosorption: an interplay between marine algae and potentially toxic elements—a review. Mar Drugs 16(2):65
31. Singh A, Sharma R, Pant D, Malaviya P (2021) Engineered algal biochar for contaminant remediation and electrochemical applications. Sci Total Environ 774:145676
32. Senthilkumar R, Reddy Prasad DM, Govindarajan L, Saravanakumar KS, Naveen Prasad BS (2020) Synthesis of green marine algal-based biochar for remediation of arsenic (V) from contaminated waters in batch and column mode of operation. Int J Phytoremediation 22(3):279–286
33. Jung K-W, Jeong TU, Kang HJ, Ahn KH (2016) Characteristics of biochar derived from marine macroalgae and fabrication of granular biochar by entrapment in calcium-alginate beads for phosphate removal from aqueous solution. Bioresour Technol 211:108–116
34. Kadimpati KK et al (2024) Design of innovative hybrid biochar prepared from marine algae and magnetite: insights into adsorption performance and mechanism. Chem Eng Res Des 201:218–227
35. Hung C, Chen C, Huang C, Cheng J, Dong C (2022) Algae-derived metal-free boron-doped biochar as an efficient bioremediation pretreatment for persistent organic pollutants in marine sediments. J Clean Prod 336:130448
36. Amin M, Chetpattananondh P (2019) Biochar from extracted marine Chlorella sp. residue for high efficiency adsorption with ultrasonication to remove Cr(VI), Zn(II) and Ni(II). Bioresour Technol 289:121578

37. Wang H, Zhao H, Yan Q (2020) Adsorption and Fenton-like removal of chelated nickel from Zn-Ni alloy electroplating wastewater using activated biochar composite derived from Taihu blue algae. Chem Eng J 379:122372
38. Nautiyal P, Subramanian KA, Dastidar M (2016) Adsorptive removal of dye using biochar derived from residual algae after in-situ transesterification: alternate use of waste of biodiesel industry. J Environ Manag 182:187–197
39. Petrie B, Barden R, Kasprzyk-Hordern B (2015) A review on emerging contaminants in wastewaters and the environment: current knowledge, understudied areas and recommendations for future monitoring. Water Res 72:3–27
40. Ali MEM, Abd El-Aty AM, Badawy MI, Ali RK (2018) Removal of pharmaceutical pollutants from synthetic wastewater using chemically modified biomass of green alga Scenedesmus obliquus. Ecotoxicol Environ Saf 151:144–152
41. Gupta VK, Nayak A, Agarwal S (2015) Bioadsorbents for remediation of heavy metals: current status and their future prospects. Environ Eng Res 20(1):1–18
42. Laurens LML et al (2012) Algal biomass constituent analysis: method uncertainties and investigation of the underlying measuring chemistries. Anal Chem 84(4):1879–1887
43. Nakarmi KJ et al (2022) Synthesis of biochar from iron-free and iron-containing microalgal biomass for the removal of pharmaceuticals from water. Environ Res 214:114041
44. Kuo H-C, Lin YG, Chiang CL, Liu S-H (2021) FeN@N-doped graphitic biochars derived from hydrothermal-microwave pyrolysis of cellulose biomass for fuel cell catalysts. J Anal Appl Pyrolysis 153:104991
45. Kousha M, Daneshvar E, Dopeikar H, Taghavi D, Bhatnagar A (2012) Box–Behnken design optimization of Acid Black 1 dye biosorption by different brown macroalgae. Chem Eng J 179:158–168
46. Okoro HK et al (2022) Nanomaterial-based biosorbents: adsorbent for efficient removal of selected organic pollutants from industrial wastewater. Emerg Contam 8:46–58
47. de Oliveira Brito SM, Andrade HMC, Soares LF, de Azevedo RP (2010) Brazil nut shells as a new biosorbent to remove methylene blue and indigo carmine from aqueous solutions. J Hazard Mater 174(1–3):84–92
48. Daneshvar E et al (2015) Application of response surface methodology for the biosorption of Acid Blue 25 dye using raw and HCl-treated macroalgae. Desalin Water Treat 53(6):1710–1723
49. Azimi A, Azari A, Rezakazemi M, Ansarpour M (2017) Removal of heavy metals from industrial wastewaters: a review. ChemBioEng Rev 4(1):37–59
50. Daneshvar E et al (2019) Hexavalent chromium removal from water by microalgal-based materials: adsorption, desorption and recovery studies. Bioresour Technol 293:122064
51. Bordoloi N, Goswami R, Kumar M, Kataki R (2017) Biosorption of Co(II) from aqueous solution using algal biochar: kinetics and isotherm studies. Bioresour Technol 244:1465–1469
52. Ge Y, Zhu S, Chang JS, Jin C, Ho SH (2020) Immobilization of Hg(II) on high-salinity Spirulina residue-induced biochar from aqueous solutions: sorption and transformation mechanisms by the dual-mode isotherms. Environ Pollut 265:115087
53. Poo KM, Son EB, Chang JS, Ren X, Choi YJ, Chae KJ (2018) Biochars derived from wasted marine macro-algae (Saccharina japonica and Sargassum fusiforme) and their potential for heavy metal removal in aqueous solution. J Environ Manag 206:364–372

Production of Biofuel from Macroalgae/Seaweed

16

C. Madhuri, N. Rajesh, T. Chandrasekhar, V. Anu Prasanna, and K. Riazunnisa

Abstract

To meet the energy demand of the world, existing fossil fuels were insufficient because of their nonrenewability and producing environmental pollution. Hence, there is an urgent need to replace fossil fuels. Marine macroalgae (green, red, and brown macroalgae) have attracted attention as an alternative source of renewable biomass for producing both fuels with zero carbon dioxide emission and biocompatible compounds with medicinal and industrial applications. Microalgae can grow easily and efficiently by utilizing the carbon resources in the water for higher biofuel production. The existence of elevated levels of structural polysaccharides and low lignin contents make them attractive for biofuel production. Thermochemical and microbiological processes are the two main processes used for converting macroalgae into biofuels. The development of algae-based biofuel emits zero carbon dioxide, in turn is eco-friendly in nature and benefits producers economic and social advancement. This chapter focuses on the biofuel production from different macroalgae species.

Keywords

Biodiesel · Biogas · Bio-oil · Bioethanol · Transesterification

C. Madhuri · N. Rajesh · K. Riazunnisa (✉)
Department of Biotechnology and Bioinformatics, Yogi Vemana University, Kadapa, Andhra Pradesh, India

T. Chandrasekhar
Department of Environmental Science, Yogi Vemana University, Kadapa, Andhra Pradesh, India

V. Anu Prasanna
Department of Animal Science, Yogi Vemana University, Kadapa, Andhra Pradesh, India

M. Rafatullah et al. (eds.), *Multidisciplinary Applications of Marine Resources*,
https://doi.org/10.1007/978-981-97-5057-3_16

16.1 Introduction

Petroleum derivatives, for example, coal, oil, and combustible fuels, are significant energy sources, and around 87% of worldwide carbon dioxide is transmitted because of the anthropogenic activities by the use of coal, oil, and petroleum gas. The rising population with more utilization combined with high speed of improvement have prodded higher double dealing of petroleum derivatives prompting the heightening costs and resultant ozone depleting substances (GHGs) presenting issues for the planet's climatic strength biofuels are arising as a promising option in contrast to fluid energizes. Understanding the capability of biomass, various advancements have developed the change of biomass into energizers, prominently called biofuels derived from inexhaustible plants or other natural squanders, they have the benefits of chopping down fossil fuel byproducts and releasing oil. Approximately, 80% of India's energy needs are satisfied by the use of biomass energy as kindling, farming deposits, cow fertilizer cake, and other normal feedstocks. Over three decades, food crops like sugarcane and corn were used to produce biofuel from the first-generation feedstock. However, this method was met with resistance resulting from the low stock and food crop conflict [1]. The main sustainable power sources that can straightforwardly swap petroleum derivatives for existing as well as future energy deficiencies are biomass as well as biofuel. The sources are ecologically positive as well as inexhaustible. Biofuels are of three ages arranged in light of their sources.

16.1.1 First-Generation Biofuels

Sugar, starch, and vegetable fats oils were the plant-based biomass used for biofuel production. This makes up most of the cheap and plentiful nonfood compounds that plants can produce. The probability of using cellulosic and heterogeneous biomass suggests lower costs. Anyway, the environmental consequence elevated from biofuel burning exceptionally disturbs the carbon cycle, which is associated with the start of oil subsidiaries. Besides, the exhaustion of different present biomass without appropriate pay achieved epic biomass lack, emerging environmental issues like deforestation and biodiversity misfortune. In a request for possible and down-to-earth choices rather than nonenvironmentally friendly power sources, past examinations have nitty gritty the abilities to overwhelm green growth construed biomass for the improvement of a superior structure [2].

These feedstocks are not maintainable in circumstances of increasing food requests as well as "food versus fuel" discussions. In addition, certain impediments such as low profit from speculation, reliance on nonrenewable energy sources for creation processes, and the more exorbitant cost of feed crops downsize the monetary plausibility for the biofuel generation from the original feedstock.

16.1.2 Second-Generation Biofuels

These biofuels were produced to overcome food scarcity. Lignocellulosic materials include waste leaves, bagasse, jatropha, agricultural waste, groundnut fodder, and grass that are utilized in place of materials for second-generation biofuels. Natural materials from timberland litter, leaves, and wood have more quantity of carbohydrate sources which becomes unrefined substances for the production of biofuel and they are efficiently accessible, so one can work on the profit from speculation with lower dependence on food crops. In any case, the withdrawal of regular natural matter thus expands the utilization of manures and nitrogenous enhancements subsequently expanding the chances of NO_2 discharges.

16.1.3 Third-Generation Biofuels

The basis of these biofuels was laid by the drawbacks of previous generation biofuels, like low returns on investment and complex manufacturing procedures. Photosynthetic microorganisms like algae, cyanobacteria, and microalgae are the sources of these biofuels. Algae are an excellent option for producing biofuel because of their harsh environmental tolerance and low area needs. They can also reduce CO_2 emissions. Almost every kind of water, including freshwater, saltwater, and even industrial waste fluids, can support the growth of algae. Regarding growth and oil content, algae grow 20–30 times quicker than crops that produce food, and they have an oil content that is around 30 times higher than traditional feedstocks from the first and second generations [3].

16.2 Macroalgae

Algal biomass can serve as an initial source for biofuel production. To generate bioenergy, macro- and microalgae must meet the following requirements. They ought to be:

1. Extremely productive.
2. Simple to harvest.
3. They ought to be resilient to water currents in an open ocean.
4. The cost at which they are produced ought to be comparable to or less than that of the other accessible sources [4].

Owing to the great scale of variety in the chemical composition of algae, there is a capability for producing different biofuels from them, like biodiesel, bioethanol, biohydrogen, as well as biomethane. However, there is one thing in common with all algae: they have a lot of water. Since this is typically an unwanted component of any fuel type, low H_2O content technologies—like fast pyrolysis, which produces bio-oils, for instance—become sparingly difficult due to the expenses associated with drying the material [5]. Choosing species or strains with high oil-like precursor content is the initial step to produce algal biofuels. It is well known that macroalgae contain important quantities that can be fermented to produce biogas. In comparison to microalgae, they are more cost-effective to produce biofuels [6]. A variety of eukaryotic, photosynthetic, and aquatic organism types can be seen in macroalgae. Seaweeds are multicellular macroalgae species that resemble plants in certain aspects. The macroalgae structures have a very varied common look. While some have mats and leaves, the main characteristic is long blades. Furthermore, certain macroalgal species have air bladders that allow them to float on the water's surface and stand erect. Macroalgae is categorized into Rhodophyta, Phaeophyta, and Chlorophyta. Macroalgae have intricate life cycles that differ between species. They can reproduce both sexually and asexually. Macroalgae are now widely grown commercially for use in food and fuel applications. They have a high proportion of starches, which can be transformed into a variety of liquid and gaseous fuels [7].

16.2.1 Chemical Composition of Macroalgae

Dry matter content in macroalgae is typically only 10–15%. About 60% of the dried macroalgae is made up of starches. Starches are mostly derived from the cell walls of macroalgae. Some have starch in addition to cellulose (especially red and green). There are sporadic variations in the debris content and biochemical arrangement. Full-scale green growth differs from microalgae and lignocellulosic biomass because of sources like mangan, fucoidan, fucose, alginate, laminarin, mannitol, carrageenan, ulvan, and uronic corrosive. Furthermore, acceptability for downstream handling such as anaerobic absorption or the aging system is provided by a lignin content which is low by 0.03 g/kg dry matter, which eliminates the need for costly pretreatment. Based on an assessment of 20 macroalgae species, *Palmaria palmata* was the species with highest concentration of carrageenan and acquired a yield of 354 mg/g. The more carrageenan content enables more bioethanol yield. However, earthy-colored green growth, the most developed type of green growth worldwide, is high in alginate. It contains enormous source of laminarin, mannitol upon harvest, and very little ordinary sugar. The brown as well as red green growth indicates a poor lipid production rate (less than 5%), which is too low to consider a full switch to biodiesel. While most green growth types are rich in starch (20–30%), a few are higher in cellulose. Furthermore, compared to biomass, which has a 1–1.5% halogen concentration and a 0.5–11% soluble base metal content, macroalgae have these greater levels. The shape of macrophyta is the primary determinant of the supplement content (N, P), with ecological supplement contamination coming in second [8].

The fundamental algal phyla are *Rhodophyta, Phaeophyta,* and *Chlorophyta.* There are currently 221 recognized species of macroalgae utilized by mankind, with 66% of them being used as food. Most of algal biomass originates from species with five genera, *Laminaria, Undaria, Eucheuma, Gracilaria, and Porphyra,* representing 76% of the complete weight for refined macroalgae. An ocean growth feedstock for biofuel creation ought to have the accompanying qualities for the selection: (a) promptly multiply vegetatively; (b) have a rapid development rate; (c) impervious to go after microbes, growths, epiphytes, and slow eaters; (d) impervious to harm and expulsion by the tide, flows, and tempests; (e) having a high calorific or heating value; and (f) low debris, sulphur, and nitrogen content [9].

16.3 Biofuel Conversion Methods

There are different methods to convert biomass into biofuels which are mainly categorized into biological and thermochemical methods. The most commonly involved methods for the generation of biofuels are fermentation, liquefication, pyrolysis, anaerobic absorption, and transesterification. Algal biomass might influence the hydrolysis step that requires additional time which influences the biofuel yield. It may be reduced by appropriate pretreatment. Pretreatment breaks the obligation of particles and depolymerizes the complications, subsequently expanding the solubilization. The solubilized tests can be handily utilized in the transformation cycle and help in biofuel production. Different pretreatment strategies, for example, physical, synthetic, organic, and mechanical techniques are utilized for solubilization of the complicated substrates in macroalgae. For extraction of biofuel from macroalgae, an underlying drying step is required [10]. This prompts two gatherings of cycles:

1. Biofuel production methods from dry macroalgae
   - Combustion
   - Transesterification
   - Pyrolysis
2. Biofuel production strategies for wet macroalgae
   - Aging to bioethanol or biobutanol
   - Anaerobic processing (Fig. 16.1)

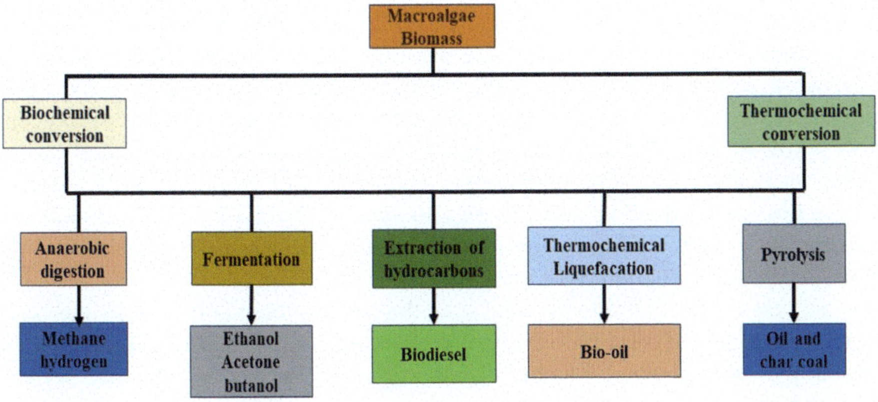

Fig. 16.1 Conversion methods for biofuels from macroalgae

16.3.1 Direct Combustion

The usage of macroalgal biomass for biofuel production by using combustion has been concentrated by numerous scientists. It is not suggested to utilize feedstock with high debris, dampness, soluble base metal (sodium and potassium), and halogen content for direct combustion. The higher warming worth (HHV) of macroalgal biomass is lower than that of the earthbound energy crops (14–20 MJ/kg). The combustion productivity of ocean growth biomass for the age of intensity is exceptionally low and is normally somewhere in the range of 15% and 30%. Nonetheless, the utilization of kelp biomass for direct burning interaction is not suggested because of the gamble of consumption, fouling, and particulate emanations. The combustion waste deposits are financially suitable, albeit specialized issues like debris taking care of and pipe gas emissions (NO_x and SO_2) remain [11].

16.3.2 Pyrolysis

Pyrolysis includes the thermochemical disintegration of natural designs into an enthusiastically helpful noncondensable combination of gases, condensable fluids, and strong buildups without any oxygen/follow oxygen air and in the presence/nonappearance of an impetus. In most cases, this process involves three crucial steps such as getting rid of moisture, breaking down organic structures, and slowly breaking down any remaining solids.

Nonreactant pyrolysis alludes the warm exploitation of natural designs and happens modestly at 623–973 K without any impetus. The temperature and the warming rate affect the yields at high temperatures and are trailed by the expansion of pyrolytic fumes. The downstream buildup produces a dim thick bio-oil or pyrolysis oil as well as nonvolatile strong biochar. Pyrolysis, in particular, carbonization and traditional pyrolysis at warming steps of 0.1–1 K s^{-1} in min-hours, and quick pyrolysis through quick and blaze pyrolysis at 1–200 K s^{-1} less than 0.5–0.5 s, was noticed. Microalgal strains through slow pyrolysis yield biochar, CH_4, and CO_2 vaporous items [12]. The temperature goes from 350 to 650 °C where the significant fuels are delegated strong (ex: biochar), gas (ex: syngas), and fluid (ex: bio-oil). Temperature, molecule size, as well as warming rate pyrolysis, are arranged into slow, quick, and glimmer pyrolysis. Bio-oil and biochar were produced by pyrolysis and gasification of various biomass in the presence as well as absence of oxygen. Slow pyrolysis is preferred for the production of large quantities of biofuel due to longer residence time, slow temperature rise, and slow degradation of biomass favours biochar production over bio-oil and syngas. Pyrolysis can be rationalized into three responses. During the initial step, drying out of biomass happens followed by the essential production of biochar with the extra biomass where a few unstable parts and gas are lost. During the final step, the optional burn is framed by the deterioration of essential roast brought by a carbon material. In addition, the temperature of pyrolysis impacts biochar's properties like surface region and surface utilitarian gatherings. As a result of the aliphatic alkyl and ester groups in the natural mixtures breaking down at high temperatures, the porosity and surface area of biochar increase, removing the materials that restrict pores. While the biochar supplied at low pyrolysis temperatures is hydrophilic, the biochar is hydrophobic and thermally stable. At higher temperatures, surface functional groups like lactone, carboxyl, pyridine, and phenol are introduced and chemical bonds are rearranged. Algae bio-oil of high quality is produced through the rapid pyrolysis process. Quick pyrolysis of a macroalgae *Scenedesmus* species was completed at 500 °C with fume of 2 s and 55% bio-oil was noticed and in *C. vulgaris*, 53% of bio-oil was gotten at the quick pyrolysis temperature of 500 °C [13].

16.3.3 Simultaneous Saccharification and Fermentation (SSF)

SSF combines enzymatic hydrolysis and maturation into a single process for the production of bioethanol from biomass. It is generally preferred over the SHF cycle due to the restraint of cellulase that can be limited and more bioethanol production rate can be accomplished because of fast transformation of glucose into ethanol by yeast. Also, the SSF process represented a few benefits generally safe from defilement, little starting osmotic pressure of maturing specialist, and high energy effectiveness. The biomass from *K. alvarezii* could be used for the production of bioethanol and achieved 53 g/L of bioethanol. Bioethanol delivered *S. japonica* through SSF and pretreatment methods, utilizing different maturation mediums, for example, sodium citrate, deionized H_2O, and fluid hydrolysate showed inconsequential contrasts. Bioethanol convergence of 6.65 g/L was acquired, which related to glucose comparable concentration of 13.01 g/L, or SSF yield of 67.41% because of the complete accessible glucan of the pretreated *S. japonica* [14].

16.3.4 Anaerobic Digestion

Marine biomass has an estimated energy potential of over 100 EJ year, which is significantly more than the energy potential of terrestrial biomass (22 EJ year) or municipal solid waste (7 EJ year) for methane production. The promotion of kelp has been accounted to bring about an extensive variety of biomethane yields brought in species. Eutrophic circumstances can increase biomass yield from kelp cultivating, albeit harming kelp epiphytes development. Close to lessening supplement load, the assortment, and conceivable development, of kelp is probably going to moderate this eutrophication and have other financial advantages. Contingent upon ocean growth types and species, items like unsaturated fats, oils, normal colors, cell reinforcements, and high-esteem natural parts can be separated with natural "extras" went for the gold. The accumulation from ocean growth following dissolvable extraction is more appropriate for biogas production than new or float kelp substrates alone. Manageability is supposed to increment with ideas like BioEnergy with Carbon Catch and Capacity (BECCS). Microalgae development with the expulsion of CO_2 from biomethane might be one more elective technique for manageable carbon catch [15].

16.3.5 Liquefaction

Aqueous liquefaction has been utilized to handle an extensive assortment of biomass including biowaste (excrement, sewage muck, metropolitan waste, and food processing waste), and lignocellulose (wood and yield buildups). Because of its high (10-half) inborn debris and dampness content, macroalgae is undeniably fit for HTL, especially because of its high salt substance, which can create some issues of slagging and fouling in pyrolysis and combustion; however, it has been recommended to chemically affect biounrefined creation under HTL conditions. Handling of kelp of the macrocystis yielded 19.2% (on a dry, debris free premise (daf)) of bioraw petroleum. The sugars in macroalgal biomass produce oils, made transcendently out of phenolic compounds. The macroalgae *Enteromorpha (Ulva)* and *Laminaria* yields 10–25% are observed. These yields are especially less than those commonly noticed for microalgae; this has been attributed to bring down lipid levels in macroalgal feedstocks as for microalgae, and high starch levels rather than prompting the arrangement of fluid stage products. Extraction of broken down organics from the fluid stage items effectively expands the general yield of biomass yet brings down its general quality by expanding heteroatom content and draining higher warming worth (HHV). It was estimated that the presence of unsaturated fats and proteins prompted an increase in biounrefined nitrogen levels. The event of optional responses between liquefaction items was affirmed: various pinnacles were seen in GC/MS of the biounrefined produced by the feedstock blend that was absent in the oils shaped from the singular green growth [16] (Fig. 16.2).

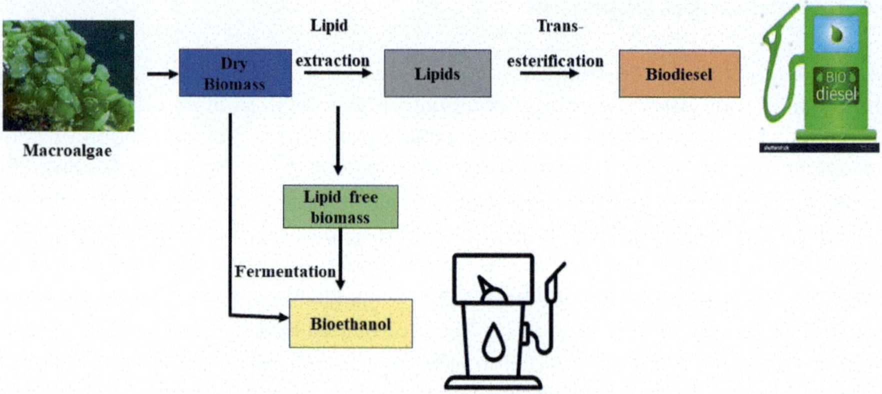

Fig. 16.2 Flow chart for production of biofuels from macroalgae

16.4 Bioethanol

Bioethanol delivered from algal sources has acquired significance in view of expanded intensity of vaporization as well as octane rating in examination with the ordinary energizes. Bioethanol shown as a perfect and environmentally friendly fuel, produced from different feedstock is the best option in contrast to gas fuel. Bioethanol produced from original feedstock when delivered in enormous amounts might prompt food deficiency around the world though the development of bioethanol from the lignocellulose materials isn't viewed as suitable on the modern scale [17]. Bioethanol is broadly created utilizing different compound and natural methodologies. The natural methodology includes the aging of biomass utilizing microorganisms under anaerobic circumstances. Aging is an old innovation in the world, which suggests the conversion of starch into corrosive or liquor.

The bioprocessing of carbohydrate containing feedstock is principally completed in two stages. The initial stage includes the polysaccharide hydrolysis into fermentable sugars followed by their transformation to bioethanol utilizing reasonable microbes. Furthermore, downstream handling incorporates purging and centralization of bioethanol by the refining process. A significant limit of the creation cycle is the lower convergence of bioethanol in the maturation stock. Subsequently, a lot of energy is expected for the partition and cleansing of bioethanol from the maturation stock [18]. The bioethanol production from *Gracilaria* species provides 0.236 g of bioethanol from 1 g of dry macroalgae yield. An ideal bioethanol convergence of 37 g/L was accomplished after 12 h of maturation, relating to a 90.6% hypothetical yield. After maturation by *S. cerevisiae* at 40 °C for 48 h, the bioethanol change pace of the compound hydrolysate from *Sargassum* species reached 89%, which was uniquely higher than the hypothetical yield of 51% in light of glucose as substrate. Consequently, it very well may be presumed that the aging of glucose or galactose after hydrolysis can be matured by the yeast of *S. cerevisiae*. The cellulosic residue of *K. alvarezii* could be used for the production of bioethanol and achieved 53 g/L of bioethanol using the SSF process. Bioethanol production of 6.65 g/L was acquired, where glucose comparable concentration of 13.01 g/L, or SSF yield of 67.41% from pretreated *S. japonica*. Bioethanol production from *Gelidium amansii* with bioethanol change yields 76.9% after 24 h. A limit of 6.2% of bioethanol was acquired under sonication in 3 h versus just 4.9% bioethanol under brooding even after 48 h [14]. The production of biofuel from *Padina tetrastromatica* and *Sargassum swartzii* with four-hour enzyme treatment and fermentation by *S. cerevisiae* and *Z. mobilis* acquired a 95.12% of yield [17]. However, 22% of biofuel yield was achieved by the fermentation of *Ulva fasciata* by *Saccharomyces cerevisiae, Pseudomonas,* with the help of the chemical hydrolysis pretreatment method [19]. *Chaetomorpha linum* used for biofuel production through fermentation and saccharification *by S. cerevisiae* methods along with prehydrolysis treatment achieved 18% of yield [20] (Table 16.1).

Table 16.1 Macroalgae used for the production of bioethanol

Type of species	Conversion methods	Microorganism used	Pretreatment methods	Biofuel yield	References
Padina tetrastromatica and *Sargassum swartzii*	Fermentation	*Saccharomyces cerevisiae* and *Zymomonas mobilis*	4 h of enzyme treatment	95.12%	[17]
Phaeophyceae (Saccharina japonica and Undaria pinnatifida) and Rhodophyceae (Eucheuma, Kappaphycus, and Pyropia)	Fermentation & saccharification	*Saccharomyces cerevisiae* and *Escherichia coli*	Acid hydrolysis with sulfuric acid	64%	[21]
Ulva fasciata	Fermentation	*Saccharomyces cerevisiae SH02, Pseudomonas* sp. *SH03*	Chemical hydrolysis	22%	[19]
Jania rubens, Ulva lactuca, Sargassum latifolium	Saccharification, fermentation	*Trichoderma asperellum*	Hydrolysis	0.29 g	[22]
Gelidium elegans	Fermentation	*Saccharomyces cerevisiae,*	Hydrolysis, purification	90%	[23]
G. sesquipedale, U. lactuca	Fermentation	*S. cerevisiae* yeast	Enzymatic hydrolysis	67%	[24]
Laminaria digitata	Saccharification, fermentation	*P. angophorae*	2 M HCl, 30 min, 5% load	13.2%	[25]
Chaetomorpha linum	Saccharification, fermentation	*S. cerevisiae*	Prehydrolysis	18%	[20]
Sargassum sagamianum	Thermal liquefication	*P. stipitis*	–	10.0%	[26]
Gracilaria salicornia	Fermentation	*E. coli KO11*	2% sulfuric acid, 120 °C, 30 min	7.91%	[27]

16.5 Biodiesel

Biodiesel is one of the most broadly utilized biofuels in the transport business. Biodiesel production expenses can be reduced for commercialization by investigation of nonconsumable savvy feedstocks. Algal species with high lipid content are accessible for biodiesel production which can decrease the expense of production of biodiesel [14]. The macroalgae was washed with seawater at first to clean the visible epiphytes and sand particles. The macroalgae was again washed with water and dried five to multiple times followed by stove drying at 55–70 °C for 22 h. The powder was used for the production of biofuel. Hexane is famous for oil extraction because of its incredible destructiveness, low oily repeating impacts, edge of boiling over, and high solidness.

According to the available literature, the bond elute process was used to extract triglycerides using the same solvent combination. Roughly 6.58% of petrol extraction yield was accomplished using n-hexane as a dissolvable. The algal oil extraction of 6.38% utilizes n-hexane-ether in dissolvable blend. While applying n-hexane-ethyl liquor and n-hexane 2-isopropanol as a dissolvable combination, the oil extraction yield was seen to be 5.23% and 5.57%, respectively. Hexane is the most generally involved dissolvable for the expulsion of oil acquired from microalgae and macroalgae. It is considerably less hazardous, has higher particular towards essential lipid portions, and insignificant liking towards nonlipid pollutants. Hexane has been utilized as a dissolvable compound for oil reclamation from freshwater green growth for biodiesel creation [17]. *Enteromorpha compressa* produces 90.6% of biodiesel yield, when the macroalgae pretreated with NaOH, and methanol in the ratio of 9:1 by undergoing the conversion method called as transesterification [28]. *Chaetomorpha antennina* was treated with chloroform and ethanol in the ratio of 1:20 and transesterification method and achieved 2.1 mL/10 g biomass of yield [29]. *Gracilaria corticata* undergoes transesterification and acquired 2 mL/1.0 g biomass of yield [30] (Table 16.2).

Table 16.2 Macroalgae used for production biodiesel and biohydrogen

Type of biofuel	Type of species	Pretreatment methods	Conversion method	Biofuel yield	References
Biodiesel	*Padina tetrastromatica* and *Sargassum swartzii*	Soxhlet extraction	–	92%	[17]
	Laminaria japonica	Lipid extraction	Transesterification	48.30%	[31]
	Ulva fasciata	6 h Temperature—80–100 °C	Catalytic transesterification	88%	[32]
	Enteromorpha compressa	NaOH, Methanol–oil ratio—9:1	Base transesterification	90.6%	[28]
	Chaetomorpha antennina	Chloroform-ethanol solvent—1:20	Transesterification	2.1 mL/10 g	[29]
	Gracilaria corticata	Hexane-ether solvent—1:20 (w/v)	Transesterification	2 mL/10 $g_{biomass}$	[30]
Biohydrogen	*Laminaria japonica*	Temperature—160 °C, Time—30 min	Microwave	89%	[33]
	Laminaria digitate	Temperature—140 °C, Time—20 min	Hydrothermal	44.0 ± 1.2 mL/g VS	[34]
	Chaetomorpha antennina	Microwave power—0.36 kW Ammonium dodecyl sulfate—0.0035	Surfactant-aided microwave pretreatment	74.5 mL/g COD	[34]
	Ulva reticulate	Power—0.36 kWH_2O_2 dosage—24 mg H_2O_2/g	Microwave-H_2O_2 alkali pretreatment	87.5 mL H_2/g COD	[35]

16.6 Biogas

Biogas is a sustainable power source made essentially out of methane (55–65%) and carbon dioxide (35–45%). Carbon dioxide lessens the energy worth of methane and can be eliminated by cleaning. The cleaning system produces biogas, a maintainable, accommodating substitute for regular gaseous petrol in addition to applications like cooking, power production, etc. Methane for energy recuperation lessens its negative environmental impact factor. Methane is a greenhouse gas with a half-existence of roughly multidecade as well as an unnatural weather change potential 20-crease more noteworthy than CO_2. A reciprocal supplement thick digestate is produced in farming as a dirt conditioner [36]. Biogas alludes with a combination of gases delivered through the anaerobic processing of biomass. It could be agrarian squanders, plant material, sewage, fertilizer, or food waste. Algal biomass which contains more lipid content can produce biogas and might be hypothetically determined to be around 287–611 L/Kg biomass. The biogas yield would be profoundly dependent upon the algal strain selection and working temperatures. For instance, *Chlamydomonas reinhardtii* strain produces around 580 L/Kg at temperatures around 28–31 °C though *Chlorella Scenedesmus* produces around 611 L/Kg at temperatures around 45 °C. In addition, it has been seen that thermophilic assimilation yields biogas of around six to multiple times more than mesophilic processing [3]. *Chaetomorpha antennina* produces 215 mL/g of biogas when treated with 1 N NaOH and 80 °C temperature [37]. *Laminaria digitata* produces 235.4 mL of biogas when oven dried at the temperature of 70 °C for 72 h [38] (Table 16.3).

Table 16.3 Macroalgae used for production of biogas and bio-oil

Type of biofuel	Type of species	Pretreatment methods	Conversion method	Biofuel yield	References
Biogas	*Palmaria palmata*	Semi continuous anaerobic digestion	Anaerobic digestion	320 mL CH_4/g VS	[39]
	Chaetomorpha antennina	Disperser g force—1613 g, Treatment time—30 min, Ozone dosage—0.00049 g O_3/g TS	Ozone disperser pretreatment	0.20 g COD/g COD	[28]
	Chaetomorpha antennina	Disperser g-force of 1613 g, Temperature—80 °C, NaOH—1 N, pH—11	Thermo-chemo disperser	215 mL/g VS	[37]
	Laminaria digitata	Temperature—104 °C	Heat	97.66%	[38]
	Laminaria digitata	Temperature—70 °C,Time −72 h	Oven drying	235.4 ± 14.1 mL/g VS	[40]
Bio-oil	*Saccharina japonica*	Temperature—450 °C	Fixed bed reactor pyrolysis	47%	[41]
	Ulva lactuca	Temperature—500 °C	Microwave pyrolysis	18.4 wt.%	[42]
	Porphyra tenera	Temperature—500 °C	Packed tube reactor pyrolysis	47.4 wt.%	[43]
	Laminaria japonica	Temperature—500 °C	Packed tube reactor pyrolysis	45.8 wt.%	[43]
	Undaria pinnatifida	Temperature—500 °C	Packed tube reactor pyrolysis	37.5 wt.%	[43]

16.7 Biohydrogen

Hydrogen is the perfect and cleanest type of biofuel with energy worth of 140 KJ/g. In spite of the advantages, it contains high production cost. Generally, hydrogen production from petroleum products by reformation of methane is utilized. However, because of exhaustion of nonrenewable energy sources, elective production from biomass has been proposed overall [11]. Throughout the long term, numerous motivations are acquainted with increased gas production and the nature of hydrogen gas. Impetuses, for example, nickel, potassium mineral-based impetus, dolomite, concrete furnace dust, and so on are being utilized for hydrogen production from biomass. A few promoters like Li, Ca, and Na were added to lessen coke development. The realization was observed that the biomass likewise has specific metals which holds after the pyrolysis method. Thus, biochar can be eventually utilized as an impetus without adding any promoters. The principal job of impetus in biohydrogen production is the capacity to improve the water-gas shift response and to lessen tar content during the process of biomass gasification. It helps in molding the premise of biochar biomass utilization in hydrogen production by using biomass gasification.

Cotton scorch along with nickel showed the best return of hydrogen, i.e., 92.08 mg/g of biomass. It was inferred that advanced stacking of nickel starduced H_2 with utmost molar division as well as furthermore existence of basic metals favors H_2 production. Algal biochar from *Sargassum*'s impetus was used for hydrogen development from the pyrolysis of crude green growth, which gave a negligible measure of hydrogen, i.e., 3 mmol/g of Sargassum [13]. *Laminaria digitate* acquired 2 mL/g of biohydrogen yield with pretreatment methods like hydrothermal process [34]. *Laminaria japonica* treated biomass converted into biohydrogen fuel biohydrogen and yields 89% fuel [33]. *Chaetomorpha* antennina produced 87.5 mL of biohydrogen fuel when treated with H_2O_2 alkali pretreatment [34] (Fig. 16.3).

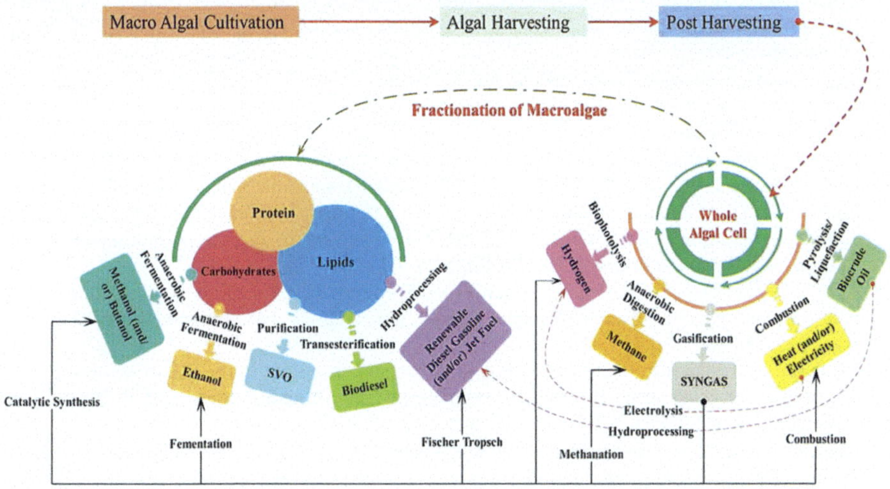

Fig. 16.3 Production of biofuel refinery from macroalgae (Adopted from Godvin Sharmila et al. [10])

16.8 Conclusion

The most promising use of the biomass derived from macroalgae species is the production of third-generation biofuels, a more advanced kind of biofuel. Because they are adaptable plants, they may grow in various aquatic conditions, including ones with high salinity or waste content. Facilities that produce macroalgae can be located in regions that are unsuitable for the development of agroecosystems or forests. There was a discussion of the significance of renewable energy sources such as bioethanol, biodiesel, biogas, and biohydrogen. Nonetheless, it is crucial to diversify the production of biofuels from algae biomass to enhance the overall energy balance. The prospects for algal oil will be improved by advancements in the processes used to extract oil from biomass and separate algae biomass from water. Algal-based fuel production may become viable, widely adaptable, and affordable over the next 7–10 years, but only if we keep raising awareness of these amazing species and strengthening our ability to modify them with the express purpose of fostering the growth of the new energy sector. Algal biomass may be crucial for dealing with the tension between food production and biofuel production in the future. Regarding fuel requirements, bioenergy derived from macroalgae will be a better substitute for fastest-growing economies like India.

References

1. Ramachandra TV, Hebbale D (2019) Bioethanol from macroalgae: prospects and challenges. Renew Sustain Energy Rev 117:109479. https://doi.org/10.1016/j.rser.2019.109479
2. Mahmood T, Hussain N, Shahbaz A, Mulla SI, Iqbal HMN, Bilal M (2023) Sustainable production of biofuels from the algae-derived biomass. Bioprocess Biosyst Eng 46:1077–1097. https://doi.org/10.1007/s00449-022-02796-8
3. Anto S, Mukherjee SS, Muthappa R, Mathimani T, Deviram G, Kumar SS, Verma TN, Pugazhendhi A (2020) Algae as green energy reserve: technological outlook on biofuel production. Chemosphere 242:125079. S0045-6535(19)32318-5. https://doi.org/10.1016/j.chemosphere.2019.125079
4. Rajkumar R, Yaakob Z, Takriff MS (2014) Potential of the micro and macro algae or biofuel production: a brief review. Bioresources 9(1):1606–1633
5. Griffiths G, Hossain AK, Sharma V, Duraisamy G (2021) Key targets for improving algal biofuel production. Clean Technol 3:711–742. https://doi.org/10.3390/cleantechnol3040043
6. Gao G, Burgess JG, Wu M, Wang S, Gao K (2020) Using macroalgae as biofuel: current opportunities and challenges. Bot Mar 63(4):355–370. https://doi.org/10.1515/bot-2019-0065
7. Jalilian N, Najafpour GD, Khajouei M (2020) Macro and micro algae in pollution control and biofuel production—a review. ChemBioEng Rev 7:1–17
8. Chen H, Zhou D, Luo G, Zhang S, Chen J (2015) Macroalgae for biofuels production: progress and perspectives. Renew Sustain Energy Rev 47:427–437. https://doi.org/10.1016/j.rser.2015.03.086
9. Milledge J, Harvey PJ, John. (2016) Potential process 'hurdles' in the use of macroalgae as feedstock for biofuel production in the British Isles. J Chem Technol Biotechnol 91:2221–2234. https://doi.org/10.1002/jctb.5003
10. Godvin Sharmila V, Dinesh Kumar M, Pugazhendi A, Bajhaiya AK, Gugulothu P, Rajesh Banu J (2021) Biofuel production from macroalgae: present scenario and future scope. Bioengineered 12(2):9216–9238. https://doi.org/10.1080/21655979.2021.1996019

11. Sudhakara K, Mamatc R, Samykanoc M, Azmic WH, Ishakd WFW, Yusaf T (2018) An overview of marine macroalgae as bioresource. Renew Sustain Energy Rev 91:165–179. https://doi.org/10.1016/j.rser.2018.03.100
12. Kumar G, Shobana S, Chen W-H, Bach Q-V, Kim S-H, Atabani AE, Chang J-S (2017) A review of thermochemical conversion of microalgal biomass for biofuels: chemistry and processes. Green Chem 19:44–67. https://doi.org/10.1039/C6GC01937D
13. Chi NTL, Anto S, Ahamed TS, Smita Kumar S, Shanmugam S, Melvin Samuel S, Mathimani T, Brindhadevi K, Pugazhendhi A (2021) A review on biochar production techniques and biochar based catalyst for biofuel production from algae. Fuel 287:119411. https://doi.org/10.1016/j.fuel.2020.119411
14. Tan IS, Lam MK, Foo HCY (2020) Advances of macroalgae biomass for the third generation of bioethanol production. Chin J Chem Eng 28:502. https://doi.org/10.1016/j.cjche.2019.05.012
15. Milledge JJ, Nielsen BV, Maneein S, Harvey PJ (2019) A brief review of anaerobic digestion of algae for bioenergy. Energies 12:1166. https://doi.org/10.3390/en12061166
16. Raikova S, Allen MJ, Chuck CJ (2019) Hydrothermal liquefaction of macroalgae for the production of renewable biofuels. Biofuel Bioprod Biorefin 13:1483. https://doi.org/10.1002/bbb.2047
17. Jeyakumar N, Hoang AT, Nižetić S, Balasubramanian D, Kamaraj S, Pandian PL, Sirohi R, Nguyen PQP, Nguyen XP (2022) Experimental investigation on simultaneous production of bioethanol and biodiesel from macro-algae. Fuel 329:125362. https://doi.org/10.1016/j.fuel.2022.125362
18. Dave N, Selvaraj R, Varadavenkatesan T, Vinayagam R (2019) A critical review on production of bioethanol from macroalgal biomass. Algal Res 42:101606. https://doi.org/10.1016/j.algal.2019.101606
19. Hamouda RA, Sherif SA, Ghareeb MM (2018) Bioethanol production by various hydrolysis and fermentation processes with micro and macro green algae. Waste Biomass Valor 9:1495–1501. https://doi.org/10.1007/s12649-017-9936-7
20. Schultz-Jensen N, Thygesen A, Leipold F, Thomsen ST, Roslander C, Lilholt H, Bjerre AB (2013) Pretreatment of the macroalgae *Chaetomorpha linum* for the production of bioethanol—comparison of five pretreatment technologies. Bioresour Technol 140:36–42. https://doi.org/10.1016/j.biortech.2013.04.060
21. Ravanal MC, Camus C, Buschmann AH, Gimpel J, Olivera-Nappa Á, Salazar O, Lienqueo ME (2019) Production of bioethanol from brown algae. In: Advances in feedstock conversion technologies for alternative fuels and bioproducts, pp 69–88. https://doi.org/10.1016/B978-0-12-817937-6.00004-7
22. Soliman RM, Younis SA, El-Gendy NS, Mostafa SSM, El-Temtamy SA, Hashim AI (2018) Batch bioethanol production via the biological and chemical saccharification of some Egyptian marine macroalgae. J Appl Microbiol 125(2):422–440. https://doi.org/10.1111/jam.13886
23. Hessami MJ, Cheng SF, Ambati RR, Yin YH, Phang SM (2019) Bioethanol production from agarophyte red seaweed, *Gelidium elegans*, using a novel sample preparation method for analysing bioethanol content by gas chromatography. 3 Biotech 9:25. https://doi.org/10.1007/s13205-018-1549-8
24. Amamou S, Sambusiti C, Monlau F, Dubreucq E, Barakat A (2018) Mechano-enzymatic deconstruction with a new enzymatic cocktail to enhance enzymatic hydrolysis and bioethanol fermentation of two macroalgae species. Molecules 23:174. https://doi.org/10.3390/molecules23010174
25. Adams JMM, Toop TA, Donnison IS, Gallagher JA (2011) Seasonal variation in Laminaria digitata and its impact on biochemical conversion routes to biofuels. Bioresour Technol 102:9976–9984. https://doi.org/10.1016/j.biortech.2011.08.032
26. Yeon JH, Lee SE, Choi WY, Kang DH, Lee HY, Jung KH (2011) Repeated-batch operation of surface-aerated fermentor for bioethanol production from the hydrolysate of seaweed Sargassum sagamianum. J Microbiol Biotechnol 21:323–331. https://doi.org/10.4014/jmb.1010.10057

27. Wang X, Liu X, Wang G (2011) Two-stage hydrolysis of invasive algal feedstock for ethanol fermentation. J Integr Plant Biol 53:246–252. https://doi.org/10.1111/j.1744-7909.2010.01024.x
28. Kraan S (2013) Mass cultivation of carbohydrate rich macro algae, a possible solution for sustainable biofuel production. Mitig Adapt Strateg Glob Change 18(1):27–46
29. Sankaran R, Cruz RAP, Pakalapati H et al (2019) Recent advances in the pretreatment of microalgal and lignocellulosic biomass: a comprehensive review. Bioresour Technol 298:122476
30. Atadashi IM, Aroua MK, Aziz AA (2010) High quality biodiesel and its diesel engine application: a review. Renew Sustain Energy Rev 14(7):1999–2008
31. Xu X, Kim JY, Yu Ri O, Park JM (2014) Production of biodiesel from carbon sources of macroalgae, Laminaria japonica. Bioresour Technol 169:455–461. https://doi.org/10.1016/j.biortech.2014.07.015
32. Abomohra AE-F, El-Naggar AH, Baeshen AA (2018) Potential of macroalgae for biodiesel production: screening and evaluation studies. J Biosci Bioeng 125(2):231–237
33. Jung K-W, Kim D-H, Shin H-S (2011) Fermentative hydro gen production from Laminaria japonica and optimization of thermal pretreatment conditions. Bioresour Technol 102(3):2745–2750
34. Hinks J, Edwards S, Sallis PJ et al (2013) The steady state anaerobic digestion of Laminaria hyperborea—effect of hydraulic residence on biogas production and bacterial community composition. Bioresour Technol 143:221–230
35. Dinesh Kumar M, Yukesh Kannah R, Kumar G et al (2020) A novel energetically efficient combinative microwave pretreatment for achieving profitable hydrogen pro duction from marine macro algae (Ulva reticulate). Bioresour Technol 301:122759
36. Thompson TM, Young BR, Baroutian S (2019) Advances in the pretreatment of brown macroalgae for biogas production. Fuel Process Technol 195:10615. https://doi.org/10.1016/j.fuproc.2019.106151
37. Tamilarasan K, Kavitha S, Rajesh Banu J et al (2017) Energy- efficient methane production from macroalgal biomass through chemo disperser liquefaction. Bioresour Technol 228:156–163
38. Ding L, Cheng J, Lin R, Ding L, Cheng J, Lin R (2020) Improving biohydrogen and biomethane co-production via two-stage dark fermentation and anaerobic digestion of the pretreated seaweed Laminaria digitata. J Clean Prod 251:119666
39. Chai WS, Bao Y, Jin P et al (2021) A review on ammonia, ammonia-hydrogen and ammonia-methane fuels. Renew Sustain Energy Rev 147:111254
40. Adams JMM, Schmidt A, Gallagher JA (2014) The impact of sample preparation of the macroalgae Laminaria digitata on the production of the biofuels bioethanol and biomethane. J Appl Phycol 27(2):985–991
41. Wang X, Li C, Chen M, Mmh M, Manickam S et al (2021) Microwave-assisted pyrolysis of seaweed biomass for aromatics-containing bio-oil production. In: 7th international conference on energy materials and environment engineering
42. Bae YJ, Ryu C, Jeon JK et al (2011) The characteristics of bio-oil produced from the pyrolysis of three marine macroalgae. Bioresour Technol 102(3):3512–3520
43. Suh DJ, Choi JH, Woo HC (2014) Pyrolysis of seaweeds for bio-oil and bio-char production. Chem Eng Trans 37:121–126

17 Application of Marine Biomass for Carbon Neutrality, Sustainable Environment and Bioeconomy, Regulations, and Policy Framework

Deepshikha Datta, Sayantan Sarkar, Deep Laha,
Piyansi Bhangar, Sudipta Kumar Das, and Bimal Das

Abstract

Marine biomass is a diverse and abundant resource that holds significant potential for promoting carbon neutrality, a sustainable environment, and a thriving bioeconomy. It can be harnessed to create a wide array of valuable products, including biofuels, biomaterials, and various other items, all of which contribute to reducing greenhouse gas emissions, conserving natural resources, and generating new economic opportunities. The development of the marine bioeconomy is integral to reaching carbon neutrality and a sustainable environment. Marine biomass can play a significant role in achieving carbon neutrality and addressing climate change. This work deals with the applications of marine biomass for achieving carbon neutrality by various processes, such as restoration of seagrass and mangrove, conservation and reestablishment of salt marshes and seagrasses, cultivation of marine biomass-based carbon neutral bioproducts, such as bioplastics, development of sustainable mariculture, production of ocean-based renewable energy, seaweed cultivation, and carbon neutral shipping. Further this work also highlights on the utilization of biomass in a variety of ways to promote sustainability and support a bio-based economy by biofuel production, carbon sequestration, aquaculture, bioremediation, and economic opportunities to coastal communities. Regulatory frameworks and policies of different countries

D. Datta
Department of Chemistry, Brainware University, Kolkata, West Bengal, India

S. Sarkar · D. Laha · B. Das (✉)
Department of Chemical Engineering, National Institute of Technology, Durgapur, West Bengal, India
e-mail: bdas.che@nitdgp.ac.in

P. Bhangar · S. K. Das
Department of Biotechnology, Brainware University, Kolkata, West Bengal, India

M. Rafatullah et al. (eds.), *Multidisciplinary Applications of Marine Resources*,
https://doi.org/10.1007/978-981-97-5057-3_17

are discussed to ensure the conservation of marine resources. This chapter provides insights into the challenges and opportunities in this dynamic field.

Keywords

Marine biomass · Carbon neutrality · Bioeconomy · Sustainable environment · Regulations

17.1 Introduction

The utilization of marine biomass in the realms of carbon neutrality, sustainable environmental practices, and the burgeoning bioeconomy represents a pivotal intersection between science, industry, and policy. Due to facing the challenges of climate change and resource depletion, marine biomass is emerging as a sustainable solution with multiple benefits. Exploration of the complexity of marine biomass and the potential benefits in terms of carbon neutrality, environmental sustainability, and bioeconomy are very significant. Furthermore, the responsible utilization of these precious marine resources is guided by the intricate landscape of regulations and policy frameworks.

At its essence, marine biomass encapsulates a diverse array of organic materials derived from marine organisms, ranging from macroscopic seaweeds and microalgae to various forms of aquatic plants and animals [1]. These resources, abundant in our oceans, hold immense potential for sustainable development due to their rapid growth rates and minimal environmental impact compared to traditional land-based biomass sources [2]. The utilization of marine biomass extends beyond mere extraction for food consumption; it encompasses a broader spectrum of applications, from bioenergy production to the creation of bio-based materials. Understanding the intricacies of marine biomass involves recognizing the dynamic interactions within marine ecosystems and the potentialities they present for a more sustainable future [3]. Harnessing marine biomass for energy and materials is a promising avenue for achieving carbon neutrality. The photosynthetic prowess of marine plants, particularly macroalgae and microalgae, makes them efficient carbon dioxide absorbers, aiding in the sequestration of atmospheric carbon [4]. By integrating marine biomass into energy production, we have the potential to mitigate carbon emissions and combat climate change. The benefits extend beyond carbon neutrality to encompass the broader spectrum of environmental sustainability. Marine biomass cultivation requires minimal freshwater, no arable land, and limited synthetic inputs, mitigating the adverse environmental impacts associated with traditional agriculture. Moreover, the cultivation of marine biomass contributes to nutrient cycling and fosters biodiversity in marine ecosystems [5]. As a cornerstone of the bioeconomy, marine biomass provides a renewable resource for the production of biofuels, biomaterials, and biochemicals. The bioeconomy relies on the sustainable conversion of biological resources into valuable products, reducing dependence on finite fossil resources and fostering economic growth. The applications of marine biomass are diverse and span various sectors, including energy, agriculture, and industry. From biofuel production and wastewater treatment to the development of biodegradable materials, marine biomass emerges as a versatile resource capable of addressing multiple societal needs [6]. This chapter will explore these applications in depth, shedding light on the transformative potential of marine biomass across different industries.

In the pursuit of a responsible and sustainable utilization of marine resources, robust regulations and policy frameworks play a pivotal role [7]. The exploration of marine biomass is embarked upon; the convergence of science, industry, and policy becomes evident [8]. A future where the oceans significantly contribute to carbon neutrality, environmental sustainability, and a thriving bioeconomy can be paved through a comprehensive understanding of marine biomass and its applications, coupled with adherence to stringent regulations [9].

Approximately, a total of 550 Gt C is biomass mass across all taxa on Earth, of which about 80% (450 Gt C; SI Appendix, Table 17.1) are plants, with land plants (embryophytes) accounting for the majority. Bacteria (about 70 Gt C) make up approximately 15% of the world's biomass and are the second largest component. Other groups that make up the remaining <10% include viruses, protists, fungi, mammals, and archaea, as listed in descending order. We estimate that plants are the dominate kingdom in terms of biomass at a probability of around 90% despite the high degree of ambiguity surrounding the total biomass of microorganisms (Table 17.1). Marine biomass has a significant contribution in the total amount biomass across all taxa.

This chapter will navigate the intricate landscape of international, regional, and national regulations that govern the extraction, cultivation, and utilization of marine biomass. By understanding and adhering to these frameworks, stakeholders can ensure the preservation of marine ecosystems while unlocking the full potential of marine biomass for the benefit of society.

17.2 Marine Biomass and Carbon Neutrality

Marine biomass, the vast and diverse realm of organic matter found within our oceans, holds immense potential for achieving carbon neutrality. Its unique ability to capture and store carbon, combined with its potential for sustainable utilization, offers a promising avenue for combating climate change and building a carbon-neutral future.

17.2.1 Carbon Sequestration Through Marine Biomass

One of the key mechanisms by which marine biomass contributes to carbon neutrality is through carbon sequestration. This involves the long-term removal of carbon dioxide (CO_2) from the atmosphere and its storage in various forms within the ocean. Marine organisms, particularly phytoplankton and seagrasses, play a crucial role in this process. Through photosynthesis, they utilize sunlight and CO_2 to

Table 17.1 Estimated total biomass for taxonomic groups [10–12]

Taxon	Mass (Gt C)
Plants	450
Fish	0.7
Bacteria	70
Marine bacteria	1.3
Soil bacteria	7
Terrestrial deep	60
subsurface	7
bacteria	12
Marine deep	0.3
subsurface	12
bacteria	
Fungi	
Marine fungi	
Soil fungi	
Archaea	7
Marine archaea	0.3
Soil archaea	0.5
Terrestrial deep	4
subsurface	3
archaea	
Marine deep	
subsurface	
archaea	
Wild mammals	0.003
Marine mammals	0.004
Protists	4
Animals	2
Terrestrial	0.2
anthropods	1
Marine	0.006
anthropods	0.01
Terrestrial	0.002
nematodes	0.06
Marine nematodes	0.007
Chordates, wild	0.1
birds	0.7
Chordates,	0.2
humans	0.2
Chordates, wild	0.1
mammals	0.2
Chordates,	
livestock	
Chordates, fish	
Molluscs	
Annelids	
Cnidarians	
Viruses	

Global biomass (measured in tonnes of carbon) versus the abundance (number of individuals) of different taxonomic groups

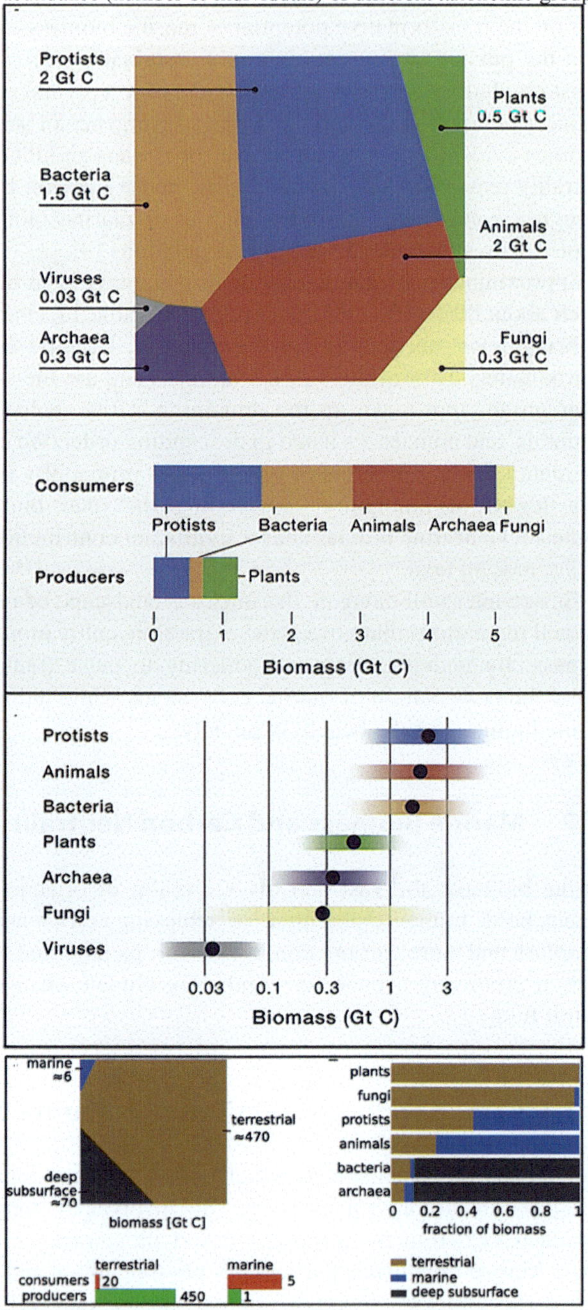

produce organic matter, effectively locking away atmospheric carbon within their biomass [13].

Estimates suggest that the ocean absorbs around 25% of CO_2 produced globally, with marine biomass acting as a significant sink for this carbon. Seagrasses, for instance, are known to store vast amounts of carbon in their meadows, burying it in the sediment for centuries. Similarly, mangrove forests trap and accumulate carbon from the atmosphere, contributing to coastal carbon sequestration [13].

17.2.2 Restoration and Conservation Efforts

Several initiatives aim to restore and conserve these crucial carbon sinks. Mangrove restoration projects are underway around the world, with the goal of re-establishing lost mangrove forests and their associated carbon sequestration potential [13]. Similarly, efforts are underway to protect and restore seagrass meadows, ensuring that their continued role as carbon sinks and vital marine ecosystems [2].

17.2.3 Reducing Greenhouse Gas Emissions

Beyond carbon sequestration, marine biomass also contributes to carbon neutrality by reducing greenhouse gas (GHG) emissions. One significant way that this is achieved is through the production of biofuels. Marine organisms, such as algae and seaweed, can be cultivated and processed to produce biofuels, such as biodiesel and bioethanol [3]. These biofuels offer a cleaner alternative to fossil fuels, as they release fewer GHGs during combustion [6].

Several countries are investing heavily in developing and deploying marine biofuels as a means to reduce their dependence on fossil fuels and transition toward a low-carbon economy [14]. In addition, marine biomass can be utilized for the production of biomaterials, such as bio-plastics and bio-composites. These biomaterials offer sustainable alternatives to traditional plastic products, which contribute significantly to GHG emissions throughout their lifecycle [4].

17.2.4 Mitigating Climate Change and Global Carbon Neutrality Efforts

The potential of marine biomass to combat climate change extends beyond individual contributions in carbon sequestration and emission reduction. It offers a holistic approach that can be woven into existing global efforts toward carbon neutrality. By actively investing in research and development of marine biotechnologies, governments and private sectors can unlock the full potential of this resource and accelerate the transition to a low-carbon future [14]. Imagine a future powered by clean biofuels derived from ocean-grown algae, where sustainable biomaterials replace plastic pollution, and restored mangrove forests and seagrass meadows act as natural carbon capture machines [2]. This vision is not a distant dream; it is a tangible path paved with scientific advancements, policy frameworks that incentivize responsible utilization, and international collaboration. International forums such as the Paris Agreement already recognize the potential of marine ecosystems in climate change mitigation. By integrating robust marine biomass strategies into national action plans, countries can significantly contribute to global carbon neutrality goals. This could involve implementing policies that support sustainable marine biofuel production, incentivize mangrove restoration projects, and promote research on efficient carbon sequestration technologies [15]. Furthermore, fostering international cooperation in technology sharing and resource management is crucial to maximize the impact of marine biomass. Collaborative efforts can optimize infrastructure for biofuel production, facilitate knowledge exchange about best practices in sustainable harvesting, and establish robust regulations to prevent any harmful environmental impacts [16].

The journey toward carbon neutrality demands multifaceted solutions. Marine biomass, with its unique ability to cleanse the air, power our lives, and protect our planet, offers a beacon of hope in this challenge. By embracing its potential and fostering responsible utilization, we can chart a course toward a cleaner and more sustainable future, where the ocean becomes not just a victim of climate change, but a powerful ally in mitigating its effects [4]. The application of marine biomass plays a pivotal role in maintaining carbon neutrality by acting as a natural carbon sink. Marine organisms, such as algae and seaweed, undergo photosynthesis, absorbing substantial amounts of carbon dioxide from the atmosphere. Utilizing these organisms for various purposes, such as biofuel, bio-electricity, and biogas production, not only reduces reliance on fossil fuels but also contributes to carbon sequestration. In addition, the cultivation of marine biomass promotes sustainable practices, preventing the release of stored carbon back into the environment. By integrating marine biomass into diverse sectors, we not only harness renewable resources but actively engage in a natural carbon mitigation strategy, thereby nurturing a balanced and carbon-neutral ecosystem. Embracing the potential of marine biomass applications aligns with global efforts to combat climate change and ensures a sustainable future with minimized carbon footprints (Fig. 17.1).

17.3 Applications for a Sustainable Environment

Marine biomass, with its diverse and abundant offerings, presents a myriad of applications for promoting environmental sustainability [14]. This section explores the multifaceted applications of marine biomass, focusing on its utilization for habitat restoration, pollution control, and marine ecosystem preservation. Drawing on a wealth of research, including studies by leading experts in the field, we delve into the innovative ways in which marine biomass contributes to a sustainable environment.

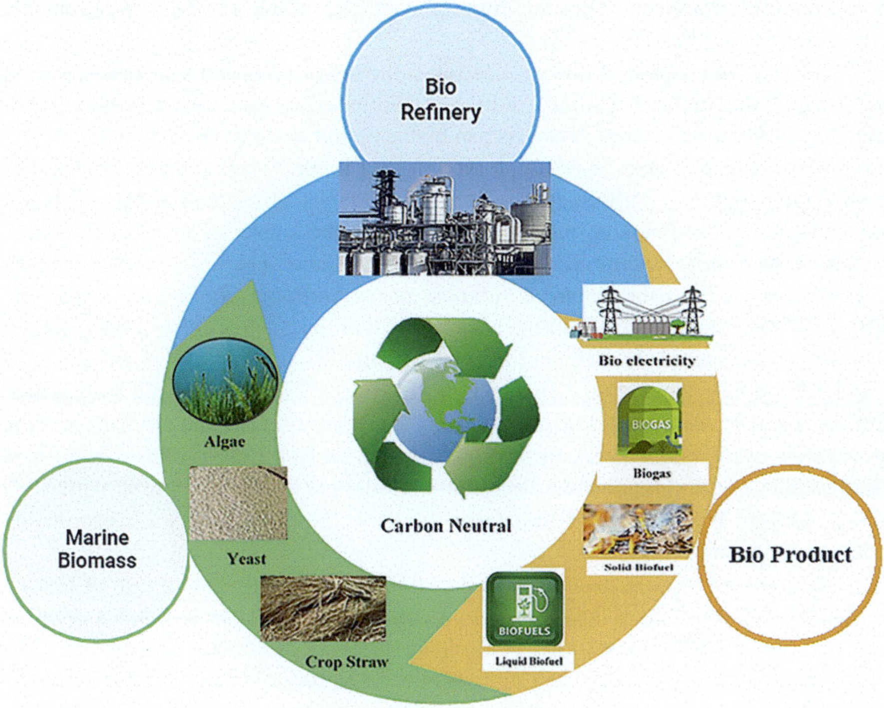

Fig. 17.1 Application of marine biomass to maintain carbon neutrality

17.3.1 Application of Marine Biomass to Promote Environmental Sustainability

The application of marine biomass for environmental sustainability is a cornerstone in the quest for a healthier planet. The chapter explores the utilization of marine biomass in creating valuable products, such as biomaterials, that can replace traditional, resource-intensive materials. By delving into the eco-friendly attributes of marine biomass-derived products, this section establishes a connection between the bioeconomy and sustainable environmental practices [17].

At the heart of this endeavor lies the recognition of marine biomass as a versatile resource capable of revolutionizing traditional practices that often strain the environment. The research underscores how marine biomass serves as a linchpin in the creation of valuable biomaterials, offering a sustainable alternative to resource-intensive materials commonly used in various industries [17].

The versatility of marine biomass is manifest in its ability to generate biomaterials that can seamlessly replace conventional materials, thereby reducing the ecological footprint associated with resource extraction and processing. By scrutinizing the eco-friendly attributes of these marine biomass-derived products, the study establishes a pivotal connection between the emerging bioeconomy and sustainable environmental practices. This is aligning with the SDG 12, i.e., "Responsible Consumption and Production" [5]

One significant aspect of this application is the potential to mitigate deforestation and habitat destruction. Unlike traditional biomaterials that often require land-intensive cultivation, marine biomass can be harvested without disturbing terrestrial ecosystems. For instance, seaweed, a type of marine biomass, can be cultivated in aquatic environments without encroaching on valuable terrestrial habitats. This not only safeguards biodiversity but also contributes to the conservation of critical ecosystems which indicates the SDG 15, i.e., "Life on Land" [1].

Furthermore, the role of marine biomass in fostering circular economies is significant. The cyclic nature of marine biomass utilization, from cultivation to product creation and eventual decomposition or recycling, aligns with the principles of a circular economy. This contrasts starkly with linear models that result in resource depletion and environmental degradation. By incorporating marine biomass into the production cycle, industries can minimize waste and pollution, contributing to a more sustainable and regenerative economic model [17]. This is aligning with SDG 8, i.e., "Decent Work and Economic Growth," as well as SDG 12, i.e., "Responsible Consumption and Production."

It is important to raise awareness and encourage widespread adoption of marine biomass-based products. Education campaigns and collaborative initiatives can play a pivotal role in shifting consumer preferences toward eco-friendly alternatives. The widespread acceptance of marine biomass-derived products can catalyze a significant reduction in the demand for resource-intensive materials, steering industries toward more sustainable practices [18].

Therefore, the application of marine biomass for environmental sustainability is a transformative journey toward a healthier planet. Different research works illuminate how this not only addresses current environmental challenges but also lays the foundation for a more sustainable and regenerative future. By focusing on biomaterials and their eco-friendly attributes, the study exemplifies the potential of marine biomass to redefine industrial practices, mitigate environmental impact, and foster a harmonious coexistence between human activities and the natural world [5].

17.3.2 Marine Biomass Utilization for Habitat Restoration

Habitat restoration stands as a critical application of marine biomass, contributing to the preservation and rejuvenation of ecosystems. Different researches showed the pivotal role of marine biomass, particularly seagrasses and mangroves, in habitat restoration projects [2]. By drawing attention to the symbiotic relationship between marine biomass and habitat restoration, this section emphasizes the interconnectedness of environmental health and biodiversity. The intricate characteristics of seagrasses and mangroves are highlighted as pivotal components in creating resilient and thriving coastal ecosystems [13].

Nutrient-rich aquaculture effluent, once a threat to coastal ecosystems, can be channeled into constructed wetlands. Bacteria degrade organic matter, fungi immobilize heavy metals, and plants uptake excess nutrients, filtering the water and creating a haven for marine life. Shellfish, the ocean's natural filters, play a starring role. Oysters, mussels, and clams efficiently filter suspended particles and pollutants from the water, improving water quality and creating a fertile substrate for the return of seagrass meadows and diverse marine communities [1]. Coral larval propagation techniques nurture coral larvae in controlled environments, boosting their survival rates and accelerating reef regeneration. Scientists are also exploring cryopreservation and assisted fertilization to preserve genetic diversity and enhance coral resilience. Microbiome manipulation holds immense promise. By understanding and promoting beneficial microbes, scientists aim to improve coral resistance to stress and accelerate their recovery from bleaching events. Seaweed-based biopolymers offer remarkable versatility [19]. Kelp and alginate can be woven into biodegradable mats and ropes, used to stabilize eroded coastlines and create artificial reefs teeming with life. Bio-bricks crafted from oyster shells and mussel byssus, a naturally occurring adhesive, offer a sustainable alternative to traditional concrete. These bio-bricks provide sturdy shelters for marine life and promote shoreline stabilization [1].

17.3.3 Pollution Control Through Marine Biomass

Marine biomass emerges as a potent tool in pollution control, offering sustainable solutions to mitigate the impact of pollutants on marine ecosystems. Marine biomass, including seaweeds and microalgae, can absorb and neutralize pollutants. This section explores the potential of cultivating marine biomass in polluted areas, acting as a natural filter and promoting the restoration of ecosystems affected by various pollutants. By highlighting the capacity of marine biomass to contribute to water quality improvement, the chapter underscores its role as a catalyst for sustainable environmental practices [14].

Some of the recent studies spotlight the extraordinary capabilities of organisms, such as seaweeds and microalgae, functioning as nature's own filters to neutralize and absorb an array of pollutants. In this ecological orchestra, these microscopic janitors showcase a remarkable repertoire of techniques [20]. Seaweeds, epitomized by the intriguing "spaghetti kelp" (*Laminaria Hyperborea*), feature porous cell walls adept at absorbing heavy metals and organic contaminants from the water. Simultaneously, microalgae wield specialized enzymes and molecules, skillfully breaking down toxins into harmless compounds. These are not mere theoretical notions; projects worldwide underscore the tangible power of cultivating these natural biofilters. In China, kelp forests adjacent to mariculture farms have significantly diminished nutrient levels, effectively mitigating harmful eutrophication [21]. Similarly, Norway embraces microalgae-based systems to treat wastewater from fish farms, successfully reducing nitrogen and phosphorus pollution [22].

The benefits of this purification process extend beyond mere detoxification. As these marine organisms cleanse the water, they concurrently oxygenate it, cycle essential nutrients, and foster critical habitats. This cascading effect sets the stage for natural ecological recovery, weaving a tapestry of marine life resilience [22]. Harnessing this cleansing symphony presents a sustainable solution to a pressing problem. By fostering responsible stewardship and implementing targeted interventions, we can leverage the insights gleaned from different research papers to reclaim polluted waters. This not only aligns with global goals for pollution control, but also envisions a future where pristine oceans are not the exception but the norm [5]. In essence, pollution control through marine biomass encapsulates a harmonious interplay between nature's ingenuity and human intervention. As we explore this story of change, we understand that the power of marine biomass in fighting pollution is more than just science—it represents a hope for a healthier and long-lasting marine environment [18].

17.3.4 Marine Ecosystem Preservation

Preserving marine ecosystems is a fundamental goal in the broader context of environmental sustainability. The chapter is shedding light on the intricate relationship between marine biomass utilization and ecosystem preservation. It examines how sustainable mariculture practices, integrated with conservation efforts, contribute to the maintenance of biodiversity and the protection of marine habitats [5]. In addition, the cultivation of seaweed and its role in carbon-neutral shipping are explored as means to preserve delicate marine ecosystems while advancing economic activities in a sustainable manner [19].

Traditionally, extractive fishing practices often resemble plundering a treasure chest, depleting stocks and disrupting the delicate equilibrium of marine ecosystems. Sustainable mariculture, in contrast, offers a harmonious alternative. The meticulously cultivated shellfish farms suspended in the sun-dappled water, not just replenishing fish stocks but also filtering pollutants and providing nurseries for countless marine species. This closed-loop approach minimizes nutrient and waste discharge, fostering biodiversity and nurturing the very foundation of the ocean's food web [23]. The marine giants not only provide food and shelter for myriad creatures, but also act as potent carbon capture machines. Their rapid growth absorbs vast amounts of atmospheric CO_2, transforming it into organic matter and locking it away for millennia [24]. Cultivating kelp forests strategically becomes a natural climate-change mitigation strategy, offsetting carbon emissions from shipping and other ocean-based activities. Sustainable utilization and ecosystem preservation cannot exist in isolation. Protecting vital marine habitats, from coral reefs to seagrass meadows, requires robust frameworks that integrate conservation efforts with responsible resource management [19]. Marine protected areas, meticulously zoned and monitored, become sanctuaries for threatened species and breeding grounds for future generations. By implementing quotas, catch limits, and gear restrictions, we can prevent overfishing and allow depleted stocks to recover [23]. Understanding the intricate web of ecological processes is crucial for navigating the path of sustainable utilization. Oceanographers map complex currents, fisheries scientists model population dynamics, and marine biologists unravel the intricate food webs that bind the ocean's ecosystems together. This scientific knowledge informs policy decisions, guides mariculture practices, and identifies crucial areas for protection [25].

Therefore, from the restoration of degraded habitats to pollution control and the preservation of marine ecosystems, marine biomass emerges as a versatile and indispensable resource. These applications not only contribute to the broader goals of carbon neutrality and bioeconomy development, but also serve as tangible steps toward ensuring a resilient and sustainable future for our planet. Thus, marine biomass plays a significant role on the development of sustainable environment through significant contribution in climate change mitigation, ensuring carbon neutrality, generation of biofuels for reducing environmental pollution, and waste valorization by its conversion to useful products, as shown in Fig. 17.2.

17.4 Bioeconomy and Economic Opportunities

17.4.1 Marine Bioeconomy and Its Potential to Drive Economic Growth in Coastal Communities

Using marine bio-resources as the source or target of biotechnological applications is known as marine biotechnology which can relate to the knowledge of marine bioeconomy. Therefore, this comprehensive knowledge of marine biotechnology encompasses that both marine biotechnologies encompass both conventional methods such as aquaculture and contemporary techniques such as bioremediation, biofuel generation, and fish genetic manipulation [29]. Numerous significant and broad advancements have already been made in the realms of medicine, cosmetics, nutraceuticals, food production, and environmental industrial uses by this discipline. When compared to terrestrial resources, the marine environment and the full extent of its abundance of bio-resources were unexplored for a very long period [29–31]. Still, things are shifting as we now have a better understanding of the marine environment and new tools to research marine creatures and ecosystems because of scientific and technological advancements, particularly in the field of "omics" technologies. Particularly, metagenomics sequencing is increasingly being utilized to investigate microbial ecosystems and find genes with novel functions [31]. Marine microorganisms appear to have special potential for marine biotechnology and are showing biodiversity's hitherto unrealized boundaries [32].

Conventional biofuel biorefineries use microorganisms that ferment biomass sugars to produce bioethanol. Specifically, *S. cerevisiae* has been optimized to yield substantial amounts of ethanol at the industrial scale. Nonetheless, these industrial strains originate from the earth, and salts typically significantly reduce their activity [33]. On the other hand, marine fermentation makes use of resources found in the sea, such as seawater, marine yeast, and seaweed. Therefore, it is essential to isolate and identify any novel marine strains [33]. Genetic engineering was employed in the early stages of attempts to create effective halotolerant bacteria. Limtong et al. [34] hybridized *S. cerevisiae* with *Zygosaccharomyces rouxii* to produce a

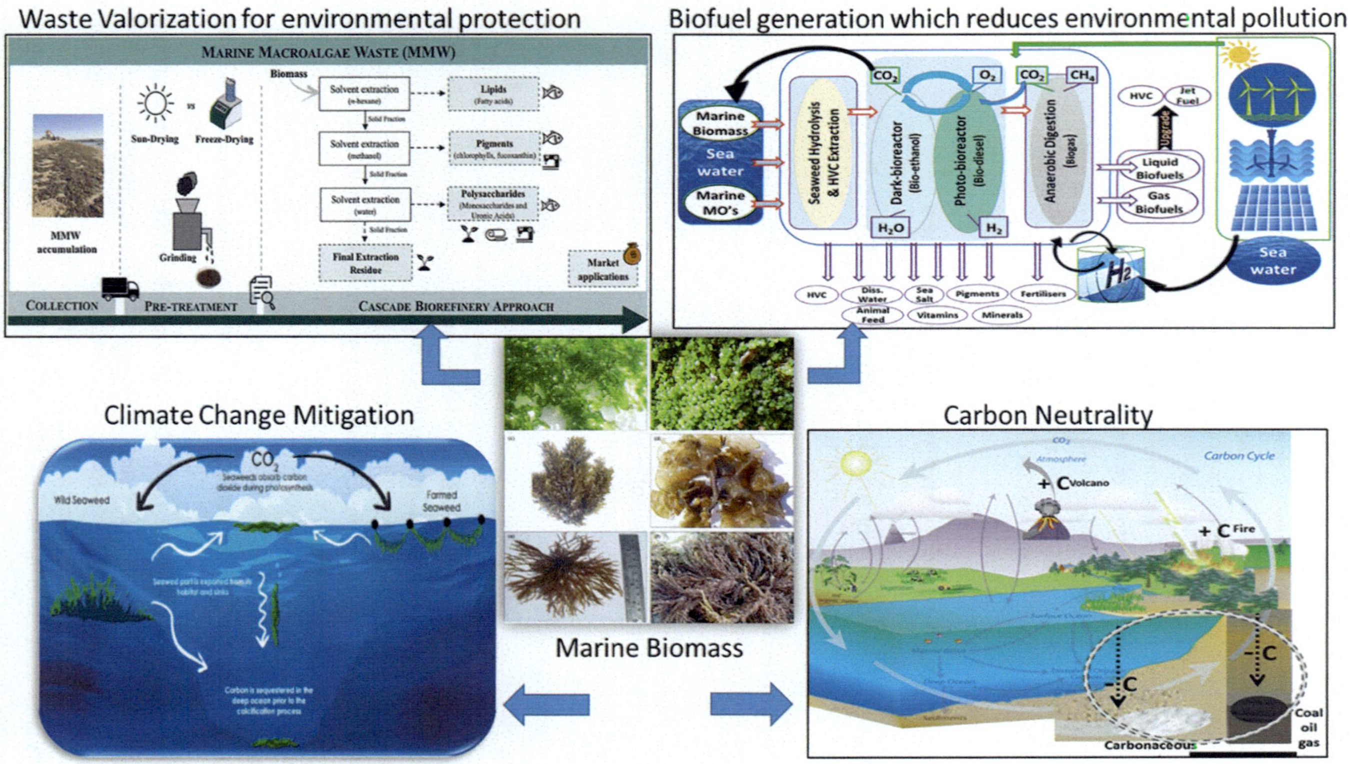

Fig. 17.2 Application of marine biomass for the development of sustainable environment [26–28]

high-ethanol-fermenting, halotolerant microbe. It can also be observed that Europe leads to the production of biofuel from marine biomass followed by North America and Asia, as shown in Table 17.2.

17.4.2 Businesses and Industries that Have Leveraged Marine Biomass for Sustainable Economic Development

One of such industries where entrepreneurship may yield significant rewards in a supportive setting is aquaculture. In order for entrepreneurship to flourish, a fundamental shift in mindset and culture is required to dispel the myth that commercialization will restrict academic freedom and get over the reluctance to commercialize findings. Scientists will discover a fit between their academic pursuits and their commercial endeavors when this occurs [55]. It could be observed that more than 78% of biofuel companies are located in USA; however, the maximum macroalgae production both wild harvest (75%) and cultivated is observed in China (28%), as shown in Fig. 17.3.

Case Study No. 1
Aquaculture business expansion is limited by hatcheries producing grouper seeds in an unsustainable manner

The highly valued and sought-after grouper species in Sabah has seen its population decline due to overfishing and excessive exploitation practices. Landings from capture fisheries fall well short of what the market will bear. Due to female infertility and other factors, the hatchery output of giant grouper (Epinephelus lanceolatus), one of the most valuable species, seed is inadequate and unreliable. Resolution aspects such as hormone therapy, maturity diets, and broodstock training were the subject of research. The hybridization of a large grouper male and a tiger grouper male was an inventive strategy that led to the breakthrough. It produced seed, which made a major contribution to closing the gap between the fish's supply and demand [32].

Case Study No. 2
Skewed sex ratio as a result of groupers' protogynous state, which increases the cost of managing broodstock and the unpredictability of seed supply

Groupers are protogynous hermaphrodites, meaning that they may switch from being female to male at birth. The lack of males in the native population has hampered its aquaculture. A lot of study attention has been focused on the process by which the grouper gonad transitions from the ovary to the intersexual state and finally to the testis. The challenge for newly established hatcheries is locating the male grouper with ease. It is challenging and costly to purchase and care for the broodstock of some species, such as the tiger grouper (Epinephelus fuscoguttatus), in a hatchery. If a male is unavailable, breeding fails, resulting in financial losses and a lack of investors, and the solution in sex distinction has been studied, particularly in relation to hormones. In the end, it was determined to investigate if certain

Table 17.2 Estimated global potential of marine biomass for biofuels production

Marine biomass for biofuel production (Bioethanol)				**Marine biomass for biofuel production (Biodiesel)**			
Marine yeasts	**Source**	**Bioethanol (Max. EtOH) (g/L)**	**References**	**Microalgae**	**Source**	**Lipid content (% DW)**	
S. cerevisiae **YPS128**	Plymouth, UK	13	[35]	Desmodesmus sp. WC08	Hainan, China	31.30	[36]
S. cerevisiae **C-19**	Tokyo Bay	87.7	[37]	Chlorella sp. PKUAC 102	Guangdong, China	52.80	[36]
Candida sp.	West Coast India	12.3	[38]	Coelastrum sp. HA-1	Tianjin, China	43.20	[39]
Defluviitalea. haphyphila **Alg1**	Yellow Sea, China	10	[40, 41]	Shizochytrium sp. PKU#Mn4	Guangdong, China	51.50	[42]
P. salicaria	Mangrove, SE India	28.5	[43, 44]	**Marine biomass for the extraction of sugar for bioethanol production** [45]			
S. cerevisiae **(JN387604)**	Mangrove, *Southeast Asia*	25.1	[46]	**Seaweed/Class**	**Sugar composition**	**Sugar released (%w/w dry biomass)**	
C. albicans	Mangrove, SE India	47.3	[47]	*Laminaria japonica/ Brown*	Glucose	9.6–37.6	
S. cerevisiae **AZ65**	Caernarfon, *Wales, UK*	86.72	[48]	*Ulva lactuca/Green*	Glucose	19.4	
				Gelidium amansii/ Red	Galactose	34.6–67.5	
				Sargassum fulvellum/ Brown	Mannitol	9.6	
				Ulva pertusa/Green	Glucose	59.6	
Continent	**North America**	**South America**	**Asia**	**Europe**	**Africa**	**Oceania**	
Estimated Potential (EJ/year)	4.0–12.0	1.0–3.0	11.0–30.0	40.0–160.0	0.5–2.0	0.5–1.5	
References	[49]	[50]	[51]	[52]	[53]	[54]	

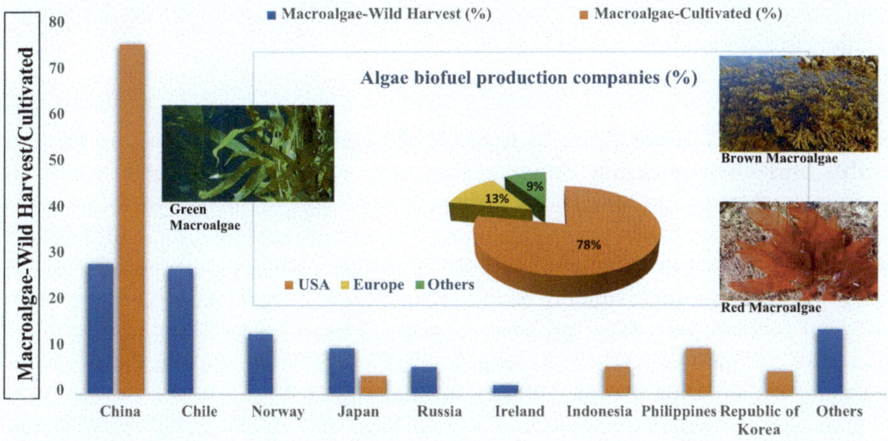

Fig. 17.3 Global distribution of algal biofuel production companies and the distribution of algal biomass [56, 57]

cues may cause sex shift and affect the ratio of sexes in the hatchery tanks. It was found that when female specimens are preparing for reproduction and there is not a single male in the tank for prolonged periods of time, the gonad of one of the female specimens in the cohort experiences transitional processes that result in sex reversal and differentiation into a functional male. Others in the tank are subsequently discouraged from going through the same process as a result of this sex change. After then, tiger grouper seed is created in the hatchery by breeding [29].

Case Study No. 3

The main issue is the marine ecology and capture fisheries are being negatively impacted by aquacultures over reliance on fish-based feed, which is also likely to jeopardize aquaculture's viability. Entrepreneurs can profit financially from the aquaculture of groupers and some other carnivorous fish species because of their high demand, market value, and popularity in the live reef food fish trade. These species' native populations have been reduced by overexploitation, which has also disrupted the coral reef environment where they are mostly taken. The use of low-value prey fish as feed is putting pressure on the long-term sustainability of capture fisheries, despite the fact that aquaculture has the potential to increase supply. It has a significant effect on the balance of the ecosystem and marine biodiversity. Because wild stocks of low-value prey fish are limited, the price of these species has increased dramatically in recent years, and the available supply is decreasing. This will probably limit aquaculture's ability to grow in the future if its reliance on prey fish does not lessen. To feed high-value groupers, the aquaculture sector has also resorted to buying large quantities of pelagic fish, such as sardines and mackerels, which is jeopardizing the food security of the less fortunate communities that these species have been sustaining. Resolution creating alternative supplies of fish oil and protein from sustainable goods is the answer to this issue. BMRI conducted studies to create diets that do not rely on fish [58].

Case Study 4

Negative view of aquaculture as a result of negative consequences on human health and environmental damage The environmental effects of effluents in receiving habitats, eutrophication, and the conversion of mangrove forests into farms have all been blamed on aquaculture. The use of pesticides and antibiotics to treat the farmed stocks has also drawn a lot of criticism, since it put consumer health at risk. In order for this critically essential business to expand and contribute more to global food security, these problems needed to be resolved. The scientific community has examined these issues from several angles in an effort to pinpoint best practices that can support aquaculture's sustainability [59].

However, the marine biomass has a great influence on enhancing the bioeconomy by the generation of economically beneficial products, such as biodiesel, bioethanol, biohydrogen, and also other value-added products and chemicals (Fig. 17.4). Fuel crisis has further indicated that methane generated from marine biomass can act as a significant substitute to mitigate this crisis.

Table 17.3 outlines specific economic indicators, such as the area of marine biomass products and the potential for benefit with overall impact on local and global economies.

17.5 Regulatory Frameworks for Marine Biomass

17.5.1 Regulatory Landscape Surrounding Marine Biomass Utilization, Highlighting the Need for Responsible Management

The both affluent and poor nations' coastal and marine ecosystems have suffered greatly as a result of persistent overfishing that ignores scientific warnings, fishing down food webs, habitat destruction, and increased pollutant loading, particularly nitrogen export. Lack of cooperation amongst nations sharing marine ecosystems, fragmentation among institutions, and adopting ecosystem-based techniques to controlling human activity in these systems to avert severe social and economic disruption is becoming increasingly necessary due to the actions of international agencies, disciplines, and slack national laws, policies, and enforcement [68]. The Global Environment Facility (GEF) has received approaches from developing countries, in record numbers, pleading for help in safeguarding the futures of their common vast marine ecosystems (LMEs) [69]. This document outlines the GEF-supported procedures that are being applied to help them adopt ecosystem-based methods for controlling human activity in these systems to prevent significant disruptions to the social and economic spheres. Developing nations have flocked to the GEF in record

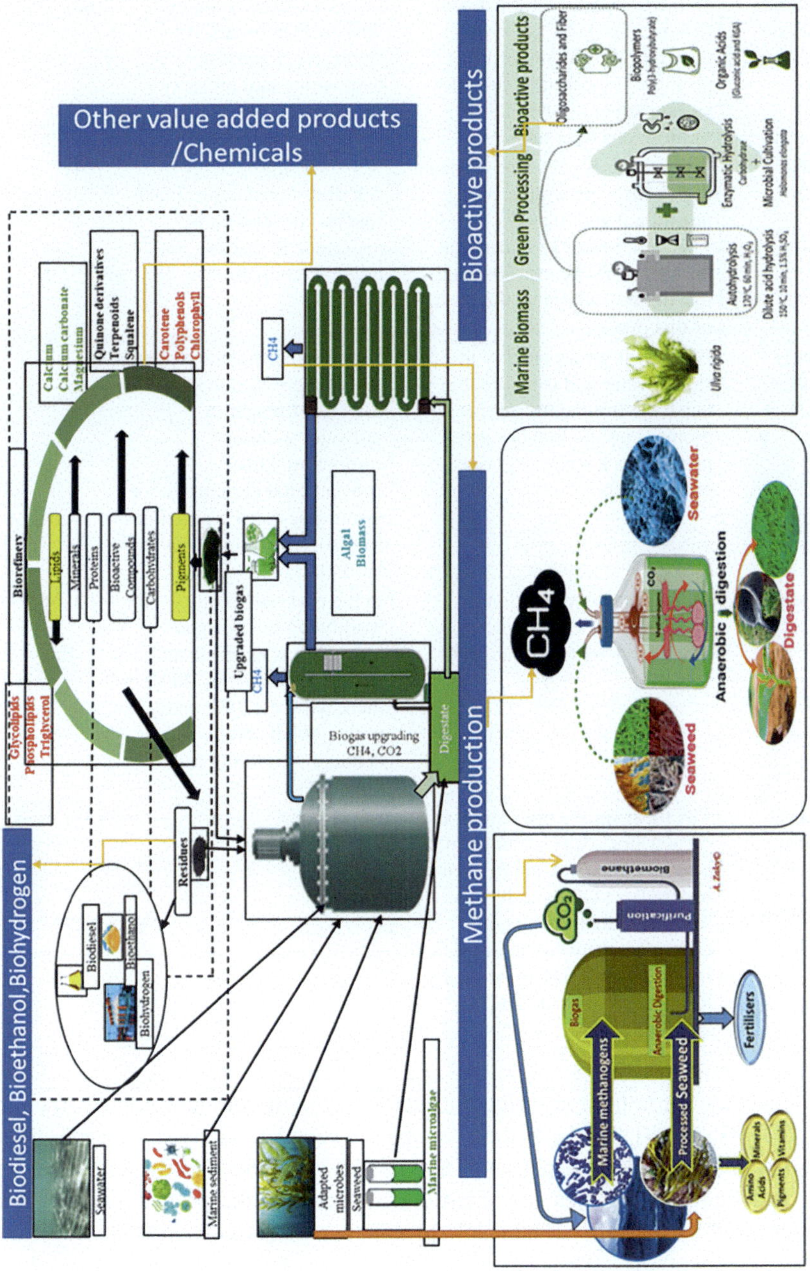

Fig. 17.4 Economic impact of marine biomass utilization [60–62]

Table 17.3 Economic benefits of marine biomass utilization

Area	Benefits	Observations	References
Aquaculture and fisheries	While the demand for fish and fisheries products has been growing, the amount of fish produced by captured fisheries has stalled at roughly 88–90 million tons over time	Fisheries directly support 90% of livelihoods. As a result, it can prevent overfishing and preserve biodiversity	[63]
Feed and fertilizer	Algae naturally replace soybeans in fish diets, providing both nutritional and financial benefits. This is because the nutritional profiles of soybeans show that fish do not fully meet the needs of this plant	Seaweed supplements to fish diets boost several fish species' development, resistance to disease, and lipid metabolism	[64]
Bioenergy production	Because it grows quickly in saline water and is resilient in harsh environments, marine biomass production has great promise as a source of non-edible biofuel	Because marine macroalgal biomass grows quickly and can be cultivated in saltwater without competing with conventional agriculture for resources, it is a promising feedstock for transportation biofuels	[65]
Macroalgal biomass	Enhances the production of biofuels in an environmentally sustainable manner than land crops	In the past half-century, the worldwide macroalgae industry has experienced rapid expansion. By 2040, there will be a 28% increase in the global energy demand, according to projections	[63, 66]
Carbon capture and storage	The most significant seagrass carbon stock was found from seagrass beds in Ujung Pancu waters is 231.43 g carbon/m^2, while the smallest seagrass carbon stocks were obtained from seagrass beds in Ahmad Rhang Manyang at 123.93 g carbon/m^2		[67]

numbers, asking for help in ensuring the survival of their common vast marine ecosystems (LMEs) [69]. To help them adopt an ecosystem-based, science-driven approach to managing human activities that impact connected freshwater basins and coastal and marine ecosystems, this article summarizes the procedures that have been supported by the GEF. Renewable products and services worth $10.6 trillion annually are in jeopardy [31].

The Council has authorized ten projects totaling 72 nations, while seven further LMEs with 54 countries are in the process of preparing GEF international

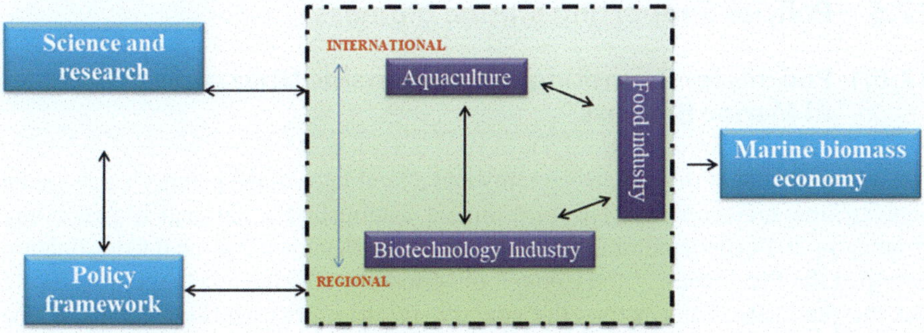

Fig. 17.5 Policy frameworks for marine biomass management

waterways projects. The nations are in the process of testing a five-module evaluation and management approach that uses place-based management to help them develop joint governance institutions that are practicable. By including stakeholders and encouraging scientific community engagement, this strategy helps to create institutions for adaptive management.

Four LMEs' comprehensive initiatives are described. For certain LMEs and the connected watersheds they affect, the significance of forging and arranging relationships between the north and south is emphasized. These well-coordinated north–south collaborations greatly strengthen the GEF's catalytic interventions, which result in investments and reforms aiding in the shift to sustainable development [70].

17.5.2 International, National, and Regional Regulations Governing Marine Resource Extraction and Cultivation

Working in the realm of the marine environment and its resources are a staggering number of national, international, and local governmental and non-governmental organizations. Numerous international legal frameworks govern the operations of these organizations. These range from broad international treaties such as the United Nations Convention on the Law of the Sea to regional accords designed to preserve and develop oceans within specific regions. The predominance of sectoral methods characterizes the administration of these waters at both the national and international levels. A number of steps have been made throughout time to enhance collaboration, coordination, and integration to attain better coherence of policies and strategies across many organizations that deal with marine and maritime management, both inside and outside the framework of the United Nations. However, the progress has not been really significant [71] (Fig. 17.5).

17.6 Policies Supporting Marine Biomass

17.6.1 Policies for Promoting the Sustainable Utilization of Marine Biomass

A key component of the Italian economy and a facilitator of the country's new green pact are the bio-economy. The government established a national bio-economy strategy in 2017 (updated in 2019) and a specially designed National Coordination Group at the Presidency of the Council of Ministers, which is in charge of carrying out the plan effectively across the country, in an effort to further maximize the opportunities brought about by the bio-economy in Italy. Over the next several years, this plan should allow the Italian bio-economy to grow by 1.25% annually in turnover and employment. More importantly, though, it should play a critical role in ensuring Italy's continued competitiveness in this important area of the European economy and Green Deal package [69].

The sustainability, health, and biomass yields of marine resources can be increased by taking a more thorough and ecologically focused approach to assessing, monitoring, and managing coastal ecosystems than we have for much of this century. The 1992 United Nations Conference on Environment and Development (UNCED) declaration was endorsed by most coastal nations worldwide. Its objectives were to develop and increase the potential of marine living resources to meet human nutritional needs as well as social, economic, and development goals, and prevent, reduce, and control degradation of the marine environment to maintain and improve its life-support and productive capacities [69]. This was a significant step toward the advancement of more ecologically based management practices. Aerosol pollutants, toxic algal blooms, nutrient loadings, habitat deterioration, and the depletion of living resources are due to pollution and overuse. When discipline-oriented ecological studies are carried out within a multidisciplinary scientific framework at the organizational level and concentrate on populations, habitats, and ecosystems at broad spatial scales, they can make a greater contribution to the attainment of resource sustainability [72]. A number of sizable marine ecosystems' primary, secondary, and tertiary driving causes for biomass output variations are explored, together with changes in the systems' overall "health." Issues with marine resources that UNCED highlighted are being dealt with. Large-scale marine ecosystem post-UNCED programs are currently being developed to move toward commercially feasible biomass yields, healthy ecosystems, and resource sustainability. International organizations are funding the programs in an attempt to link the less developed coastal nations at the edges of ocean basins with the more developed coastal nations in terms of ecological monitoring, management, and stress-reduction techniques [72].

To keep climate change below 2 °C, the European Council and Parliament have set a long-term objective to reduce GHG emissions in the European Union (EU) by 81–96% by 2052, relative to 1991 levels. The Roadmap for Moving to a Competitive Low Carbon Economy in 2050 describes the EU's climate initiatives that will help the EU become a competitive low carbon economy by that year. This aligns with the stance taken by global leaders in the Cancun and Copenhagen Agreements to implement long-term low-carbon development plans and cost-effective approach necessitates a 40% domestic GHG emission reduction for 2030 compared to 1990 levels, and an 80% decrease for 2050, according to the study of several scenarios. With the percentage of low-carbon technologies in the power mix expected to rise from around 45% in 2020 to 60% in 2030, 75–80% in 2050, and almost 100% in 2050, the energy sector should play a significant role in accomplishing these targets [69]. The Europe 2020 Strategy [COM (2010) 2020] for employment and growth seeks to both escape the financial crisis and equip the EU economy for the challenges of the next 10 years. Europe 2020 lays forth a plan to achieve inclusive growth (jobs, productivity, and social and territorial cohesion), sustainable growth (resource-efficient and low-carbon economy), and smart growth (research and innovation).

Seven "flagship initiatives" that tackle each major issue are used to help the plan reach its objectives: (1) sustainable growth (an industrial policy for the globalization period and resource efficient Europe); (2) inclusive growth (an agenda for new skills and jobs and European platform against poverty); and (3) smart growth (digital agenda for Europe, innovation union, and youth on the move) [59].

17.6.2 Supports of Marine Biomass Development and Research

Seaweed is one kind of marine biomass that has the potential to be converted into sustainable energy in line with the United Nations Sustainable Development Goals (UN SDGs) amidst the global energy crisis and environmental concerns. Seaweed shows itself as an underutilized renewable energy source in the urgent need to switch to sustainable energy sources. The history, present, and future directions of seaweed biomass conversion are thoroughly examined. A thorough analysis is provided, assessing the benefits, drawbacks, and developments of thermochemical processes and other conversion techniques. Seaweed is a great option for producing bioenergy because of its quick growth and high glucose content. Seaweed conversion contributes to waste management, nutrient recycling, and carbon capture—all of which are beneficial to many UN SDGs [69]. There are difficulties with large-scale seaweed cultivation and energy conversion, such as technical obstacles, financial viability, and environmental issues. It is clear how valuable marine biomass—particularly seaweed—is for producing sustainable energy, underscoring the complexity of this resource. It is advised to implement strict sustainability guidelines, comprehensive life-cycle evaluations, and advancements in conversion technology. The focus of future efforts should be on comprehensive feasibility studies and interdisciplinary cooperation to solve the complex issues surrounding this energy transition and bring it into line with the UN's sustainability goals [70]. The various polices based on biofuel are listed in Table 17.4.

Table 17.4 Different policies associated with biofuel

Country	Biofuel policy	References
Brazil	National fuel Alcohol Program	[73]
	Diesel Substitution Program—2002	[74]
	Diesel Substitution Program—2004	[75]
	National Biofuels Policy—2017	[75]
China	National High Technology Research and Development Initiative (Plan 863)	[75]
	Pilot Testing Program for Bioethanol—2002	[75]
	Renewable Energy Law	[76]
	Medium and Long-Term Development Plan for Renewable Energy	[76]
	Working Plan for Promotion of Biodiesel—2009	[76]
	Biodiesel Industrial Development Policy—2013	[76]
Russia	Energy Policy	[75]
	Authorized technical standard for 5% blending of ethanol with petrol—2002	[75]
	Energy Strategy for Russia till (2020–2003)	[75]
	Authorized 10% of blending of ethanol with Petrol—2004	[75]
	Authorized 5% of biodiesel blending with diesel—2005	[75]
	Energy Strategy for Russia till (2030)—2009	[75]
	Draft Russian Energy Strategy till (2035)—2015	[75]
South Africa	White Paper on Energy Policy	[77]
Chile	National Energy Security Policy—2006	[78]
Costa Rica	National Commission on Biofuels—2006	[78]
Argentina	Biofuels Law (26,093)—2006	[78]
Malaysia	National Biofuel Policy of 2006	[79]
Indonesia	National Energy Policy—2006	[77]
Philippines	Biofuels Act—2006	[77]
Canada	EcoEnergy for Biofuel Programs—2008	[80]
	Renewable Fuel Legislation—2010	[80]
USA	Energy Independence and Security Act of 2007	[80]
India	Ethanol Bending Program—2002	[81]
	National Biodiesel Mission—2003	[79]
	National Policy on Biofuels—2009	[81]
	National Policy on Biofuels—2018	[82]

17.7 Challenges and Considerations

17.7.1 Challenges and Considerations Associated with the Application of Marine Biomass

Seaweeds and kelp, two marine biomass sources, offer enormous potential for producing fuels and high-value goods. They are good options for converting the vast amounts of solar energy into usable energy and biological element-based goods because of their quick growth rates, simplicity in collecting, and cheap preproduction costs [83]. Nowadays, seaweeds are utilized to make industrial gums, fertilizers, cosmetics, and long- and short-chain compounds with industrial and therapeutic use. For any kind of biomass, the selective separation of high-value products and fuels is a significant problem [83]. For both terrestrial and marine biomass systems, thermochemical and biochemical conversion techniques have undergone extensive testing [83]. The literature contains reports on the pyrolysis, gasification, and combustion of algae and lignocellulosic feedstocks in a laboratory setting [83]. Anaerobic digestion and fermentation methods for commercial use have tremendous worldwide function; an alternative to the aforementioned techniques has been reported, hydrothermal conversion of wet, and entire biomass. Recent reports have described the extraction of proteins and high-value compounds from wet algae using microwave techniques and other approaches [84]. Future technologies would have to concentrate on making the most of the organic materials found in biomass to produce a range of valuable bio-based goods and/or fuels. The future of exploiting marine biomass will lie in the combination and integration of several technologies, even though a single approach might not be as appealing. This issue will cover issues, such as biomass harvesting, fractionation/extraction, upgrading (and conversion) into fuels and coproducts, and systems analysis, such as life systems, with an emphasis on the latest developments in marine biomass use bioeconomics for using marine biomass sustainably [84].

A long, homogeneous shoreline is necessary for large-scale macroalgal cultivation, both geographically and legally leading to a spatial challenge for finding a favorable cite. These rules, though, should also safeguard the ecosystem and guarantee that environmental sustainability can coexist with farming, harvesting, and processing. These objectives may restrict the range of seaweed species that can be used as well as the appropriate locations for seaweed farming [85]. Local policies for industrial sectors are different from that of agricultural sectors. However, there is no explicit employment model or policy for the macroalgal sector that specifies whether these fields fall under the category of industry or agriculture. The production of seaweed is a priority area according to the Indian government. In general, legal support is provided through professional training and financial support from agencies, such as the State Bank of India, the National Fisheries Development Board, and the Aquaculture Foundation of India, an NGO with headquarters in Chennai, Tamil Nadu. In addition, unlike other seaweed-growing nations, such as the Philippines or Indonesia, there are no government guidelines accessible to support institutional development, corporate ventures, or research in the seaweed industry [86]. The weather and environment have an additional impact on where large-scale seaweed farms are located. Algae are sensitive to a variety of growth circumstances, such as salinity, pH, turbidity, temperature, and concentrations of nutrients. Seaweed production can be rapidly impacted by the distinct seasonal pattern, variations, and natural disasters (high tides, hailstorms, cyclones, and tsunami) of the Indian subcontinent through a variety of changes in the quantity and quality of the water around the cultivation area. Moreover, flooding that occurs during the wet season can contaminate and pollute seaweed. For example, frequent storms and sedimentation present challenges to the commercial cultivation of seaweeds, especially for the production of agarophytes [87]. There are still many issues that need to be resolved, such as selecting the right species and using controlled cultivation and harvesting techniques [88], finding technically sound and efficient ways to increase biomass yields, and creating affordable production streams, low-cost transportation systems appropriate for the current global markets [89, 90] (Fig. 17.6).

17.7.2 Biodiversity Conservation, Invasive Species Management, and Potential Conflicts with Other Ocean Uses

The control of invasive species, possible conflicts with other ocean users, and biodiversity protection are essential components of marine ecosystem management. In the Preservation of Biodiversity Ocean, biodiversity is essential for preserving the resilience, productivity, and balance of ecosystems [83]. It encompasses a wide range of environments, animals, and genetic diversity. The main risks to marine biodiversity include overfishing, pollution, habitat degradation, and climate change [91]. These problems may result in the extinction of important species, disturbance of food webs, and general deterioration of ecosystems [84]. Establishing boundaries for particular areas where human activity is prohibited to preserve marine animals and environments. Putting laws into place to guarantee that fishing is done in a way that has the fewest detrimental effects possible on non-target species and their ecosystems [84]. Managing invasive species and non-native species that have the potential to destroy the local ecosystem, economic, or public health when they are brought into a new area are known as invasive species [92, 93]. The equilibrium of ecosystems can be upset by invasive species ability to out compete with native species and modify environments. Ship ballast water discharge is a frequent vector for invasive organisms in the maritime environment [94].

Invasive species should be recognized and dealt with as soon as possible to stop their establishment and spread [91].

Biosecurity procedures: putting policies in place to lessen the possibility of bringing in invasive species, including ship ballast water.

17.8 Future Prospects and Innovations

17.8.1 Emerging Technologies and Innovations in the Field of Marine Biomass Utilization

When it comes to tackling environmental issues, advancing a circular economy, and managing resources sustainably, emerging technologies and ideas in the field of marine biomass utilization are essential [94]. Advances in microalgae cultivation for the production of biofuel, dietary supplements, and medicines are known as "algae and seaweed farming." Lipids, proteins, and carbohydrates abound in algae. Large-scale seaweed farming is used in food items, biofuels, and bioplastics, among other uses [92, 93].

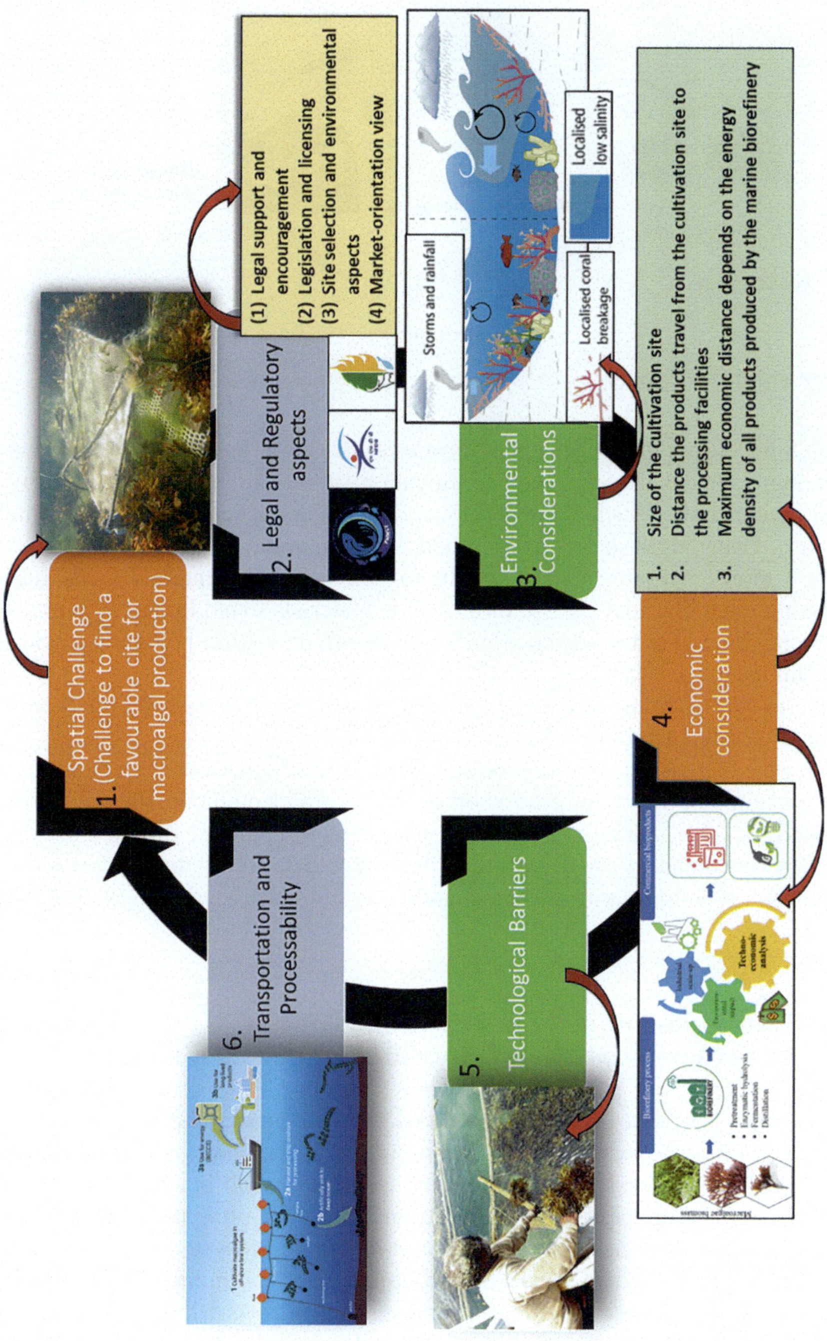

Fig. 17.6 Challenges in marine biomass application [85–90]

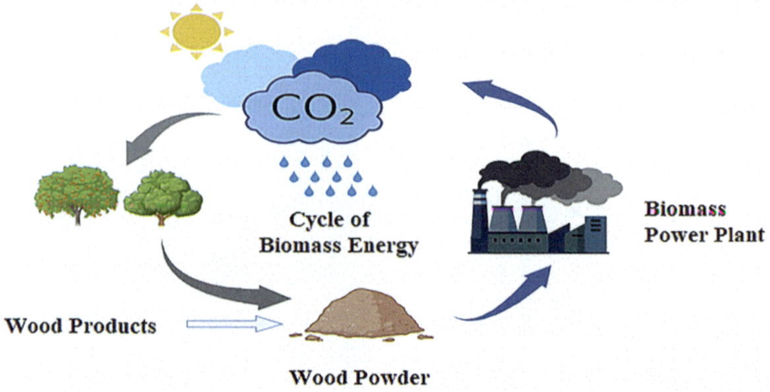

Fig. 17.7 Cycle of biomass energy

Searching aquatic life for bioactive substances that could have industrial, medicinal, or nutraceutical uses, genetic modification of marine creatures to improve desired characteristics, such as higher biomass output or resilience to environmental stress [95]. Using algae to produce biofuels, such biodiesel and bioethanol, which provide a more environmentally friendly substitute for conventional fossil fuels [83]. Using microbial fuel cells to turn organic materials found in marine biomass into power. Value-of-waste uses marine life to absorb or degrade pollutants to clean up contaminated areas [96].

Value-to-Waste: transformation of marine waste—such as fish by-products—into useful goods, including collagen, fish oil, and feed for animals. Including the use of marine biomass in circular economy models, in which waste from one operation is converted into a resource for another [91, 94–98]. Combining the production of seaweed or other marine plants with fish farming to minimize resource use and lessen environmental effect using AUVs to gather data for studies and resource management, monitor marine biomass, and carry out underwater surveys [96, 97] (Fig. 17.7).

17.8.2 Developments in Regulations and Policies to Accommodate New Opportunities and Challenges

Regulations and policies must be updated and adjusted on a regular basis due to the dynamic nature of the businesses pertaining to marine resources and the environment. In an effort to protect vital maritime ecosystems and biodiversity, several nations are expanding the size and quantity of MPAs. Adopting more comprehensive management strategies that take into account entire ecosystems as opposed to concentrating just on certain species or resources [94, 97]. Limiting the quantity of fish can be taken to protect sustainable fisheries and avoid overfishing. Fisheries that follow sustainable methods are acknowledged and supported by certification schemes, such as the Marine Stewardship Council . Application of both national and international regulations limits the spread of exotic species and regulates ballast water discharge. Creating plans promptly detect and control invasive species before they appear [91–93, 96].

17.9 Conclusion

A comprehensive, scientifically grounded, cooperative strategy that values innovation and upholds the inherent worth of marine ecosystems is necessary for the future management of marine resources. Dedication to sustainability, technological advancement, and international collaboration will be essential to guarantee the health and resilience of our seas for future generations as civilizations continue to navigate complex environmental concerns. Several promising approaches must be considered to effectively valorize marine waste, including the development of mechanical technology for biomass harvesting, the creation of chemical and biological pipelines to conserve and process marine biomass, market research to maximize potential use in various industries, the use of communication strategies to increase consumer acceptance of greener technologies or improve public health through enhanced waste-derived food and feed ingredients as well as biomedicals, and the encouragement of public and private investment for technological advancement and innovation, with a focus on resolving potential legal roadblocks. In addition, the integration of environmentally friendly concepts into industrial processes is crucial. Despite notable advancements in research and innovation related to marine-derived biomass, there is an unfulfilled demand for technological scaling up and the creation of economic prospects to support sustainability and a circular economy. The conversion of trash into valuable goods requires close cooperation between the research and industrial sectors, as well as governing agencies. The European Green Deal, Europe's blueprint for a sustainable economy, represents the most current policy effort reflecting this commitment to governance and the promotion of resource efficiency on the continent. As waste valorization firms participate, the utilization of bioactive chemicals valorized from waste and side streams will significantly contribute to global environmental sustainability, promoting a circular, blue, and green economy. These businesses should cultivate a socially conscious image and offer communities, particularly those relying on fishing and aquaculture, an alternative source of income. In summary, irrespective of the time of year or location, future research should focus on streamlining side streams for processing into valuable goods and simultaneously ensure a sustainable supply and product quality.

References

1. Renzi JJ, He Q, Silliman BR (2019) Harnessing positive species interactions to enhance coastal wetland restoration. Front Ecol Evol 7(4):1–14. https://doi.org/10.3389/fevo.2019.00131
2. Fourqurean JW, Duarte CM, Kennedy H et al (2012) Seagrass ecosystems as a globally significant carbon stock. Nat Geosci 5(7):505–509. https://doi.org/10.1038/ngeo1477

3. Mata TM, Martins AA, Caetano NS (2010) Microalgae for biodiesel production and other applications: a review. Renew Sustain Energy Rev 14(1):217–232. https://doi.org/10.1016/j.rser.2009.07.020
4. Srivastava RK, Shetti NP, Reddy KR, Kwon EE, Nadagouda MN, Aminabhavi TM (2021) Biomass utilization and production of biofuels from carbon neutral materials. Environ Pollut 276:116731. https://doi.org/10.1016/j.envpol.2021.116731
5. Abelson A, Halpern BS, Reed DC et al (2016) Upgrading marine ecosystem restoration using ecological-social concepts. Bioscience 66(2):156–163. https://doi.org/10.1093/biosci/biv171
6. Wijffels RH, Barbosa MJ, Eppink MHM (2010) Microalgae for the production of bulk chemicals and biofuels. Biofuels Bioprod Biorefin 4(3):287–295. https://doi.org/10.1002/BBB.215
7. Bilawal KM, Wang S, Zhang X et al (2023) Technology advancement and international law in marine policy, challenges, solutions and future prospective. Front Mar Sci 10:1–17
8. Hamon KG, Kreiss CM, Pinnegar JK et al (2021) Future socio-political scenarios for aquatic resources in Europe: an operationalized framework for marine fisheries projections. Front Mar Sci 8(3):1–21. https://doi.org/10.3389/fmars.2021.578516
9. Khan N, Sudhakar K, Mamat R (2021) Role of biofuels in energy transition, green economy and carbon neutrality. Sustainability 13(22):12374
10. Bar-On YM, Phillips R, Milo R (2018) The biomass distribution on earth. Proc Natl Acad Sci U S A 115(25):6506–6511
11. Whitman WB, Coleman DC, Wiebe WJ (1998) Prokaryotes: the unseen majority. Proc Natl Acad Sci U S A 95:6578–6583
12. Bar-On YM, Milo R (2019) The biomass composition of the oceans: a blueprint of our blue planet. Cell 179(7):1451–1454
13. Alongi DM (2012) Carbon sequestration in mangrove forests. Carbon Manag 3(3):313–322. https://doi.org/10.4155/cmt.12.20
14. Gao K, McKinley KR (1994) Use of macroalgae for marine biomass production and CO2 remediation: a review. J Appl Phycol 6(1):45–60. https://doi.org/10.1007/BF02185904
15. John RP, Anisha GS, Nampoothiri KM, Pandey A (2011) Micro and macroalgal biomass: a renewable source for bioethanol. Bioresour Technol 102(1):186–193. https://doi.org/10.1016/j.biortech.2010.06.139
16. Jones DOB, Yool A, Wei CL et al (2014) Global reductions in seafloor biomass in response to climate change. Glob Chang Biol 20(6):1861–1872. https://doi.org/10.1111/gcb.12480
17. Rudovica V, Rotter A, Gaudêncio SP et al (2021) Valorization of marine waste: use of industrial by-products and beach wrack towards the production of high added-value products. Front Mar Sci 8(6):39. https://doi.org/10.3389/fmars.2021.723333
18. Gaspard S, Ncibi MC (2013) Biomass for sustainable applications. Royal Society of Chemistry, Cambridge
19. Bayraktarov E, Saunders MI, Abdullah S et al (2016) The cost and feasibility of marine coastal restoration. Ecol Appl 26(4):1055–1074. https://doi.org/10.1890/15-1077
20. Paz S, Rubio C, Frías I et al (2019) Toxic metals (Al, Cd, Pb and Hg) in the most consumed edible seaweeds in Europe. Chemosphere 218:879–884. https://doi.org/10.1016/j.chemosphere.2018.11.165
21. Xu S, Yu Z, Zhou Y, Yue S, Liang J, Zhang X (2023) The potential for large-scale kelp aquaculture to counteract marine eutrophication by nutrient removal. Mar Pollut Bull 187(22):114513. https://doi.org/10.1016/j.marpolbul.2022.114513
22. Wicker RJ (2023) Bioprospecting for nordic photosynthetic consortia for wastewater treatment, carbon capture, and value creation
23. Pauly D, Palomares MLD (2001) Fishing down marine food webs: an update. Waters Peril 279(2):47–56. https://doi.org/10.1007/978-1-4615-1493-0_4
24. Rose DJ, Hemery LG (2023) Methods for measuring carbon dioxide uptake and permanence: review and implications for macroalgae aquaculture. J Mar Sci Eng 11(1):175. https://doi.org/10.3390/jmse11010175

25. Tonin S (2018) Citizens perspectives on marine protected areas as a governance strategy to effectively preserve marine ecosystem services and biodiversity. Ecosyst Serv 34:189–200. https://doi.org/10.1016/j.ecoser.2018.03.023
26. Yong WTL, Thien VY, Rupert R, Rodrigues KF (2022) Seaweed: a potential climate change solution. Renew Sustain Energy Rev 159(2):112222
27. Pardilhó S, Cotas J, Pacheco D et al (2023) Valorisation of marine macroalgae waste using a cascade biorefinery approach: exploratory study. J Clean Prod 385:135672
28. Johnston KG, Abomohra A, French CE, Zaky AS (2023) Recent advances in seaweed biorefineries and assessment of their potential for carbon capture and storage. Sustainability 15(17):13193
29. Zhu Y, Romain C, Williams CK (2016) Sustainable polymers from renewable resources. Nature 540(7633):354–362. https://doi.org/10.1038/nature21001
30. Thomas JE, Sinha R, Strand Å et al (2022) Marine biomass for a circular blue-green bioeconomy? A life cycle perspective on closing nitrogen and phosphorus land-marine loops. J Ind Ecol 26(6):2136–2153. https://doi.org/10.1111/jiec.13177
31. Scarlat N, Dallemand JF, Monforti-Ferrario F, Nita V (2015) The role of biomass and bioenergy in a future bioeconomy: policies and facts. Environ Dev 15(2015):3–34. https://doi.org/10.1016/j.envdev.2015.03.006
32. Ng HS, Kee PE, Yim HS, Chen PT, Wei YH, Chi-Wei Lan J (2020) Recent advances on the sustainable approaches for conversion and reutilization of food wastes to valuable bioproducts. Bioresour Technol 302(135):122889. https://doi.org/10.1016/j.biortech.2020.122889
33. Torres MD, Kraan S, Domínguez H (2019) Seaweed biorefinery, 18(2):335
34. Limtong S, Deejing S, Yongmanitchai W, Santisopasri W (1998) Construction of high ethanol fermenting halotolerant hybrid by intergeneric protoplast fusion of Saccharomyces cerevisiae and Zygosaccharomyces rouxii. J Nat Sci 32(1997):213–223
35. Kostas ET, White DA, Du C, Cook DJ (2016) Selection of yeast strains for bioethanol production from UK seaweeds. J Appl Phycol 28(2):1427–1441
36. Chandel AK, Singh OV (2011) Weedy lignocellulosic feedstock and microbial metabolic engineering: advancing the generation of biofuel. Appl Microbiol Biotechnol 89(5):1289–1303
37. Obara N, Ishida M, Hamada-Sato N, Urano N (2012) Efficient bioethanol production from scrap paper shredder by a marine Saccharomyces cerevisiae derived C-19. Stud Sci Technol 1:127–132
38. Khambhaty Y, Upadhyay D, Kriplani Y, Joshi N, Mody K, Gandhi MR (2013) Bioethanol from macroalgal biomass: utilization of marine yeast for production of the same. Bioenergy Res 6(1):188–195
39. Yanagisawa M, Nakamura K, Ariga O, Nakasaki K (2011) Production of high concentrations of bioethanol from seaweeds that contain easily hydrolyzable polysaccharides. Process Biochem 46(11):2111–2116
40. Ji S, Wang B, Lu M, Li F (2016) Defluviitalea phaphyphila sp. nov, a novel thermophilic bacterium that degraded brown algea. Appl Environ Microbiol 82:868–877
41. Ji SQ, Wang B, Lu M, Li FL (2016) Direct bioconversion of brown algae into ethanol by thermophilic bacterium Defluviitalea phaphyphila. Biotechnol Biofuels 9(1):1–10
42. Ren M, Ogden K, Lian B (2013) Effect of culture conditions on the growth rate and lipid production of microalgae Nannochloropsis gaditana. J Renew Sustain Energy 5(6):063138
43. Senthilraja P, Kathiresan K, Saravanakumar K (2012) Comparative analysis of bioethanol production from whey by different strains of immobilized. J Yeast Fungal Res 2(8):113–116
44. Kathiresan K, Saravanakumar K, Senthilraja P (2011) Bio-ethanol production by marine yeasts isolated from coastal mangrove sediment. Int Multidiscip Res J 1(1):19–24
45. Wei N, Quarterman J, Jin YS (2013) Marine macroalgae: an untapped resource for producing fuels and chemicals. Trends Biotechnol 31(2):70–77
46. Saravanakumar K, Senthilraja P, Kathiresan K (2013) Bioethanol production by mangrove-derived marine yeast, Sacchromyces cerevisiae. J King Saud Univ Sci 25(2):121–127

47. Obara N, Okai M, Ishida M, Urano N (2015) Bioethanol production from mixed biomass (waste of Undaria pinnatifida processing and paper shredding) by fermentation with marine-derived Saccharomyces cerevisiae. Fish Sci 81(4):771–776
48. Zaky AS, French CE, Tucker GA, Du C (2020) Improving the productivity of bioethanol production using marine yeast and seawater-based media. Biomass Bioenergy 139(5):105615
49. Park H, Lee CG (2016) Theoretical calculations on the feasibility of microalgal biofuels: utilization of marine resources could help realizing the potential of microalgae. Biotechnol J 11(11):1461 1470
50. International Energy Agency, Renewables 2020. Paris (2020)
51. Bauen A, Berndes G, Junginger M, Vuille F, Londo M (2009) Bioenergy—a sustainable and reliable energy source a review of status and prospects
52. Darda S, Papalas T, Zabaniotou A (2019) Biofuels journey in Europe: currently the way to low carbon economy sustainability is still a challenge. J Clean Prod 208:575–588
53. Silas K, Kwaji HB, Gutti B (2015) Microalgae as potential feedstock for biodiesel production—a review. Int J Recent Res Phys Chem Sci 2(3):1–9
54. Hemer MA, Manasseh R, McInnes KL, Penesis I, Pitman T (2018) Perspectives on a way forward for ocean renewable energy in Australia. Renew Energy 127:733–745. https://doi.org/10.1016/j.renene.2018.05.036
55. Wu Y, Shen X, Tang C-J, Chen Z-L, Hu Q, Shi W (2002) Synthesis of natural fragrant molecules cis -3-methyl-4-decanolide and Aerangis lactone. General enantioselective routes to β,γ- cis -disubstituted γ-lactones and γ,δ- cis -disubstituted δ-lactones. J Org Chem 67(11):3802–3810. https://doi.org/10.1021/jo025540o
56. Singh J, Gu S (2010) Commercialization potential of microalgae for biofuels production. Renew Sustain Energy Rev 14(9):2596–2610
57. Murphy F, Devlin G, Deverell R, McDonnell K (2013) Biofuel production in Ireland-an approach to 2020 targets with a focus on algal biomass. Energies 6(12):6391–6412
58. Chandel AK, Garlapati VK, Singh AK, Antunes FAF, da Silva SS (2018) The path forward for lignocellulose biorefineries: bottlenecks, solutions, and perspective on commercialization. Bioresour Technol 264(4):370–381. https://doi.org/10.1016/j.biortech.2018.06.004
59. Duda AM, Sherman K (2002) A new imperative for improving management of large marine ecosystems. Ocean Coast Manag 45(11):797–833. https://doi.org/10.1016/S0964-5691(02)00107-2
60. Faisal S, Zaky A, Wang Q, Huang J, Abomohra A (2022) Integrated marine biogas: a promising approach towards sustainability. Fermentation 8(10):1–28. https://doi.org/10.3390/fermentation8100520
61. Leandro T, Teles M, Gomes-Dias JS et al (2023) Ulva rigida valorization into poly(3-hydroxybutyrate), organic acids and functional ingredients. Mar Drugs 21(10):537. https://doi.org/10.3390/md21100537
62. Zaky AS (2021) Introducing a marine biorefinery system for the integrated. Processes 9:1841
63. Narwal S, Kaur M, Yadav DS, Bast F (2024) Sustainable blue economy: opportunities and challenges. J Biosci 49(1):18
64. Güroy D, Güroy B, Merrifield DL, Ergün S, Tekinay AA, Yiğit M (2011) Effect of dietary Ulva and Spirulina on weight loss and body composition of rainbow trout, Oncorhynchus mykiss (Walbaum), during a starvation period. J Anim Physiol Anim Nutr (Berl) 95(3):320–327
65. Fernand F, Israel A, Skjermo J, Wichard T, Timmermans KR, Golberg A (2017) Offshore macroalgae biomass for bioenergy production: environmental aspects, technological achievements and challenges. Renew Sustain Energy Rev 75:35–45
66. Pravin R, Baskar G, Rokhum SL, Pugazhendhi A (2023) Comprehensive assessment of biorefinery potential for biofuels production from macroalgal biomass: towards a sustainable circular bioeconomy and greener future. Chemosphere 339(8):139724
67. Uly W, Haridhi HA, Dewiyanti I (2024) Estimation of seagrass blue carbon stocks in Ahmad Rhang Manyang and Ujung Pancu 02006, 1–8

68. Costanza R, de Groot R, Braat L et al (2017) Twenty years of ecosystem services: how far have we come and how far do we still need to go? Ecosyst Serv 28:1–16. https://doi.org/10.1016/j.ecoser.2017.09.008
69. Fava F, Gardossi L, Brigidi P, Morone P, Carosi DAR, Lenzi A (2021) The bioeconomy in Italy and the new national strategy for a more competitive and sustainable country. New Biotechnol 61(9):124–136. https://doi.org/10.1016/j.nbt.2020.11.009
70. Arrow K, Bolin B, Costanza R, Dasgupta P, Folke C, Holling CS (1996) Economic growth, carrying capacity, and the environment. Environ Dev Econ 1(1):104–110. https://doi.org/10.1017/S1355770X00000413
71. Grip K (2017) International marine environmental governance: a review. Ambio 46(4):413–427. https://doi.org/10.1007/s13280-016-0847-9
72. McCormick K, Kautto N (2013) The bioeconomy in Europe: an overview. Sustainability 5(6):2589–2608. https://doi.org/10.3390/su5062589
73. Scheffran J (2010) The global demand for biofuels: technologies, markets and policies. In: Biomass to biofuels: strategies for global industries
74. de CGP (2010) Biofuels: policies, standards and technologies. World Energy Council
75. Saravanan AP, Pugazhendhi A, Mathimani T (2020) A comprehensive assessment of biofuel policies in the BRICS nations: implementation, blending target and gaps. Fuel 272(1):117635
76. Hao H, Liu Z, Zhao F et al (2018) Biofuel for vehicle use in China: current status, future potential and policy implications. Renew Sustain Energy Rev 82:645–653
77. Jull C (2007) The legal framework for bioenergy
78. Charlotta Jull PCR (2007) Recent trends in the law and policy of bioenergy production, promotion and use, pp 1–60
79. Ram Mohan MP, Thomas Phillippe GT, Shiju MV (2006) Biofuel laws in Asia: instruments for energy access, security, environmental protection and rural empowerment. Asian Biotechnol Dev Rev 8(2):51–75
80. Das S (2020) The National Policy of biofuels of India—A perspective. Energy Policy 143:111595
81. Chandel AK, Sukumaran RK (2017) Sustainable biofuels development in India
82. Kothari R, Vashishtha A, Mohan H (2020) Assessment of Indian bioenergy policy for sustainable environment and its impact for rural India: strategic implementation and challenges. Environ Technol Innov 20:101078
83. Holden NM, Neill AM, Stout JC, Brien DO, Morris MA (2023) Biocircularity: a framework to define sustainable, circular bioeconomy. Circ Econ Sustain 3(1):77–91. https://doi.org/10.1007/s43615-022-00180-y
84. Johnston P, Everard M, Santillo D, Robèrt K (2007) Discussion articles reclaiming the definition of sustainability. Environ Sci Pollut Res 14(1):60–66
85. Ingle KN, Traugott H, Golberg A (2020) Challenges for marine macroalgal biomass production in Indian coastal waters. Bot Mar 63(4):327–340
86. Mantri VA et al (2017) An appraisal on commercial farming of Kappaphycus alvarezii in India: success in diversification of livelihood and prospects. J Appl Phycol 29(1):335–357
87. Ganesan M, Eswaran K, Reddy CRK (2017) Farming of agarophytes in India—a long-time sustainability for the industry and preserving wild stocks. J Appl Phycol 29(5):2239–2248
88. Golberg A, Liberzon A (2015) Modeling of smart mixing regimes to improve marine biorefinery productivity and energy efficiency. Algal Res 11:28–32
89. Chanakya HN, Mahapatra DM, Ravi S, Chauhan VS, Abitha R (2012) Sustainability of large-scale algal biofuel production in India. J Indian Inst Sci 92(1):63–98
90. Balina K, Romagnoli F, Blumberga D (2017) Seaweed biorefinery concept for sustainable use of marine resources. Energy Procedia 128:504–511
91. Mace GM (2019) The ecology of natural capital accounting. Oxford Rev Econ Policy 35(1):54–67. https://doi.org/10.1093/oxrep/gry023
92. Srinivasan UT, Carey SP, Hallstein E et al (2008) The debt of nations and the distribution of ecological impacts from human activities. PNAS 105(5):1768–1773. https://doi.org/10.1073/pnas.0709562104

93. Chiesura A, De Groot R (2003) Critical natural capital: a socio-cultural perspective. Ecol Econ 44(2–3):219–231
94. Srinivasan R, Mishra A (2008) Okra (Hibiscus esculentus) and fenugreek (Trigonella foenum graceum) mucilage: characterization and application as flocculants for textile effluent treatments. Chin J Polym Sci 26(6):679–687. https://doi.org/10.1142/s0256767908003424
95. Costanza R, de Groot R, Braat L et al (1997) The value of the world's ecosystem services and natural capital. Nature 387(6630):253–260. https://royalroads.on.worldcat.org/oclc/4592801201
96. Havlík P, Schneider UA, Schmid E et al (2011) Global land-use implications of first and second generation biofuel targets. Energy Policy 39(10):5690–5702. https://doi.org/10.1016/j.enpol.2010.03.030
97. Dauber J, Jones MB, Stout JC (2010) The impact of biomass crop cultivation on temperate biodiversity. GCB Bioenergy 2(6):289–309. https://doi.org/10.1111/j.1757-1707.2010.01058.x
98. Walzberg J, Lonca G, Hanes RJ, Eberle AL, Carpenter A, Heath GA (2020) Do we need a new sustainability assessment method for the circular economy? A critical literature review. Front Sustain 1(1):1–20. https://doi.org/10.3389/frsus.2020.620047

Biomonitoring of Mangroves as an Indicator of Sustainable Coastal Environment

18

Indahayu Ab. Rahman, Harlina Ahmad, Sazlina Salleh, and Widad Fadhullah

Abstract

Coastal environments worldwide face increasing threats from natural stressors and human-driven impacts. Mangroves have emerged as critical indicators of environmental sustainability through their role in climate change mitigation, coastal protection, water quality enhancement, and sustainable resource management. This review explores the role of mangroves as an interceptor of microplastics, their function in climate change mitigation, and biomonitoring applications using bioindicators in the mangroves. A case study analyzing the water quality and heavy metals analyses in mangrove water, sediments, and biological organisms was also explained under the biomonitoring applications. In the face of a changing climate, it is imperative to understand how the mangroves function paves the way toward sustainability.

I. Ab. Rahman
Environmental Technology Section, School of Industrial Technology, Universiti Sains Malaysia, Penang, Malaysia

H. Ahmad · W. Fadhullah (✉)
Environmental Technology Section, School of Industrial Technology, Universiti Sains Malaysia, Penang, Malaysia

Renewable Biomass Transformation Cluster, School of Industrial Technology, Universiti Sains Malaysia, Penang, Malaysia
e-mail: widad@usm.my

S. Salleh
Centre for Policy Research, Universiti Sains Malaysia, Penang, Malaysia

Centre for Marine and Coastal Studies, Universiti Sains Malaysia, Penang, Malaysia

M. Rafatullah et al. (eds.), *Multidisciplinary Applications of Marine Resources*,
https://doi.org/10.1007/978-981-97-5057-3_18

Keywords

Mangrove coastal management · Climate change · Biomonitoring · Water quality · Biological indicators

18.1 Introduction

Coastal environments mostly extend seaward from high-water mark to the edge of the continental shelf, including soft shore sediments, rocky shores and reefs, embayment, deltas, estuaries, and mangroves ecosystems [1]. Multiple stressors in coastal environments are related to its hydrodynamic features in response to tidal cycles and the expanding industrialization and urbanization due to human population growth and economic needs [2]. Biomonitoring techniques are crucial to assessing the impact of pollution on mangrove ecosystems and monitoring their health. Biomonitoring is an environmental monitoring technique that uses certain organisms or bioindicators which can provide information on the changes in coastal environments [3] that were affected by anthropogenic activities.

Mangroves have a promising ability to be used for biomonitoring in coastal environments [4–8], as these plants have unique biological characteristics as they can readjust according to the environmental condition. Although they comprise only 1% of the world's tropical forests, mangrove ecosystems provide significant services to coastal regions and the communities residing there [9]. Being a resourceful ecosystem, mangrove in coastal systems provides a wide range of services and functions, as this ecosystem plays critical roles in maintaining the health and productivity of the coastal environment, including habitat for local fauna and flora, carbon sequestration, and protection from natural disasters (storm surges and coastal erosion) [7].

Degradation of the coastal environment occurs due to climate-induced factors, such as increased storminess threat, and sea-level rise [10] together with anthropogenic-induced nutrients, toxic chemicals, and pathogens [5] in greater amounts than found naturally in the environments. These impacts alter or intensify biophysical changes of mangroves along the coastline. Species extirpation and habitat collapse in mangrove environments may increase in response to climate changes and man-made activities. The degradation of this natural environment has been accelerated, since the dumping of urban and sewage waste in the nearby areas of sea and river sides and unauthorized land use areas is wildly being exploited [11].

One of the serious toxic chemical pollutions in the mangrove ecosystem is heavy metal (HM) pollution. Studies have found that these metals generally sink, together with the sediments and accumulated mainly in root tissues, in numerous mangrove species [12–14]. Mangroves can take up HMs and enter the aquatic systems mainly through weathering and erosion of rocks [15], but the increases of anthropogenic sources such as disposal of industrial effluents, sewage sludge, and agricultural activities may impose the pressure of contamination of HM in this ecosystem [3, 11, 14, 16]. According to a study conducted by Nguyen et al. [17], the total mangrove forest biomass decline from 1,750,000 tons (no HM pollution scenario) to 850,000 tons (current HM pollution scenario) and 350,000 tons (double HM pollution scenario) under three scenario simulations to predict mangrove dynamics under different levels of HM pollution in combined environmental conditions of salinity and elevation.

The purpose of this chapter is to provide an overview of the importance of mangroves through biomonitoring their functions and services toward achieving sustainability. In this chapter, the mangrove ecosystem characteristics and function were briefly explained focusing upon its sustainable role as an interceptor of microplastics. One of its crucial ecosystem services offered; carbon sequestration role is covered which is then followed by biomonitoring applications in mangroves including one example of a related case study on preliminary biomonitoring assessment of water quality and HM accumulation in the mangrove ecosystem located in Kuala Gula, Perak Malaysia. Considering its multiple function and services, it is imperative to evaluate the current challenges and limitations for a sustainable mangrove ecosystem management in the future.

18.2 Mangrove Ecosystems' Characteristics, Function, and Its Role in Carbon Sequestration

18.2.1 Characteristics and Function

Mangroves are plant communities that flourish in the intertidal areas of tropical and subtropical regions along the coast [18]. They are vital in safeguarding coastlines, capturing and storing carbon, and serving as homes for diverse marine and land-dwelling species. Mangrove roots have prominent and unique features than other vegetation by having pneumatophores, stilt roots, and prop roots which specializes in breathing in the soil adjacent to the water and for better wind and wave resistance [19]. Mangrove ecosystems recorded the highest diversity of more than 50 species in the Southeast Asian regions including in Malaysia [20].

Some of the examples of widespread mangrove distribution in Malaysia are Matang Forest Reserve, Kuala Selangor Nature Park, Tanjung Tuan Forest Reserve, and Bako National Park. Some of the dominant genera of mangrove species are Bakau (*Rhizophora*), Berembang and Perepat (*Sonneratia*), Api-api (*Avicennia*), and Tumu (*Bruguiera*) [21]. These different genera are easily distinguished upon its above ground roots; *Rhizophora* have prop roots, *Avicennia* have pneumatophores, while *Bruguiera* have root knees. Another difference between these genera refers to their ability to secrete salt. *Avicennia* are able to exclude most of the salt from the seawater through its salt glands which categorize them as secreters. In contrast, the non-secreters such as *Rhizophora, Sonneratia,* and *Bruguiera* remove salt through the transpiration process [22]. The zones of mangrove are also different; the genera found at the zone closest to sea are *Avicennia* and *Sonneratia*, while the middle zone (behind *Avicennia* and *Sonneratia*) is generally *Rhizophora*. *Bruguiera* grows in the area furthest from the sea which is behind *Rhizophora* [23].

One of the emerging functions of mangroves is as an intercept in microplastics. Mangroves have always been recognized as interceptors of nutrients, pollutants, and sediments prior to reaching the oceans [24]. Initial focus area for microplastics study was in the ocean as the endpoint or final destination of microplastic accumulation. However, the distribution of the aquatic environment involves the freshwater, environment, coastal areas, marine, and the ocean compartment. Microplastics which are commonly less than 5 mm can easily be trapped among the roots of mangroves before entering the adjacent marine counterpart [25]. Among the dominant microplastic found in the mangrove ecosystems is polystyrene, polypropylene, and polyethylene which are related to laundry waste, domestic discharges, fishing, tourism, and shipping activities [26]. Microplastic in the mangrove seawater is commonly floating and deposited on the sea surfaces due to its lower density than seawater ($<1.02\ g/cm^3$).

The interception rates of microplastics in the mangrove ecosystems are influenced by the river flow velocity, ecological indicators, and the sediment porosity. The influence of river flow is dominant during the wet season, whereby rainfall causes higher water flow which subsequently releases the deposited microplastic from the sediments and leaf litters area. The total above biomass was found to block more microplastics compared to the total coverage area of mangrove trees. This factor is dependent upon the arching root system of the mangroves which trapped more of the larger microplastics. Mangrove sediments trapped more of the microplastics in their pores compared to the sediment organic matter content [19]. Higher porosity allows more microplastics to flush out to the ocean particularly for the larger microplastics, but lower porosity trapped more of the microplastics in the mangrove sediments.

In general, microplastic abundance was found higher in densely populated areas and urbanized localities compared to those in less populated and urbanized areas [25, 26]. Using mangroves in Penang as a case study, estuarine sediment and coastal bottom sediments were reported to have the highest abundance of microplastics in mangrove areas. This factor could be likely due to the potential of aerial roots trapping those microplastics which settled at the estuarine and coastal bottom sediments. Microplastics were also found in deeper layers of the sediment up to 142 cm depth due to infiltration and bioturbation process by benthic organisms.

Microplastics abundance was reported to be higher in the high tide zone sediments than in the low and mid tide zones. This factor is driven by the tide velocity which may disrupt microplastics from the sediment. It could also be impacted by the presence of aerial roots trapping these microplastics during high tide events. Other factors influencing microplastic abundance is the particulate organic content [27]. A study by Mohamed et al. [28] reported the abundance of microplastics near the mouth of the river in Kuala Gula mangroves. This area received reduced currents which allows microplastics to be deposited. Fibers and fragments were highest which may be related to the laundry washing of the communities in the area.

18.2.2 Mangroves and Its Role in Carbon Sequestration

Mangroves are collectively categorized as the blue carbon ecosystem alongside seagrasses and wetlands. This so-called blue carbon ecosystems can keep and store carbon over a long period of time if photosynthesis process exceeds the respiration and decomposition capacity [29]. Mangrove sediments are areas with anoxic conditions which inhibit decomposition, and therefore, the leaf litter and mangrove residues are buried over a lengthy period of time in these sediments. The carbon dioxide from the atmosphere is captured and stored as carbon within the leaves and mangrove trees and also the sediments. This carbon may be contributed by autochthonous (internally derived) and allochthonous input (externally derived). Root production of mangroves is one of the autochthonous inputs for mangrove carbon [30]. This whole process portrays the role of mangroves in sequestering carbon within its constituents.

The carbon capture ability of mangroves was calculated using estimates of the aboveground biomass and the belowground biomass, loss of ignition, organic matter concentrations, and net primary productivity. The geomorphological settings in the mangroves together with the physical factors such as tide and waves as well as hydrological and biogeochemical properties control the carbon inputs in the aboveground and belowground biomass [31]. Apart from that, satellite image approach was also used to obtain a better geographical coverage and longer temporal monitoring of the mangroves [32]. This approach provides an alternative in mapping and monitoring mangrove ecosystems worldwide. Land use and land cover connection with mangroves is another critical factor to consider in studying the carbon sequestration ability in mangroves. Among the largest loss of carbon storage capacity is linked to the conversion of mangroves into aquaculture ponds (more than 80%) as reviewed by Sasmito et al. [33].

Mangrove carbon sequestration capability is dependent upon the size and age of the plants. Mature forests have greater biomass, more litter, and residues produced which accumulate in the thick rich sediments of mangroves than in younger and new plants. A study by [34] reported that planted mangroves with a diameter breast height (DBH) class size of 11–20 cm have a better ability to sequester than other size classes. Younger mangrove plants of about 9–14 years with an average DBH of 6.88 cm in the Philippines were only able to sequester about 51.1 tons per hectare of total biomass [35]. Indonesia with the mangrove distribution of 3.3 million hectares accounts for 950.5 Mg C ha^{-1} of potential CO_2 absorption by mangrove stands [36]. Among the regions in Indonesia, Papua accounts for 56% of carbon capture potential [37] followed by Sumatra and Kalimantan (17.2–17.5%).

Carbon capture was found to be highest in the mangroves than the terrestrial forest [38]. This finding was contributed by the anaerobic conditions and waterlogged sediment of mangroves. The role of aboveground biomass was 40% higher than belowground biomass in capturing and storing carbon in planted mangrove [34]. The carbon in the aboveground biomass was reported to be contributed mostly by the stem and bark of the mangrove plants compared to the branches and leaves [39].

The ability of mangroves in storing carbon is species specific, whereby fast-growing species from the genera *Avicennia* are better candidates than *Rhizophora* and *Bruguiera*.

18.3 Biomonitoring Applications

18.3.1 Biological Indicators

Biological indicators or bioindicators use living organisms to measure and evaluate the quality of their environment [40]. The reactions exhibited by these living organisms are observed and quantified in response to alterations in their surroundings. The extent of harmful effects on organisms and ecosystems caused by combinations of pollutants in various stress scenarios can be assessed by applying the concept of bioindicators [41]. Thus, by studying the health and behavior of these organisms, researchers can acquire valuable insights into the overall state of the ecosystem and detect possible environmental stressors. For example, diatoms have been used to indicate the changes in water quality [42], meiofauna in sediments pollution [43], and molluscs for the presence of HMs [44].

Diatoms (Bacillariophyceae) are crucial as primary producers in the mangrove ecosystems [45]. They are valuable bioindicators, because they represent the conditions at the sampling site, are widely available, and are simple to collect [46]. They serve as valuable model organisms for evaluating the ecological well-being of environments impacted by human activities in mangroves and marine and freshwater ecosystems. The suitability of diatom assemblages for bioindication purposes and their ability to accurately reflect site-specific water quality depends on how they respond to variations in water quality, even when nutrient concentrations are low. This is primarily due to their short life cycles, abundance, and quick response to environmental changes. When nutrient levels increase, there is a direct correlation with a boost in the primary productivity of a selected species. For instance, the high abundance of *Rhizosolenia* spp. is known to be associated with high ammonia habitats, thus recognizing it as a negative indicator of water quality [47]. Owing to that, diatoms are frequently utilized to assess the extent of eutrophication in water due to their rapid sensitivity to water contamination. Enrichment by nutrients and other pollutants could favor the proliferation of motile diatoms through various mechanisms. Initially, mobile diatoms can release extracellular enzymes, allowing them to utilize large molecules attached to solid surfaces or sediments. Motile diatoms typically exhibit larger sizes than sessile diatoms, allowing them to accumulate more nutrients [48]. Furthermore, it should be noted that motile diatoms can swiftly transition from nutrient-poor microenvironments to areas with a higher concentration of nutrients [49].

The benthic invertebrate fauna indicates sediment changes in the marine environment. However, their response times are slower than primary producers, who can quickly convert sporadic events into observable changes. For example, the meiofauna population in mangrove sediments is highly abundant and diverse, and the community structure is shaped by the mangrove sediment characteristics, making them an ideal bioindicator. For example, the genus *Terschellingia* (Nematoda) are tolerant to organic contamination, particularly six polychlorinated biphenyls and low oxidized muddy sediment [43], hence are commonly found in disturbed mangroves. Besides nematodes, molluscs play a crucial role as bioindicators, providing valuable insights into mangrove environments' ecological health and resilience. Molluscs, as stationary filter feeders, are highly affected by variations in water quality, sediment conditions, and the overall suitability of their habitat. Bivalves and gastropods have been used to determine the quantity of hazardous metals, since they accumulate these compounds in their tissues over time. However, metal accumulation in mollusc tissues differs between species (feeding habits), HM elements, and habitats (sediment characteristics) [44]. For instance, three bivalves (*Meretrix meretrix*, Macoma *birmanica*, and *Sanguilonaria (Soletellina acuminata*) from the same habitat recorded significant concentrations of polycyclic aromatic hydrocarbon (PAH) congeners in body tissues, mainly due to the capabilities of the organism to metabolize and biotransformation PAHs [50].

The selection of bioindicators is a complex task that requires careful consideration of various criteria to ensure the accuracy and relevance of monitoring initiatives. It is crucial to select bioindicators with different tolerance levels to these pollutants to identify areas of concern within the ecosystem. Identifying certain species, recognized for their capacity to accumulate pollutants, has proven to be a valuable method for assessing contamination levels. Ideally, the species selected should be accessible and identifiable to facilitate efficient and cost-effective monitoring to enable assessments across various spatial and temporal scales. These species serve as effective indicators, providing valuable insights into the impact of human activities on the quality of water and sediment in mangrove ecosystems. Table 18.1 presents a summary of organisms that are commonly used as bioindicators in mangrove ecosystems. Through their responses to environmental changes, bioindicators serve as early warning systems, helping identify shifts in ecological dynamics before irreversible damage occurs. Ultimately, it is a valuable tool for measuring these changes and conveying a complex message in a simplified and practical way [41].

18.3.2 Case Study: Preliminary Assessment of Water Quality and Heavy Metals Accumulation in Mangrove Ecosystem, Kuala Gula, Perak, Malaysia

Contamination of marine environment with toxic HMs is one of the important worldwide problems, because they are non-degradability, bioaccumulation, and biomagnification, particularly in mangrove environments along coastal areas throughout the world [56, 57]. Kuala Gula is known as a small fishing village and an important bird sanctuary in Malaysia and the Asian region. This village has been experiencing this pollution, since it lies at the northern end of the Matang Mangrove Forest Reserve.

Table 18.1 Summary of organisms used as bioindicators in mangrove ecosystems

Organism	**Ecological role**			**Reference**
Foraminifera	**Habitat**	**Application**	**Sensitivity and Responses**	
Sorites orbiculus/ *Peneroplis planatus*/ *Coscinospira* spp.	Seaward intertidal flats	• Monitoring of water nutrient • Monitoring of mangrove in relation to salinity changes • Reconstruction of paleoecology	• Abundance in lower salinity and and organic matter sites • Abundance in clear and low nutrients sites (species designates good water quality) • Positive impact on growth and productivity by producing calcium carbonate (5.8–77.9%)	[51]
Genus: *Quinqueloculina*	Swamp	• Monitoring in high nutrient environment	• Positive impact on growth and productivity in hypoxic conditions (5.1–5.6 mg/L) • Hypersaline conditions (45.0–45.9 ppt)	
Ammonia tepida	Landward swamp	• Monitoring and evaluating the impact of elevated salinity on the environment	• Positive impact on growth and productivity in hypersaline conditions (45.0–45.9 ppt) • Reported to have high survival rate in stresses environment (tolerant species)	
Macroinvertebrate	**Habitat**	**Application**	**Sensitivity and responses**	**References**
Polychaeta	Intertidal zone/ mangrove forest	• Monitoring the water quality (nutrient enrichment) • Used as proxy to detect aquaculture effluent	• Distribution of species (type) are due to polychaetes characteristic and local environmental condition • Essential in nitrogen cycling process and influences sediment structure by varying burrowing activities upon detection of stressed environment	[52, 53]
Leptuca leptodactyla (fiddler crabs)	Mangrove ecosystems and beaches	• Bioindicators to verify the presence of rare-earth elements (REE)	• REE were detected in crabs and in sediments and the concertation are related to seasonal variability	[54]

Diatom	Habitat	Application	Sensitivity and responses	Reference
Thalassionema nitzschiodes (neritic planktonic species)	Downstream	• Indicator of high nutrients and salinity condition	• Positive impact on abundance in high temperature environment (25–30 °C), high salinity range (24–29 ppt), high pH range (7.3–7.8) and DO (3.9–5.5 mg/L)	[46]
Cyclotella meneghiniana/Cyclotella atomus Hustedt	Freshwater with polyhaline characteristics	• Bioindicator in poor water quality environments	• Persist well in environmental conditions with high temperature range (27–30.4 °C), high conductivity range (40.1–43.4 mS/cm), high salinity range (20–28 ppt) and high phosphate content (0.04–0.05 mg/L)	
Nitzschia amphibia	Upstream/freshwater	• Study on the oxygenation dynamics and ecological response to oxygen depletion in the mangrove habitat	• Increase in abundance in high nitrate (0.3–0.4 mg/L) and high phosphate content (0.04–0.06 mg/L) and oxygen-depleted conditions (DO: 3–4.5 mg/L) but has high range of oxygen saturation percentage (25–61%)	
Marine sponges	**Habitat**	**Application**	**Sensitivity and responses**	**Reference**
Halichondria melanadocia (Breadcrumb sponge)	Attached to prop roots of *Rhizophora* mangle	• Determination of the rate of anthropogenic waste (microplastic)	• Efficient filter feeders hence suitable to detect microplastics in the water column • The ability of sponges to bioaccumulate microplastics in their tissues	[55]

The study was represented with four selected sampling areas along the coastal environment, including at the river mouth (P1), middle of coastal water (P2), aquaculture area (P3), and near the estuary which was adjacent to a mangrove forest reserve (P4). Water samples were collected to obtain water quality status by comparing with the Malaysia's National Water Quality Standard (NWQS). The water physio-chemical characteristics were recorded in-situ. The sediment samples also were collected using a grab sampler at respective sampling areas to determine HM accumulation. Besides, several biological samples such as *Arius tenuispinis* or known as catfish, *Valamugil seheli*, commonly known as blue spot mullet, mud crabs (*Scylla* sp.), red blood cockles (*Anadara granosa*), and mussels were also collected to be analyzed for HMs level.

Table 18.2 shows the physico-chemical parameters of Kuala Gula Mangrove. The changing physico-chemical characteristics of water of the sampling points were influenced by discharging domestic and aquaculture wastes as one-third of the population was involved in fishery and aquaculture industries and ongoing land reclamation in the study area [28]. Most of the recorded parameters were still acceptable, especially for pH values fall within the range between 7.19 and 7.63 which categorized in class I. According to DOE Malaysia, a pH value ranged from 6.5 to 8.5 is acceptable for domestic water supply and practically no treatment necessary.

In this study case, the Total Suspended Solid (TSS) of four points in Kuala Gula Mangrove varied from 196.67 to 231.11 mg/L and it categorized in class IV which indicates that the study area was polluted. The reading level was quite different between Kuala Sepetang Mangrove [56], whereas the highest reading averagely 49.4 mg/L during that year. The DO level was also found to be low in this case study; the lowest DO value was P1, 3.32 mg/L. Nitrate–nitrogen (NO_3–N) and ammoniacal–nitrogen (NH_3–N) concentrations were found high except for phosphate according to NWQS. Excessive usage of feeding stock from aquaculture activities will contribute to high concentration of nutrients in water. Remnant bait, fish stock excretion, biological debris, and dissolved nutrient salts are examples of aquaculture pollutants which negatively affects the sustainability mangrove environment [58].

TSS has strong relationship with HM concentration [5]. HMs that enter the waterbody were distributed by suspended chemical solids and stored within the waterbed bottom sediments. Figure 18.1 presents the HM concentration in four

Table 18.2 Physico-chemical characteristics of Kuala Gula Mangrove

Parameters	P1	P2	P3	P4
Temperature (°C)	30.4	32.33	30.63	30.07
pH	7.21	7.34	7.19	7.63
TSS (mg/L)	196.67	206.67	198.89	231.11
DO (mg/L)	3.32	5.70	4.73	4.77
BOD (mg/L)	2.27	9.73	2.47	6.29
NO_3–N (mg/L)	22.70	24.40	71.03	24.87
NH_3–N (mg/L)	2.11	2.31	1.72	2.45
PO_4^3 (mg/L)	0.76	1.25	0.37	1.55

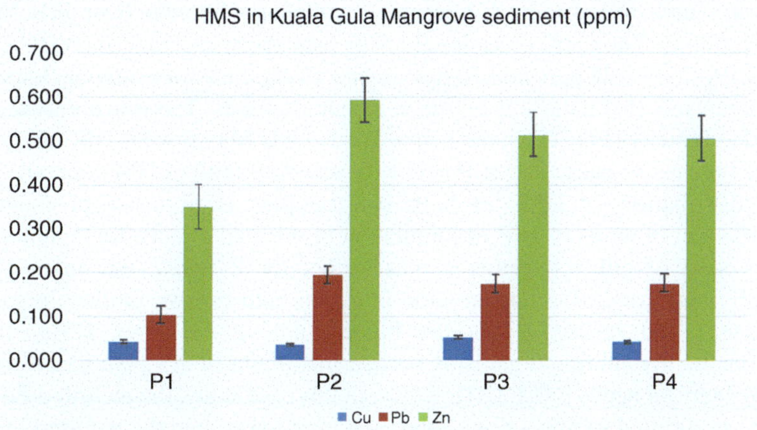

Fig. 18.1 Mean concentration and associated standard deviation of heavy metals in sediments collected at Kuala Gula Mangrove

Table 18.3 Average of HMs concentration in collected biological samples from Kuala Gula Mangrove

	Heavy metal concentration (ppm)		
Samples/species	Cu	Pb	Zn
Scylla sp.	0.48	0.02	1.750
Andara granosa	0.05	0.06	0.990
Mussels	0.17	0.07	1.520
Arius tenuispinis			
Intestine	0.011	0.024	1.614
Gill	0.083	0.037	1.965
Muscle	0.000	0.035	0.201
Valamugil seheli			
Intestine	0.043	0.076	0.643
Gill	0.035	0.042	0.681
Muscle	0.045	0.060	0.959

sediments samples collected from four representative sampling points. The distribution of three types of HMs, copper (Cu), lead (Pb), and zinc (Zn) varied in each sampling point. From the results, it was found that the highest mean concentration of HMs was Zn > Pb > Cu. Midstream (P2) and aquaculture area (P3) were recorded among the highest Zn concentration, 0.570 ppm and 0.513 ppm, respectively. Compared to regional study, Zn content in all the sediment samples was found to be lower than in other mangrove ecosystem [59, 60].

The average of HMs concentration in collected biological samples and selected organs (for fish samples only) such as intestine, gill, and muscle is presented in Table 18.3. Like sediment samples, Zn content was found to be the highest in all selected biological samples with the range from 0.201 to 1.750 ppm. It has been observed that the samples accumulated Zn the most are *Scylla* sp. as the whole individual sample followed by *A. tenuispinis*, which was found in their gills. Yet, the concentration of Zn is still considered lower than those specified in WHO (2004) and USEPA (1999), 25 ppm and 16 ppm, respectively. Zn is an essential element for living organisms that natural constituent in human foods. Too much uptake of Zn can lead to toxicity [61]. Thus, the fish in Kuala Gula Mangrove is still safe to consume by humans, because it is still below the standard limit of HM content.

The distribution of HMs in both fish samples is in order of magnitude Zn > Pb > Cu. In terms of HM accumulation in selected organs, for *A. tenuispinis* were in order of gill > intestine > muscle, but for *V. seheli*, the order is muscle > gill > intestine. Bioconcentration of Zn in both bivalve mollusc organisms (*A. granosa* and mussels) was the highest among the three HMs studied. Theoretically, live marine organisms needed maximum of 2.7 ppm of Zn to meet their biochemical needs [60, 62]. However, in this case study, the recorded Zn accumulation in selected samples was slightly 0.8 ppm lower than the required level.

From this case study, it is evident that all these samples have the criteria to be the potential bioindicator to point out pollution especially the current status of HMs in the area of Kuala Gula Mangrove. Facing the challenges of the world, this area had been exposed to a variety of pollutants and contaminants, such as HMs. Therefore, it is crucial to protect the coastal environment by having competent and decent policies, legislation, and regulation on mangrove ecosystem managements.

18.4 Challenges, Limitations, and Future Directions

18.4.1 Pollution in Mangrove Ecosystems

The main challenges present in the mangrove ecosystems are related to pollutions. Mangrove ecosystems constantly experience pollution from direct and indirect sources, such as municipal waste, aquaculture, mariculture, industries, and run-off from urban centers.

Some of the most endangered plant species in the world can be found in mangrove forests. However, many mangrove areas are extensively deforested to make land available for aquaculture and coastal development. In the last 50 years, mangrove populations have lost over one-third of their total size [7]. In a review by Akram et al. [7], they noted that the global mangroves have decreased by 5245.24 km^2 in 24 years. Among the 64 species of mangrove plants recorded, 12 species were declared threatened by the International Union for Conservation of Nature (IUCN) Red List. Akram et al. [7] also noted that the threats to mangroves were severe threats (coastal development, increase in aquaculture/agriculture land and harvesting of timber), moderate threats (climate change, eutrophication, and hydrological alteration), and low-level threats (disease, tourism, and pollution). However, the delicate balance of mangrove ecosystems is increasingly threatened by pollution, posing a significant environmental challenge. Pollution from land-based and marine sources introduces harmful substances, such as chemicals, hydrocarbons, and HMs, into these fragile habitats. Although facing detrimental threats, the mangrove ecosystems provide crucial services for mitigating climate change and serve multiple roles, such as contaminant and sediment filtering and protecting coastlines [14, 63].

Research has shown that the mangrove ecosystem has the remarkable capacity to withstand the presence of HMs in its surroundings [14, 64, 65]. In addition, the plants within this ecosystem play a crucial role in restricting the movement of contaminants in their environment. Roots, leaves, and stems serve as entry points for pollutants, with various transport mechanisms facilitating their movement within the plant. For example, in a study conducted in a mangrove forest contaminated with HMs, it was found that the physicochemical properties of the pollutants, such as their solubility and particle size, influenced their uptake by mangrove trees [64].

It is crucial to acknowledge that the processes of HM accumulation can differ among various mangrove species and are affected by environmental factors, such as sediment properties, water quality, and the specific concentrations of metals in the ecosystem [3]. For example, metal pollution loading in the *Avicennia marina* exhibits significant variation among locations and leaves, suggesting the potential for biological effects on the mangrove-associated biota in these areas [66]. In addition, Alam [66] observed a strong positive correlation between metals found in sediments and leaves of *A. marina*. This correlation was particularly evident for Mn, Co, Zn, and Pb. These findings suggest that *A. marina* can accumulate these metals as the sedimentary metal levels increase. Furthermore, this implies that the leaves could be a valuable bioindicator for assessing environmental metal concentrations. The accumulation of metal loads in leaves differs among species, depending on the importance of metals for plant growth and the availability of metals in sediment [14].

Furthermore, the build-up of HMs in mangrove plants can have ecological consequences, as mangroves function as habitats and nourish diverse organisms in coastal ecosystems. Understanding these variations is crucial for assessing the potential use of different plant species as bioindicators or phytoremediators in polluted environments (Table 18.4). The mangrove ecosystems are recognized for their exceptional ability to retain HMs, such as chromium (Cr), copper (Cu), zinc (Zn), manganese (Mn), and cadmium (Cd) [64]. However, excess loading can exceed the binding capacity of the sediment. This natural wastewater treatment feature enables them to remove these pollutants effectively. They can store metals, transferring these elements from the sediment and accumulating them in their bark tissues and roots [42]. These capabilities can be utilized as a method for capturing and eliminating pollutants. In addition, HMs found in the tissues can indicate the level of HM concentration in the surrounding ecosystems.

HM contamination in mangrove systems is recognized to have a detrimental impact on root growth, leading to substantial reductions in cell division activity and causing damage to the nucleoli in the tip meristem. This is frequently considered a detoxification strategy, sequestering metals in roots and reducing their translocation to above-ground tissues. As such, *Rhizophora apiculata*, known for its prop roots, can accumulate HMs in its root tissues [17]. In addition, the plants have distinct root systems and employ viviparous reproduction to adapt to intertidal environments.

Table 18.4 Various species of mangroves have been utilized in scientific research to monitor the presence of heavy metals in aquatic environments

Species	Application	Section	Response	Source
Avicennia marina (Grey mangrove)	Heavy metal pollution monitoring	Roots and leaves	• Metal accumulation occurs with root uptake via apoplastic and symplastic pathways • Shows increased metal tolerance in its root tissues due to its high tolerance to metals via exclusion of metals from acropetal transport • A positive relationship between the sediments with leaf tissues for selected heavy metals	[66, 67]
Laguncularia racemosa (White mangrove)		Leaf	• Impacts the leaves morphological characteristics can indicate the presence of potentially toxic elements (PTEs) • This species is known to be tolerant to high PTEs and has the potential of the species for green remediation of contaminated mangroves	[68]
Bruguiera gymnorrhiza (Black mangrove)	Hydrocarbon (oil) pollution monitoring	Leaf and stem (seedlings)	The impact of oil on seed germination and plant growth varies based on sediment type: Sandy Sediment: • Spent oil negatively affects growth parameters (stem diameter, height, biomass) due to the larger sediment size • Fresh oil adversely affects growth in terms of leaf length, width, and biomass Muddy Sediment: • Physiological parameters remain unaffected by both spent and fresh oil	[69]
Rhizophora apiculata	Heavy metal pollution	Roots, Leaf and stem	• Chromium impacted the leave and stem except the root number • Chromium and nutrients affected the plant's growth, in high chromium. The number of stomata increased under chromium stress, especially in salty conditions	[17]

The complex root system provides a large surface area for metal adsorption, contributing to metal sequestration in the root zone [70].

Gaining insight into the uptake and transport mechanisms of pollutants in mangrove trees can also aid in the identification of potential pollution hotspots within mangrove ecosystems. In addition, HM pollution also hindered plant growth by altering the photosynthetic pigments and associated enzyme activities, thus hindering carbon assimilation and growth of the plant. The acquired knowledge can be utilized to develop precise approaches for mitigating pollution and restoring areas that have been adversely impacted. In addition, examining the impact of pollutants on individual trees can offer valuable insights into the overall health and resilience of mangrove forests, which play a crucial role in safeguarding coastal areas and preserving biodiversity.

18.4.2 Future Directions

Mangrove conservation and preservation by ecologist and biologist was initially focused upon estimating the species diversity and abundance. The scope of the mangrove study has always been consistently biased toward biodiversity as the top priority [20]. This intention is still relevant until now to continue upgrading the database for areas with less coverage and for future use in modelling and predictions. With the emergence of sustainability and ecosystem service-oriented management, an ecosystem-based approach and nature-based solutions integrating interdisciplinary approaches emerged in the current agenda.

The tsunami incidence was the triggering factor of the importance of mangroves as a 'green belt' in Indonesia since 2004 [36]. The continuous conflict between reaching the equilibrium between environmental and socio-economic gains has led to the confounding estimation of the monetary value of the mangrove's ecosystem services. This effort was done to bridge the communication gap between multiple stakeholders in mangrove ecosystem, such as scientist, researchers, managers, officers, non-governmental organizations, policy makers, communities, and the public.

The yield of mangroves as nursery grounds of fishes were higher (US$ 37,500/ha/year serving) than the yield of shrimp farming (US$1220/ha/year) in mangrove converted areas in Buri [71]. This finding indicates that it is more profitable to conserve mangroves than to convert into other type of land uses. Similarly, in another area of 4354.11 hectares of mangroves in Bintan Island, Indonesia, the economic value of selling fish resources from the area is Rp. 225,971,478/ha/year [72]. This value indicates that with the existing mangrove area, it is able to sustain the livelihood of the fishermen. Payments for ecosystem services are one of the means for protecting and restoring mangrove ecosystem services. A study by Hasan-Basri et al. [73] reported that the respondents involved in their study are willing to pay up to RM 18.31 for mangrove conservation program in Kuala Perlis, Malaysia.

Multiple effort has been carried out since the 1970s to conserve mangroves through mangrove reforestation, mangrove restoration, rehabilitation, and mangrove afforestation. The intention of these efforts was marginally done through direct planting [20]. Community engagements are also flourishing locally but are limited to small scale and region specific. It is apparent that the acknowledgement of the importance of mangrove conservation in climate change mitigation has increased the motivation to enhance mangrove restoration efforts. This factor could be because of more economic and financing incentives provided in relation to the blue carbon initiatives than mangrove restoration due to biodiversity purposes. The way to move forward in the mangrove conservation agenda is through price-based mechanism, such as carbon credits. It is suggested that a carbon price of US$ 6–10 US per tCO_2 would be enough to support mangrove conservation [74].

18.5 Conclusion

The use of mangrove ecosystems for biomonitoring applications in coastal environments is very appreciated as these ecosystems have been recognized as the most productive intertidal ecosystems on earth. Their unique features compared to other vegetation help them in safeguarding coastline, capturing and storing carbon, serving as nursery home for many types of marine species. With these kinds of features, mangroves have an emerging function in intercepting microplastics which are commonly found floating and deposited on the sea surface due to its lower density than seawater (<1.02 g/cm^3). Microplastic abundance was reported to be higher in estuarine sediments and coastal bottom sediments of mangrove areas due to the potential of aerial roots trapping those microplastics. Carbon capture was profoundly to be highest in the mangrove forests than the terrestrial forests. A mature mangrove forest has greater capability in carbon sequestration as it has more litter and residues produced which is accumulates in the thick rich sediments of mangroves than in younger and new plants.

While this ecosystem has been providing the earth in many aspects, studying the health and behavior of living organisms in mangrove ecosystems or known as biological indicators or bioindicators can help to evaluate the quality status of the environment. Diatoms, benthic invertebrates' fauna, and marine sponges that attached to prop roots of *Rhizophora* are valuable bioindicators, because they represent the conditions at the sampling sites and highly affected by the variations in water quality, sediment conditions, and the overall suitability of their habitat. Bivalve and gastropods have been used to determine the quantity of hazardous metals in their tissues over time.

One case study was explained regarding the application of some organism as bioindicators of HMs in Kuala Gula Mangrove, Perak Malaysia. Although the concentration obtained for the HMs studied (Cu, Pb and Zn) are considered below documented toxic levels, however, the fast growth of urbanization and industrialization levels in mangrove areas requires continuous vigilance and monitoring of this sensitive and important ecosystem to ensure that HM pollution is at a minimal level.

In general, the main challenges present in the mangrove ecosystems are related to pollution. About 12 species from 64 recorded species of mangrove plants were declared as threatened by IUCN Red List. The build-up of HMs in mangrove plants can have ecological consequences as excessive toxic loading in this ecosystem will affect the natural wastewater treatment features as they are unable to remove those pollutants effectively. Numerous efforts have been carried out for mangrove conservation and preservation by ecologists and biologists since the 1970s. The intention is still relevant until now but with different approaches: an ecosystem-based approach and a nature-based solution integrating interdisciplinary approach to fit the emergence of sustainability and ecosystem service-oriented management.

Acknowledgements I would like to acknowledge the funding by Sultan Ismail Nasiruddin Shah Terengganu Scholarship 2022. This review is also contributed by Lim Angel, Lee Weng Yan, Elisa Rani Charles Johnson and Nur Intan Erinna Rosshadi.

References

1. Cooley SR, Schoeman DS, Bopp L, Boyd P, Donner S, Ito S, Kiessling W, Martinetto P, Ojea E, Racault M, Rost B, Skern-Mauritzen M, Ghebrehiwet DY (2023) Oceans and coastal ecosystems and their services. Cambridge University Press, Cambridge
2. Omar H, Misman MA, Linggok V (2018) Characterizing and monitoring of mangroves in Malaysia using Landsat-based spatial-spectral variability. IOP Conf Ser Earth Environ Sci 169(1):012037
3. Wilda R, Hamdan AM, Rahmi R (2021) A review: the use of mangrove for biomonitoring on aquatic environment. IOP Conf Ser Mater Sci Eng 980(1):012083
4. Diganta MTM, Saifullah ASM, Siddique MAB, Mostafa M, Sheikh MS, Uddin MJ (2023) Macroalgae for biomonitoring of trace elements in relation to environmental parameters and seasonality in a sub-tropical mangrove estuary. J Contam Hydrol 256:104190
5. Azratul ANMD, Akbar JB, Kamaruzzaman BY, Sheikh HI, Jalal KCA, Noor Faizul HN (2017) Biomonitoring selected heavy metal concentration in Nerita Sp. collected from Tanjung Lumpur mangrove Forest. Environ Ecosyst Sci 1(1):4–7
6. Edu EAB, Edwin-Wosu NL, Inegbedion A (2015) Bio-monitoring of Mangal sediments and tissues for heavy metal accumulation in the mangrove forest of cross river estuary. Insight Ecol 4(1):46–52
7. Akram H, Hussain S, Mazumdar P, Chua KO, Butt TE, Harikrishna JA (2023) Mangrove health: a review of functions, threats, and challenges associated with mangrove management practices. Forests 14(9):1–38
8. Blasco F, Saenger P, Janodet E (1996) Mangroves as indicators of coastal change. Catena 27(3–4):167–178
9. Suman DO (2019) Chapter 31—Mangrove management: challenges and guidelines. In: Perillo GME, Wolanski E, Cahoon DR, Hopkinson CS (eds) Coastal wetlands: an integrated ecosystem approach, 2nd edn. Elsevier, Amsterdam, pp 1055–1079
10. Gopalakrishnan S, Landry CE, Smith MD (2018) Climate change adaptation in coastal environments: modeling challenges for resource and environmental economists. Rev Environ Econ Policy 12(1):48–68
11. Sarath NG, Puthur JT (2021) Heavy metal pollution assessment in a mangrove ecosystem scheduled as a community reserve. Wetl Ecol Manag 29(5):719–730
12. Wenjiao Z, Peng L (1996) Accumulation and distribution of Cu, Pb, Zn and Cd in Avicennia marina mangrove community of Futian in Shenzhen. Oceanologia et Limnologia Sinica. Environmental Science Research Center, Xiamen University, Xiamen, 27(4), pp 388–393
13. Wang Y, Qiu Q, Xin G, Yang Z, Zheng J, Ye Z, Li S (2013) Heavy metal contamination in a vulnerable mangrove swamp in South China. Environ Monit Assess 185(7):5775–5787

14. Yan Z, Sun X, Xu Y, Zhang Q, Li X (2017) Accumulation and tolerance of mangroves to heavy metals: a review. Curr Pollut Rep 3(4):302–317
15. Barakat A, El Baghdadi M, Rais J, Nadem S (2012) Assessment of heavy metal in surface sediments of Day River at Beni-Mellal region, Morocco. Res J Environ Earth Sci 4:797–806
16. Sun Z, Chen J, Wang X, Lv C (2016) Heavy metal accumulation in native plants at a metallurgy waste site in rural areas of northern China. Ecol Eng 86:60–68
17. Nguyen KL, Nguyen HA, Richter O, Pham MT, Nguyen VP (2017) Ecophysiological responses of young mangrove species Rhizophora apiculata (Blume) to different chromium contaminated environments. Sci Total Environ 574:369–380
18. Taillardat P, Marchand C, Friess DA, Widory D, David F, Ohte N, Nakamura T, Vinh VT, Thanh-Nho N, Ziegler AD (2020) Respective contribution of urban wastewater and mangroves on nutrient dynamics in a tropical estuary during the monsoon season. Mar Pollut Bull 160:111652
19. Jiao M, Ren L, Wang Y, Ding C, Li T, Cao S, Li R, Wang Y (2022) Mangrove forest: an important coastal ecosystem to intercept river microplastics. Environ Res 210:112939
20. Gerona-Daga MEB, Salmo SG (2022) A systematic review of mangrove restoration studies in Southeast Asia: challenges and opportunities for the United Nation's decade on ecosystem restoration. Front Mar Sci 9:1–23
21. Giesen W, Wulffraat S (2006) Mangrove guidebook for Southeast Asia
22. Marchand M (2008) Mangrove restoration in Vietnam key considerations and a practical guide
23. Costa LD, Budiastuti MTHS, Sutrisno J, Sunarto (2019) The diversity of plant species in a mangrove forest in the Coast of Metinaro, Timor-Leste. IOP Conf Ser Earth Environ Sci 256(1):012026
24. Suratman MN (2008) Carbon sequestration potential of mangroves in Southeast Asia. In: Bravo F, Jandl R, LeMay V, von Gadow K (eds) Managing forest ecosystems: the challenge of climate change. Springer Netherlands, Dordrecht, pp 297–315
25. Tan E, Mohd Zanuri NB (2023) Abundance and distribution of microplastics in tropical estuarine mangrove areas around Penang, Malaysia. Front Mar Sci 10:1–20
26. Talukdar A, Kundu P, Bhattacharjee S, Dey S, Dey A, Biswas JK, Chaudhuri P, Bhattacharya S (2023) Microplastics in mangroves with special reference to Asia: occurrence, distribution, bioaccumulation and remediation options. Sci Total Environ 904:166165
27. Zhang L, Wang S, Jian Q, Zhang P, Lu Y, Liu H (2022) Tidal variation shaped microplastic enrichment patterns in mangrove blue carbon ecosystem of northern Beibu gulf, China. Front Mar Sci 9:927884
28. Mohamed CAR, Nur Shahruddin A, Pradit S, Loh PS, Nitiratsuwan T, Kobkeatthawin T, Noppradit P, Quynh Le TP, Oeurng C, Sok T, Lee CW, Bong CW, Lu X, Anshari GZ, Kandasamy S, Wang J (2023) Depth profiles of microplastic in sediment cores in the mangrove area of Kuala Gula mangrove, Malaysia. J Mar Sci Eng 11(6):1223
29. Zhu JJ, Yan B (2022) Blue carbon sink function and carbon neutrality potential of mangroves. Sci Total Environ 822:153438
30. Bouillon S, Dahdouh-Guebas FS, Rao AVV, Koedam N, Dehairs F (2003) Sources of organic carbon in mangrove sediments: variability and possible ecological implications. Hydrobiologia 495(1):33–39
31. Arnaud M, Krause S, Norby RJ, Dang TH, Acil N, Kettridge N, Gauci V, Ullah S (2023) Global mangrove root production, its controls and roles in the blue carbon budget of mangroves. Glob Chang Biol 29(12):3256–3270
32. Pham TD, Xia J, Ha NT, Bui DT, Le NN, Tekeuchi W (2019) A review of remote sensing approaches for monitoring blue carbon ecosystems: mangroves, sea grasses and salt marshes during 2010–2018. Sensors 19(8):1933
33. Sasmito SD, Taillaedat P, Clendenning JN, Cameron C, Friess DA, Murdiyarso D, Hutley LB (2019) Effect of land-use and land-cover change on mangrove blue carbon: a systematic review. Glob Chang Biol 25(2):4291–4302
34. Aye WN, Tong X, Li J, Tun AW (2023) Assessing the carbon storage potential of a young mangrove plantation in Myanmar. Forests 14(4):1–13

35. USAID (2017) Assessment of carbon sequestration potential of mangrove forests in the Talon-Talon and Mampang treevolution sites of Zamboanga City
36. Sidik F, Lawrence A, Wagey T, Zamzani F, Lovelock CE (2023) Blue carbon: a new paradigm of mangrove conservation and management in Indonesia. Mar Policy 147:105388
37. Maulana MI, Auliah NL, Onrizal O (2021) Potential carbon storage of Indonesian mangroves. IOP Conf Ser Earth Environ Sci 782(3):2–6
38. Cui X, Liang J, Lu W, Chen H, Liu F, Lin G, Xu F, Luo Y (2018) Stronger ecosystem carbon sequestration potential of mangrove wetlands with respect to terrestrial forests in subtropical China. Agric For Meteorol 249:71–80
39. Rozainah MZ, Nazri MN, Sofawi AB, Hemati Z, Juliana WA (2018) Estimation of carbon pool in soil, above and below ground vegetation at different types of mangrove forests in peninsular Malaysia. Mar Pollut Bull 137:237–245
40. Aguirre-Rubí JR, Ortiz-Zarragoitia M, Izagirre U, Etxebarria N, Espinoza F, Marigómez I (2019) Prospective biomonitor and sentinel bivalve species for pollution monitoring and ecosystem health disturbance assessment in mangrove–lined Nicaraguan coasts. Sci Total Environ 649:186–200
41. Desrosiers C, Leflaive J, Eulin A, Ten-Hage L (2013) Bioindicators in marine waters: benthic diatoms as a tool to assess water quality from eutrophic to oligotrophic coastal ecosystems. Ecol Indic 32:25–34
42. Chowdhury R, Favas PJC, Jonathan MP, Venkatachalam P, Raja P, Sarkar SK (2017) Bioremoval of trace metals from rhizosediment by mangrove plants in Indian Sundarban wetland. Mar Pollut Bull 124(2):1078–1088
43. Michelet C, Zeppilli D, Hubas C, Baldrighi E, Cuny P, Dirberg G, Militon C, Walcker R, Lamy D, Jezequel R, Receveur J, Gilbert F, El Houssainy A, Dufour A, Heimburger-Boavida L, Bihannic I, Sylvi L, Vivier B, Michaud E (2021) First assessment of the benthic meiofauna sensitivity to low human-impacted mangroves in French Guiana. Forests 12(3):338
44. Hong AH, Hargan KE, Williams B, Nuangsaeng B, Siriwong S, Tassawad P, Chaiharn C, Huertos ML (2020) Examining molluscs as bioindicators of shrimp aquaculture effluent contamination in a southeast Asian mangrove. Ecol Indic 115(11):106365
45. Hassan H, Salleh S, Wong NLW (2021) Gut content of mangrove gastropod, Cerithidea obtusa (Lamarck, 1822) from Kuala Selangor Nature Park, Selangor and Tanjung Piai National Park, Johor, Peninsular Malaysia. IOP Conf Ser Earth Environ Sci 870(1):012047
46. Costa-Böddeker S, Thuyên LX, Schwarz A, Huy HĐ, Schwalb A (2017) Diatom assemblages in surface sediments along nutrient and salinity gradients of Thi Vai estuary and can Gio mangrove Forest, southern Vietnam. Estuar Coasts 40(2):479–492
47. Hemraj DA, Hossain MA, Qin Q, Ye JG, Leterme SC (2017) Plankton bioindicators of environmental conditions in coastal lagoons. Estuar Coast Shelf Sci 184:102–114
48. Hilaluddin F, Yusoff FM, Toda T (2020) Shifts in diatom dominance associated with seasonal changes in an estuarine-mangrove phytoplankton community. J Mar Sci Eng 8(7):528
49. Fu M, Wang Z, Pu X, Qu P, Li Y, Wei Q, Jiang M (2016) Response of phytoplankton community to nutrient enrichment in the subsurface chlorophyll maximum in Yellow Sea cold water mass. Acta Ecol Sin 36(1):39–44
50. Zuloaga O, Prieto A, Usobiaga A, Sarkar SK, Chatterjee M, Bhattacharya BD, Battacharya A, Alam Md. A, Satpathy KK (2009) Polycyclic aromatic hydrocarbons in intertidal marine bivalves of Sunderban mangrove wetland, India: an approach to bioindicator species. Water Air Soil Pollut 201(1–4):305–318
51. El-Menhawey W, Kholeif SEA, Elshanawany R, Ibrahim MIA (2020) Benthic foraminifera as bio-indicator of mangrove ecosystem quality: a case study from Abu Ghoson area, Red Sea coast, Egypt. Reg Stud Mar Sci 40:101506
52. Abdullah AM, Akhtar A, Kamal AHM, AftabUddin S, Islam MS, Sharifuzzaman SM (2022) Assessment of benthic macroinvertebrates as potential bioindicators of anthropogenic disturbance in Southeast Bangladesh coast. Mar Pollut Bull 184:114217
53. Miri M, Seyfabadi J, Ghodrati SM, Rahimian H, Valipour M (2023) Polychaete diversity and functional trait composition in subtropical mangrove ecosystems. Diversity 15(9):1–20

54. Lavezzo B, Kinoshita A, Figueiredo AMG, Pinheiro MMF, Santana W (2020) Detection of rare-earth elements using fiddler crabs Leptuca leptodactyla (Crustacea: Ocypodidae) as bioindicators in mangroves on the coast of São Paulo, Brazil. Sci Total Environ 738:139787
55. Celis-Hernández O, Ávila E, Ward RD, Rodríguez-Santiago MA, Aguirre-Téllez JA (2021) Microplastic distribution in urban vs pristine mangroves: using marine sponges as bioindicators of environmental pollution. Environ Pollut 284:117391
56. Dias HQ, Nayak GN (2016) Geochemistry and bioavailability of mudflats and mangrove sediments and their effect on bioaccumulation in selected organisms within a tropical (Zuari) estuary, Goa, India. Mar Pollut Bull 105(1):227–236
57. Xie Z, Zhu G, Xu M, Zhang H, Yi W, Jiang Y, Liang M, Wang Z (2022) Risk assessment of heavy metals in a typical mangrove ecosystem—a case study of Shankou mangrove National Natural Reserve, southern China. Mar Pollut Bull 178(2):113642
58. Peng Y, Li X, Wu K, Peng Y, Chen G (2009) Effect of an integrated mangrove-aquaculture system on aquacultural health. Front Biol China 4(4):579–584
59. Hossain M (2000) Distribution of zinc in Sepang mangrove Forest environment, Malaysia. Khulna Univ Stud 2:149–157
60. Billah M, Kamal AHM, Bin Idris MJ, Ismail J, Bhuiyan MKA (2014) Cu, Zn, Fe, and Mn in mangrove ecosystems (sediment, water, oyster, and macroalgae) of Sarawak, Malaysia. Zool Ecol 24(4):380–388
61. Tchounwou PB, Yedjou CG, Patlolla AK, Sutton DJ (2012) Heavy metal toxicity and the environment. Exp Suppl 101:133–164
62. Pequegnat JE, Fowler SW, Small LF (1969) Estimates of the zinc requirements of marine organisms. J Fish Res Board Canada 26(1):145–150
63. Li R, Tan H, Zhu Y, Zhang Y (2017) The retention and distribution of parent, alkylated, and N/O/S-containing polycyclic aromatic hydrocarbons on the epidermal tissue of mangrove seedlings. Environ Pollut 226:35–142
64. Joseph P (2019) Heavy metal contamination in representative surface sediments of mangrove habitats of Cochin, southern India. Environ Earth Sci 78(15):1–11
65. Shen X, Chai R, Li M, Cheng S, Tam NFY, Han J (2021) Does combined heavy metal stress enhance iron plaque formation and heavy metal bioaccumulation in Kandelia obovata? Environ Exp Bot 186:104463
66. Alam MR (2022) Metal(loid) accumulation in the leaves of the grey mangrove (Avicennia marina): assessment of robust sampling requirements and potential use as a bioindicator. Environ Res 211:113065
67. Robin SL, Marchand C, Ham B, Pattier F, Laporte-Magoni C, Serres A (2021) Influences of species and watersheds inputs on trace metal accumulation in mangrove roots. Sci Total Environ 787:147438
68. Pimentel VC (2023) Laguncularia racemosa leaves indicate the presence of potentially toxic elements in mangroves. Sci Rep 13(1):1–15
69. Zhang CG, Leung KK, Wong YS, Tam NFY (2007) Germination, growth and physiological responses of mangrove plant (Bruguiera gymnorrhiza) to lubricating oil pollution. Environ Exp Bot 60(1):127–136
70. Yap CK, Al-Mutairi KA (2023) Potentially toxic metals in the tropical mangrove non-salt secreting Rhizophora apiculata: a field-based biomonitoring study and phytoremediation potentials. Forests 14(2):237
71. Sala E (2013) A general business model for marine reserves. PLoS One 8(4):e58799
72. Arkham MN, Riadi S, Wahyudin Y, Krisnafi Y (2023) Economic value of mangrove ecosystem services in the coastal area of Bintan Island, Indonesia. Wetl Ecol Manag 1–19
73. Hasan-Basri B, Samdin Z, Ghani ANA (2020) Willingness to pay for conservation of mangrove forest. J Ekon Malaysia 54(3):89–99
74. Jakovac CC, Latawiec AE, Lacerda E, Lucas IL, Korys AK, Iribarrem A, Malaguti GA, Turner KR, Luisetti T, Strassburg BBN (2020) Costs and carbon benefits of mangrove conservation and restoration: a global analysis. Ecol Econ 176:106758

Inventive Applications of Marine Resources in Cosmetic Production: A Review

19

K. Harinisri Ram and B. Thamarai Selvi

Abstract

The utilization of marine resources in the cosmetics industry is a burgeoning field with vast potential. Marine-derived compounds, such as algae extracts, marine peptides, and sea minerals, offer unique benefits for skincare and cosmetics products. These resources provide natural sources of antioxidants, moisturizers, and anti-aging agents. Marine ingredients are increasingly incorporated into formulations for sunscreen, anti-wrinkle creams, and skincare treatments. In addition, the sustainable harvesting of marine resources aligns with eco-friendly practices in the cosmetics industry. As the demand for natural and effective cosmetic solutions grows, this chapter critically reviews marine resources that hold promise for innovative and environmentally conscious beauty products.

Keywords

Anti-aging agents · Cosmetics · Marine organisms · Marine minerals · Natural compounds

19.1 Introduction

Self-care has evolved into an essential and crucial aspect of contemporary lifestyles, propelling the beauty industry to explore various innovative concepts and novel elements within cosmetic and skin care products. In recent decades, the ultra-effective cosmetic industry has channeled its curiosity into the bountiful and enigmatic realm of marine resources, uncovering a treasure trove from the depths of our ocean. The

---

K. Harinisri Ram (✉) · B. Thamarai Selvi
Department of Microbiology, Sri Ramakrishna College of Arts and Science for Women, Coimbatore, India

M. Rafatullah et al. (eds.), *Multidisciplinary Applications of Marine Resources*, https://doi.org/10.1007/978-981-97-5057-3_19

worldwide cosmetic and cosmeceutical industry is poised to reach a staggering $463.5 billion by 2027 [1], signifying the future for this flourishing sector.

With the anticipation of growth in the cosmetics industry, numerous products have been developed to meet customer needs while prioritizing safety and avoiding adverse effects. However, it is worth noting that many synthetic chemicals employed in these products can be expensive and pose potential health risks associated with side effects. For instance, certain skin-whitening compounds can not only irritate the skin but also lead to more severe issues, such as skin damage or even cancer [2]. Consequently, in recent years, there has been a burgeoning demand for cosmetic products that harness the power of natural ingredients. Natural ingredients offer several advantages, including environmental friendliness, minimal side effects, and enhanced safety. Consequently, the cosmeceutical industry is actively scouring for active compounds derived from natural sources, with a particular focus on marine resources.

The vastness, mystery, and beauty of Earth's oceans have perpetually captivated humanity throughout history. Beyond their enchantment and allure, these marine environments harbor invaluable resources that are indispensable to human survival and well-being. Beneath the oceanic waves lie surprising ecosystems, ranging from minuscule microalgae to significant kelp forests. These entities have evolved over millennia, adapting to unique and challenging environments. Among the myriad gifts of the sea, perhaps, none are as intriguing and versatile as the wealth of marine organisms and compounds that have revolutionized the cosmetic world.

The oceans, covering over 70% of the Earth's surface, are a source of immense inspiration and intrigue. Yet, their importance transcends aesthetics and curiosity. These marine ecosystems have gifted humanity with a remarkable array of compounds and organisms that have unlocked new horizons in the field of cosmetics. From vibrant coral reefs to the depths of the abyss, marine life has yielded a multitude of ingredients that offer not only cosmetic benefits but also a renewed commitment to sustainability and eco-consciousness.

One of the standout features of marine-derived cosmetics is their environmentally friendly nature. The extraction and cultivation of these resources often follow sustainable practices that ensure the long-term health of our oceans. Furthermore, the natural origins of these ingredients reduce the likelihood of adverse reactions, making them suitable for a wider range of consumers. In addition to their environmental and safety advantages, marine-derived cosmetic ingredients have demonstrated remarkable efficacy in addressing various skincare concerns. From seaweed-based moisturizers that hydrate and rejuvenate the skin to sea sponge extracts that promote collagen production, the potential applications are vast and exciting.

This chapter embarks on a captivating journey to delve into the profound significance of marine resources and the extraordinary cosmetics derived from them. By harnessing the wonders of the ocean, the beauty industry continues to evolve, offering products that not only enhance our appearance but also contribute to the preservation of our precious marine ecosystems and shedding spotlight light on the symbiotic relationship between our oceans and the beauty industry, revealing the fascinating interplay between nature's bounty and human innovation.

19.2 The Significance of Marine Resources in the Cosmetic Industry

Over 70% of the Earth's surface is covered by the ocean, which plays a vital role in sustaining life on our planet. It is home to an astonishing array of flora and fauna, many of which are still unexplored and enigmatic. Besides their ecological significance, marine organisms have provided humans with a wealth of resources, including food, medicine, and industry. Among these resources, marine-derived ingredients have established their niche in the cosmetic industry, contributing to the quality and effectiveness of numerous skincare and beauty products.

Marine resources are highly valued for their unique and potent properties, which are utilized in cosmetics to provide exceptional benefits to the skin and hair. From the depths of the ocean, scientists and cosmetic formulators have discovered bioactive compounds, such as seaweeds, algae, and plankton. These organisms have evolved over millennia to withstand the harsh conditions of their underwater habitats, and in doing so, they have developed remarkable adaptations that can enhance the health and appearance of human skin.

One of the most renowned marine resources in the world of cosmetics is seaweed. Seaweeds are a diverse group of marine algae that thrive in various aquatic ecosystems. They are rich in essential vitamins, minerals, and antioxidants, making them an ideal ingredient for skincare products [3]. Seaweed extracts are highly effective in hydrating, nourishing, and rejuvenating the skin, helping combat the effects of ageing and environmental damage. As consumers increasingly seek natural and sustainable alternatives in their beauty routines, seaweed-based cosmetics have gained popularity for their efficacy and eco-friendly credentials.

Marine-derived oils and extracts have become popular ingredients in cosmetics. Marine oils, such as omega-3-rich fish oils, have countless benefits, such as anti-inflammatory and moisturizing properties, which make them valuable additions to skincare formulations. These oils can soothe irritated skin, reduce redness, and support overall health. Marine extracts from organisms such as microalgae and kelp contain bioactive compounds, which contribute to collagen production, firmness, and elasticity in the skin.

Marine resources are also a source of inspiration for innovation in the cosmetic industry. Many scientists explore the vast genetic diversity of marine organisms, leading to the discovery of novel compounds with transformative potential. For instance, identifying unique enzymes found in deep-sea bacteria has enabled the development of advanced skincare products that concentrate on specific skin concerns, such as hyperpigmentation and acne.

Marine-derived cosmetics are not only used to enhance beauty, but they also have ecological significance. The sustainable harvesting and cultivation of marine resources for cosmetic production promote responsible and environmentally friendly practices. By investing in these resources, the cosmetic industry can contribute to the conservation of marine ecosystems, fostering a harmonious relationship between human needs and the preservation of our oceans.

19.3 The Evolution of Cosmetic Utilization from the Ocean

There is substantial evidence to suggest that the pursuit of enhancing our beauty dates back at least 7000 years, and even further into the past. It comes as no surprise that people worldwide have looked toward the sea as a source of beauty-enhancing elements. There exists a specific term for utilizing seawater for therapeutic purposes: Thalassotherapy. While historical sources vary regarding the origins of this concept, the practice predates these recorded instances.

Many ancient beauty products included simple or harmful ingredients, but our methods of harnessing beauty from the sea have advanced and improved over time.

Historical records suggest that Cleopatra, inspired by the Queen of Sheba, established her spa to reap the benefits of the mineral-rich Dead Sea and its therapeutic mud, as shown in Fig. 19.1. Many people have widely used mud for its hydrating, firming, and nourishing properties, and soothing various skin ailments (Fig. 19.1).

Ancient Mesopotamians used seaweed as a lip colorant, and during China's Qin Dynasty, an Empress outlined her skincare regimen, which featured ingredients derived from jellyfish and seaweed.

Furthermore, there is documented evidence of pearl powder made from fresh and saltwater being consumed orally or applied topically to enhance skin health in China as far back as 320 AD. Pearl powder also found its place in Ayurvedic medicine in India, serving as a love potion, and was utilized in the Philippines and Europe to promote skin paleness and firmness [4] (Fig. 19.2).

Fig. 19.1 Dead sea and mud mask

Fig. 19.2 Pearl powder

19.4 Marine Organisms: A Key Focus in Cosmetic Innovation

Marine organisms have perpetually captivated the imagination of scientists and skincare experts, thanks to their extraordinary and multifaceted attributes. These remarkable life forms, inhabiting the profound depths of the ocean, have undergone millennia of evolution to thrive in their underwater habitats. This extended period of adaption has yielded an abundance of bioactive compounds and naturally occurring ingredients that have become integral components in the formulation of cosmetics.

The allure of marine organisms lies in their ability to produce unique and potent substances as a means of survival in the challenging aquatic ecosystems they call home. From resilient corals and resilient algae to elusive sea creatures, such as sponges and jellyfish, each species has developed specialized mechanisms and chemical defenses to thrive amidst the harsh conditions of the ocean floor.

These marine bioactive compounds have not only piqued the curiosity of the scientific community but have also been harnessed by skincare experts and cosmetic manufacturers. The diverse properties of these compounds offer a wide array of benefits for skin care, including hydration, anti-ageing, and skin revitalization. As such, these natural ingredients have found their way into numerous cosmetic formulations, where their unique properties contribute to enhancing the efficacy and appeal of skincare products.

In essence, the fascination with marine organisms is not merely a scientific endeavor but also a practical one, as these incredible life forms continue to provide a wealth of natural ingredients that elevate the world of cosmetics and skincare to new heights.

1. *Marine Sponges*: Marine sponges are commonly found in various shapes and colors and are known as renewable resources. Marine sponges are a wealthy supply of natural ingredients with various chemical, bioactive metabolites, and antimicrobial compounds [5]. They contain compounds with potent antioxidants, such as flavonoids, polyphenols, and anti-inflammatory properties, making them valuable additions to skincare products. These compounds help combat free radicals, reduce skin inflammation, and promote a healthy complexion by reducing wrinkles and skin damage, skin whitening, and cleaning purposes. Sponges produce more than 35 carotenoids and tend to have photoprotective and antioxidant functions (Fig. 19.3).
2. *Sea Cucumber:* Sea cucumbers are marine organisms that contain beneficial compounds, such as peptides and glycosaminoglycans. These compounds have the potential to increase the production of collagen in the skin [6]. Sea cucumber is majorly found in anti-ageing, antioxidant products, anti-inflammatory and collagen-based skincare products and their derivatives are also found in ointments, toothpaste, body lotions, skin cleaners, and so on [7]. In addition to peptides and glycosaminoglycans, sea cucumbers consist of sulphated polysaccharides, mucopolysaccharides, and collagen widely used and safe compounds in the cosmetic industry [8]. There are great sources of vitamins, such as A and B2, and minerals, such as magnesium, iron, zinc, etc. which are widely used in skin treatments and hydration [9]. The compounds extracted from sea cucumber are also used for the wound-healing process and for reducing scars by tissue regeneration (Fig. 19.4).
3. *Seaweeds:* Seaweed, which includes diverse types, such as Kelp, Nori, and Algae, is perhaps one of the most well-known, non-toxic, biodegradable marine ingredients in cosmetics. It is rich in essential vitamins such as A, C, E, and K and derivatives of vitamin B, such as B1, B2, Niacine, B12, and so on. These vitamins help prevent photoaging, provide antioxidants, reduce pigmentation, and promote hair and nail growth even help with medical conditions, such as

Fig. 19.3 Phylum *Porifera*—Marine Sponges

Fig. 19.4 Phylum *Echinodermata*—Sea Cucumber

melanogenesis [10], skin cancer, and clotting properties. Polysaccharides and mineral-rich seaweeds are majorly used as moisturizers, facial cleaners, masks, and anti-aging products [11]. Seaweed helps detoxify the skin, purify, and restore natural radiance, and protects from ultraviolet (UV) damage. Other bioactive compounds such as polyphenols, carotenoids, etc. are produced by different red algae, brown algae, and green algae (Fig. 19.5).

4. *Corals:* Coral and its compounds have found a place in the cosmetics industry due to their unique properties and potential benefits for skincare. It is crucial to emphasize responsible and sustainable sourcing practices to prevent harm to fragile coral ecosystems. Coral calcium, rich in minerals, such as calcium and magnesium, is derived from coral reefs and is often used in cosmetics to provide essential minerals to the skin, contributing to skin vitality and health. These are also used as exfoliants to remove dead skin cells and unclog pores; trace elements in corals help improve skin texture, rejuvenation, and antioxidant compounds [12]. Coral compounds enhance cosmetic product colors, particularly in lip and nail products (Fig. 19.6).
5. *Microalgae*: Microalgae a unicellular, photosynthetic organism and its bioactive compounds have gained prominence because of their potential benefits for skincare and beauty products [1]. These tiny aquatic organisms, including species, such as *Spirulina*, *Chlorella*, and *Dunaliella*, offer a wide range of advantages when incorporated into cosmetic formulations. Microalgae contain essential nutrients, such as vitamins (A, B, C, and E), minerals (iron, calcium, and magnesium), amino acids, and fatty acids. These nutrients nourish the skin and help maintain its health and vitality. Certain species contain pigments, such as phycocyanin and fucoxanthin, which can help fade dark spots, and promote even skin other products lipsticks, and eyeshadow are used. Polysaccharides, humectants, and collagen enhance and lock moisture in the skin. Carotenoids, a natural sunscreen, used to protect the skin from UV rays and are commercialized as sunscreens. Many products such as scrubs, peels, and anti-inflammatories

Fig. 19.5 Phylum *Chlorophyta*—Seaweeds

Fig. 19.6 Phylum *Cnidaria*—Corals

to reduce redness are released by industries. In addition, mycosporine amino acids by cyanobacteria and other microalgae act as photo protectants to the skin [13] (Fig. 19.7).

6. *Bacteria:* Many marine bacteria produce bioactive compounds with natural antibacterial and anti-inflammatory properties. These compounds can be effectively incorporated into skincare products for combating acne, soothing irritated skin, and reducing redness [8]. In addition to this, some marine organisms can produce collagen, peptides, and antioxidants that are widely used in anti-ageing, anti-wrinkles, moisturizing, serums, masks, and hydration. Furthermore, some compounds from marine bacteria can serve as natural preservatives in cosmetic products. These natural alternatives help extend the shelf life of cosmetic products. Marine bacterial compounds help in the reduction of dark spots and higher pigment production (Fig. 19.8).
7. *Fungi:* Fungi that thrive in marine environments offer a range of advantages when integrated into cosmetic formulations. Marine fungi produce bioactive compounds, such as polyphenols, flavonoids, and peptides that possess potent antioxidant properties. These antioxidants help protect the skin from free radicals, preventing premature ageing and maintaining the skin's youthful appearance. Hydroquinone compounds act as antioxidants derived from fungi found in mangroves and seawater [14]. Tyrosinase inhibitors produced from marine fungi are used for hyperpigmentation, melasma, and skin lightening [15] (Fig. 19.9).
8. *Phytoplankton:* Phytoplankton extracts contain essential fatty acids, such as omega-3 and omega-6, which are important for maintaining healthy skin [16]. These fatty acids contribute to nourishing and preserving the natural barrier of the skin, contributing to its overall health. In addition, phytoplankton extracts are known for their antioxidant properties, which can help protect the skin from environmental stressors and oxidative damage (Fig. 19.10).

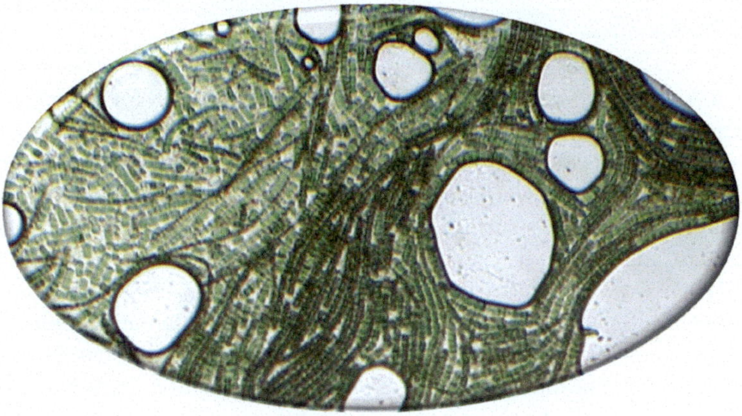

Fig. 19.7 Phylum *Haptophyta*—Microalgae

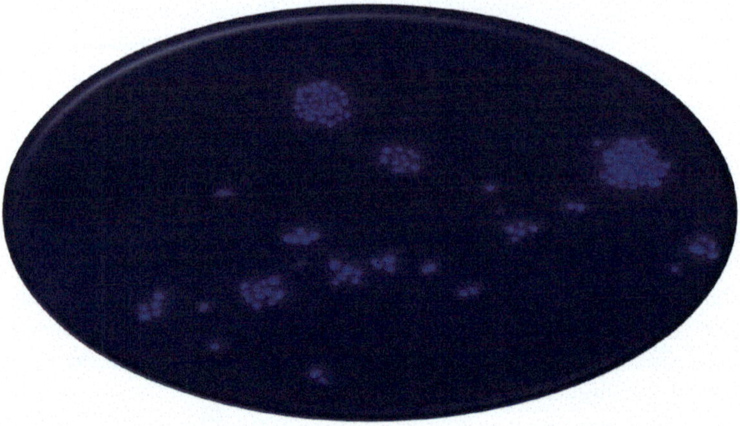

Fig. 19.8 Phylum *Marine Prokaryotes*—Bacteria

Fig. 19.9 Phylum *Ascomycota, Microsporidia,* etc.—Fungi

9. *Jellyfish:* Jellyfish-derived collagen is highly valuable for its potential to enhance skin elasticity and firmness. Collagen is a key structural protein in the skin that supports its youthful appearance. Mucus and peptides extracted from various organisms are widely applied in moisturizing creams, creating a protective layer on the skin's surface. In addition to providing hydration, these peptides are also used as anti-ageing and skin regeneration products [12] (Fig. 19.11).
10. *Sea Fennel:* Sea fennel is highly valued for its high vitamin C content, which possesses powerful antioxidant properties that can help brighten the skin, promote an even skin tone, and reduce wrinkles. In addition, stem cells found in sea fennel can aid in the regeneration of keratin in the skin, resulting in improved skin tone.

Fig. 19.10 Phylum *Cyanobacteria, Dinophyta,* etc.—Phytoplankton Bloom

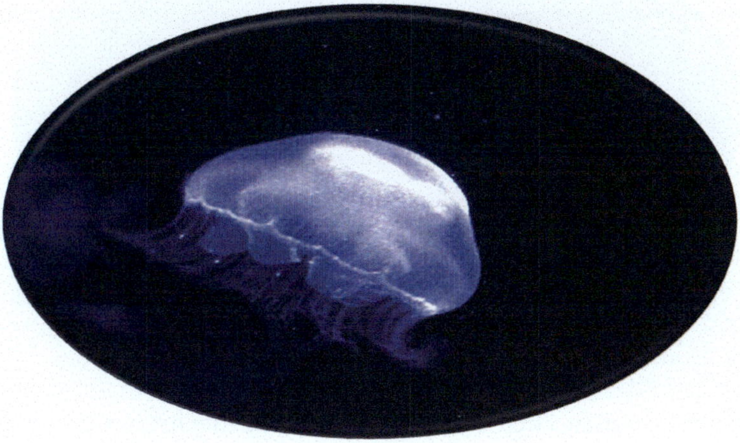

Fig. 19.11 Phylum *Cnidaria*—Jellyfish

11. *Sea Mud and Sea Water:* Sea mud is known to contain fine particles that help exfoliate the skin gently. This process removes dead skin cells and impurities detoxify the skin. This exfoliation leads to a smoother and brighter complexion. Many mud-based products, including facial masks, creams, and shampoos, are produced because of mineral-sourced mud [17] (Fig. 19.12).

Sea water is rich in minerals, such as magnesium, calcium, potassium, and sodium. These minerals can help nourish the skin and promote overall skin health. Sea water is a natural source of hydration. Cosmetic products often include seawater to help maintain the skin's moisture balance, providing a refreshing and

Fig. 19.12 Sea mud and Sea water

Fig. 19.13 Shark fish and Turtles

hydrating sensation. Deep seawater treats skin diseases because of its high mineral content [1].

12. *Shark Fish and Turtles:* Shark liver oil is a prized marine ingredient, because it contains a high amount of squalene, a natural emollient that moisturizes and nourishes the skin. It is also available in lip balms, deodorants, sun lotion, lipsticks, shampoo, perfumes, and so on.

Marine turtle oils are rich in vitamin E and commonly used in cosmetics, such as lotions, skin creams, nail creams, etc. [12] (Fig. 19.13).

19.5 Integration of Marine Natural Elements in Cosmetics Formulations

The beauty industry has long been intrigued by the untapped potential of marine natural elements as key ingredients in cosmetics. These exceptional components, sourced from the depths of our oceans, have become a focal point of skincare innovation. Over millions of years, these elements have adapted in response to the unique challenges posed by their underwater environments, resulting in a rich reservoir of bioactive compounds and natural substances that now play a pivotal role in cosmetic formulations.

The marine natural elements are fascinating due to their unique and diverse properties. They possess the ability to thrive in extreme conditions and produce potent substances as a means of adaptation and survival. These substances have garnered attention not only for their remarkable resilience but also for their remarkable skincare benefits. Various elements such as amino acids, polysaccharides, peptides, and so on offer a wide range of skin care advantages, from moisturization and revitalization to anti-ageing properties. These marine elements explore their incredible potential and the transformative impact they have had on the cosmetic industry.

1. *Mycosporines:* Mycosporines and mycosporine-like amino acids (MAAs) are natural cyclohexanone or cycloheximide chromophores with nitrogen compounds that act as sunscreens for marine organisms that live in shallow waters, where there is high UV exposure. These compounds are commonly found in marine sources, such as cyanobacteria [18], fungi, seaweeds, corals, lichens, and so on. Mycosporines have properties such as water solubility and low molecular mass, and they are typically colorless. Mycosporins-like amino acids also provide osmatic balance to the cells when cells undergo oxidative stress [19]. These compounds are used for their photoprotective properties, which help to shield the skin from the harmful effects of UV rays, including sunburn, premature ageing, and skin damage. Organisms that produce Mycosporins-like amino acids when exposed to higher levels of UV radiation [20]. The production of amino acids varies based on various factors, such as organism type, light, temperature, nutrients, salt concentration, and more. Mycosporine amino acid exhibits antioxidant properties, effectively scavenging free radicals generated by UV radiation and environmental pollutants. This antioxidant activity helps protect the skin from oxidative stress and reduces the risk of DNA damage, which can lead to skin ageing and skin cancer. They also have anti-inflammatory, skin hydration, anti-cancer, and wound-healing properties. Mycosporins-like amino acids can stabilize skin product formulations, extend product life, act as natural preservatives, and are eco-friendly. It is important to note that while MAAs offer natural photoprotection, they are typically not used as standalone sunscreens in cosmetic products. Instead, they are often integrated into sunscreen formulations or combined with other UV filters for comprehensive protection. In addition, thorough safety assessments and testing are essential to ensure the efficacy and safety of cosmetic products containing MAAs, especially regarding their UV protection capabilities (Fig. 19.14).

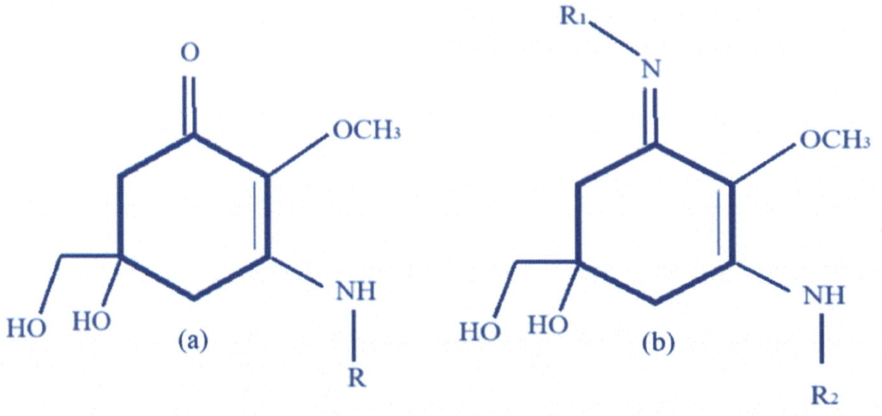

Fig. 19.14 (**a**) Mycosporine. (**b**) Mycosporine-like aminoacid

2. *Carotenoids:* Carotenoids are tetraterpene pigments that occur naturally in photosynthetic organisms, such as algae, plants, and cyanobacteria, and non-photosynthetic organisms, such as archaea, bacteria, fungi, and animals. Carotenoid pigments are classified into two types: carotenes and xanthophylls [21]. They create the vibrant colors found in many marine life forms and have numerous applications in the cosmetic industry. Carotenoids, such as astaxanthin and fucoxanthin, are widely produced by brown algae, and red algae [22] which provide vibrant red, orange, and yellow hues in cosmetics. These pigments are used to impart natural color to products, such as lipsticks, lip glosses, eyeshadows, and nail polishes. These pigments, extracted from marine sources, offer an eco-friendly alternative to synthetic colorants and impart advantages through cosmetic products. Studies done in vivo and in vitro have shown that fucoxanthin extracted from marine sources has photoprotective properties [21]. Carotenoids are potent antioxidants, capable of neutralizing free radicals generated by UV radiation and environmental pollutants, thus reducing premature ageing and maintaining skin health. Carotenoids offer a diverse range of benefits in cosmetic production, from providing natural color to offering antioxidant protection and anti-ageing effects. They can help enhance skin tone, giving a healthy and radiant complexion. When used responsibly and ethically, these natural pigments enhance the appeal and efficacy of cosmetics while meeting the demands of consumers seeking more eco-conscious and skin-friendly beauty products (Fig. 19.15).
3. *Polysaccharides:* Polysaccharides are complex carbohydrates made up of long chains of sugar molecules. In the production of cosmetics, polysaccharides obtained from various sources such as marine, plant, and microbial are utilized for their diverse properties and applications. Marine-based polysaccharides are classified as sulfated and non-sulfated and are used in various industrial applications. Sulfated polysaccharides such as carragenans, fucoidans, galactans, etc. are negatively charged with sulfate cross-links widely and are produced by red

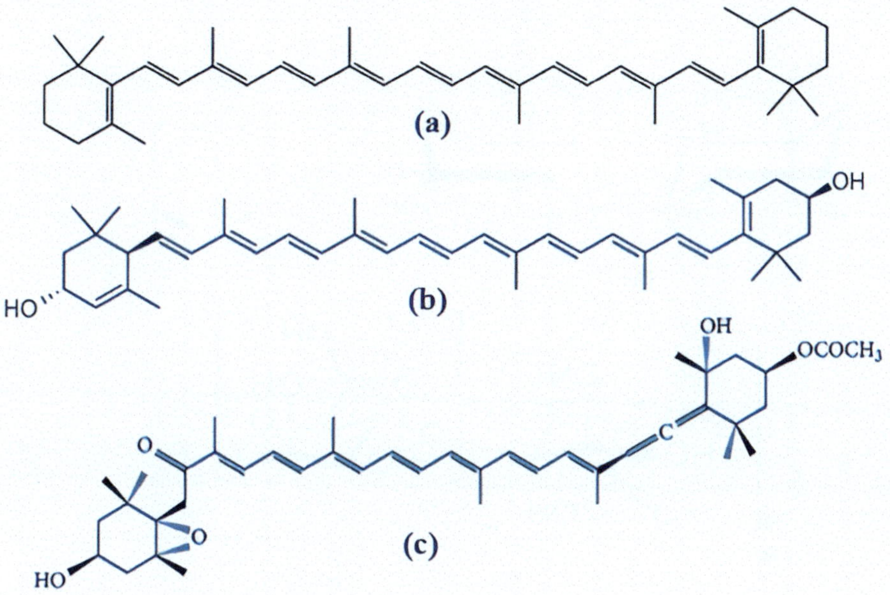

Fig. 19.15 (**a**) Carotene. (**b**) Xanthophyll. (**c**) Fucoxanthin

algae, brown algae, and green algae [23]. Non-sulfated polysaccharides, such as alginates and agars, which originate from algae, are also used in cosmetic production. Polysaccharides derived from red algae have wound-healing properties by enhancing cell multiplication and migration at wound sites [24]. Carrageenan is a common thickening and gelling agent found in moisturizers, lotions, and hair care products due to its emollient and texture-enhancing properties. Brown algae polysaccharides have various properties, such as skin whitening, antioxidant activity, moisturizing, and water retention when formulated in skin care products [25]. Alginic acid is the precursor, where salts combine with it to form an alginate. Alginate is often used in skincare products, such as masks and peels. It forms a gel-like texture when mixed with water, which can be applied to the skin to hydrate, soothe, and deliver the active ingredients effectively. Fucoidan and hyaluronic acid bioactivity contrast based on their molecular weight [1]. Fucoidan has antioxidant and anti-inflammatory properties, making it suitable for skin care products targeting skin health. It is used in serums, creams, and masks to combat oxidative stress, soothe the skin, and reduce redness. Hyaluronic acid is a powerful humectant, capable of holding up to 1000 times its weight in water and a key ingredient in moisturizers and serums, maintains skin suppleness and reduces the appearance of fine lines. Polysaccharides align with the growing consumer preference for natural and clean beauty products. They are sourced from a variety of eco-friendly and sustainable sources, making them appealing to environmentally conscious consumers (Fig. 19.16).

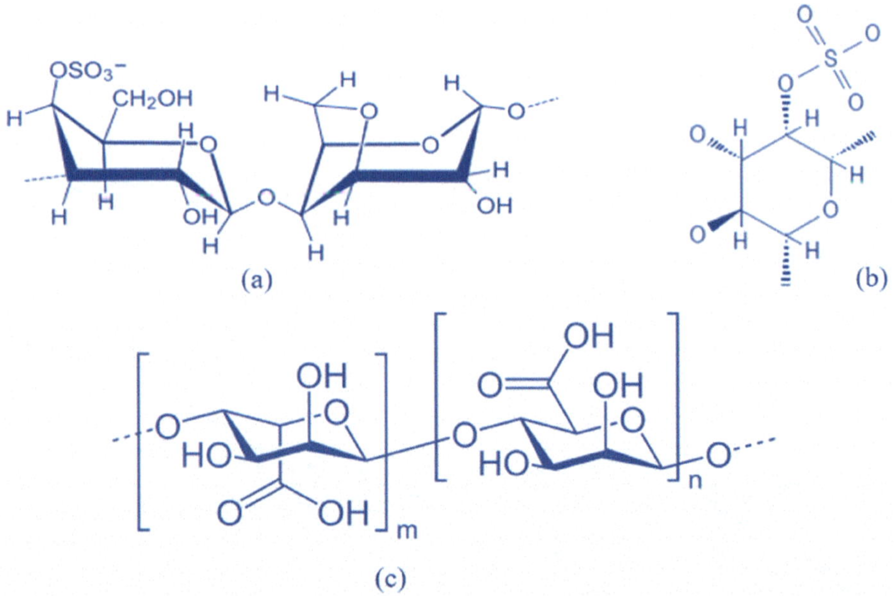

Fig. 19.16 (**a**) Carrageenan. (**b**) Fucoidan. (**c**) Alginic acid

4. *Terpenoids:* Terpenoids and terpenes are hydrocarbon groups mostly found in microalgae and sponges. Terpenoids are known for their biological activities, and their incorporation into cosmetic formulations offers a range of potential benefits. Some marine terpenoids have anti-inflammatory properties that can soothe and calm-irritated skin. They may be particularly useful in products designed for sensitive or inflamed skin types. Marine terpenoids can help in preventing and treating acne and other skin conditions related to microbial imbalances. They are often incorporated into skincare products targeting blemishes and acne-prone skin. Terpenoids found in algae and sponges, often possess antioxidant properties. These compounds can help protect the skin from oxidative stress caused by free radicals, thereby reducing the risk of premature ageing and maintaining skin health by stimulating collagen production and preventing fine lines and wrinkles [26]. It is important to note that using marine-based terpenoids in cosmetics is a growing field, and research continues to uncover new applications and benefits (Fig. 19.17).
5. *Proteins and Peptides:* Compounds have a wide range of applications and are highly valued for their ability to address various skin concerns. Collagen is one of the crucial components majorly found in many cosmetics because of its anti-ageing, moisturizing, and humectant properties [27]. Peptides derived from marine sources can stimulate collagen production, improve skin texture, increase elasticity, and reduce the signs of ageing. Certain marine proteins also possess strong antioxidant properties, which protect the skin from oxidative stress caused by free radicals, UV radiation, and pollution. In addition, they have wound-

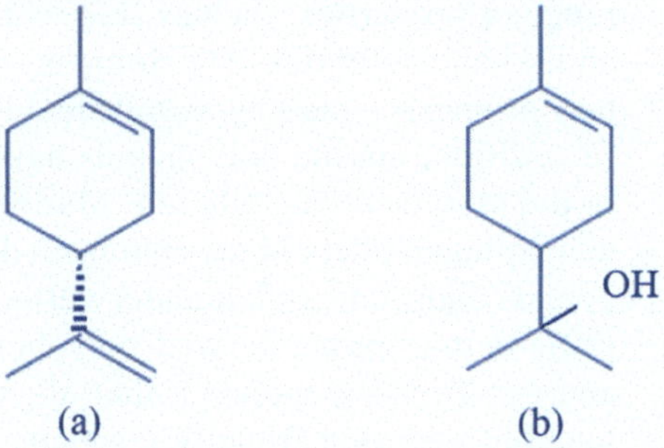

Fig. 19.17 (**a**) Terpenes. (**b**) Terpenoids

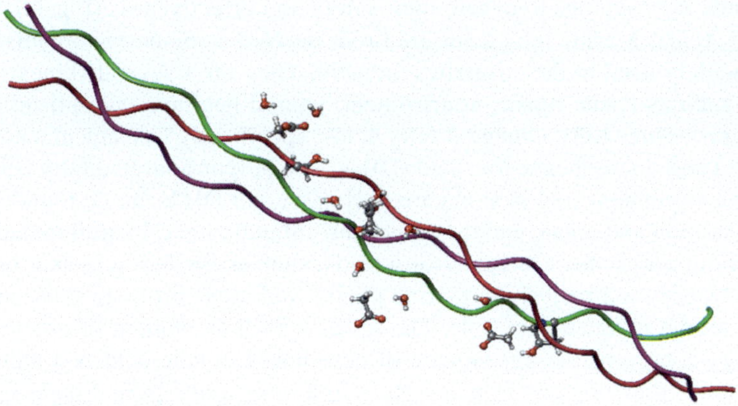

Fig. 19.18 Collagen triple-helical structure

healing properties and can reduce hyperpigmentation. Marine collagens are found in marine organisms, such as fish, sponges, sea cucumbers, jellyfish, and so on [1]. Proteins and peptides from marine organisms represent a promising avenue for the cosmetic industry to provide effective and innovative skincare solutions. These marine-derived ingredients offer natural, eco-friendly alternatives that align with the growing demand for sustainable and skin-friendly beauty products (Fig. 19.18).

6. *Polyphenolic Compounds:* The following are types of secondary bioactive metabolites that are categorized based on their phenol rings and structural elements [28]. These polyphenol metabolites are classified into two major groups: flavonoids and non-flavonoid compounds. Flavonols (Catechin), Flavones, and Flavanones are some of the categories of flavonoid compounds, whereas phenolic acid and tannins are categories of non-flavonoids [29]. Marine organisms such as micro and macroalgae, sponges, fungi, cyanobacteria, and others produce polyphenols in large quantities [30]. Various large compounds derived from marine algae such as phlorotannin, phenolic acids, etc. protect against ultraviolet B rays which are included in sunscreen products [21]. Polyphenolic compounds extracted from seaweed have a wide

range of properties, such as skin whitening, UV protection, anti-ageing, and antioxidant activities [31]. Some marine polyphenols possess anti-inflammatory properties, making them valuable for calming-irritated or inflamed skin. In addition, marine polyphenols have been shown to protect collagen fibers in the skin from degradation. Marine polyphenols have been shown to protect collagen fibers in the skin from degradation. Brown algae phenolic compounds show anti-inflammatory effects that can help treat allergic responses [32]. Using marine-derived polyphenols in cosmetics products is a natural and eco-friendly approach that aligns with the increasing demand for sustainable and skin-friendly cosmetic products. However, it is important to carry out thorough safety assessments and testing before using these compounds in cosmetics to ensure their safety and effectiveness (Fig. 19.19).

7. *Fatty Acids:* Marine fatty acids are lipids derived from marine organisms and are widely used in the cosmetics industry. They offer several skincare benefits, such as tissue repair, nourishment, wound-healing, and anti-inflammatory properties [33]. Omega-3 fatty acids, specifically eicosapentaenoic acid (EPA) and docosahexaenoic acid (DHA), are polyunsaturated fatty acids that play a significant role in skin care [14]. EPA and DHA can be found in fish-like salmon and algae, and they have anti-inflammatory, hydration, and skin-soothing properties. Omega-6 fatty acids, such as linoleic acid and the alpha variation, found majorly in marine plants, and algae promote skin moisture, and soothe irritated skin [34]. The lipids with high content of omega-6 and omega-3 prevent the appearance of eczemas. Linoleic acid is a significant ceramide that reduces skin hardness and sensitivity [35]. Single-cell oils from yeast and molds are more than fish oil and have higher antioxidant properties [1]. Fish liver oils are used in cosmetic formulations as an emulsifier and thickening agent. Some marine fatty acids, especially those from algae and certain fish oils, can inhibit melanin production and regulate sebum production (Fig. 19.20).

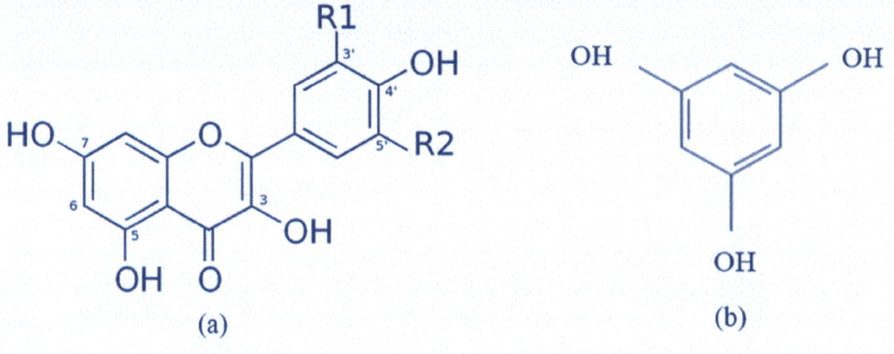

Fig. 19.19 (**a**) Flavonol. (**b**) Tannins

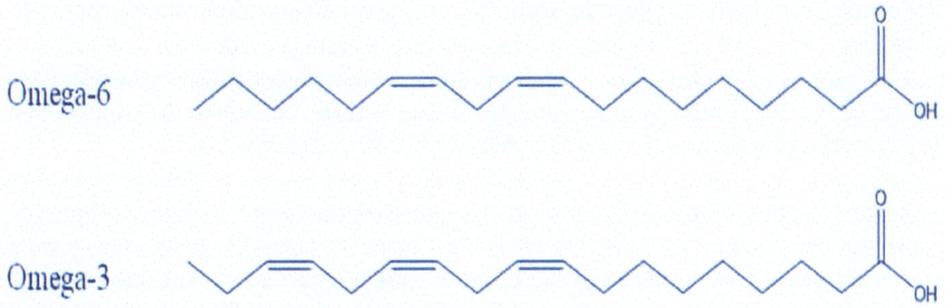

Fig. 19.20 Polyunsaturated fatty acids—Omega fatty acids

19.6 Cosmetic Formulations Procured from Marine Natural Compounds

In recent decades, marine cosmetic products have emerged as a revolutionary trend in the skincare and beauty industry, captivating consumers with their promise of natural efficacy, sustainability, and innovative formulations. Sourced from the rich biodiversity of the ocean, these products harness the power of marine ingredients, such as seaweeds, algae, marine minerals, and microorganisms to provide a wide range of skincare benefits. The allure of marine cosmetic products lies in their deep-rooted connection to the ocean's vast and diverse ecosystems. For centuries, coastal communities around the world have recognized the therapeutic properties of seawater and marine botanicals, incorporating them into traditional skincare rituals. Today, advancements in marine biotechnology and sustainable harvesting practices have unlocked the full potential of marine ingredients, paving the way for their widespread adoption in modern skincare formulations.

One of the key distinguishing features of marine cosmetic products is their ability to deliver a potent blend of vitamins, minerals, antioxidants, and other bioactive compounds directly to the skin. Seaweeds and algae, for example, are rich sources of essential nutrients, such as vitamins A, C, and E, as well as minerals, such as magnesium, calcium, and iodine. These nutrients work synergistically to nourish, hydrate, and revitalize the skin, promoting a healthy and radiant complexion. Marine ingredients often exhibit unique skincare benefits that set them apart from traditional cosmetic ingredients. Certain types of algae contain polysaccharides and peptides that have been shown to improve skin elasticity reduce inflammation and enhance collagen production. Marine minerals such as magnesium and calcium help maintain skin barrier function and regulate sebum production, making them ideal for balancing oily or acne-prone skin.

Another compelling aspect of marine cosmetic products is their sustainability and eco-friendliness. Unlike many synthetic and plant-based ingredients, marine resources can be harvested in a renewable and environmentally responsible manner, minimizing ecological impact. In addition, ongoing research into marine biotechnology has led to the development of innovative extraction methods that maximize the yield of bioactive compounds while minimizing waste and energy consumption. Marine cosmetics offer a glimpse into the future of skincare, where the treasures of the ocean become essential allies in achieving healthy, radiant skin.

1. *Sunscreen:* Photoprotection refers to the measures and mechanisms employed to protect the skin and other biological tissues from the harmful effects of exposure to UV radiation, typically from the sun. UV radiation is classified as ultraviolet C, ultraviolet B, and ultraviolet A. It can cause various forms of damage to the skin, such as premature ageing, hyperpigmentation caused by ultraviolet A sunburn, and oxidative stress by ultraviolet B rays with increased risk of skin cancer [36]. Prolonged exposure to UV rays affects and causes skin effects, such as sunburns, ageing, precancerous and cancerous lesions, and so on [37]. Primary and secondary protecting factors are involved in the photoprotection process. Primary elements involve sunscreens, which encompass physical barriers responsible for reflecting and dispersing light and chemical barriers that soak up light. Secondary elements encompass antioxidants, osmolytes, and DNA repair enzymes, contributing to reducing skin damage triggered by direct UV exposure [38]. Chemical sunscreens are composed of organic compounds which block ultraviolet A and ultraviolet B rays. Physical sunscreens are composed of compounds zinc oxide and titanium oxide. The effectiveness of sunscreen is estimated by the reflective index, particle size, film thickness, and base dispersion. Sunscreen with antioxidants such as vitamin C, vitamin E, and polyphenols limits sunburns, erythema, photoaging, and anti-inflammatory properties [37]. The sun protection factor (SPF) is used to measure the amount of time the sunscreen will protect the skin from UV rays and varies with their exposure time and skin type (Fig. 19.21).
   Marine-derived compounds such as MAAs, astaxanthin, and marine-derived oils are potential photoprotectors and reduce skin damage. MAAs absorb ultraviolet A and B radiation and act as a physical barrier with antioxidant properties. Seaweed extracts from marine algae reflect or scatter radiation and form a protective layer on the skin. Carotenoids protect skin from oxidative damage and neutralize free radical generation. Various sources such as corals, microalgae, lichens, and so on contribute to UV rays protection and they are not standalone sunscreens and are often used in combination with traditional sunscreen agents to enhance their efficacy.
2. *Moisturizers and Hydration:* The primary cellular component found in the dermis is the fibroblast, residing within a complex network of collagen and elastic fibers. Surrounding this fibroblast are various elements, including an extracellular matrix comprising glycoproteins, hyaluronic acid, glycosaminoglycans, and additional components, such as water and salts, which collectively create a gel-like structure. Among these, glycosaminoglycans (GAGs) play a crucial role as proteoglycans, responsible for retaining a significant portion of the skin's water content [39]. External factors make the skin lose its natural elasticity and moisturizing properties. The skin layers consist of hygroscopic natural moistur-

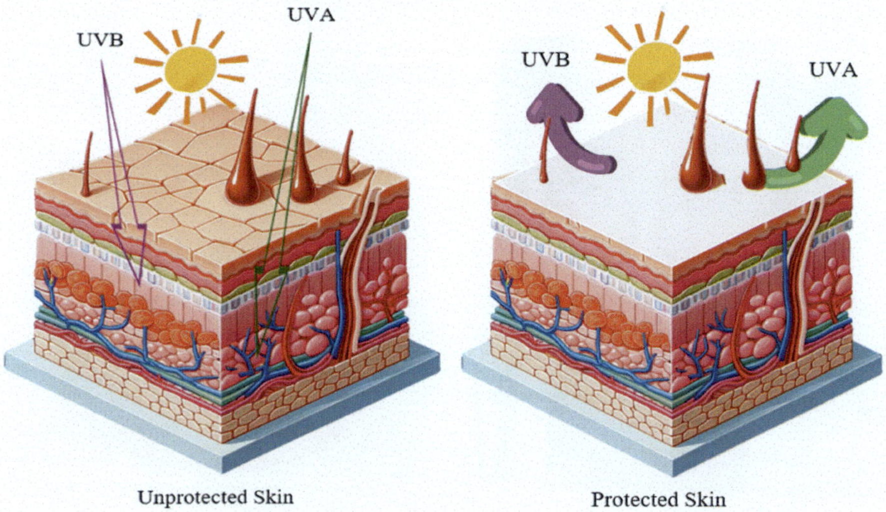

Fig. 19.21 Sunscreen working mechanism on skin

izing factors that prevent water loss [40]. Natural moisturizing factors consist of 40% of amino acids, 12% of carboxylic acid, 8% of urea, sugars, minerals, and so on. The mentioned compounds act as humectants on topical application which improves the skin texture [41]. Many molecules such as fatty acids and polysaccharides are produced by marine algae, seaweeds, and so on and are potentially used in cosmetic production. Polysaccharides and oligosaccharides can form hydrogen bonds with keratin, aiding in the retention of moisture [42]. Seaweed polysaccharides have a potential moisturizing effect, and their effect is more significant than that of hyaluronic acids. Beyond seaweeds, many marine bacteria produce exopolysaccharides that also have hydrating properties when incorporated in cosmetics [39].

Alginates natural polysaccharides derived from brown seaweeds can bind to water molecules creating a hydrating and moisturizing effect when applied to the skin. They form a film on the skin's surface, locking in moisture and preventing water loss, which helps keep the skin hydrated and supple. Hyaluronic acid is a powerful humectant that attracts and retains water in the skin. When applied topically, it can hold water molecules, providing immediate hydration and plumping effects. Squalene is a natural emollient that mimics the skin's natural lipids. It forms a protective barrier on the skin's surface, enhancing skin's suppleness and softness, and antioxidant properties. These compounds are valued for their ability to promote skin health and maintain hydration, making them popular additions to skincare products designed to address dryness, dehydration, and overall skin comfort (Fig. 19.22).

3. *Skin-Whitening Agents*: Skin-whitening products often target hyperpigmentation, dark spots, and uneven skin tone caused by factors, such as sun exposure, age, and hormonal changes. The skin, comprising the epidermis and dermis,

Fig. 19.22 Moisturizers hydrating mechanism on skin

serves as a protective barrier within the human body. The epidermis, situated on the skin's outermost layer, primarily consists of melanocytes and keratinocytes. Melanocytes facilitate the provision of melanin to keratinocytes via the dendritic transfer of melanin bodies. This process allows keratinocytes to create melanin caps, thereby diminishing UV-induced DNA damage to the epidermal layer [43]. Two forms of melanin exist eumelanin, characterized by dark brown insoluble polymers, and non-melanin, comprising reddish sulfur-containing polymers. The enzyme tyrosinase plays a crucial role in melanin synthesis. It catalyzes the conversion of tyrosine into dopaquinone, which, through a series of intricate steps, ultimately leads to melanin formation. Marine organisms such as kojic acid by fungi represent significant reservoirs capable of inhibiting tyrosinase [39]. Microorganisms in marine environments, such as microalgae and bacteria, are capable of producing active compounds that possess skin-whitening properties. For instance, marine yeast can produce Astaxanthin, which has been discovered to offer protection against age spots when applied to the skin [14]. Crustaceans-derived dehydrotyrosine has the property of a tyrosinase inhibitor which is much significant than commercially available whitening agents [44].

Marine polyphenols from algae and seaweed can inhibit the production of melanin, by inhibiting the activity of tyrosinase involved in melanin synthesis responsible for skin darkening, helping to lighten existing dark spots and prevent new ones from forming. Alginates can also help in skin whitening by promoting the exfoliation of pigmented skin cells. They assist in the removal of dead skin cells, revealing fresher and brighter skin. Consulting with a dermatologist or skincare professional is advisable to determine the most appropriate and safe approach to address specific skin concerns related to skin whitening (Fig. 19.23).

Fig. 19.23 Skin-whitening mechanism on skin

4. *Anti-ageing Products:* Marine natural compounds have gained recognition in the cosmetic and skincare industries for their potential anti-aging properties. These compounds, sourced from various marine organisms and environments, offer unique mechanisms of action to address signs of ageing and promote youthful skin. Skin ageing results from a combination of internal and external factors. Internally, the ageing process is primarily influenced by age-related changes. Externally, environmental stressors such as elevated temperatures, smoke, pollutants, and UV radiation also play a significant role [45]. The visible signs of skin ageing predominantly stem from diminished levels of collagen, elastic fibers, and hyaluronic acid, which manifest as wrinkles, dryness, reduced elasticity, sagging, and a rough texture [46]. Numerous compounds sourced from marine organisms exhibit a remarkable ability to battle cellular oxidative damage and thwart the ageing of the skin. For instance, Fucoxanthin, extracted from brown seaweed, has demonstrated its capacity to shield keratinocytes from oxidative harm [47]. Marine-sourced peptides have stimulated bioactivity, and collagen production and reduced oxidative damage in skin layers.
   Marine collagen and elastin are essential proteins that improve skin elasticity, reduce the appearance of fine lines and wrinkles, and promote a firmer complexion. When applied topically, they may help replenish lost collagen and elastin in the skin. Marine antioxidants, such as marine polyphenols, carotenoids, and vitamins, protect the skin from oxidative stress and free radical damage. They neutralize harmful free radicals generated by UV radiation and environmental pollutants, reducing premature ageing, and promoting skin health. marine-derived growth factors can promote skin cell renewal and collagen production leading to smoother, firmer, and more youthful-looking skin (Fig. 19.24).

Fig. 19.24 Anti-ageing mechanism on skin

5. *Colorants:* Marine natural compounds can serve as pigments and colorants in cosmetics and skincare products. These compounds are often used to provide natural and sustainable alternatives to synthetic colorants. Vibrant and appealing colors offer a delightful sensory experience and are commonly found in cosmetics, such as nail polish and lipstick. Traditional industrial pigments are typically synthesized using chemicals like benzene and toluene, which can raise concerns about toxicity and may not be acceptable to many consumers. As a result, natural pigments have gained widespread popularity in cosmetic products due to their safety and stability. These natural pigments include anthocyanins, carotenoids, and chlorophyll [48]. Marine microorganisms are increasingly recognized as valuable sources of cost-effective, innovative, stable, and safe biological pigments, and numerous studies have explored their potential in this regard. Marine microbes produce novel, stable, and safe pigments that show better water solubility properties. Certain marine invertebrates, such as sponges and corals, display vivid and striking colors, which could be linked to the presence of photosynthetic pigments produced by symbiotic microorganisms. Seaweeds such as red algae produced acid-resistant melanin with water solubility properties [49]. Pigments produced from cyanobacteria and microalgae are used as colorants in eye shadow, natural dyes, lipsticks, and eyeliners [39]. Cosmetic products with pink and purple pigments can be created using natural pigments sourced from red microalgae [50].
   Phycobiliproteins are water-soluble pigments that exhibit bright and vibrant colors, including red, blue, and green. These pigments are often used in natural cosmetics and skin care products to provide various shades of color. Astaxanthin is a red–orange pigment that acts as a natural antioxidant and colorant. Astaxanthin is often used for its ability to impart a healthy and youthful glow to the skin. Marine minerals are often used as colorants in cosmetics and skincare

products. Mica, for example, is a mineral pigment that can provide shimmer and shine to makeup and skincare items. Guanine, a compound found in fish scales, is often used in cosmetics and nail polishes to create a pearly and iridescent sheen. It provides a reflective and glittery effect, making it popular in products, such as nail polishes and eyeshadows. Marine natural compounds used as pigments work by interacting with light to produce various colors and effects. These natural pigments offer a sustainable and eco-friendly alternative to synthetic colorants in cosmetics and skincare products (Fig. 19.25).

6. *Lipid Compounds:* Marine natural compounds can serve as lipid conditioning agents in cosmetics and skincare products, helping to improve skin and hair health by providing essential lipids and moisture. These compounds often come from marine sources, such as algae, fish, and seaweed. Optimum sebum secretion nourishes the skin and prevents the skin from dryness. Excessive sebum production from the sebaceous gland makes skin oily, and prone to acne, whereas less production of sebum makes the skin very dry caused itchy and scaly [51]. Lipase is widely used in the production of cleansing facial oil. Fatty acids are widely produced by various sources, such as sponges, fish, algae, and seaweeds and commercially included in cosmetic products. Many marine organisms produce alkali-resistant lipases, collagen which increases the skin lipids, and treat dry skin [39].

   Algal oils are rich in essential fatty acids, such as omega-3 and omega-6. These fatty acids help improve the skin's lipid barrier function, preventing moisture loss and maintaining skin hydration. In haircare products, algal oils can provide nourishment to the hair shaft, reducing dryness and frizz. These fatty acids, spe-

Fig. 19.25 Colorants cosmetic products

cifically EPA and DHA, help enhance the skin's lipid barrier, reducing transepidermal water loss and improving skin moisture levels. Ceramides are essential lipids found naturally in the skin's outermost layer. Marine-derived ceramides replenish the skin's ceramide levels, improving the skin's lipid barrier. Marine squalane from shark liver is a natural emollient that mimics the skin's sebum. It helps maintain skin moisture balance, leaving the skin soft and supple. Alginates can also contribute to lipid conditioning. They assist in forming a protective barrier on the skin's surface, preventing moisture loss. They are often incorporated into cosmetics and skincare products to address dryness, dehydration, and overall lipid balance for a healthier and more radiant appearance (Fig. 19.26).

7. *Aromatic Compounds:* Fragrances are scented substances that are detectable through the sense of smell. Olfactory cells located on the mucous membrane within the nasal cavity can be activated by odor molecules, leading to the generation of olfactory signals. The aroma of perfume travels through the nasal cavity and triggers these olfactory cells, which transmit the fragrance to the brain's olfactory center. This process results in the formation of olfactory sensations and can evoke pleasurable experiences [52]. Fragrances are short-fatty acids and alcohol esters. The inherent fragrant compounds produced by certain marine microorganisms hold potential as a novel perfume source. Conversely, established perfume components can be generated via biological processes, offering more environmentally friendly and sustainable alternatives compared to traditional chemical production methods [39].

   Marine natural compounds are not commonly used as aromatics in cosmetics and skincare products. Aromatics typically refer to fragrances and scents that are added to products to enhance their sensory appeal. These fragrances are often

Fig. 19.26 Fatty acid-infused cosmetic products

Fig. 19.27 Aromatic products

derived from botanical sources, essential oils, or synthetic compounds, rather than marine sources.

However, certain marine-derived ingredients may have a mild natural scent, but they are primarily used for their functional properties rather than their aromatic qualities. For example, seaweed extracts or marine minerals may have a subtle marine or oceanic scent due to their marine origin, but this scent is usually not the primary focus when formulating cosmetics or skincare products.

In general, the aromatic properties of marine natural compounds are not widely harnessed for fragrance purposes in cosmetics and skincare. Instead, marine ingredients are typically valued for their skincare benefits, such as hydration, anti-aging properties, or antioxidant effects. If a specific marine ingredient is used for its aroma, it is often used for its unique and mild scent, rather than as a primary fragrance component in cosmetic or skincare formulations (Fig. 19.27).

19.7 Challenges

The commercialization of cosmetic products derived from marine sources comes with specific questions and hurdles. These challenges primarily pertain to the sustainability of these products and encompass aspects, such as production capacity, ecological impact, economic viability, and waste generation. When incorporating marine-derived compounds into cosmetic formulations, additional challenges arise concerning standardization, effectiveness, and traceability. One set of challenges revolves around obtaining easy access to marine resources and developing efficient screening processes to identify promising bioactive elements. Ensuring the sustainable production of these bioactive compounds and gaining a thorough understanding of their mechanisms of action pose another set of hurdles. Market-related issues, including processes, costs, and the establishment of partnerships, also contribute to the complexities of utilizing marine-derived ingredients in cosmetic products. Furthermore, marine-derived compounds encounter the issue of production volume, since they are typically found in limited amounts and can be difficult to isolate [53].

Obtaining marine raw materials presents significant challenges, particularly in accessing deep oceanic environments. Despite advancements in robotics and engineering, accessing the deepest oceanic locations remains a formidable task. Consequently, research efforts primarily focus on shallow coastal waters, where depths are more accessible, typically less than 30 m. However, this preference for shallow waters leads to gaps in our understanding of biodiversity and chemodiversity, as deeper regions remain largely unexplored.

Furthermore, the utilization of marine-derived products faces obstacles due to limited taxonomic knowledge and the abundance of undiscovered and undescribed species. Identifying and characterizing marine organisms, particularly those with potential cosmetic applications, require extensive taxonomic expertise. In addition, the vast number of undiscovered marine species presents a challenge in harnessing the full potential of marine biodiversity for cosmetic purposes [54] (Fig. 19.28).

Similar to terrestrial organisms, marine organisms exhibit significant biological and biochemical diversity, especially regarding the quantity and diversity of metabolites that they produce, which are influenced by their specific environmental conditions. Establishing a sustainable source for harnessing natural products from marine organisms can be challenging. This difficulty arises because the desired compounds

Fig. 19.28 Challenges faced during the Production of cosmetics

may be present in minimal quantities within the raw material, a quantity that can be affected by external factors. In addition, isolating these compounds presents considerable challenges. Consequently, large quantities of raw materials are often required to obtain meaningful quantities of the desired compounds [55].

Ensuring the consistency of extracts and transitioning cultivations from laboratory to industrial scale pose major challenges for industries dealing with natural products. Natural products tend to adjust their metabolite levels and growth rates in response to environmental factors as a protective mechanism. Various factors such as temperature, light availability, level of oxygen, mixing parameters, level of nutrients, contamination risk, formation biomass film, and loss of transparency in the bioreactor significantly influence the growth of macro- and microalgae. The scale-up process is vital in process development but often leads to significant productivity losses due to the substantial differences in optimal conditions between laboratory and industrial scales [54].

Contamination of raw materials and the resulting extracts poses a notable concern when integrating natural ingredients into cosmetic formulation and production processes. Biological contamination caused by water and air pollution, salt, and nutrient accumulation leads to cell growth strikes and protein pile-affecting contamination. Chemical contamination such as heavy metals creates toxicity in the environment.

19.8 Regulations

Regulatory standards in the cosmetic industry govern the manufacturing and branding of products intended for consumer use. These guidelines ensure product safety, prevent health risks, and provide consumers with essential information regarding product benefits. The regulation of cosmetic products varies across different countries. The Drugs & Cosmetics Act of 1940, along with its accompanying rules from 1945, assigns diverse responsibilities to central and state regulators for overseeing the regulation of drugs and cosmetics. Its objective is to ensure consistent enforcement of the Act and its associated rules, thereby safeguarding the safety, rights, and welfare of patients by regulating drugs and cosmetics. In the United States, the Food and Drug Administration oversees cosmetics through two key laws—the Federal Food, Drug, and Cosmetic Act and the Fair Packaging and Labelling Act. Brazil's cosmetic sector is regulated by three authorities: the Ministry of Health, the Brazilian Health Regulatory Agency, and the Hygiene, Perfume, Cosmetics, and Sanitizing Products Management. China's regulatory framework involves three major authorities: the State Administration for Market Regulation, the National Medical Products Administration, and the General Administration of Customs. In Canada, cosmetic regulations fall under the Cosmetic Regulation Act (1977) and the Food and Drugs Act (1985), which have undergone amendments over time. In Japan, cosmetic products are governed by the Pharmaceutical and Medical Devices Law, overseen by the Ministry of Health, Labor, and Welfare [1, 56].

The Scientific Committee on Consumer Safety periodically reviews and revises the "Notes of Guidance for the Testing of Cosmetic Ingredients and Their Safety Evaluation." This document offers guidance to both public authorities and the cosmetic industry regarding the safety assessment of cosmetic ingredients. It outlines various toxicological testing methodologies, including fundamental toxicity assessments, aimed at evaluating diverse human health-related toxicological endpoints.

Cosmetic product packaging serves as a multifaceted medium, not only attracting consumers and conveying product information but also safeguarding, transporting, and storing the product. To prevent misleading labels and ensure packaging safety, each country has enacted regulations. In terms of labeling, there are broad similarities in requirements across countries, including standardized label displays and information specifications, along with the adoption of the International Nomenclature of Cosmetic Ingredients for ingredient labelling. In addition, packaging plays a crucial role in protecting, transporting, and storing products, necessitating compliance with climatic conditions and protection against biological, physical, and chemical hazards to ensure product stability and integrity [57]. Disclosure of ingredients on their labels allows consumers to make informed decisions about product usage and helps individuals with allergies or sensitivities avoid potential allergies.

Marine cosmetics, particularly those containing natural ingredients, may be susceptible to microbial contamination. Regulatory agencies establish guidelines for microbial safety, requiring manufacturers to implement appropriate preservation methods and conduct microbial testing to ensure product safety and stability. Marine-derived ingredients may contain trace amounts of heavy metals, which can pose health risks if present in high concentrations. Regulatory authorities set limits for heavy metal content in cosmetic products and require manufacturers to conduct rigorous testing to ensure compliance with these limits.

Cosmetic regulations often require stability testing to assess the shelf-life and efficacy of products. Stability testing evaluates factors such as pH, viscosity, and microbial growth over time to ensure that products remain safe and effective throughout their intended shelf life.

The issue of animal testing in cosmetics has been extensively debated over the years, prompting heightened consumer awareness and growing demands for animal welfare considerations in product development. European countries have implemented bans on cosmetic products and ingredient testing on animals. In the United States, eight states have enacted laws prohibiting animal testing, with other states largely adopting alternative testing methods. Regulations in Japan and Canada are less stringent, while China lacks mandatory regulations for animal testing. Brazil has banned testing in ten states [56].

Regulatory agencies scrutinize product claims to ensure that they are supported by scientific evidence. Manufacturers must substantiate claims related to the efficacy and safety of marine cosmetics through appropriate testing and documentation. While regulations vary between countries and regions, efforts are underway to harmonize cosmetic regulations globally. Organizations such as the International Cooperation on Cosmetics Regulation work to align regulatory requirements across different jurisdictions to facilitate international trade and ensure consistent safety standards for cosmetic products, including marine cosmetics.

19.9 Prospects of Marine-Based Beauty Products

The beauty and skincare industry has been witnessing a remarkable shift toward natural and sustainable ingredients in recent years. Among these, marine-based cosmetic products have emerged as a promising trend with substantial growth potential. These products, derived from various marine organisms and resources, offer a range of benefits and hold exciting prospects for the future of the beauty industry.

The ocean is teeming with diverse marine life, from algae and seaweeds to sponges and microorganisms, which produces a plethora of bioactive compounds. These compounds, such as antioxidants, peptides, and fatty acids, have demonstrated remarkable skincare properties. As research continues to uncover new marine-derived ingredients, the future of marine-based cosmetics looks promising in terms of innovation and product development.

The beauty industry is becoming increasingly environmentally conscious because of the ecotoxicity created by synthetic ingredients [58], and marine-based cosmetic products align well with this trend. Sustainable harvesting and cultivation of marine resources can help protect fragile marine ecosystems while providing a renewable source of ingredients. Ethical sourcing and responsible practices will likely be integral to the future growth of this sector.

Many marine ingredients have demonstrated potent anti-ageing and skin rejuvenation properties. Compounds such as collagen and hyaluronic acid from marine sources can help improve skin elasticity and hydration, making them valuable assets in the fight against wrinkles and fine lines. As consumers continue to seek natural alternatives to synthetic skincare products, marine-based options are well-positioned to meet these demands. The ocean is a source of natural UV-absorbing compounds, such as MAAs found in marine algae. Incorporating these natural sunscreens into cosmetic formulations can enhance sun protection capabilities. In the future, marine-based products may play a significant role in sun care routines, reducing the reliance on conventional chemical sunscreens.

Marine ingredients, such as seaweed extracts and sea minerals, have shown anti-inflammatory and wound-healing properties. This makes them valuable for addressing various skin conditions, including acne and eczema. The ongoing research in this area may lead to the development of specialized marine-based skincare solutions.

The future of skincare is likely to be more personalized, with products tailored to individual skin types and concerns. Marine-based cosmetics offer a wide range of ingredients that can be combined to create customized skincare regimens. This personalization trend may drive the demand for marine-derived ingredients in beauty products. The demand for natural and organic cosmetics is on the rise, not only in established markets but also in emerging economies. Marine-based cosmetic products have the potential to tap into this growing global market. As awareness of their benefits spreads, the market for these products is likely to expand further.

The beauty industry's investment in research and development is expected to increase significantly in the marine-based cosmetics sector. This will lead to the discovery of new bioactive compounds, innovative extraction methods, and advanced formulations that maximize the potential of marine ingredients. In recent years, waste generation, especially plastic from cosmetic and personal products, has accumulated in the marine environment, creating major land and aquatic pollution. To avoid the use of plastics in cosmetic product packaging, many efforts are being made to focus on marine resources and create various compostable packaging solutions [1].

As marine-based cosmetics gain popularity, consumer education will play a crucial role. Consumers will seek information about the sustainability of marine sourcing, the benefits of specific ingredients, and how to incorporate these products into their skincare routines. Brands that prioritize transparent communication and education will have a competitive edge. In conclusion, the future of marine-based cosmetic products looks promising and vibrant.

19.10 Summary

The utilization of marine resources in cosmetic production represents a burgeoning field at the intersection of skincare innovation and sustainability. This review explores the inventive applications of marine ingredients in cosmetic formulations, highlighting their diverse benefits and potential for the beauty industry. Marine resources offer a wealth of bioactive compounds, ranging from seaweeds and algae to marine microorganisms, each boasting unique skincare properties. These ingredients are rich in vitamins, minerals, antioxidants, and peptides, providing nourishment, hydration, and anti-aging benefits to the skin. Moreover, marine-derived compounds exhibit soothing and anti-inflammatory properties, making them suitable for sensitive skin types. The review underscores the significance of sustainable sourcing and extraction methods in harnessing marine resources for cosmetic production. Sustainable harvesting practices ensure the preservation of marine ecosystems while maintaining a renewable supply of raw materials. Advanced extraction techniques optimize the yield of bioactive compounds, minimizing waste and energy consumption. The challenges and opportunities associated with incorporating marine ingredients into cosmetic formulations. Factors such as variability in metabolite levels, scale-up challenges, and regulatory considerations necessitate careful formulation and product development processes. However, the potential benefits of marine cosmetics, including efficacy, natural origin, and consumer appeal, outweigh these challenges, driving innovation in the beauty industry. The regulatory landscape governs marine cosmetics, highlighting differences in regulations between regions, such as Europe, the United States, and Asia. While some regions have implemented bans on animal testing and strict regulations on ingredient safety, others exhibit less stringent requirements, underscoring the need for global harmonization and adherence to ethical principles in cosmetic production.

In conclusion, the review emphasizes the transformative potential of marine resources in cosmetic production. By harnessing the diverse benefits of marine ingredients and adopting sustainable practices, the beauty industry can innovate and thrive while promoting environmental stewardship and consumer well-being.

References

1. Fonseca S, Amaral MN, Reis CP, Custódio L (2023) Marine natural products as innovation cosmetic ingredients. Mar Drugs 21(3):170
2. Jesumani V, Du H, Aslam M, Pei P, Huang N (2019) Potential use of seaweed bioactive compounds in skincare—a review. Mar Drugs 17(12):688
3. Sudhakar K, Mamat R, Samykano M, Azmi WH, Ishak WFW, Yusaf T (2018) An overview of marine macroalgae as bioresource. Renew Sustain Energy Rev 91:165–179
4. Siegel H (2018) In depth: a short history of beauty from the sea
5. Zhang H, Dong M, Chen J, Wang H, Tenney K, Crews P (2017) Bioactive secondary metabolites from the marine sponge genus *Agelas*. Mar Drugs 15(11):351
6. Olivera-Castillo L, Grant G, Kantún-Moreno N, Barrera-Pérez HA, Montero J, Olvera-Novoa MA, Carrillo-Cocom LM, Acevedo JJ, Puerto-Castillo C, Solís VM, Pérez-Vega JA, Gil-Zamorano J, Hernández-Garibay E, Fernández-Herrera MA, Pérez-Tapia M, Medina-Contreras O, Villanueva-Toledo JR, Rodriguez-Canul R, Dávalos A (2020) A glycosaminoglycan- rich fraction from sea cucumber *Isostichopus badionotus* has potent anti-inflammatory properties in vitro and in vivo. Nutrients 12(6):1698
7. Motuhi S-E, Mehiri M, Payri CE, La Barre S, Bach S (2016) Marine natural products from New Caledonia—A review. Mar Drugs 14(3):58
8. Alves A, Sousa E, Kijjoa A, Pinto M (2020) Marine-derived compounds with potential use as cosmeceuticals and nutricosmetics. Molecules 25(11):2536
9. Siahaan EA, Pangestuti R, Munandar H, Kim S-K (2017) Cosmeceuticals properties of sea cucumbers: prospects and trends. Cosmetics 4(3):26
10. Searle T, Al-Niaimi F, Ali FR (2020) The top 10 cosmeceuticals for facial hyperpigmentation. Dermatol Ther 33(6):e14095
11. López-Hortas L, Flórez-Fernández N, Torres MD, Ferreira-Anta T, Casas MP, Balboa EM, Falqué E, Domínguez H (2021) Applying seaweed compounds in cosmetics, cosmeceuticals and nutricosmetics. Mar Drugs 19(10):552
12. Uppala L (2015) A review on active ingredients from marine sources used in cosmetics. SOJ Pharm Pharma Sci 2:1–3
13. Gupta A, Singh AP, Singh VK, Singh PR, Jaiswal J, Kumari N, Upadhye V, Singh SC, Sinha RP (2023) Natural sun-screening compounds and DNA-repair enzymes: photoprotection and photoaging. Catalysts 13(4):745
14. Corinaldesi C, Barone G, Marcellini F, Dell'Anno A, Danovaro R (2017) Marine microbial-dervied molecules and their potential use in cosmeceutical and cosmetic products. Mar Drugs 15(4):118
15. Agarwal S, Adholeya A, Barrow CJ, Deshmukh SK (2018) Marine fungi: an untapped bioresource for future cosmeceuticals. Phytochem Lett 23:15–20
16. Galloway AWE, Winder M (2015) Partitioning the relative importance of phylogeny and environmental conditions on phytoplankton fatty acids. PLoS One 10(6):e0130053
17. Abdel-Fattah A, Pingitore NE Jr (2009) Low levels of toxic elements in Dead Sea black mud and mud-derived cosmetic products. Environ Geochem Health 31:487–492
18. Geraldes V, Jacinavicius FR, Genuário DB, Pinto E (2020) Identification and distribution of mycosporine-like amino acids in Brazilian cyanobacteria using ultrahigh-performance liquid

chromatography with diode array detection coupled to quadrupole time-of-flight mass spectrometry. Rapid Commun Mass Spectrom 34 Suppl 3:e8634
19. Oren A, Gunde-Cimerman N (2007) Mycosporines and mycosporine-like amino acids: UV protectants or multipurpose secondary metabolites. FEMS Microbiol Lett 269(1):1
20. Chrapusta E, Kaminski A, Duchnik K, Bober B, Adamski M, Bialczyk J (2017) Mycosporine-like aminoacids: potential health and beauty ingredients. Mar Drugs 15(10):326
21. Pangestuti R, Siahaan EA, Kim S-K (2018) Photoprotective substances derived from marine algae. Mar Drugs 16(11):399
22. Mohamed S, Hashim SN, Rahman HA (2012) Seaweeds: a sustainable functional food for complementary and alternative therapy. Trends Food Sci Technol 23(2):83–96
23. Muthukumar J, Chidambaram R, Sukumaran S (2021) Sulfated polysaccharides and its commercial application in food industries—A review. J Food Sci Technol 58(7):2453
24. Veeraperumal S, Qiu H-M, Zeng S-S, Yao W-Z, Wang B-P, Liu Y, Cheong K-L (2020) Polysaccharides from Gracilaria lemaneiformis promote the HaCaT keratinocytes wound healing by polarised and directional cell migration. Carbohydr Polym 1:116310
25. Jesumani V, Du H, Pei P, Zheng C, Cheong K-L, Huang N (2019) Unravelling property of polysaccharides from Sargassum sp. as an anti-wrinkle and skin whitening property. Int J Biol Macromol 1:216
26. Elena Balboa M, Li Y, Ahn B, Eom SH, Domínguez H, Jiménez C, Rodriguez J (2015) Photodamage attenuation effect by a tetraprenyltoluquinol chromane meroterpenoid isolated from *Sargassum muticum*. J Photochem Photobiol B Biol 148:51–58
27. Alves AL, Marques ALP, Martins E, Silva TH, Reis RL (2017) Cosmetic potential of marine fish skin collagen. Cosmetics 4(4):39
28. Ignat I, Volf I, Popa VI (2011) A critical review of methods for characterisation of polyphenolic compounds in fruits and vegetables. Food Chem 126(4):1821
29. Jayawardhana HHACK, Jayawardena TU, Sanjeewa KKA, Liyanage NM, Nagahawatta DP, Lee HG, Kim JI, Jeon YJ (2023) Marine algal polyphenols as skin protective agents: current status and future prospectives. Mar Drugs 21(5):285
30. Mateos R, Pérez-Correa JR, Domínguez H (2020) Bioactive properties of marine phenolics. Mar Drugs 18(10):501
31. Kalasariya HS, Pereira L (2022) Dermo-cosmetic benefits of marine macroalgae-derived phenolic compounds. Appl Sci 12(23):11954
32. Vo T-S, Ngo D-H, Kim S-K (2012) Marine algae as a potential pharmaceutical source for anti-allergic therapeutics. Process Biochem 47(3):386
33. Ziboh VA, Miller CC, Cho Y (2000) Metabolism of polyunsaturated fatty acids by skin epidermal enzymes: generation of anti-inflammatory and antiproliferative metabolites. Am J Clin Nutr 71:361S
34. Huang T-H, Wang P-W, Yang S-C, Chou W-L, Fang J-Y (2018) Cosmetics and therapeutic applications of fish oils's fatty acids on the skin. Mar Drugs 16(8):256
35. Bialek A, Bialek M, Jelinska M, Tokarz A (2015) Fatty acid profile of new promising unconventional plants oils for cosmetic use. Int J Cosmet Sci 38(4):382
36. Rai R, Srinivas CR (2007) Photoprotection. Indian J Dermatol Venereol Leprol 73(2):73
37. Rai R, Shanmuga SC, Srinivas CR (2012) Update on photoprotection. Indian J Dermatol 57(5):335
38. Gabros S, Nessel TA, Zito PM (2023) Sunscreens and photoprotection. StatPearls
39. Ding J, Baochuan W, Chen L (2022) Application of marine microbial natural products in cosmetics. Front Microbiol 13:892505
40. Pimentel FB, Alves RC, Rodrigues F, Beatriz P. P Oliveria M (2017) Macroalgae-derived ingredients for cosmetic industry—An update. Cosmetics 5(1):2
41. Couteau C, Coiffard L (2016) Seaweed application in cosmetics. In: Seaweed in health and disease prevention, pp 423–441
42. Bedoux G, Hardouin K, Burlot AS, Bourgougnon N (2014) Bioactive components from seaweeds: cosmetic applications and future development. Adv Bot Res 71:345–378
43. Costin G-E, Hearing VJ (2007) Human skin pigmentation: melanocytes modulate skin color in response to stress. FASEB J 21(4):976

44. Deering RW, Chen J, Sun J, Ma H, Dubert J, Barja JL, Seeram NP, Wang H, Rowley DC (2016) N-acyl dehydrotyrosines, tyrosinase inhibitors from the marine bacterium Thalassotalea sp. PP2-459. J Nat Prod 79(2):447
45. Zhang S, Duan E (2018) Fighting against skin aging: the way from bench to bedside. Cell Transplant 27(5):729
46. Brunt EG, Burgess JG (2018) The promise of marine molecules as cosmetic active ingredients. Int J Cosmet Sci 40(1):1
47. Zheng J, Piao MJ, Keum YS, Kim HS, Hyun JW (2013) Fucoxanthin protects cultured human keratinocytes against oxidative stress by blocking free radicals and inhibiting apoptosis. Biomol Ther (Seoul) 21(4):270–276
48. Karla R, Conlan XA, Goel M (2020) Fungi as a potential source of pigments: harnessing filamentous fungi. Front Chem 8:369
49. Ganesh Kumar C, Sahu N, Narender Reddy G, Prasad RBN, Nagesh N, Kamal A (2013) Production of melanin pigment from *Pseudomonas stutzeri* isolated from red seaweed *Hypnea musciformis*. Lett Appl Microbiol 57(4):295
50. Malis Arad S, Yaron A (1992) Natural pigments from red microalgae for use in foods and cosmetics. Trends Food Sci Technol 3:92–97
51. Endly DC, Miller RA (2017) Oily skin: a review of treatment options. J Clin Aesthet Dermatol 10(8):49–55
52. Silva Teixeira CS, Cerqueira NMFS, Silva Ferreira AC (2016) Unravelling the olfactory sense: from the gene to odor perception. Chem Senses 41(2):bjv075
53. Martins A, Vieira H, Gaspar H, Santos S (2014) Marketed marine products in the pharmaceutical and cosmeceutical industries: tips for success. Mar Drugs 12(2):1066
54. Costa JP, Custódio L, Reis CP (2023) Exploring the potential of using marine-derived ingredients: from the extraction to cutting-edge cosmetics. Mar Drugs 21(12):620
55. Matos GS, Pereira SG, Genisheva ZA, Gomes AM, Teixeira JA, Rocha CMR (2021) Advances in extraction methods to recover added-value compounds from seaweeds: sustainability and functionality. Foods 10(3):516
56. Ferreira M, Matos A, Couras A, Marto J, Ribeiro H (2022) Overview of cosmetic regulatory frameworks around the world. Cosmetics 9(4):72
57. Briasco B, Capra P, Cozzi AC, Mannucci B, Perugini P (2016) Packaging evaluation approach to improve cosmetic product safety. Cosmetics 3(3):32
58. Juliano C, Magrini GA (2017) Cosmetic ingredients as emerging pollutants of environmental and health concern. A mini-review. Cosmetics 4(2):11

The manufacturer's authorised representative in the EU is Springer Nature Customer Service Centre GmbH, Europaplatz 3, 69115 Heidelberg, Germany. If you have any concerns regarding our products, please contact ProductSafety@springernature.com

Printed and bound by CPI Group (UK) Ltd, Croydon, CR0 4YY
01/10/2026
02225926-0017